Lecture Notes in Control and Information Sciences

Volume 461

AF786300

About this Series

This series aims to report new developments in the fields of control and information sciences—quickly, informally and at a high level. The type of material considered for publication includes:

1. Preliminary drafts of monographs and advanced textbooks
2. Lectures on a new field, or presenting a new angle on a classical field
3. Research reports
4. Reports of meetings, provided they are
   (a) of exceptional interest and
   (b) devoted to a specific topic. The timeliness of subject material is very important.

More information about this series at http://www.springer.com/series/642

M. Kanat Camlibel · A. Agung Julius
Ramkrishna Pasumarthy
Jacquelien M.A. Scherpen
Editors

Mathematical Control Theory I

Nonlinear and Hybrid Control Systems

Springer

Editors
M. Kanat Camlibel
Johann Bernoulli Institute for Mathematics and Computer Science
University of Groningen
Groningen
The Netherlands

A. Agung Julius
Department of Electrical, Computer and Systems Engineering
Rensselaer Polytechnic Institute
Troy, NY
USA

Ramkrishna Pasumarthy
Department of Electrical Engineering
Indian Institute of Technology
Chennai, Tamil Nadu
India

Jacquelien M.A. Scherpen
Engineering and Technology Institute Groningen
University of Groningen
Groningen
The Netherlands

ISSN 0170-8643 ISSN 1610-7411 (electronic)
Lecture Notes in Control and Information Sciences
ISBN 978-3-319-20987-6 ISBN 978-3-319-20988-3 (eBook)
DOI 10.1007/978-3-319-20988-3

Library of Congress Control Number: 2015942816

Mathematics Subject Classification: 34H05, 34H15, 47N70, 70Q05, 93B05, 93B52, 93C05, 93C10, 93C15, 93C35

Springer Cham Heidelberg New York Dordrecht London

Printed on acid-free paper

Springer International Publishing AG Switzerland is part of Springer Science+Business Media (www.springer.com)

Foreword

This volume is offered to Prof. Dr. Arjan van der Schaft in celebration of his birthday. It contains papers by his collaborators, including a number of former Ph.D. students and postdoctoral fellows.

This is the first of a series of two books, appearing in connection with the workshop "Mathematical systems theory: from behaviors to nonlinear control" dedicated to the 60th birthdays of Arjan van der Schaft and Harry Trentelman, both at the Johann Bernoulli Institute for Mathematics and Computer Science, and at the Jan C. Willems Center for Systems and Control at the University of Groningen.

Preface

It is our great pleasure to present this book in celebration of the 60th birthday of Arjan van der Schaft. Arjan received his M.Sc. degree in Mathematics with honors in 1979 from the University of Groningen. Subsequently, he pursued his doctoral degree, also in Mathematics at the University of Groningen, under the tutelage of the late Jan C. Willems. His doctoral thesis, System Theoretic Descriptions of Physical Systems, was completed in 1983. Quite remarkably, he started his academic career as an Assistant Professor in Applied Mathematics at the University of Twente in 1982, before his doctoral thesis was written. We have to note, however, that at this point, Arjan had already published seven journal papers on control theory. At Twente, Arjan's academic career went all the way up to Full Professor at the Chair of Mathematical Systems and Control Theory. In 2005, Arjan's academic career came to a full circle, when he returned to the University of Groningen as Full Professor.

Over the past 30 years, Arjan's footprint in the field of systems and control theory has been deep and extensive. The books "Nonlinear Dynamical Control Systems (with Henk Nijmeijer), "L_2-gain and Passivity in Nonlinear Control", sole-authored by Arjan, and "An Introduction to Hybrid Dynamical Systems" (with Hans Schumacher), all had great impact in the field. Arjan's impact on H_∞ control for nonlinear systems is witnessed by his paper "L_2-gain Analysis of Nonlinear Systems and Nonlinear State-Feedback H_∞ Control", which was recognized as the Dutch research paper in international technical sciences journals with the largest number of citations during the period 1994–1998. Furthermore, Arjan is one of the founders (with Bernhard Maschke) of port-Hamiltonian systems theory; a comprehensive and influential theory for mathematical modelling, analysis, simulation and control of complex multiphysics systems. This theory offers new paradigms for control (energy-shaping, interconnection-shaping, control by interconnection), and has been applied to many areas: from robotics, mechatronics, power systems, to chemical reaction networks. The systems and control community recognizes Arjan's excellence in research and academic leadership. He was inaugurated as a Fellow of the IEEE in 2002, invited as a keynote speaker at the International Congress of Mathematicians in 2006, was rewarded with the SICE Takeda Best

Paper Prize (with Noboru Sakamoto) in 2008, and was awarded the three-yearly Certificate of Excellent Achievements from the IFAC Technical Committee on Nonlinear Systems in 2013.

Looking back, we recognize that Arjan's scientific legacy is not only his hundreds of peer-reviewed papers and half a dozen technical books, but also around three dozen young researchers (Ph.D. students and post-doctoral researchers) whose careers benefitted from his tutelage and collaboration. To celebrate this milestone in Arjan's academic life, we present this book to our colleague and teacher, Arjan van der Schaft, with affection and admiration and our best wishes for several decades more of top-level scientific productivity.

Groningen, The Netherlands — M. Kanat Camlibel
New York, USA — A. Agung Julius
Chennai, India — Ramkrishna Pasumarthy
Groningen, The Netherlands — Jacquelien M.A. Scherpen
May 2015

Contents

Contributors

Barbara M. Bakker Department of Pediatrics and Systems Biology Centre for Energy Metabolism and Ageing, University of Groningen, University Medical Center Groningen, Groningen, The Netherlands

M. Kanat Camlibel Johann Bernoulli Institute for Mathematics and Computer Science, University of Groningen, Groningen, The Netherlands

Jorge Cortés Department of Mechanical and Aerospace Engineering, UC San Diego, La Jolla, CA, USA

J.C. Engwerda Department of Econometrics and Operations Research, Tilburg University, Tilburg, The Netherlands

Sasanka V. Gottimukkala Hightech Solutions B.V. Apeldoorn, Apeldoorn, The Netherlands

Jun-ichi Imura Graduate School of Information Science and Engineering, Tokyo Institute of Technology, Meguro, Tokyo, Japan

Takayuki Ishizaki Graduate School of Information Science and Engineering, Tokyo Institute of Technology, Meguro, Tokyo, Japan

Bayu Jayawardhana Engineering and Technology Institute Groningen, University of Groningen Groningen, The Netherlands

Dimitri Jeltsema Delft Institute of Applied Mathematics, Delft University of Technology, Delft, The Netherlands

A. Agung Julius Department of Electrical, Computer, and Systems Engineering, Rensselaer Polytechnic Institute, Troy, NY, USA

Yu Kawano Graduate School of Informatics, Kyoto University, Sakyo-ku, Kyoto, Japan

Krishna Chaitanya Kosaraju Department of Electrical Engineering, Indian Institute of Technology Madras, Chennai, India

Alessandro Macchelli Department of Electrical, Electronic, and Information Engineering, University of Bologna, Bologna, Italy

Riccardo Marino Department of Electronic Engineering, University of Rome Tor Vergata, Roma, Italy

Sonia Martínez Department of Mechanical and Aerospace Engineering, UC San Diego, La Jolla, CA, USA

Jonathan C. Mayo-Maldonado Vision, Learning and Control Group, School of Electronics and Computer Science, University of Southampton, Southampton, UK

Henk Nijmeijer Department of Mechanical Engineering, Eindhoven University of Technology, Eindhoven, The Netherlands

Romeo Ortega Laboratoire des Signaux et Systèmes, Gif-sur-Yvette, France

Ramkrishna Pasumarthy Department of Electrical Engineering, Indian Institute of Technology Madras, Chennai, India

Shodhan Rao Ghent University Global Campus, Yeonsu-gu, Incheon, South Korea

Paolo Rapisarda Vision, Learning and Control Group, School of Electronics and Computer Science, University of Southampton, Southampton, UK

P.V. Reddy GERAD, HEC Montréal, Montréal, Canada

H. Rodríguez-Cortés Departamento de Ingeniería Industrial y Operaciones, Instituto Tecnológico Autónomo de México, México, D.f. México

Noboru Sakamoto Faculty of Science and Engineering, Nanzan University, Showa-ku, Nagoya, Japan

Jacquelien M.A. Scherpen Engineering and Technology Institute Groningen, Jan C. Willems Center for Systems and Control, ENTEG-DTPA, University of Groningen, Groningen, The Netherlands

Kurt Schlacher Institute of Automatic Control and Control Systems Technology, Johannes Kepler University, Linz, Austria

J.M. Schumacher Department of Econometrics and Operations Research, Tilburg University, Tilburg, The Netherlands

Markus Schöberl Institute of Automatic Control and Control Systems Technology, Johannes Kepler University, Linz, Austria

Ward Sikkema Engineering and Technology Institute Groningen, University of Groningen, Groningen, The Netherlands

Tjerk W. Stegink Johann Bernoulli Institute for Mathematics and Computer Science, University of Groningen, Groningen, The Netherlands

Erik Steur Institute for Complex Molecular Systems and Department of Mechanical Engineering, Eindhoven University of Technology, Eindhoven, The Netherlands

Stefano Stramigioli University of Twente, Enschede, The Netherlands

Viswanath Talasila MSRIT, Bangalore, India

Le Quang Thuan Department of Mathematics, Quy Nhon University, Quy Nhon, Binh Dinh, Vietnam

Patrizio Tomei Department of Electronic Engineering, University of Rome Tor Vergata, Rome, Italy

Harry L. Trentelman Johann Bernoulli Institute for Mathematics and Computer Science, University of Groningen, Groningen, The Netherlands

Daniele Zonetti Laboratoire des Signaux et Systèmes, Gif-sur-Yvette, France

Hans Zwart Department of Applied Mathematics, University of Twente, Enschede, The Netherlands; Dynamics and Control, TU/e, Eindhoven, The Netherlands

Chapter 1
A Port-Hamiltonian Formulation of a Wireless Communication System

Viswanath Talasila and Ramkrishna Pasumarthy

Abstract In this chapter we model the traffic dynamics in a wireless communication system (characterized by a set of routers exchanging data with each other) in the port-Hamiltonian framework. Communication systems are characterized by elements which produce significant time delays (by design) in their response, unlike (say) an idealized circuit or mechanical element. Furthermore, the communication between two routers (compositionality) involves losses due to the characteristics of radio signal propagation. In this paper we study the type of Dirac structure used to model a communication element (a router), we analyze the stability properties of a router and finally we study the compositionality properties that evolve under lossy interconnections.

1.1 Introduction

This chapter is dedicated to Arjan's 60th birthday. We were both extremely fortunate to be his Ph.D. students and have admired him, over the years, for his deep technical insights and fundamental contributions to systems and control theory. His work has influenced our own professional interests and growth in many ways.

Arjan's discussions and work about *compositions of mathematical structures* over the years have been an important contribution to systems (and control) theory. This has led to some really interesting results in various allied disciplines. For example, the use of the compositionality property in spatial discretization was a wonderful concept of his. Recently, when working with wireless communication systems the first author was studying if there was an elegant way to describe the interconnection structures in wireless networks. This motivated the authors to model wireless communication systems in a port-Hamiltonian setting. It is indeed impressive that

V. Talasila (✉)
MSRIT, MSR Nagar, Bangalore, India
e-mail: viswanath.talasila@msrit.edu

R. Pasumarthy
Department of Electrical Engineering, Indian Institute of Technology, Chennai, India
e-mail: ramkrishna@ee.iitm.ac.in

M.K. Camlibel et al. (eds.), *Mathematical Control Theory I*,
Lecture Notes in Control and Information Sciences 461,
DOI 10.1007/978-3-319-20988-3_1

the port-Hamiltonian theory of Arjan and Bernhard Maschke can be used across so many engineering disciplines.

On a personal level, Arjan's influence is deep. His generosity, patience and kindness are attributes that we have tried to emulate over the years. From both the authors: Arjan, thank you for some beautiful memories. And of course, happy birthday!

The design and deployment of a wireless communication system involves multiple aspects: modeling the radio signal propagation characteristics, router design, network layer routing protocols, the design of the antenna, and so on. There is a standard for describing these multiple subsystems and their interconnections, and this is the OSI (open systems interconnection) model, [11]. The OSI model partitions the functions of a communication systems into seven abstraction layers. These include, among others, the data link layer, the physical layer, and the network layer. The workings of each of these seven layers, and their interactions completely define a wireless communication system. For example, the physical layer may use a full duplex transmission mode, the MAC layer in the data link layer controls the permissions to transmit data; and the network layer implements various protocols to decide on message routing. The internal functioning of the OSI architecture is not of clear relevance (at least not directly) from a dynamical systems viewpoint, which is one of the themes of this book. What is more relevant is the traffic flow dynamics in communication systems, the way that congestion occurs in a network, the analysis of network latencies, and of course stability issues. The MAC layer and network layer protocols can be considered to be more relevant from a control viewpoint, and as such is not the focus of this chapter—which focuses entirely on modeling the traffic dynamics in a specific component—the wireless router.

Network traffic is usually considered as a sequence of arrival of data packets in a time interval, called a *point process*, [3], which is characterized by a set of packet arrival times $\{t_1, t_2, \ldots\}$. Some traffic models rely on the use of stochastic processes to represent, for example, the packet arrival times $\{t_1, t_2, \ldots\}$. The most well-known stochastic process is based on assuming that the packet arrival data is independent and the packets are exponentially distributed—this is the (memoryless) Poisson process traffic model. An important drawback of the Poisson model is that it cannot capture traffic burstines, [5], as the Poisson process assumes that the arrival rate is constant. One approach to handle this is to consider the basic process as a Poisson process and modulate the arrival rate—this leads to a Markov modulated Poisson process, [7]. Another modeling approach that considers inter-arrival times is the Markov model, which models the events in a traffic network by a finite number of states, [7]. A general problem with models which are based on inter-arrival times is that since they consider the arrival of each data packet as a separate event, there is a significant overhead. This has motivated the use of fluid flow models to describe network traffic, which characterizes traffic flow by the flow rate. Since flow rate changes much less frequently than individual packet arrival times, the computational burden is reduced [1]. The assumption is that the changes in flow rates capture the different events that occur in the network [7]. The use of graph theoretic tools to model large-scale communications network is, to the best of the authors' knowledge, surprisingly limited, except some works such as [2, 12].

There are related issues surrounding network traffic modeling, such as congestion handling, optimal router buffer sizing, channel capacity [13], channel/link utilization, network latency, etc. Network congestion can occur due to a variety of reasons—e.g., by improper sizing of buffers in network routers [14], traffic flow exceeding actual network capacity, etc. Buffers are used to prevent packet drops and to maintain link/channel utilization.[1] The sizing of router buffers is an important area of research in network communications, [14, 23]. Too large a buffer size and network latency can be significantly degraded, too small a buffer size and we can expect channel/link utilization to be poor and even lead to buffer instability.

Existing network traffic modeling techniques do not capture the inherent physical nature of the interconnection in the data flow. The design of various advanced network protocols rely primarily on the type of traffic flow (e.g., Poisson process), [9], or the type of MAC protocols used, [4], but not on the interconnection structure itself. There is need for a network-based modeling technique which can capture the interconnection structure of each node in a communication network, as well as to model the interconnection structure of the entire network, while capturing the underlying fundamental physical laws. The framework of port-Hamiltonian systems, [19–21] and energy based modeling, [17] is very well suited to this task. The concept here is that systems from different domains (mechanical, electrical, chemical, etc.) can be interconnected to form a network, and this networked system can then be modeled (and controlled) by using energy as the unifying theme, [16–19, 21]. A complex physical system can be viewed as an interconnection of simpler subsystems. The interconnection, in this framework, then results in the total energy being equal to the sum of energies of individual subsystems. A fundamental result is that any (power conserving) interconnection of (multiple) port-Hamiltonian systems results in another port-Hamiltonian system, [8, 19]. Recently, the port Hamiltonian framework has been extended to modeling dynamics on graphs, [22]; an extension of this to communication networks may provide interesting results in stability and control.

1.2 Models of Wireless Communication Systems

Our objective in this work is to model the traffic flow in communication systems, from an energy viewpoint, more specifically in a Hamiltonian setting. We abstract the working methodology of the router to a set of simple differential equations—from a traffic flow viewpoint. Note that data flow in a communication network is typically considered in time slots or intervals. In this paper we make the assumption that the data flow can be described by differential equations, and thus we model the traffic flow as continuous linear time invariant systems. We focus on the data inflow/outflow, the influence of the buffer on traffic dynamics, and the analogy of dissipation in such

[1] A routers link is utilized as long as the router is sending data on that link. Link utilization is important, for, e.g., if a particular router is given a certain channel bandwidth and it is unable to use that, it would be a significant waste of expensive resource.

systems. While our focus is on router traffic, the same (or similar) dynamics can also be used for other communication devices which have buffering capability.

Let $\omega(t)$ denote the number of buffered packets in a router at time t. Let ω_ξ denote the size of the buffered packets in the router at time t. Next, let μ denote the number of transmitted packets by the router at time t, and finally μ_ξ denotes the size of transmitted packets by the router at time t. We note the following constitutive relationships:

$$\omega_\xi = \frac{\omega}{q}, \quad \mu_\xi = \frac{\mu}{s} \tag{1.1}$$

with q and s being unit-less scaling constants.

1.2.1 Router Configuration—Transmitted Data Fed Back to Receiver

First we consider the simplest possible router configuration, where the transmitted data is directly fed back to the receiver side. Further we make the following assumptions:

- transmission is instantaneous and takes place at a fixed rate
- reception (through the looping back) is also instantaneous (i.e., no delays)
- no buffer overflows (infinite buffer assumed)
- no transmission or reception losses

Then we obtain the two differential equations governing the router dynamics as

$$\frac{\mathrm{d}\omega}{\mathrm{d}t} = -\mu, \quad \frac{\mathrm{d}\mu_\xi}{\mathrm{d}t} = \omega_\xi$$

The first differential equation, $\frac{\mathrm{d}\omega}{\mathrm{d}t} = -\mu$ expresses that the rate of change in the number of buffered packets, $\omega(t)$ is a function of the number of packets that have been transmitted (out of the buffer). The second differential equation, $\frac{\mathrm{d}\mu_\xi}{\mathrm{d}t} = \omega_\xi$ expresses that the rate of change in the size of the transmitted packets is equal to the size of the buffered packets—in other words, all the buffered packets are instanteneously transmitted each time. In a Hamiltonian formulation we can rewrite the above equations as follows. First, let (ω, μ_ξ) be the energy states and (ω_ξ, μ) be the energy co-states. Then we define the Hamiltonian as $H(\omega, \mu) = \frac{1}{2}\left(\frac{\omega^2(t)}{q} + \mu_\xi^2(t)s\right)$. Then we have

$$\begin{bmatrix} \dot{\omega} \\ \dot{\mu_\xi} \end{bmatrix} = \begin{bmatrix} 0 & -1 \\ 1 & 0 \end{bmatrix} \begin{bmatrix} \frac{\partial H}{\partial \omega} = \omega_\xi \\ \frac{\partial H}{\partial \mu_\xi} = \mu \end{bmatrix} \tag{1.2}$$

In fact, these are equations of a harmonic oscillator, and in this case we can compute the time rate of the Hamiltonian as

$$\frac{\mathrm{d}H}{\mathrm{d}t} = \omega_\xi\,(-\mu) + \mu\omega_\xi = 0$$

which is indeed what we expect from a harmonic oscillator system. The main assumption made in the set of Eq. (1.2) is that buffering and data transmission are instantaneous. In actual practice (always occurs during network congestion), usually only a part of the buffered data is transmitted, and the router continues buffering the remainder of the data. Clearly, the assumptions we made above are unrealistic for wireless systems. The following subsections will incorporate additional criteria to make the models more realistic.

1.2.2 Router Configuration—No Reception, Only Transmission

In the second case we assume the router is only transmitting data and is not receiving any data. We assume that the router has some data already buffered, and this data is being transmitted.

The dynamics then has the following form:

$$\frac{\mathrm{d}\omega}{\mathrm{d}t} = -c_1\mu, \quad \frac{\mathrm{d}\mu_\xi}{\mathrm{d}t} = c_2\omega_\xi$$

The interpretation of $\frac{\mathrm{d}\mu_\xi}{\mathrm{d}t} = c_2\omega_\xi$ is simply that the buffer does not empty its contents instantaneously, and in each time slot it can empty only $c_2\omega_\xi$ amount of data. Similarly, the interpretation of $\frac{\mathrm{d}\omega}{\mathrm{d}t} = -c_1\mu$ is that the rate of change of buffered packets is always a fraction (bounded from above by 1) of the total number of transmitted packets. If we consider the Hamiltonian, as before, $H(\omega,\mu) = \frac{1}{2}\left(\frac{\omega^2(t)}{q} + \mu_\xi^2(t)s\right)$, we obtain the Hamiltonian dynamics as

$$\begin{bmatrix} \dot{\omega} \\ \dot{\mu}_\xi \end{bmatrix} = \begin{bmatrix} 0 & -c_1 \\ c_2 & 0 \end{bmatrix} \begin{bmatrix} \frac{\partial H}{\partial \omega} = \omega_\xi \\ \frac{\partial H}{\partial \mu_\xi} = \mu \end{bmatrix} \tag{1.3}$$

These dynamics are the same as the dynamics of a harmonic oscillator only when $c_1 = c_2 = 1$. The resulting time rate of change of the Hamiltonian is

$$\frac{\mathrm{d}H}{\mathrm{d}t} = \omega_\xi\,(-c_1\mu) + \mu c_2\omega_\xi = (c_2 - c_1)\,\mu\omega_\xi \tag{1.4}$$

The sign of $\frac{\mathrm{d}H}{\mathrm{d}t}$ then depends on the relative values of c_1 and c_2. As an illustrative simulation, we consider a router with 1000 kb of buffered data; and a 10 kbps transmission rate; the router does not receive any data. In this example, when data transmission just starts, $c_1 = 1$ and then it keeps decreasing. $c_2 = 0$ at the beginning and and converges to 0.5 (toward the end of the data transmission). The left plot in

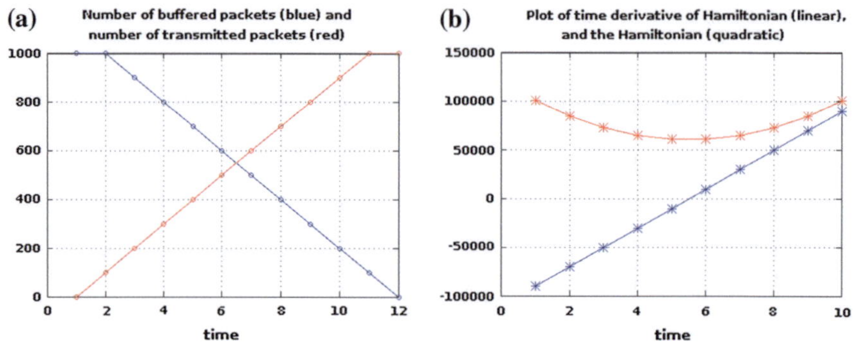

Fig. 1.1 Simulation response for a system with buffered data and only data transmissions. **a** Buffered (*decreasing curve*) and transmitted (*increasing curve*) packets; starting with 1000 buffered packets and 0 transmitted packets. The buffer is empty after 12 time units. **b** Rate of change of the Hamiltonian; also shown is the quadratic Hamiltonian function itself (the Hamiltonian values are scaled for easy comparison). Minimum energy corresponds to $c_2 = c_1$

Fig. 1.1 shows the buffered and transmitted packets; after about 12 time units the buffer is emptied. The right plot of Fig. 1.1 plots the time rate change of the Hamiltonian from Eq. 1.4; the total energy (Hamiltonian) is also shown in this plot (the quadratic curve). The interpretation of $\frac{\mathrm{d}H}{\mathrm{d}t}$ being negative (i.e., $c_2 < c_1$) is that there are more buffered packets than transmitted packets,[2] when $\frac{\mathrm{d}H}{\mathrm{d}t}$ becomes positive (i.e., $c_2 > c_1$) there are more transmitted packets than buffered packets.[3]

1.2.3 Router Configuration—No Transmission, Only Reception

In the third case we assume that the router is only receiving data with no data transmissions. The dynamics then has the following form:

$$\frac{\mathrm{d}\omega}{\mathrm{d}t} = 0 + \eta, \quad \frac{\mathrm{d}\mu_\xi}{\mathrm{d}t} = 0$$

where η is the number of data packets being received by the router at time t. If we consider the Hamiltonian, as before, $H(\omega, \mu) = \frac{1}{2}\left(\frac{\omega^2(t)}{q} + \mu_\xi^2(t)s\right)$, we obtain the Hamiltonian dynamics as

[2]We will see in Sect. 1.3 that this implies that the buffer utilization is improving, whereby buffer utilization can be defined as simply the amount of buffer space that is being utilized.

[3]In Sect. 1.3 we will see that this implies that the buffer utilization is becoming poorer, and is not a desirable condition in wireless communication.

$$\begin{bmatrix} \dot{\omega} \\ \dot{\mu}_\xi \end{bmatrix} = \begin{bmatrix} 0 & 0 \\ 0 & 0 \end{bmatrix} \begin{bmatrix} \frac{\partial H}{\partial \omega} = \omega_\xi \\ \frac{\partial H}{\partial \mu_\xi} = \mu \end{bmatrix} + \begin{bmatrix} 1 \\ 0 \end{bmatrix} \eta$$

Then

$$\frac{\mathrm{d}H}{\mathrm{d}t} = \omega_\xi \eta > 0, \text{ if } \eta > 0$$

In reality, the energy cannot indefinitely increase. A router does not have an infinite buffer size, and once the routers buffer capacity limit is reached it begins to discard data packets. This phenomena shall be modeled in a dissipation framework in Sect. 1.2.6. In any case, we have the implication that $H(\omega, \mu_\xi)$ is bounded from above.

1.2.4 Router Configuration—Both Transmission and Reception Enabled

In the last case, we assume the router is both receiving and transmitting data. The dynamics then has the following form:

$$\frac{\mathrm{d}\omega}{\mathrm{d}t} = -c_1\mu + \eta, \quad \frac{\mathrm{d}\mu_\xi}{\mathrm{d}t} = c_2\omega_\xi$$

where η is the number of data packets being received by the router at time t. If we consider the Hamiltonian, as before, $H(\omega, \mu) = \frac{1}{2}\left(\frac{\omega^2(t)}{q} + \mu_\xi^2(t)s\right)$, we obtain the Hamiltonian dynamics as

$$\begin{bmatrix} \dot{\omega} \\ \dot{\mu}_\xi \end{bmatrix} = \begin{bmatrix} 0 & -c_1 \\ c_2 & 0 \end{bmatrix} \begin{bmatrix} \frac{\partial H}{\partial \omega} = \omega_\xi \\ \frac{\partial H}{\partial \mu_\xi} = \mu \end{bmatrix} + \begin{bmatrix} 1 \\ 0 \end{bmatrix} \eta \tag{1.5}$$

Then

$$\frac{\mathrm{d}H}{\mathrm{d}t} = (c_2 - c_1)\,\mu\omega_\xi + \omega_\xi\eta$$

In Fig. 1.2 we simulate the traffic of a router with varying input data rates and fixed transmission rates. Note that the minimum energy corresponds to $c_2 = c_1$. In Fig. 1.2c the number of buffered packets, $\omega(t)$, remains constant since input data rate is equal to transmission rate. The buffer utilization (which means the amount of buffer space utilized, ideally this should be fairly high) for the two cases when η is greater than or equal to the transmission rate is very good. In the first case when η is smaller than the transmission rate, the buffer utilization is poor. However, when η is much greater than the transmission rate, the buffer will quickly fill up, see Fig. 1.2e and this can lead to router instability; this will be studied in Sect. 1.3.

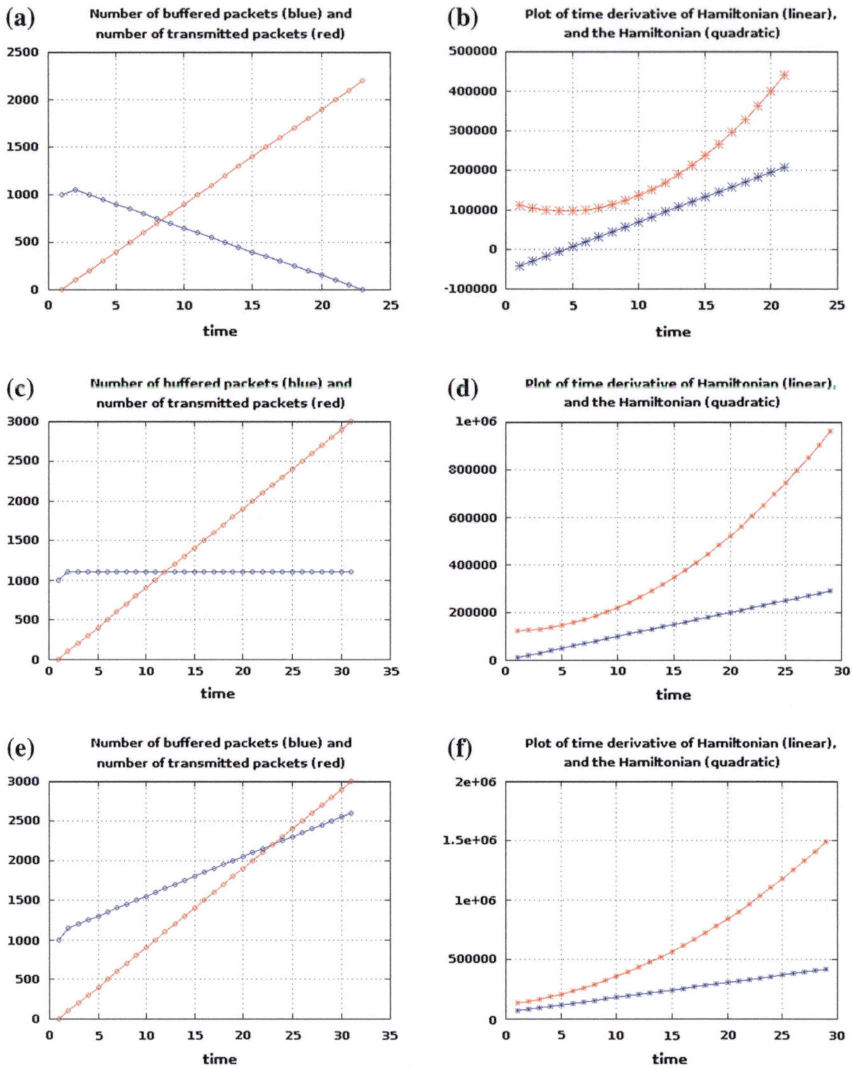

Fig. 1.2 Simulation response for a system with varying input data rates and fixed transmission rates. **a** Input data rate = 50kbps, Transmission rate = 100 kbps. Buffered data (decreasing) and transmitted data (*increasing*). **b** Rate of change of the Hamiltonian and the quadratic Hamiltonian. **c** Input data rate = 100 kbps, Transmission rate = 100 kbps. Buffered data (*constant*) and transmitted data (*increasing*). **d** Rate of change of the Hamiltonian and the quadratic Hamiltonian. **e** Input data rate = 150 kbps, Transmission rate = 100 kbps. Buffered data (*increasing, smaller slope*) and transmitted data (*increasing*). **f** Rate of change of the Hamiltonian and the quadratic Hamiltonian

1.2.5 Dirac Structures

We have claimed that Eq. (1.5) is a port-Hamiltonian model. In this section we will prove this claim by showing that the space of flows and efforts corresponding to the interconnection structure $\begin{bmatrix} 0 & -c_1 \\ c_2 & 0 \end{bmatrix}$ is a Dirac structure. We consider the dynamics as in Eq. 1.5 and rewrite it as

$$\begin{bmatrix} \dot{\omega} \\ \dot{\mu}_\xi \end{bmatrix} = \underbrace{\begin{bmatrix} 0 & -1 \\ 1 & 0 \end{bmatrix}}_{J} \begin{bmatrix} \frac{\partial H}{\partial \omega} = \omega_\xi \\ \frac{\partial H}{\partial \mu_\xi} = \mu \end{bmatrix} + \underbrace{\begin{bmatrix} 0 & 1 - c_1 \\ c_2 - 1 & 0 \end{bmatrix}}_{B} \begin{bmatrix} \frac{\partial H}{\partial \omega} = \omega_\xi \\ \frac{\partial H}{\partial \mu_\xi} = \mu \end{bmatrix} + \begin{bmatrix} 1 \\ 0 \end{bmatrix} \eta$$

The matrix $J := \begin{bmatrix} 0 & -1 \\ 1 & 0 \end{bmatrix}$ is the standard skew-symmetric Poisson structure (and hence Dirac). We denote the new matrix $\begin{bmatrix} 0 & 1 - c_1 \\ c_2 - 1 & 0 \end{bmatrix}$ by $\mathfrak{B}$ and call this the Buffer structure. The resulting energy balance equation will have three terms, as follows:

$$\frac{\mathrm{d}H}{\mathrm{d}t} = 0 + (c_2 - c_1)\,\mu\omega_\xi + \omega_\xi \eta$$

The 0 power contribution is due to the skew-symmetric Poisson structure. The $\omega_\xi \eta$ power contribution is the supply rate. Finally, the $(c_2 - c_1)\mu\omega_\xi$ is the *buffered power* in the system. Note that this is *not* the dissipated power; there is no loss of energy/power in the system. Instead, because the router is unable to transmit the entire received data in a single time slot, some data are left buffered in the router. This is given by $(c_2 - c_1)\mu\omega_\xi$.

Recall the definition of a Dirac structure from [21].

Definition 1.1 A constant Dirac structure on an l-dimensional linear space, $\mathfrak{F}$ is a linear subspace $\mathfrak{D} \subset \mathfrak{F} \times \mathfrak{F}^*$ such that $\mathfrak{D} = \mathfrak{D}^\perp$, where $\perp$ is defined with respect to a bilinear form on $\mathfrak{F} \times \mathfrak{F}^*$.

It follows that $\langle e | f \rangle = 0, \forall (e, f) \in \mathfrak{D}$, [21].

Set $f_s := \begin{bmatrix} \dot{\omega} \\ \dot{\mu}_\xi \end{bmatrix}$, $e_s := \begin{bmatrix} \frac{\partial H}{\partial \omega} \\ \frac{\partial H}{\partial \mu_\xi} \end{bmatrix}$, $e := \frac{\partial H}{\partial \omega}$ and $f := \eta$. Denote $x = [\omega, \mu_\xi]^T$. Then we have the energy balance

$$\begin{aligned} \frac{\mathrm{d}H(x)}{\mathrm{d}t} &= \frac{\partial H}{\partial x}\dot{x} = -\, e_s^T f_s = \begin{bmatrix} \frac{\partial H}{\partial \omega} \, \frac{\partial H}{\partial \mu_\xi} \end{bmatrix} \begin{bmatrix} \dot{\omega} \\ \dot{\mu}_\xi \end{bmatrix} \\ &\Rightarrow e_s^T f_s + (c_2 - c_1)\omega_\xi \mu + \frac{\partial H}{\partial \omega}\eta = 0 \\ &\Rightarrow \underbrace{e_s^T f_s}_{\text{stored power}} + \underbrace{(c_2 - c_1)\omega_\xi \mu}_{\text{buffered power}} + \underbrace{ef}_{\text{supplied power}} = 0 \end{aligned}$$

Fig. 1.3 The port-Hamiltonian structure which includes a buffer element; note that dissipation has not yet been considered

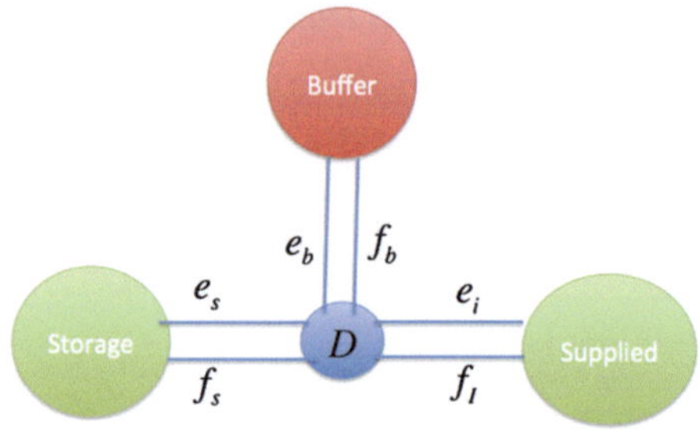

Thus $\langle\langle\cdot,\cdot\rangle\rangle = 0$ and this bilinear form defines a Dirac structure for the matrix operator $J + \mathfrak{B}$ on the space $\mathfrak{F} \times \mathfrak{F}^*$. There is an additional term, $(c_2 - c_1)\omega_\xi \mu$, which is not seen in the usual energy balance equations for standard physical systems. Denote $(c_2 - c_1)$ by $\tilde{c}$, so we have $\tilde{c}\omega_\xi \mu$, which corresponds to the *buffered energy* in the system. This motivates a modified port-Hamiltonian interconnection structure, e.g., see p. 14 of [21] for the standard port-Hamiltonian interconnection structure, as follows (Fig. 1.3).

1.2.6 Including Dissipation

The analogy of electrical or mechanical dissipation for a communication system (e.g., a single router) is related to congestion control. Wireless technologies have become very popular in the past decade. This has led to a proliferation of devices connected over a WiFi channel (such as 802.11). Inevitably, this has led to network congestion. Even in wired communication systems congestion is a commonly observed problem.

Consider Fig. 1.4 where there are six routers, each modeled as a port-Hamiltonian system as in Eq. 1.5. Router 1, denoted by $\sum_1$, is transmitting to router 3, $\sum_3$; and router 2, $\sum_2$, is transmitting to router 4, $\sum_4$. This example is strongly motivated from the example presented in [6]; we present the same example in a port-Hamiltonian context. Denote, as in [6], λ_i to be the sending rate of router i, and $\tilde{\lambda}_i$ to be the

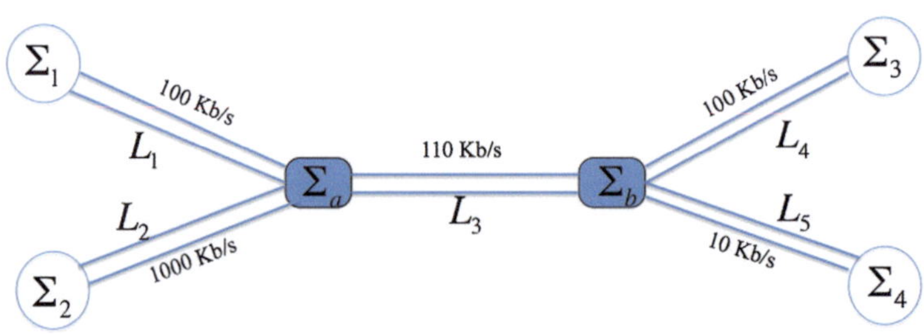

Fig. 1.4 Six routers, modeled as PH systems Σ_i, with channel capacities indicated

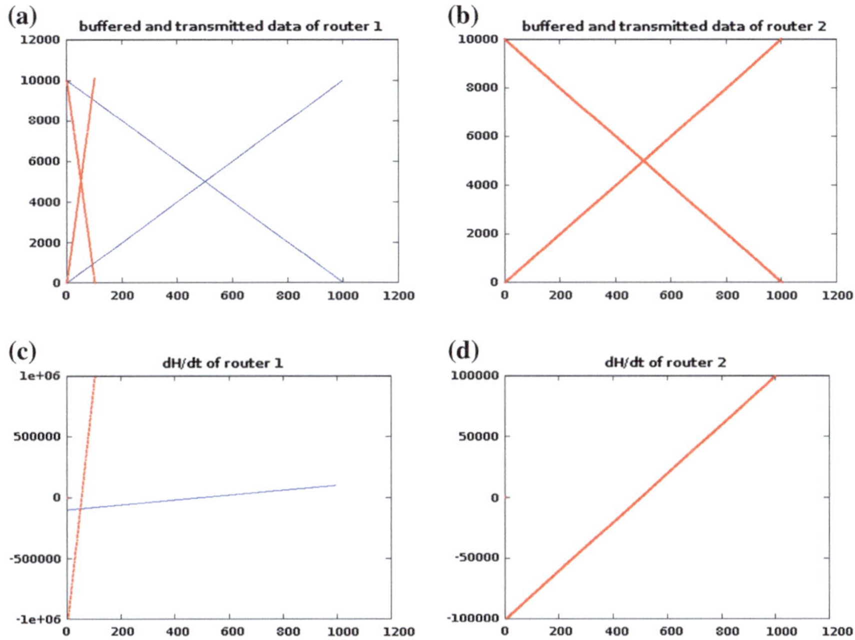

Fig. 1.5 Simulation of network inefficiency when upstream routers do not have feedback from downstream routers. **a** *Router1* Without feedback (*light*), with feedback (*dark*). Both buffered and transmitted packets are shown. **b** *Router2* Without feedback (*light*), with feedback (*dark*). Both buffered and transmitted packets are shown. **c** Energy rate change for router 1, with (*dark*) and without (*light*) feedback. **d** Energy rate change for router 2, with (*dark*) and without (*light*) feedback

actual outgoing rate of router i. Assuming there is no feedback from the network the total throughput is just 20 kbps. This is because source 2 is limited to 10 kbps because of link 5. And because source 1 is competing with source 2 (which is trying to transmit at 1000 kbps) on link 3, source 1 can only transmit at 10 kbps. Thus the total throughput is 20 kbps. However, if source 2 knew (via feedback) that it cannot exceed 10 kbps, it would not attempt to transmit at 1000 kbps, and it would simply transmit at 10 kbps. Then source 1 could have transmitted at 100 kbps. This is a form of network inefficiency and it can lead to the phenomena of *congestion collapse*, [6, 10]—which simply means that the achieved throughput $\rightarrow 0$ as the offered load $\rightarrow \infty$.

In Fig. 1.5 we perform simulations to model the effect of feedback on the router traffic dynamics. In Fig. 1.5a we see the buffered and transmitted packets of router 1 under conditions of no feedback and when feedback is present. Note that when feedback is not present (router 1 operates at 10 kbps throughput); when feedback is present (router 1 operates at 100 kbps) and Fig. 1.5a shows a significantly faster response time in emptying the buffer (the two dark curves until $t = 150$. When feedback is absent, Fig. 1.5a shows a much slower response time in emptying the buffer (takes about 1000 time units). Router 2, in Fig. 1.5b has the same response times since it continues to operate at 10 kbps throughout with or without feedback. We observe in Fig. 1.5c that the slope for $\frac{\mathrm{d}H}{\mathrm{d}t}$ is much sharper with feedback and with the sign changing very quickly, indicating that buffer utilization is reducing faster and leading to reduced buffer stability; we will study this formally in Sect. 1.3. To implement congestion control each node (router) requires feedback from the downstream network. Feedback may be in the form of missing packets (downstream); if missing packets are detected the sources reduce their transmission rate. In congestion, control dropping packets (often deliberately) is a form of feedback control, whereby the upstream nodes detect the dropped packets and react by reducing their transmission rates. Note that a router may also drop packets if the buffer is filled up. In this paper we model the dropping of packets in the dissipation framework of port-Hamiltonian systems. This dissipation is not *always present* in the system dynamics; and is usually observed only during congestion and occurs at discrete time intervals. To model this we first have the following dynamics for the buffered packets:

$$\frac{\mathrm{d}\omega}{\mathrm{d}t} = -c_1\mu - R_\omega\omega + \eta$$

where the parameter $R_\omega(t)\omega(t)$ models the number of dropped packets (from the buffer) at time t. We then have the corresponding port-Hamiltonian dynamics:

$$\begin{bmatrix} \dot{\omega} \\ \dot{\mu_\xi} \end{bmatrix} = \underbrace{\begin{bmatrix} 0 & -1 \\ 1 & 0 \end{bmatrix}}_{J} \begin{bmatrix} \frac{\partial H}{\partial \omega} = \omega_\xi \\ \frac{\partial H}{\partial \mu_\xi} = \mu \end{bmatrix} + \underbrace{\begin{bmatrix} 0 & 1 - c_1 \\ c_2 - 1 & 0 \end{bmatrix}}_{\mathfrak{B}} \begin{bmatrix} \frac{\partial H}{\partial \omega} \\ \frac{\partial H}{\partial \mu_\xi} \end{bmatrix} - \underbrace{\begin{bmatrix} R_\omega & 0 \\ 0 & 0 \end{bmatrix}}_{R} \begin{bmatrix} \frac{\partial H}{\partial \omega} \\ \frac{\partial H}{\partial \mu_\xi} \end{bmatrix} + \begin{bmatrix} 1 \\ 0 \end{bmatrix} \eta \tag{1.6}$$

This corresponds to the standard port-Hamiltonian input-output model with dissipation, with $J - R = \begin{bmatrix} -R_\omega & -1 \\ 1 & 0 \end{bmatrix}$. Denote: $e := \frac{\partial H}{\partial \omega}$, $f := \eta$. We then have the energy balance as

$$\frac{\mathrm{d}H}{\mathrm{d}t} = \underbrace{e_s^T f_s}_{\text{stored power}} + \underbrace{\tilde{c}\mu\omega_x i}_{\text{buffered power}} + \underbrace{ef}_{\text{supplied power}} - \underbrace{R_\omega\omega_\xi^2}_{\text{dissipated power}} = 0 \tag{1.7}$$

With dissipation included we obtain the complete structure of the port-Hamiltonian system for the class of communication systems considered in this paper and is as shown below (Fig. 1.6).

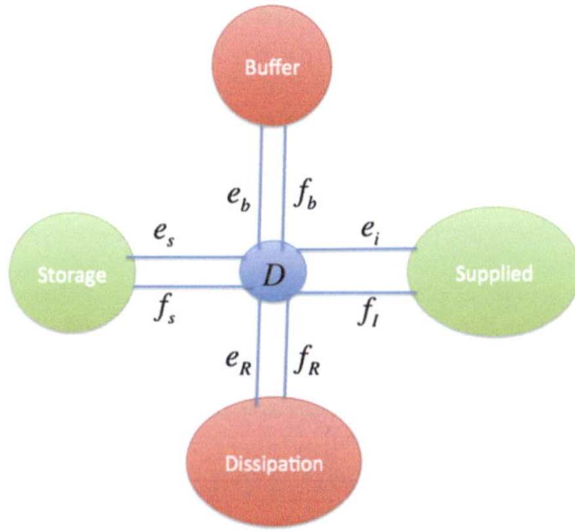

Fig. 1.6 The port-Hamiltonian structure which includes a buffer element; note that dissipation has not yet been considered

1.3 Formal Characterization of Stability, Buffer Utilization, and System Latency

A fundamental problem in wireless communication networks is that of designing a scheduling policy that is guaranteed to stabilize the node buffers (i.e., a router in a network) for a given arrival process with some specified arrival rate. Related problems involve guaranteed maximum latency (i.e. delay) and buffer utilization (discussed below). While we do not discuss scheduling policies in this chapter, the concept of stability, latency, and buffer utilization will be analyzed below. A typical definition of stability for wireless networks is as follows, see [15].

Definition 1.2 (*Buffer/Router Stability*) Let $\omega(t)$ be the number of packets in a routers buffer at time t. The buffer is called stable if

$$\lim_{B\to\infty}\lim_{t\to\infty}\sup \Pr\{\omega(t) > B\} = 0$$

The above definition simply says that a queue is stable if its asymptotic buffer overflow probability goes to zero as the buffer size, B, goes infinitely large, [15]. In other words the queue length (in the buffer) should not go to infinity! The network is called stable if all node buffers are stable.

Scheduling algorithms should also consider end-to-end delays (latency). Network delay is characterized as acceptable if,

$$\Pr\left(\frac{X^{\text{agg}}(t)}{\lambda}\right) > \tau_d = 0$$

where $X^{\text{agg}}(t)$ denotes the aggregated queue length, λ denotes the arrival rate, and τ_d denotes the service time allowed to a particular task. (Assuming packets are arriving at a constant rate and all queues are First In First Out), [23]. The above definition is simply saying that the delay in the network is considered acceptable if the probability that the latency exceeds a delay threshold is zero. Another way to measure the latency is, related to the definition in [3], as follows:

Definition 1.3 (*Latency*) The packet latency can be defined as the ratio of the buffer size to the channel/link capacity, i.e.,

$$\tau = \frac{B}{C} \quad \text{and} \quad \tau_{\text{current}} = \frac{\omega(t)}{\mu(t)/s}$$

where B is the buffer sizing required to keep the link fully utilized (see Definition 1.4 and the discussion following that), μ/s is the number of packets transmitted in a unit interval of time, and C is the channel/link capacity (in bits per sec). Note that τ is the maximum latency possible when the buffer is full, and τ_{current} is the current latency experienced by a user.

Buffers add queuing delay, and this increases packet latency. Thus as the requirement on the current buffer size increases, the queuing delay (latency) proportionately increases.

A different problem exists in the utilization of the buffer.

Definition 1.4 (*Buffer Utilization*) (optimal) Buffer utilization is defined as the amount of buffer sizing required in order to keep the link fully utilized while ensuring latencies satisfy service times, during congestion.

While buffer utilization may be simply defined as the amount of buffer space occupied; buffer utilization is strongly linked to the channel/link utilization. The link or channel utilization must remain high always, especially during congestion. This in turn implies that a router's buffer must never be empty, or else the link utilization can go down. Thus during congestion we can expect that the desired relation between the buffer size and channel capacity to be as follows: $c(t) \rightarrow C \Rightarrow \omega(t) \rightarrow \tilde{B}$, where $c(t)$ is the current channel utilization and C is the maximum channel capacity, $\omega(t)$ captures the current buffer usage, and $\tilde{B}$ is the congestion buffer threshold ($\tilde{B} < B$). Note that B is the buffer size which is calculated via a standard rule set forth in [23] and is theoretically high enough to ensure complete link utilization during congestion, see the footnote below. For example, advanced congestion control mechanisms prefer the buffer size, $\tilde{B}$, to be as high as possible (during congestion) so that link utilization remains high—but below[4] the typical buffer size B obtained via the [23] rule, so as to keep latencies under control.

Definition 1.5 (*Link Utilization*) Link utilization can be defined as the ratio of the current data transfer rate to the link capacity, i.e.,

---

[4]How much *below* is a complicated question. One standard buffer sizing rule used is based on the round trip time of a flow and channel capacity, and is given by $B = RTT \times C$, [23]. Studies have shown that this rule leads to large buffer sizing, and can lead to unacceptable latencies; and these studies, [3] have provided new mechanisms for router buffer sizing (with $\tilde{B} < B$), which can provide acceptable latencies while ensuring link utilization is high.

$$LU = \frac{\mu(t)/s}{C}$$

where μ/s is the number of packets transmitted in a unit interval of time.

Ideally, during congestion, we require $LU = 1$.

To summarize, the four parameters—buffer utilization, link utilization, latency, and stability—are strongly linked with each other; in the following subsection we analyze these parameters in relation to our port-Hamiltonian model, Eq. 1.5. Note that while we ignore the port Hamiltonian model with dissipation below, the same result can be easily extended to the case with dissipation.

1.3.1 Analysis of Stability, Buffer Utilization, Link Utilization, and System Latency in a Port-Hamiltonian Setting

In this section we analyze stability, latency, and buffer utilization in a port-Hamiltonian setting. Specifically, we will study the time rate of change of the Hamiltonian, $\frac{\mathrm{d}H}{\mathrm{d}t}$ and analyze these properties. The analysis below studies how the energy balance equation directly influences these four crucial performance parameters of a communication system:

Remark 1.6 Consider the system dynamics as in Eq. 1.5 and the associated energy balance equation $\frac{\mathrm{d}H}{\mathrm{d}t} = (c_2 - c_1)\,\mu\omega_\xi + \omega_\xi\eta$. Assuming that the input $\eta(t)$ is constant, we then have the following:

(1) $c_2 = 0 \Rightarrow \frac{\mathrm{d}H(t)}{dt} > 0$ if $\eta(t) > \mu(t)$, and leads to decreasing buffer stability, optimal link utilization, and poor latency
(2) $c_2 = 0 \Rightarrow \frac{\mathrm{d}H(t)}{\mathrm{d}t} \leq 0$ if $\eta(t) \leq \mu(t)$, and leads to increasing buffer stability, low link utilization, and low latency
(3) $c_1 = 0$ and $\eta(t) > 0 \Rightarrow \frac{\mathrm{d}H}{\mathrm{d}t} > 0$ leads to buffer instability and zero link utilization
(4) $c_2 = c_1 \Rightarrow \frac{dH}{dt} > 0$, leads to increasing buffer stability, reduced buffer utilization
(5) For $0 < c_2 < c_1$, a large negative $\frac{\mathrm{d}H}{\mathrm{d}t}$ leads to improved buffer utilization but reduced buffer stability, whereas a smaller negative $\frac{\mathrm{d}H}{\mathrm{d}t}$ leads to reduced buffer utilization and increased buffer stability.
(6) For $0 < c_1 < c_2$, a large positive $\frac{\mathrm{d}H}{\mathrm{d}t}$ leads to reduced buffer utilization but increased buffer stability, whereas a smaller positive $\frac{\mathrm{d}H}{\mathrm{d}t}$ leads to increased buffer utilization and reduced buffer stability.

We now analyze the statements made in Remark 1.6. The time rate of the Hamiltonian is $\frac{\mathrm{d}H}{\mathrm{d}t} = (c_2 - c_1)\,\mu\omega_\xi + \omega_\xi\eta$.

- $c_2 = 0$ and $\eta(t) > \mu(t)$: Clearly, we have that $\frac{\mathrm{d}H}{\mathrm{d}t}$ is an increasing function. This has the implication that the number of packets in the buffer, $\omega(t)$, will eventually exceed the buffer capacity and will lead to instability, i.e., $\lim_{t\to\infty} \omega(t) > B$; where B is the buffer capacity threshold. The link utilization, $LU = \frac{\mu(t)/s}{C} \to 1$ only if sufficient buffered data is always available. In the current case we have that $\omega(t) > \mu(t)$ (at least after a finite time) implying that the requirement for buffered space is continuously increasing (and in finite time $\omega(t) \to \tilde{B}$)—indicating high buffer utilization and thereby $LU \to 1$.

- $c_2 = 0$ and $\eta(t) < \mu(t)$: Clearly we have that $\frac{\mathrm{d}H}{\mathrm{d}t}$ is a decreasing function. This has the implication that $\mu(t) > \omega(t)$. Thus $\omega(t) < B, \forall t$ and the router queue will be stable. Further, the link utilization $LU < 1$, since $\omega(t)$ is a decreasing function. Thus, though $\mu(t) > \omega(t)$ we observe that $\mu(t)$ is either a constant or a decreasing function and hence $LU < 1$. Further, we have $\omega(t)/\mu(t) < 1$, which will lead to poor buffer utilization.
- $c_2 = c_1 = 1$: This corresponds to the situation where the router does not buffer any incoming packets, it instantaneously transmits any incoming data. Then $\frac{\mathrm{d}H}{\mathrm{d}t} = \eta\omega_\xi$. This is an ideal situation from a user perspective, since the user will not experience any delays in receiving the data. The problem with this is that the buffer remains underutilized and link utilization will be poor.
- $c_2 - c_1 < 1$: Denoting $\tilde{c}$ for $(c - 1)$ we have: $\frac{\mathrm{d}H}{\mathrm{d}t} = \tilde{c}\mu\omega_\xi + \eta\omega_\xi$, with $\tilde{c} < 0$. Assume that $\eta(t)$ is a (nonzero) constant. The larger $\tilde{c}$ is, the more negative is $\frac{\mathrm{d}h}{\mathrm{d}t}$, the smaller $\tilde{c}$ is, the less negative $\frac{\mathrm{d}H}{\mathrm{d}t}$ is. The interpretation of a large $\tilde{c}$ is that c is fairly small, thus the number of transmitted packets at time t, is *small*. This can cause issues from a QoS perspective, where the user may experience significant latencies; however, the buffer utilization improves—and the system dynamics moves closer to instability. The interpretation of a smaller $\tilde{c}$ is exactly the opposite: decreased latency (and increased QoS) and poor buffer utilization—but system dynamics has improved stability.

1.4 Compositionality Results in Communication Systems

A fundamental result in the area of port-Hamiltonian systems is that the composition of two Dirac structures is again a Dirac structure [19, 21]. If we consider the total energy to be the sum of the energies of individual physical system, then the power conserving interconnection of two port-Hamiltonian systems is again a port-Hamiltonian system [19, 21]. In this section we study the interconnection of two port-Hamiltonian systems, each representing a communication system equipped with a buffer.

In a wireless network with multiple nodes (e.g., routers), often the nodes enter into contention for channel access. When this happens, collisions occur and this can result in lost data. Further, there are other reasons for data loss—related to the physical medium—e.g., path loss, which describes the radio signal attenuation caused by free space propagation, scattering, reflection, etc. Thus, data loss in communication systems will occur in general, and especially during congestion. From a systems viewpoint we can consider the data loss (due to any of these reasons) as being a result of interconnection loss (and could be looked upon as a dissipation in the interconnection, i.e., lossy interconnection). Indeed, the following result formally captures this in the sense that when we interconnect two Dirac structures, the data loss can be captured through a lossy interconnection. The following theorem is an extension of the compositionality proof in [8].

Theorem 1.7 *Let $\mathfrak{D}_A$ denote the Dirac structure which represents router A, and let $\mathfrak{D}_B$ denote the Dirac structure representing router B. We have $\mathfrak{D}_A \subset \mathfrak{F}_1 \times \mathfrak{F}_1^* \times \mathfrak{F}_2 \times \mathfrak{F}_2^*$, and defined w.r.t. their bilinear forms. Then $\mathfrak{D}_A || \mathfrak{D}_B$ is also a Dirac structure w.r.t. the bilinear form on $\mathfrak{F}_1 \times \mathfrak{F}_1^* \times \mathfrak{F}_3 \times \mathfrak{F}_3^*$, with a lossy interconnection.*

Proof This proof follows the same spirit as that of Cervera et al. If $\mathfrak{D}_A$ and $\mathfrak{D}_B$ are Dirac structures, then they admit the following image representations:

$$\begin{aligned}\mathfrak{D}_A &= [E_1\ F_1\ E_{2A}\ F_{2A}\ 0\ 0]^T\\ \mathfrak{D}_B &= [0\ 0\ E_{2B}\ F_{2B}\ E_3\ F_3]^T\end{aligned}$$

Furthermore, for the composition of the two Dirac structures we place the following constraint:

$$e_{2A} = e_{2B}, \quad f_{2A} = -kf_{2B}$$

Then $(f_1, e_1, f_3, e_3) \in \mathfrak{D}_A || \mathfrak{D}_B \iff \exists \lambda_A, \lambda_B$ s.t.

$$[f_1\ e_1\ 0\ 0\ f_3\ e_3]^T = \begin{bmatrix} E_1 & F_1 & E_{2A} & F_{2A} & 0 & 0\\ 0 & 0 & E_{2B} & -kF_{2B} & E_3 & F_3\end{bmatrix}^T \begin{bmatrix}\lambda_A\\ \lambda_B\end{bmatrix} \iff$$

$$\forall (\beta_1, \alpha_1, \beta_2, \alpha_2, \beta_3, \alpha_3)\ s.t.$$

$$\begin{bmatrix}\beta_1\ \alpha_1\ \beta_2\ \alpha_2\ \beta_3\ \alpha_3\end{bmatrix} \begin{bmatrix} E_1 & F_1 & E_{2A} & F_{2A} & 0 & 0\\ 0 & 0 & E_{2B} & -kF_{2B} & E_3 & F_3\end{bmatrix}^T = 0$$

This gives

$$\begin{aligned}\beta_1^T f_1 + \alpha_1^T e_1 + \beta_2^T f_{2A} - k\beta_2^T f_{2A} + \alpha_2^T e_{2A} - \alpha_2^T e_{2B} + \beta_3^T f_3 + \alpha_3^T e_3 = 0\\ \Rightarrow \beta_1^T f_1 + \alpha_1^T e_1 + \beta_3^T f_3 + \alpha_3^T e_3 + \beta_2^T f_{2A}\,(1-k) = 0\end{aligned}$$

$$\iff$$

$$\forall (\alpha_1, \beta_1, \alpha_2, \beta_2, \alpha_3, \beta_3)\ s.t.$$

$$\begin{bmatrix} F_1 & E_1 & F_{2A} & E_{2A} & 0 & 0\\ 0 & 0 & -kF_{2B} & E_{2B} & F_3 & E_3\end{bmatrix}^T \begin{bmatrix}\beta_1\ \alpha_1\ \beta_2\ \alpha_2\ \beta_3\ \alpha_3\end{bmatrix} = 0$$

$$\iff \forall (\alpha_1, \beta_1, \alpha_3, \beta_3) \in \mathfrak{D}_A || \mathfrak{D}_B$$
$$\beta_1^T f_1 + \alpha_1^T e_1 + \beta_3^T f_3 + \alpha_3^T e_3 = \beta_2^T f_{2A}\,(k-1)$$

Thus $\mathfrak{D}_A||\mathfrak{D}_B = (\mathfrak{D}_A||\mathfrak{D}_B)^{\perp}_{diss}$ and is a Dirac structure with interconnection losses characterized by $\beta_2^T f_{2A}(k-1)$. □

Theorem 1.7 shows that, in communication networks, the interconnection of two communication systems (such as routers) is lossy. We proved that the resulting interconnection is again a Dirac structure *with* interconnection losses. Consider a network of N routers, each router being modeled as a port-Hamiltonian system. We have the corresponding Dirac structures $\mathfrak{D}_i, i = 1, \ldots, N$. Assume that $\mathfrak{D}_1$ is a source sending packets, at a constant rate $\eta(t)$, to the destination $\mathfrak{D}_N$. Further, assume that each $\mathfrak{D}_i, i = 2, \ldots, N-1$ also receive traffic from other sources. Let us denote the interconnection losses, captured by $\beta_2^T f_{2A}(k-1)$ in Theorem 1.7, by $\mathfrak{L}_{ij}, i \neq j, i = 1 : N-1, j = 2 : N$. Then we have that the total interconnection losses in the network is the sum of each interconnection loss, i.e., $\mathfrak{L} = \sum_{i=1,j=2}^{i=N,j=N-1} \mathfrak{L}_{ij}$. Let $\mathfrak{L}_{\text{thresh}}$ be the total packet losses a network can tolerate (in terms of Quality of Service, fairness, and stability measures). Then if $\mathfrak{L} > \mathfrak{L}_{\text{thresh}}, \forall t$ the network will experience any (or all) of these issues: degraded latency, buffer instability, and poor link utilization. In fact it is possible to show, though not in this work, that there is a cascading effect on the entire network which can lead to severe traffic congestion or even network failure.

1.5 Conclusions

In this chapter we provide a port-Hamiltonian formulation of wireless network traffic flow, under the assumption of deterministic flows. The resulting Dirac structure for such systems indicates an additional term in the power balance equation, which we term as the buffered power. The loss of packets, from the buffer, is modeled as a dissipation term in the dynamics. We analyze buffer stability, channel/link utilization, and latencies using the energy balance equation. Finally we prove that the composition of two Dirac structures, each representing a communication system with buffering capability, is again a Dirac structure *but* with lossy interconnection.

The work here opens interesting possibilities. One is a study of the various conservation laws and symmetries inherent in such systems. Another is the development of novel routing protocols, for e.g., using Casimirs. Another possibility can be the study of communication networks on graphs, in a port-Hamiltonian setting—where stability and performance of large-scale communication networks can be analyzed in the well-established energy modeling framework of port-Hamiltonian systems.

References

1. A. Adas, Traffic models in broadband networks. IEEE Commun. Mag. **35**(7), 82–89 (1997)
2. H.M. Ali, A. Busson, V. Vque, Channel Assignment Algorithms: A Comparison of Graph Based Heuristics, in *Proceedings of the 4th ACM Workshop on Performance Monitoring and Measurement of Heterogeneous Wireless and Wired Networks* (ACM, New York, 2009)
3. G. Appenzeller, I. Keslassy, N. McKeown, Sizing Router Buffers, in *SIGCOMM 04*, Portland, Oregon, USA, 30 Aug–3 Sept 2004
4. A. Azgin, Y. Altunbasak, G. AlRegib, Cooperative MAC and Routing Protocols for Wireless Ad Hoc Networks, in *IEEE GLOBECOM*, 2005
5. M. Becchi, *From Poisson Processes to Self Similarity: A Survey of Network Traffic Models*, Technical Report, Citeseer, 2008
6. J.-Y.L. Boudec, Rate Adaptation, Congestion Control and Fairness—A Tutorial, Ecole Polytechnique Fédérale de Lausanne (EPFL), 12 Sept 2014
7. B. Chandrasekaran, *Survey of Network Traffic Models*, Lecture Notes (Washington University, St. Louis, 2015)
8. J. Cervera, A.J. van der Schaft, A. Baos, Interconnection of port-Hamiltonian systems and composition of Dirac structures. Automatica **43**, 212–225 (2007)
9. M. Haenggi, On routing in random rayleigh fading networks. IEEE Trans. Wireless Commun. **4**(5), 1553–1562 (2005)
10. B. Hull, K. Jamieson, H. Balakrishnan, Mitigating Congestion in Wireless Sensor Networks, in *SenSys04*, Baltimore, Maryland, USA, 3–5 Nov 2004
11. Information Technology—Open System Interconnection—Basic Reference Model: The Basic Model, ISO/IEC 7498–1 (1994)
12. M. Iliofotou et al., Network Monitoring using Traffic Dispersion Graphs, in *IMC07*, San Diego, California, USA, 24–26 Oct 2007
13. M.T. Ivrlac, W. Utschick, J.A. Nossek, Fading correlations in wireless MIMO communication systems. IEEE J. Sel. Areas Commun. **21**(5), 819–828 (2003)
14. T. Li, D. Leith, D. Malone, Buffer sizing for 802.11 based networks. IEEE/ACM Trans. Netw. **19**(1), 156–169 (2010)
15. G. Mergen, L. Tong, Stability and capacity of regular wireless networks. IEEE Trans. Inform. Theory **51**(6), 1938–1953 (2005)
16. R. Pasumarthy, A.J. van der Schaft, Achievable casimirs and its implications on control of port-Hamiltonian systems. Int. J. Control **80**(9), 1421–1438 (2007)
17. R. Ortega, A.J. van der Schaft, A.J. Mareels, B. Maschke, Putting energy back in control. IEEE Control Syst. Mag. **21**(2), 18–33 (2001)
18. R. Ortega, A.J. van der Schaft, B. Maschke, G. Escobar, Interconnection and damping assignment passivity-based control of port-controlled Hamiltonian systems. Automatica **38**(4), 585–596 (2002)
19. A.J. van der Schaft, *L2-Gain and Passivity in Nonlinear Control* (Springer, New York, Inc., Secaucus, 1999)
20. A.J. van der Schaft, Port-Hamiltonian Systems: An Introductory Survey, in *Proceedings of the International Congress of Mathematicians*, Madrid, Spain, 2006
21. A.J. van der Schaft, D. Jeltsema, *Port-Hamiltonian Systems Theory: An Introductory Overview* (NOW Publishing Inc., 2014), 188 pp (ISBN: 978-1-60198-786-0)
22. A.J. van der Schaft, B.M. Maschke, Port-Hamiltonian systems on graphs. SIAM J. Control Optim. **51**(2), 906–937 (2013)
23. C. Villamizar, C. Song, High performance TCP in ANSNET. ACM Comput. Commun. Rev. **24**(5), 45–60 (1994)

Chapter 2
Dirac Structures and Control by Interconnection for Distributed Port-Hamiltonian Systems

Alessandro Macchelli

Abstract The aim of this work is to show how the Dirac structure properties can be exploited in the development of energy-based boundary control laws for distributed port-Hamiltonian systems. Stabilisation of non-zero equilibria has been achieved by looking at, or generating, a set of structural invariants, namely Casimir functions, in closed-loop, and geometric conditions for the problem to be solved are determined. However, it is well known that this method fails when an infinite amount of energy is required at the equilibrium (dissipation obstacle). So, a novel approach that enlarges the class of stabilising controllers within the control by interconnection paradigm is also discussed. In this respect, it is shown how to determine a different control port that is instrumental for removing the intrinsic constraints imposed by the dissipative structure of the system. The general theory is illustrated with the help of two related examples, namely the boundary stabilisation of the shallow water equation with and without distributed dissipation.

2.1 Introduction

Port-Hamiltonian systems have been introduced about 20 years ago to describe lumped parameter physical systems in a unified manner, [4, 25, 26]. For these systems, the dynamic results from the power conserving interconnection of a limited set of components, each characterised by a particular "energetic behaviour," i.e. storage, dissipation, generation and conversion. The generalisation to the infinite dimensional scenario leads to the definition of distributed port-Hamiltonian systems [13, 27], that have proved to represent a powerful framework for modelling, simulation and control physical systems described by PDEs. Distributed port-Hamiltonian systems share analogous geometric properties with their finite dimensional counterpart, and also the control development follows the same rationale.

A. Macchelli (✉)
Department of Electrical, Electronic, and Information Engineering,
University of Bologna, Viale del Risorgimento 2, 40136 Bologna, Italy
e-mail: alessandro.macchelli@unibo.it

M.K. Camlibel et al. (eds.), *Mathematical Control Theory I*,
Lecture Notes in Control and Information Sciences 461,
DOI 10.1007/978-3-319-20988-3_2

This first paragraph well summarises the scientific scenario at the time I had the luck to meet Arjan, and to start collaborating with him. It was in 2001, I have to say a life ago for me, from a scientific and personal point of view. I had been staying for 6 months at the Mathematical Department of the University of Twente as a visiting Ph.D. student, with the initial idea of working on some fancy connection between sliding-mode control and port-Hamiltonian systems. After some time spent discussing with Arjan, I completely changed the topic, and I started to look at these distributed port-Hamiltonian systems, a new line of research that Arjan and Bernhard Maschke were starting to develop at those times. Everything was so intriguing to me that I continued to work on it during a second period in Twente for a Post-Doc in 2003, and until now. What I actually am professionally, I owe it also to Arjan, to his patience and clearness in teaching, and to his support and precious suggestions. The motivating idea behind this chapter is then to frame some new results on the control of distributed port-Hamiltonian systems within the classical theory and core properties of port-Hamiltonian systems, topics that Arjan thought to me and to many other PhD students during these years, and on which he is still contributing a lot. In fact, some of the results presented here are based on some recent results by him and his students for lumped parameter systems.

Since the first time I heard about distributed port-Hamiltonian systems, the general theory has been developed a lot, and most of the current research on control and stabilisation deals with the development of boundary controllers. For example, in [14, 15, 20, 23, 24], this task has been accomplished by generating a set of Casimir functions in closed-loop that independently from the Hamiltonian function relates the state of the plant with the state of the controller, a finite dimensional port-Hamiltonian system interconnected to the boundary of the distributed parameter one. The shape of the closed-loop energy function is changed by acting on the Hamiltonian of the controller. This procedure is the generalisation of the control by interconnection via Casimir generation (energy-Casimir method) developed for finite dimensional systems [19, 25], and the result is an energy-balancing passivity-based controller that is not able to deal with equilibria that require an infinite amount of supplied energy in steady state, i.e. with the so-called "dissipation obstacle."

In finite dimensions, the dissipation obstacle has been solved within the control by interconnection paradigm by defining a *new* passive output for the original system in such a way that, in closed-loop, a new set of Casimir functions that can be employed with success in the energy-shaping procedure is present, [8, 18, 28]. More precisely, in [28], a constructive way to modify the Dirac structure of the system in order to obtain a new interconnection structure that is associated to the same state evolution, but with potentially different Casimir functions is provided. Among such larger set of structural invariants, it is then possible to find the "right" Casimir functions to be employed in the control by interconnection synthesis.

Even if inspired by [28], the approach proposed here is quite different. Starting from the geometrical properties of those energy-shaping control techniques that are not limited by the dissipation obstacle [11, 12], the conditions that the Casimir functions *should* respect to obtain the same results within the control by interconnection paradigm are deduced. Then, for the given plant, new Dirac and resistive structures that allow to have not only the same state evolution, but also the previously determined Casimir functions in closed-loop are computed. At the end, the result is a new control port and, similarly to [28], the final closed-loop system is characterised by the desired set of invariants, and the limits of the "classical" control by interconnection are clearly removed. It is worth noting that, for distributed port-Hamiltonian systems, the key point is the formulation of the interconnection structure in infinite dimensions in terms of a Dirac structure on a Hilbert space, [6, 7].

This chapter is organised as follows. In Sect. 2.2, a short background on Dirac structures on Hilbert spaces and infinite dimensional port-Hamiltonian systems is given. In Sect. 2.3, the control by interconnection and the control by energy-shaping are discussed from a geometrical point, i.e. the applicability of the methods is related to the properties of the Dirac structure of the system that has to be stabilised. Then, in Sect. 2.4, the problem of defining a new control port that allows to overcome the dissipation obstacle within the control by interconnection paradigm is discussed. Then, in Sect. 2.5, the general methodology is illustrated with the help of an example, namely the shallow water equation with and without dissipation. Conclusions and ideas about future research activities are reported in Sect. 2.6.

2.2 Background

2.2.1 Dirac Structures

A Dirac structure is a linear space which describes internal power flows, and the power exchange between the system and the environment. Denote by $\mathcal{F} \times \mathcal{E}$ the space of power variables, with $\mathcal{F}$ an n-dimensional linear space, the space of flows (e.g., velocities and currents) and $\mathcal{E} \equiv \mathcal{F}^*$ its dual, the space of efforts (e.g., forces and voltages), and by $\langle e, f\rangle$ the power associated to the port $(f, e) \in \mathcal{F} \times \mathcal{E}$, where $\langle \cdot, \cdot \rangle$ is the dual product between f and e.

Definition 2.1 Consider the space of power variables $\mathcal{F} \times \mathcal{E}$. A (constant) Dirac structure on $\mathcal{F}$ is a linear subspace $\mathcal{D} \subset \mathcal{F} \times \mathcal{E}$ such that $\dim \mathcal{D} = \dim \mathcal{F}$, and $\langle e, f\rangle = 0, \forall (f, e) \in \mathcal{D}$.

A Dirac structure, then, defines a power conserving relation on $\mathcal{F}\times\mathcal{E}$. As discussed in the next Proposition, different representations are possible, [3].

Proposition 2.2 *Assume that $\mathcal{F} = \mathcal{E} = \mathbb{R}^n$, which implies that $\langle e, f\rangle = e^{\mathrm{T}} f$. Then, for any Dirac structure $\mathcal{D} \subset \mathcal{F} \times \mathcal{E}$, with there exists a pair of $n \times n$ matrices F and E satisfying the conditions*

$$E F^{\mathrm{T}} + F E^{\mathrm{T}} = 0 \qquad\qquad \operatorname{rank}\big(F \mid E\big) = n \tag{2.1}$$

such that $\mathcal{D}$ can be given in kernel representation as

$$\mathcal{D} = \left\{(f, e) \in \mathcal{F} \times \mathcal{E} \mid Ff + Ee = 0\right\} \tag{2.2}$$

or in image representation as

$$\mathcal{D} = \left\{(f, e) \in \mathcal{F} \times \mathcal{E} \mid f = E^{\mathrm{T}}\lambda,\ e = F^{\mathrm{T}}\lambda,\ \lambda \in \mathbb{R}^n\right\} \tag{2.3}$$

The definition of Dirac structure can be generalised to deal with distributed parameter systems. A possible way is to assume that the space of power variables is an Hilbert space. In this respect, Dirac structures on Hilbert spaces have been introduced in [7], while their kernel and image representations in [6]. Here, we assume that the space of flows $\mathcal{F}$ is an Hilbert space, and that the space of efforts is $\mathcal{E} \equiv \mathcal{F}$. Instead of providing their formal definition, which follows the same rationale of the finite dimensional case, their kernel and image representations is directly presented in the next Proposition, [6].

Proposition 2.3 *For any Dirac structure $\mathcal{D} \subset \mathcal{F} \times \mathcal{E}$ on an Hilbert space $\mathcal{F} \equiv \mathcal{E}$, there exists linear maps $F : \mathcal{F} \to \Lambda$ and $E : \mathcal{E} \to \Lambda$ satisfying the conditions*

$$FE^* + EF^* = 0 \qquad\qquad \overline{\operatorname{ran}\begin{pmatrix} F & E \end{pmatrix}} = \Lambda$$

being Λ an Hilbert space isometrically isomorphic to $\mathcal{F}$, such that

$$\mathcal{D} = \left\{(f, e) \in \mathcal{F} \times \mathcal{E} \mid Ff + Ee = 0\right\} \tag{2.4}$$

or, equivalently, such that

$$\mathcal{D} = \left\{(f, e) \in \mathcal{F} \times \mathcal{E} \mid f = E^*\lambda,\ e = F^*\lambda,\ \forall\lambda \in \Lambda\right\} \tag{2.5}$$

Here, $\overline{\cdot}$ and $\cdot^*$ denote the closure and the adjoint of an operator, respectively, [2].

2.2.2 Port-Hamiltonian Systems

Either in the case of lumped and distributed parameter port-Hamiltonian systems, once the Dirac structure is given, the dynamics follows when the resistive structure and the port behaviour of the energy-storage elements are given. Generally speaking, the Dirac structure defines a power conserving relation between several port variables, e.g. two internal ports $(f_S, e_S) \in \mathcal{F}_S \times \mathcal{E}_S$ and $(f_R, e_R) \in \mathcal{F}_R \times \mathcal{E}_R$, which correspond to energy-storage and dissipation respectively, and an external port $(f_C, e_C) \in \mathcal{F}_C \times \mathcal{E}_C$ which is devoted to an exchange of energy with a controller. As far as the behaviour at the resistive port is concerned, let us assume that the following linear resistive relation $\mathcal{R}$ holds

$$R_f f_R + R_e e_R = 0 \tag{2.6}$$

where R_f and R_e are $n_R \times n_R$ matrices such that

$$R_f R_e^{\mathrm{T}} = R_e R_f^{\mathrm{T}} > 0 \qquad \operatorname{rank}(R_f \mid R_e) = n_R \tag{2.7}$$

Even if most of the results presented here can be applied to a more general class of systems, in this paper we refer to the family of distributed port-Hamiltonian systems that have been studied in [9, 29], i.e. to systems described by

$$\frac{\partial x}{\partial t}(t,z) = P_1 \frac{\partial}{\partial z}\big(\mathcal{L}(z)x(t,z)\big) + (P_0 - G_0)\mathcal{L}(z)x(t,z) \tag{2.8}$$

with $x \in \mathcal{X}$ and $z \in [a, b]$. Moreover, $P_1 = P_1^{\mathrm{T}} > 0$, $P_0 = -P_0^{\mathrm{T}}$, $G_0 = G_0^{\mathrm{T}} \geq 0$, and $\mathcal{L}(\cdot)$ is a bounded and continuously differentiable matrix-valued function such that $\mathcal{L}(z) = \mathcal{L}^{\mathrm{T}}(z)$ and $\mathcal{L}(z) \geq \kappa I$, with $\kappa > 0$, for all $z \in [a, b]$. For simplicity, $\mathcal{L}(z)x(t,z) \equiv (\mathcal{L}x)(t,z)$. The state space is $\mathcal{X} = L_2(a, b; \mathbb{R}^n)$, and is endowed with the inner product $\langle x_1 \mid x_2 \rangle_{\mathcal{L}} = \langle x_1 \mid \mathcal{L}x_2 \rangle$ and norm $\|x_1\|^2_{\mathcal{L}} = \langle x_1 \mid x_1 \rangle_{\mathcal{L}}$, where $\langle \cdot \mid \cdot \rangle$ denotes the natural L_2-inner product. The selection of this space for the state variable is motivated by the fact that $H(\cdot) = \frac{1}{2}\|\cdot\|^2_{\mathcal{L}}$ is the energy function.

To define a distributed port-Hamiltonian system, the PDE (2.8) has to be "completed" by a well-defined boundary port. More precisely, given $\mathcal{L}x \in H^1(a, b; \mathbb{R}^n)$, the boundary port variables are the vectors $f_C, e_C \in \mathbb{R}^n$ given by

$$\begin{pmatrix} e_C \\ f_C \end{pmatrix} = \frac{1}{\sqrt{2}} \begin{pmatrix} W \\ \tilde{W} \end{pmatrix} \begin{pmatrix} P_1 & -P_1 \\ I & I \end{pmatrix} \begin{pmatrix} (\mathcal{L}x)(b) \\ (\mathcal{L}x)(a) \end{pmatrix} \tag{2.9}$$

where W and $\tilde{W}$ are full rank $n \times 2n$ matrices such that $W \Sigma W^{\mathrm{T}} = \tilde{W} \Sigma \tilde{W}^{\mathrm{T}} = 0$, and $W \Sigma \tilde{W}^{\mathrm{T}} = I$, being

$$\Sigma = \begin{pmatrix} 0 & I \\ I & 0 \end{pmatrix}$$

As discussed in [9, 11], it is possible to verify that $\dot{H}(x(t,\cdot)) \leq e_C^{\mathrm{T}}(t) f_C(t)$, and that (2.8) is characterised by a Dirac structure on the space of flows $\mathcal{F}_S \times \mathcal{F}_R \times \mathcal{F}_C$, with $\mathcal{F}_S = L_2(a, b; \mathbb{R}^n)$, $\mathcal{F}_R = L_2(a, b; \mathbb{R}^r)$, and $\mathcal{F}_C = \mathbb{R}^n$, being $r = \operatorname{rank} G_0$. The couple of operators $F : \mathcal{F} \to \Lambda$ and $E : \mathcal{E} \to \Lambda$ introduced in Proposition 2.3 are given by

$$F = \big(F_S \; F_R \; F_C\big) \qquad E = \big(E_S \; E_R \; E_C\big) \tag{2.10}$$

where

$$\Lambda = L_2(a, b; \mathbb{R}^n) \times L_2(a, b; \mathbb{R}^r) \times \{0\} \times \mathbb{R}^n \tag{2.11}$$

being $\{0\} \subset \mathbb{R}^n$ the set containing only the origin of $\mathbb{R}^n$. Moreover, we have that

$$\begin{aligned} F_S &= \begin{pmatrix} I \\ 0 \\ 0 \\ 0 \end{pmatrix} & F_R &= \begin{pmatrix} 0 \\ I \\ 0 \\ 0 \end{pmatrix} & F_C &= \begin{pmatrix} 0 \\ 0 \\ 0 \\ I \end{pmatrix} \\ E_S &= \begin{pmatrix} P_1 \frac{\partial}{\partial z} + P_0 \\ -G_R^{\mathrm{T}} \\ -WR\mathcal{B}_{\mathcal{J}} \\ -\tilde{W}R\mathcal{B}_{\mathcal{J}} \end{pmatrix} & E_R &= \begin{pmatrix} G_R \\ 0 \\ 0 \\ 0 \end{pmatrix} & E_C &= \begin{pmatrix} 0 \\ 0 \\ I \\ 0 \end{pmatrix} \end{aligned} \tag{2.12}$$

where $\mathcal{B}_{\mathcal{J}}(e) = \begin{pmatrix} e(b) \\ e(a) \end{pmatrix}$, with $e \in L_2(a, b; \mathbb{R}^n)$, and

$$\operatorname{dom}\begin{pmatrix} F & E \end{pmatrix} = \Big\{ (f, e) \in \mathcal{F} \times \mathcal{E} \mid e_S \text{ abs. continuous}, \\ \frac{\partial e_S}{\partial z} \in L_2(a, b; \mathbb{R}^n), \text{ and } e_C = WR\mathcal{B}_{\mathcal{J}}(e_S) \Big\} \tag{2.13}$$

It is easy to verify that the port-Hamiltonian system (2.8) is a consequence of the following port behaviour at the storage and resistive ports:

$$f_S = -\frac{\partial x}{\partial t} \qquad e_S = \frac{\delta H}{\delta x}(x) = \mathcal{L}x \qquad e_R = -\bar{G} f_R \tag{2.14}$$

where δ denotes the variational derivative, and G_R in (2.12) and $\bar{G}$ are such that $G_0 = G_R \bar{G} G_R^{\mathrm{T}}$, [27]. Note that the resistive relation is in the form (2.6) with $R_f = \bar{G}$, $R_e = I$ and $n_R = r$. Finally, simple calculations show that $F_S^* = F_S^{\mathrm{T}}$, $F_R^* = F_R^{\mathrm{T}}$, $F_C^* = F_C^{\mathrm{T}}$, $E_R^* = E_R^{\mathrm{T}}$, and

$$E_S^* = \begin{pmatrix} -P_1 \frac{\partial}{\partial z} - P_0 & -G_R & 0 & 0 \end{pmatrix} \qquad E_C^* = \begin{pmatrix} \tilde{W}R\mathcal{B}_{\mathcal{J}} & 0 & 0 & 0 \end{pmatrix}$$

with $\lambda = (\lambda_S, \lambda_R, 0, \lambda_u)$, and

$$\operatorname{dom}\begin{pmatrix} F^* \\ E^* \end{pmatrix} = \Big\{ \lambda \in \Lambda \mid \lambda_u = WR\mathcal{B}_{\mathcal{J}}(\lambda_S) \Big\} \tag{2.15}$$

2.3 Control by Interconnection and Energy-Shaping

If a port-Hamiltonian control system with Hamiltonian H_C is interconnected in power conserving way to the control port (f_C, e_C) of (2.8), the closed-loop system is again in port-Hamiltonian form, and with Hamiltonian given by the sum of the two, i.e. by $H_{cl}(x, x_C) = H(x) + H_C(x_C)$, being x_C the state variable of the controller. To use this closed-loop Hamiltonian as Lyapunov function, one has first to guarantee that this function has a minimum at the desired equilibrium with a proper choice of H_C. In both the final and infinite dimensional cases, if it is possible to find structural invariants (i.e., that do not depend on the Hamiltonian, but only on the Dirac structure) named Casimir functions of the form

$$C(x, x_C) = x_C - \Xi(x) \tag{2.16}$$

with $\Xi(x)$ some smooth well-defined functional of x, then on every invariant manifold defined by $x_C - \Xi(x) = \kappa$, with $\kappa \in \mathbb{R}$ a constant which depends on the initial plant and controller state, the closed-loop Hamiltonian may be written as, [19, 25]:

$$H_{cl}(x) = H(x) + H_C(\Xi(x) + \kappa) \tag{2.17}$$

Hence, the closed-loop equilibrium now depends on the choice of H_C, and on the invariant manifold defined by the Casimir functions the Hamiltonian H_{cl} depends on the state variable x of the plant only.

Definition 2.4 Consider a closed-loop system obtained from the power conserving interconnection at (f_C, e_C) between a couple of port-Hamiltonian systems, namely a plant with state space $\mathcal{X}$, and a (finite dimensional) controller with state space $\mathcal{X}_C \equiv \mathbb{R}^{m_C}$ for some m_C. Then, a function $C : \mathcal{X} \times \mathbb{R}^{m_C} \to \mathbb{R}$ is a Casimir function if $\dot{C} = 0$ along the trajectories of the closed-loop system for every possible choice of $H(\cdot)$ and $H_C(\cdot)$.

The applicability of the control by interconnection methodology relies then on the existence of a proper set of Casimir functions. Such property is fundamental to be able to properly shape the open-loop Hamiltonian function H, and achieve desired stability properties in closed-loop. Unfortunately, the dissipative structure of the plant may limit the number or even the existence of such structural invariants. It is well known, in fact, that a Casimir function cannot depend on the coordinates on which dissipation is present, and this implies that it is not possible to shape the closed-loop energy function along these directions. This limitation is also known as dissipation obstacle, [19].

In [1, 28], an effective way to determine the achievable Casimir functions for the closed-loop system when the plant is finite dimensional and without knowing the controller and by relying only on the Dirac and resistive structures of the plant is proposed. Such result can be generalised to infinite dimensions, [11].

Proposition 2.5 *Denote by $x \in \mathcal{X}$ the state of the plant, and by $x_C \in \mathbb{R}^{n_C}$ the state of the to-be-designed controller. Then, the achievable Casimir functions $C(x, x_C)$ associated to the Dirac structure on Hilbert space with kernel representation (2.4) and operators F and E given as in (2.12) for any kind of power conserving interconnection with the controller are such that*

$$\begin{pmatrix} 0 \; 0 \; f_C^{\mathrm{T}} \; \frac{\delta^{\mathrm{T}} C}{\delta x}(x, x_C) \; 0 \; e_C^{\mathrm{T}} \end{pmatrix}^{\mathrm{T}} \in \mathcal{D} \tag{2.18}$$

for some $(f_C, e_C) \in \mathcal{F}_C \times \mathcal{E}_C$.

Corollary 2.6 *Condition (2.18) with $C(x, x_C)$ given as in (2.16) is equivalent to*

$$-\begin{pmatrix} 0 \\ 0 \\ \frac{\delta C}{\delta x}(x, x_C) \\ 0 \end{pmatrix} = \begin{pmatrix} 0 \\ 0 \\ \frac{\delta \Xi}{\delta x}(x) \\ 0 \end{pmatrix} \in \operatorname{ran} \begin{pmatrix} E_S^* \\ E_R^* \\ F_S^* \\ F_R^* \end{pmatrix} \tag{2.19}$$

Proof The result follows from the image representation of a Dirac structure (2.5).

The constraints imposed at the resistive port (f_R, e_R) in (2.18) or, equivalently, (2.19) imply that if C is a Casimir for a specific resistive relation (2.6) that satisfies (2.7), then C is a Casimir for all the possible resistive relations, i.e. it is independent from the behaviour at the dissipative port. Thanks to this property, the dissipation obstacle is fully characterised from a geometrical point of view both in the finite and infinite dimensional cases, [4, 10–12, 25, 26]. The intrinsic limitations of the control by interconnection paradigm can be removed if the control action is explicitly thought in terms of a state-feedback law that is able to map the initial system into a new one. The target dynamics is characterised by desired Dirac structure, resistive relation, and Hamiltonian $H_d(x) = H(x) + \Xi(x)$, where now Ξ is *not* necessarily related to some Casimir function in the form (2.16). In the simplest case, i.e. when only the Hamiltonian function is shaped, in [11] it has been proved that all the admissible functions Ξ are solution of

$$\begin{pmatrix} 0 \\ \frac{\delta \Xi}{\delta x}(x) \\ 0 \end{pmatrix} \in \operatorname{ran} \begin{pmatrix} E_S^* \\ F_S^* \\ R_f E_R^* + R_e F_R^* \end{pmatrix} \tag{2.20}$$

It is easy to check that if Ξ satisfies (2.19), then also (2.20) holds, [10–12].

2.4 Overcoming the Dissipation Obstacle with a New Control Port

The motivating idea of this Section is to determine if there exists a new Dirac structure on Hilbert space $\bar{\mathcal{D}}$ with operators F and E given as in (2.12), and a resistive relation $\bar{\mathcal{R}}$ in the form (2.6) such that with the given Hamiltonian $H(x)$:

- The dynamics of the new system is the same of the original one;
- The new system is characterised by a set of Casimir functions that satisfies (2.20).

With (2.17) in mind, the second requirement implies that, for the new system, there exists a set of Casimir functions that can be employed in the control by interconnection procedure and allow to solve the dissipation obstacle. On the other hand, since the dynamics of the new system is the same of the initial one, the only difference between the twos is the behaviour at the control port. This means that a new control port $(\bar{f}_C, \bar{e}_C)$ has been determined, and when the interconnection between plant and controller takes place at $(\bar{f}_C, \bar{e}_C)$, the resulting closed-loop system is characterised by a new set of Casimir functions, that has been previously determined among the ones that allow to overcome the dissipation obstacle. With the next Proposition, a general expression for the desired Dirac structures $\bar{\mathcal{D}}$ is provided.

Proposition 2.7 *Let us consider a Dirac structure $\mathcal{D}$ on Hilbert space with kernel representation given in Proposition 2.3, where F and E are given as in (2.10). The set $\bar{\mathcal{D}} \subset \mathcal{F} \times \mathcal{E}$ defined as*

$$\bar{\mathcal{D}} = \left\{ (\bar{f}, \bar{e}) \in \mathcal{F} \times \mathcal{E} \mid \bar{F}\bar{f} + \bar{E}\bar{e} = 0 \right\} \tag{2.21}$$

with $\bar{F} : \Lambda \to \mathcal{F}$ and $\bar{E} : \Lambda \to \mathcal{E}$ a couple of linear operators such that $\bar{F} = \left(\bar{F}_S \; \bar{F}_R \; \bar{F}_C \right)$ and $\bar{E} = \left(\bar{E}_S \; \bar{E}_R \; \bar{E}_C \right)$, with $\operatorname{dom}\left(F \; E\right) = \operatorname{dom}\left(\bar{F} \; \bar{E}\right)$, and where

$$\begin{aligned} \bar{F}_S &= F_S & \bar{F}_R &= F_R + \tilde{F}_R & \bar{F}_C &= F_C \\ \bar{E}_S &= E_S + \tilde{E}_S & \bar{E}_R &= E_R + \tilde{E}_R & \bar{E}_C &= E_C + \tilde{E}_C \end{aligned}$$

is a Dirac structure iff $\overline{\operatorname{ran}\left(\bar{F} \mid \bar{E}\right)} = \Lambda$ and

$$\begin{aligned} &\tilde{E}_S F_S^* + F_S \tilde{E}_S^* + E_R \tilde{F}_R^* + \tilde{E}_R \left(F_R^* + \tilde{F}_R^* \right) + F_R \tilde{E}_R^* \\ &\qquad + \tilde{F}_R \left(E_R^* + \tilde{E}_R^* \right) + \tilde{E}_C F_C^* + F_C \tilde{E}_C^* = 0 \end{aligned} \tag{2.22}$$

Proof This result follows from Proposition 2.3.

The next Proposition provides necessary and sufficient conditions for the Dirac structure $\bar{\mathcal{D}}$ to have Casimir functions that satisfy (2.20).

Proposition 2.8 *Let us consider the Dirac structures $\mathcal{D}$ and $\bar{\mathcal{D}}$ presented in Proposition 2.7. A function $C(x, x_C)$ is a Casimir associated to $\bar{\mathcal{D}}$ that satisfies (2.20) iff*

$$\overline{\operatorname{ran}\begin{pmatrix}\tilde{E}_S^*\Phi \\ (E_R^*+\tilde{E}_R^*)\Phi \\ (F_R^*+\tilde{F}_R^*)\Phi\end{pmatrix}} \subseteq \operatorname{ran}\begin{pmatrix}(E_S^*+\tilde{E}_S^*)\Psi \\ (E_R^*+\tilde{E}_R^*)\Psi \\ (F_R^*+\tilde{F}_R^*)\Psi\end{pmatrix} \tag{2.23}$$

where $\Phi : \Lambda_\Phi \to \Lambda$ and $\Psi : \Lambda_\Psi \to \Lambda$ are a couple of linear operators such that

$$\operatorname{ran}\Phi = \operatorname{Ker} E_S^* \cap \operatorname{Ker}\left(R_f E_R^* + R_e F_R^*\right) \qquad \operatorname{ran}\Psi = \operatorname{Ker} F_S^* \tag{2.24}$$

Proof Since C satisfies (2.20), there must exists $\lambda \in \Lambda$ such that $\lambda = \Phi\lambda_\Phi$, with $\lambda_\Phi \in \Lambda_\Phi$, and that $\frac{\delta C}{\delta x} = F_S^*\lambda$. On the other hand, C is required to be a Casimir for $\bar{\mathcal{D}}$, so from (2.19) in Corollary 2.6, there must exists $\bar{\lambda} \in \Lambda$ such that

$$\bar{E}_S^*\bar{\lambda} = 0 \qquad \bar{E}_R^*\bar{\lambda} = 0 \qquad \bar{F}_R^*\bar{\lambda} = 0 \tag{2.25}$$

and $\frac{\delta C}{\delta x}(x) = F_S^*\bar{\lambda}$. This latter requirement implies that $\bar{\lambda} = \Phi\lambda_\Phi + \Psi\lambda_\Psi$, with $\lambda_\Psi \in \Lambda_\Psi$. The statement is proved once it is verified that for all λ_Φ there exists at least one λ_Ψ such that (2.25) holds, which is equivalent to require that (2.23) holds.

The next Proposition provides necessary and sufficient conditions for the port-Hamiltonian system associated to the Dirac structure $\bar{\mathcal{D}}$, with resistive structure $\bar{\mathcal{R}}$ defined later on, and Hamiltonian H to have the same state evolution of the port-Hamiltonian system with Dirac structure $\mathcal{D}$ and resistive structure $\mathcal{R}$.

Proposition 2.9 *Let us consider the Dirac structures $\mathcal{D}$ and $\bar{\mathcal{D}}$ presented in Proposition 2.7, and suppose that the resistive structure $\bar{\mathcal{R}}$ defined by*

$$\bar{R}_f \bar{f}_R + \bar{R}_e \bar{e}_R = 0 \tag{2.26}$$

is interconnected at the resistive port $(\bar{f}_R, \bar{e}_R)$ of $\bar{\mathcal{D}}$, where $\bar{R}_f$ and $\bar{R}_e$ are square matrices that satisfy conditions similar to (2.7). If the behaviour at the energy-storage port $(\bar{f}_S, \bar{e}_S)$ is as in (2.14), then the resulting state evolution is the same of the system associated to $\mathcal{D}$ iff

$$\overline{\operatorname{ran}\begin{pmatrix}\tilde{E}_S^*\bar{\Phi} \\ [\bar{R}_f(E_R^*+\tilde{E}_R^*)+\bar{R}_e(F_R^*+\tilde{F}_R^*)]\bar{\Phi}\end{pmatrix}} \subseteq \operatorname{ran}\begin{pmatrix}(E_S^*+\tilde{E}_S^*)\bar{\Psi} \\ [\bar{R}_f(E_R^*+\tilde{E}_R^*)+\bar{R}_e(F_R^*+\tilde{F}_R^*)]\bar{\Psi}\end{pmatrix} \tag{2.27}$$

where $\bar{\Phi} : \Lambda_{\bar{\Phi}} \to \Lambda$ *and* $\bar{\Psi} : \Lambda_{\bar{\Psi}} \to \Lambda$ *are a couple of linear operators such that*

$$\operatorname{ran}\bar{\Phi} = \operatorname{Ker}\left(R_f E_R^* + R_e F_R^*\right) \qquad \operatorname{ran}\bar{\Psi} = \operatorname{Ker} F_S^* \cap \operatorname{Ker} F_C^* \tag{2.28}$$

Proof Without loss of generality, assume an effort-in causality at the control ports (f_C, e_C) and $(\bar{f}_C, \bar{e}_C)$. Then, from the image representation (2.3) of a Dirac structure, and the behaviours (2.14) and (2.26) imposed at the resistive ports of $\mathcal{D}$ and $\bar{\mathcal{D}}$, respectively, we have that there must exists $\lambda = \bar{\Phi}\lambda_{\bar{\Phi}}$, with $\lambda_{\bar{\Phi}} \in \Lambda_{\bar{\Phi}}$, and

$$\bar{\lambda} \in \operatorname{Ker}\left(\bar{R}_f \bar{E}_R^* + \bar{R}_e \bar{F}_R^*\right), \quad \bar{\lambda} \in \Lambda \tag{2.29}$$

such that

$$-\frac{\partial x}{\partial t} = E_S^* \lambda = \bar{E}_S^* \bar{\lambda} \tag{2.30}$$

and $\frac{\delta H}{\delta x}(x) = F_S^* \lambda = F_S^* \bar{\lambda}$, and $e_C = F_C^* \lambda = F_C^* \bar{\lambda}$. These last two conditions are equivalent to $\bar{\lambda} = \bar{\Phi}\lambda_{\bar{\Phi}} + \bar{\Psi}\lambda_{\bar{\Psi}}$, with $\lambda_{\bar{\Psi}} \in \Lambda_{\bar{\Psi}}$. The statement is proved once it is verified that for all $\lambda_{\bar{\Phi}}$ there exists at least one $\lambda_{\bar{\Psi}}$ such that (2.29) and (2.30) hold, which is equivalent to require that (2.27) holds.

If it is possible to determine a Dirac structure $\bar{\mathcal{D}}$ and a dissipative structure $\bar{\mathcal{R}}$ such that the conditions of Propositions 2.7, 2.8 and 2.9 hold, we have determined a new control port (f_C, e_C) for the original system such that for some controller in port-Hamiltonian form the closed-loop system is characterised by a set of Casimir functions that are able to overcome the dissipation obstacle. In the next Corollary, a sufficient condition to be checked in order to have (2.23) and (2.23) satisfied is given.

Corollary 2.10 *Under the hypothesis of Propositions 2.8 and 2.9, with the further requirement that* $\bar{R}_f = R_f$ *and* $\bar{R}_e = R_e$, *conditions (2.23) and (2.27) hold if*

$$\overline{\operatorname{ran}\begin{pmatrix} \tilde{E}_S^* \Phi \\ (E_R^* + \tilde{E}_R^*)\Phi \\ (F_R^* + \tilde{F}_R^*)\Phi \end{pmatrix}} \subseteq \operatorname{ran}\begin{pmatrix} (E_S^* + \tilde{E}_S^*)\bar{\Psi} \\ (E_R^* + \tilde{E}_R^*)\bar{\Psi} \\ (F_R^* + \tilde{F}_R^*)\bar{\Psi} \end{pmatrix} \tag{2.31}$$

where Φ *and* $\bar{\Psi}$ *are defined in (2.24) and in (2.28), respectively.*

2.5 Example: Boundary Stabilisation of the Shallow Water Equation

Let us consider a rectangular open channel with a single flat reach, of length L and unitary width, which is delimited by upstream and downstream gates, and terminated by an hydraulic outfall. Moreover, it is assumed that the fluid has a unitary density;

we are in fact considering a simplified model of [5], even if all the results discussed here can be easily extended to more general cases. The dynamics is described by the shallow water equations, whose port-Hamiltonian formulation has been extensively discussed e.g. in [5, 21].

Denote by $[0,\ L]$ the spatial domain, and by $q(t,z) > 0$ and $p(t,z)$ the infinitesimal volume and kinetic momentum density, respectively. These are the state (energy) variables. Note that, due to the unitary width and fluid density assumptions, these quantities are numerically equal to the height of the fluid in the channel and to its velocity. Under the hypothesis of linearity in the internal friction forces (if present), the port-Hamiltonian formulation of the shallow water equations is in the form (2.8)

$$\frac{\partial}{\partial t}\begin{pmatrix} q \\ p \end{pmatrix} = \left[\begin{pmatrix} 0 & -1 \\ -1 & 0 \end{pmatrix}\frac{\partial}{\partial z} - \begin{pmatrix} 0 & 0 \\ 0 & D \end{pmatrix}\right]\frac{\delta H}{\delta x}(q,p) \tag{2.32}$$

where $x = (q, p)$, $D \geq 0$ models the dissipative effects, $H(q,p) = \frac{1}{2}\int_0^L (qp^2 + gq^2)\mathrm{d}z$ is the total energy of the fluid, and g is the gravity acceleration. Note that the co-energy variables are

$$\frac{\delta H}{\delta q}(q,p) = \frac{1}{2}p^2 + gq =: P(q,p) \qquad \frac{\delta H}{\delta p}(q,p) = qp =: Q(q,p)$$

which equal the hydrodynamic pressure, P, and water flow, Q, respectively. It is assumed that the controller is acting on the boundary port (f_C, e_C) defined as

$$e_C(t) = \begin{pmatrix} Q(t,0) \\ P(t,L) \end{pmatrix} \qquad f_C(t) = \begin{pmatrix} P(t,0) \\ -Q(t,L) \end{pmatrix}$$

The input is e_C. The associated Dirac structure can be written in the kernel representation (2.4), with operators F and E given in (2.10), and space Λ given in (2.11), with $n = 2$ and $r = 1$. Finally, the behaviour at the energy-storage and dissipative ports is (2.14), with $\bar{G} = D \geq 0$.

If dissipation is not present, i.e. if $D = 0$, it is possible to prove that the closed-loop system is characterised by a couple of Casimir functions in the form (2.16) that satisfy (2.18) or, equivalently, (2.19). More precisely, with the controller

$$\begin{cases} \dot{x}_C = \begin{pmatrix} 0 & 1 \\ -1 & 0 \end{pmatrix}\dfrac{\partial H_C}{\partial x_C}(x_C) + f'_C \\ e'_C = \dfrac{\partial H_C}{\partial x_C}(x_C) \end{cases}, \qquad x_C \in \mathbb{R}^2 \tag{2.33}$$

that is interconnected to the system through (f_C, e_C), i.e. $f'_C = f_C$ and $e_C = -e'_C$, the resulting closed-loop system is characterised by the following Casimir functions

$$C_1(x_C, q, p) = x_{C1} - \int_0^L p\,\mathrm{d}z \qquad C_2(x_C, q, p) = x_{C2} - \int_0^L q\,\mathrm{d}z$$

Such Casimir functions are useful to select H_C to properly shape the Hamiltonian of the closed-loop system, [11, 17].

On the other hand, when dissipation is present, i.e. when $D > 0$, no useful Casimir functions in closed-loop exist. But, it has been illustrated in [16, 17] that there exists a boundary state-feedback law thanks to which it is possible to overcome the dissipation obstacle and obtain an energy function $H(q, p) + \Xi(q, p)$ with the desired stability properties. The function Ξ satisfies (2.20), that now becomes

$$\frac{\partial}{\partial z}\frac{\delta \Xi}{\delta p}(q, p) = 0 \qquad \frac{\partial}{\partial z}\frac{\delta \Xi}{\delta q}(q, p) + D\frac{\delta \Xi}{\delta p}(q, p) = 0 \tag{2.34}$$

The same result can be obtained with the methodology discussed in this paper by relying on Corollary 2.10. In this respect, the operators Φ and $\bar{\Psi}$ are given by

$$\Phi(\lambda_q, \lambda_p) = \begin{pmatrix} D(L-z)\lambda_p + \lambda_q \\ \lambda_p \\ -D\lambda_p \\ 0 \\ 0 \\ \lambda_p \\ \lambda_q \end{pmatrix} \qquad \bar{\Psi}(\lambda_R) = \begin{pmatrix} 0 \\ 0 \\ \lambda_R \\ 0 \\ 0 \\ 0 \\ 0 \end{pmatrix}$$

with $\operatorname{dom}\Phi = \mathbb{R}^2$ and $\operatorname{dom}\bar{\Psi} = L_2(0, \ell; \mathbb{R})$. Then, it is possible to prove that conditions (2.23) and (2.27) can be satisfied by selecting $\tilde{E}_R = 0$ and

$$\tilde{E}_S^* = \begin{pmatrix} 0\,0\,0\,0\,0\,0\,0 \\ 0\,0\,0\,0\,0\,D\,0 \end{pmatrix} \quad \tilde{F}_R^* = \begin{pmatrix} 0\,0\,0\,0\,0\,-1\,0 \end{pmatrix} \quad \tilde{E}_C^* = \begin{pmatrix} 0\,D\,1\,0\,0\,0\,0 \\ 0\,0\,0\,0\,0\,0\,0 \end{pmatrix}$$

which is equivalent to have

$$\bar{E}_S = \begin{pmatrix} 0 & -\frac{\partial}{\partial z} \\ -\frac{\partial}{\partial z} & 0 \\ 0 & -1 \\ 0 & -\cdot\,|_0 \\ -\cdot\,|_L & 0 \\ -\cdot\,|_0 & D\int_0^L \cdot \\ 0 & \cdot\,|_L \end{pmatrix} \qquad \bar{F}_R = \begin{pmatrix} 0 \\ 1 \\ 0 \\ 0 \\ 0 \\ -\int_0^L \cdot \\ 0 \end{pmatrix} \qquad \bar{E}_C = \begin{pmatrix} 0 & 0 \\ D & 0 \\ 1 & 0 \\ 1 & 0 \\ 0 & 1 \\ 0 & 0 \\ 0 & 0 \end{pmatrix}$$

where $\cdot\,|_0$ and $\cdot\,|_L$ denote the value of a function in $z = 0$ and in $z = L$. With this choice, a new control port $(\bar{f}_C, \bar{e}_C)$ is defined, in which $\bar{e}_C = (\bar{e}_{C1}, \bar{e}_{C2}) = e_C$ and

$$\bar{f}_C = \begin{pmatrix} \frac{\delta H}{\delta q}(0) - 2D\int_0^L \frac{\delta H}{\delta p}(\cdot, z)\,\mathrm{d}z + D\,L\,\bar{e}_{C1} \\ -\frac{\delta H}{\delta p}(L) \end{pmatrix}$$

is the new passive output that can be used in the control by interconnection strategy to have a closed-loop system characterised by a set of Casimir functions that satisfies (2.23). In this respect, with the controller (2.33) now interconnected to the plant through the new control port, i.e. $f'_C = \bar{f}_C$ and $\bar{e}_C = -e'_C$, the resulting closed-loop system is characterised by the following Casimir functions that clearly satisfy (2.34):

$$C_1(x_C, q, p) = x_{C1} - \int_0^L \left[D(L-z)q + p\right]\mathrm{d}z \quad C_2(x_C, q, p) = x_{C2} - \int_0^L q \,\mathrm{d}z$$

Thanks to these Casimir functions, H_C can be selected to shape the Hamiltonian of the closed-loop system in the desired manner. It is possible to verify that the same control law obtained by relying on as energy-shaping approach based on trajectory matching between the open-loop system and a target one discussed e.g. in [16, 17] can be obtained within the control by interconnection paradigm.

2.6 Conclusions and Future Work

The motivating idea of the paper has been the development of a general methodology for the definition of a new control port for distributed parameter port-Hamiltonian systems with dissipation that is instrumental for the synthesis of stabilising boundary control laws able to overcome the dissipation obstacle within the control by interconnection via Casimir generation paradigm. When the interconnection between plant and controller takes place at this new control port, the same results provided by the control by energy-shaping, where the control action is explicitly determined as a state-feedback law able to shape the energy function in an appropriate manner, are recovered. Beside having established a link between these two control methodologies (i.e., between the control by interconnection via Casimir generation, and the control by energy-shaping), this result is interesting because it allows to study the properties of the closed-loop system in terms of the "interconnection of sub-systems" paradigm. This is useful, in particular, in the distributed parameter case, because it paves the way for the extension to a wider class of problems the methodologies presented e.g. in [22] that deal with the proof of the existence of solutions of systems of PDEs, and of the asymptotic/exponential stability of interconnected systems. This topic is currently under investigation.

References

1. J. Cervera, A. van der Schaft, A. Baños, Interconnection of port-Hamiltonian systems and composition of Dirac structures. Automatica **43**(2), 212–225 (2007)
2. R. Curtain, H. Zwart, *An Introduction to Infinite Dimensional Linear Systems Theory* (Springer, New York, 1995)

3. M. Dalsmo, A. van der Schaft, On representation and integrability of mathematical structures in energy-conserving physical systems. SIAM J. Control Optim. **37**, 54–91 (1999)
4. V. Duindam, A. Macchelli, S. Stramigioli, H. Bruyninckx, *Modeling and Control of Complex Physical Systems: The Port-Hamiltonian Approach* (Springer, Berlin, 2009)
5. B. Hamroun, A. Dimofte, L. Lefèvre, E. Mendes, Control by interconnection and energy-shaping methods of port Hamiltonian models. Application to the shallow water equations. Eur. J. Control **16**(5), 545–563 (2010)
6. O. Iftime, A. Sandovici, Interconnection of Dirac Structures via Kernel/Image Representation, in *Proceedings of the American Control Conference (ACC 2011)*, CA, San Francisco, USA, 2011, pp. 3571–3576
7. O. Iftime, A. Sandovici, G. Golo, Tools for Analysis of Dirac Structures on Banach Spaces, in *Proceedings of the 44th IEEE Conference on Decision and Control and European Control Conference (CDC-ECC 2005)*, 2005, pp. 3856–3861
8. D. Jeltsema, R. Ortega, J. Scherpen, An energy-balancing perspective of interconnection and damping assignment control of nonlinear systems. Automatica **40**(9), 1643–1646 (2004)
9. Y. Le Gorrec, H. Zwart, B. Maschke, Dirac structures and boundary control systems associated with skew-symmetric differential operators. SIAM J. Control Optim. **44**(5), 1864–1892 (2005)
10. A. Macchelli, Passivity-Based Control of Implicit Port-Hamiltonian Systems, in *2013 European Control Conference (ECC)*, Zürich, Switzerland, 2013, pp. 2098–2103
11. A. Macchelli, Dirac structures on Hilbert spaces and boundary control of distributed port-Hamiltonian systems. Syst. Control Lett. **68**, 43–50 (2014)
12. A. Macchelli, Passivity-based control of implicit port-Hamiltonian systems. SIAM J. Control Optim. **52**(4), 2422–2448 (2014)
13. A. Macchelli, B. Maschke, Modeling and Control of Complex Physical Systems: The Port-Hamiltonian Approach, Chapter Infinite-Dimensional Port-Hamiltonian Systems, pp. 211–271. In: Duindam et al. [4] (2009)
14. A. Macchelli, C. Melchiorri, Modeling and control of the Timoshenko beam. The distributed port Hamiltonian approach. SIAM J. Control Optim. **43**(2), 743–767 (2004)
15. A. Macchelli, C. Melchiorri, Control by interconnection of mixed port Hamiltonian systems. IEEE Trans. Autom. Control **50**(11), 1839–1844 (2005)
16. A. Macchelli, Y. Le Gorrec, H. Ramírez, H. Zwart, On the synthesis of boundary control laws for distributed port-Hamiltonian systems. IEEE Trans. Autom. Control (2014) (submitted)
17. A. Macchelli, Y. Le Gorrec, H. Ramírez, Asymptotic Stabilisation of Distributed Port-Hamiltonian Systems by Boundary Energy-Shaping Control, in *Proceedings of the 8th International Conference on Mathematical Modelling (MATHMOD 2015)*, Vienna, 2015
18. R. Ortega, L. Borja, New Results on Control By Interconnection and Energy-balancing Passivity-based Control of Port-hamiltonian Systems, in *2014 IEEE 53rd Annual Conference on Decision and Control (CDC)*, Los Angeles, California, USA, 2014, pp. 2346–2351
19. R. Ortega, A. van der Schaft, I. Mareels, B. Maschke, Putting energy back in control. IEEE Control Syst. Mag. **21**(2), 18–33 (2001)
20. R. Pasumarthy, J. van der Schaft, Achievable Casimirs and its implications on control by interconnection of port-Hamiltonian systems. Int. J. Control **80**(9), 1421–1438 (2007)
21. R. Pasumarthy, V. Ambati, A. van der Schaft, Port-Hamiltonian Formulation of Shallow Water Equations with Coriolis Force and Topography, in *Proceedings of the 18th International Symposium on Mathematical Theory of Networks and Systems (MTNS 2008)*, Blacksburg, VA, USA, 2008
22. H. Ramírez, Y. Le Gorrec, A. Macchelli, H. Zwart, Exponential stabilization of boundary controlled port-Hamiltonian systems with dynamic feedback. IEEE Trans. Autom. Control **59**(10), 2849–2855 (2014)
23. H. Rodriguez, A. van der Schaft, R. Ortega, On Stabilization of Nonlinear Distributed Parameter Port-Controlled Hamiltonian Systems via Energy Shaping, in *Proceedings of the 40th IEEE Conference on Decision and Control (CDC 2001)*, vol. 1, 2001, pp. 131–136
24. M. Schöberl, A. Siuka, On Casimir functionals for infinite-dimensional port-Hamiltonian control systems. IEEE Trans. Autom. Control **58**(7), 1823–1828 (2013)

25. A. van der Schaft, *L_2-Gain and Passivity Techniques in Nonlinear Control*, Communication and Control Engineering (Springer, New York, 2000)
26. A. van der Schaft, D. Jeltsema, Port-Hamiltonian systems theory: an introductory overview. Found. Trends® Syst. Control **1**(2–3), 173–378 (2014)
27. A. van der Schaft, B. Maschke, Hamiltonian formulation of distributed parameter systems with boundary energy flow. J. Geom. Phys. **42**(1–2), 166–194 (2002)
28. A. Venkatraman, A. van der Schaft, Energy shaping of port-Hamiltonian systems by using alternate passive input-output pairs. Eur. J. Control **16**(6), 665–677 (2010)
29. J. Villegas, H. Zwart, Y. Le Gorrec, B. Maschke, Exponential stability of a class of boundary control systems. IEEE Trans. Autom. Control **54**(1), 142–147 (2009)

Chapter 3
Energy-Aware Robotics

Stefano Stramigioli

Abstract This chapter has a tutorial nature in introducing a number of useful concepts which resulted by reasoning with power ports rather than with signals, as people usually do in control. Arjan is one of the Godfathers in this way of thinking and he has been a pioneer in bringing these concepts to a new level, introducing proper geometry, a sound system theoretic basis and divulgating these issues. This chapter shows how, by using these concepts, it is possible to address or solve certain problems in robotics, control and passivity in a simple and straightforward way. It also presents a formal proof of a claim which is often used as a conjecture and which gives theoretical arguments to counteract the statement which is often used against passivity and saying that passivity is too restrictive and stability is what should be looked for. Many of the concepts reported in the chapter have been the results of discussions with Arjan or are still issues that I am working on with Arjan. It is a great pleasure and honour to have the opportunity to contribute in this way to a recognition of the incredible career of a college and friend for which I have incredible respect from an intellectual and personal point of view.

3.1 Introduction

In many applications of robotics, a controlled robot does interact mechanically with the environment. This interaction means, in system theoretic terms, that the dynamics of the controlled system changes. This change is completely unknown in general and it is, in the opinion of the author, not meaningful in any sense to make hypothesis of linearity, structure or whatsoever of the environment and therefore of this possible change. Furthermore, this change can be discontinuous considering that for example, due to dynamic interaction, bouncing could occur and a consecutive and unpredictable contact/no-contact situation could occur. On the other hand, the robot will physically interact with the environment and the interaction will follow *physical*

S. Stramigioli (✉)
University of Twente, Enschede, The Netherlands
e-mail: S.Stramigioli@utwente.nl

M.K. Camlibel et al. (eds.), *Mathematical Control Theory I*,
Lecture Notes in Control and Information Sciences 461,
DOI 10.1007/978-3-319-20988-3_3

laws, like action and reaction and the first principle of thermodynamics of energy conservation. The first one to specifically address this issue in robotics was Neville Hogan in his famous trilogy [3]. Unfortunately, in the opinion of the author, the core message of Hogan has been often misinterpreted in the robotic literature [14]. From a more systematic and geometrical point of view, the modelling of interaction and behaviour has been presented in [10] and more extensively in [8].

This interaction can be effectively modelled with the concept of a power port known in network theory. The concept of power ports was the basis and essential element used by Paynter in the introduction of Bond Graphs [6]. In Bond Graphs the topology of energetic flows is given the main importance rather than the topology of the physical elements composing the system to be modelled.

A fundamental analysis of methods explaining the basis of bond graphs and their thermodynamical importance has been done by Breedveld [1]. The work of Arjan and Bernhard Maschke on port-Hamiltonian systems together with the deep insight of Peter Breedveld, have started in [5] a new line of research called port-Hamiltonian system theory, which gives a sound system theoretic basis to the use of port concepts in modelling and control. Arjan and Bernard Maschke have been the pioneers in this line of research and have extended these concepts very elegantly also to distributed parameter systems [13].

The implication of this theory and approach in robotics is unfortunately underestimated, but in the opinion of the author it is the only proper paradigm which can be used to control physical systems which, by their very existence, interact with a physical world where physical energy transfers dictate the way such interaction takes place. The title specifically names "energy awareness" rather than passivity, because the paradigm and ideas presented do not limit in any way the design space of control, but do give methods in order to keep track of the energy flows as a consequence of certain actions in control of robotic systems.

3.2 Why Bother About Power Ports and Energy?

A port models the mean by means of which energy can be exchanged between systems or parts of a system. It can be also used to properly model the interaction between a robot and the environment. Ports can be also used to model the interaction between the actuators of the robot and the robot itself. We can therefore model in this context, any robotic mechanism as a physical system having two multidimentional ports: one modelling the interaction and energy exchange of the robot with the (unknown) environment and one modelling the interaction and energy exchange of the robot with the actuators via which we can modify and shape the robot behaviour via control.

A port is model mathematically with the direct product of a vector space and its dual as

$$\mathcal{P} := \mathcal{V} \times \mathcal{V}^* \tag{3.1}$$

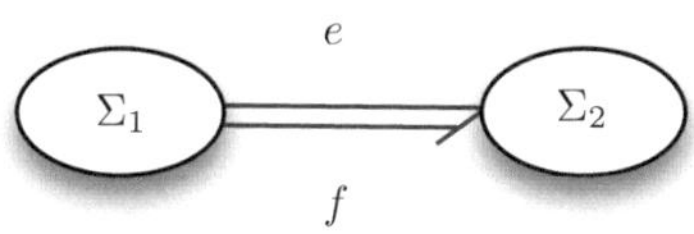

Fig. 3.1 Representation of a power-port to interconnect two systems *A* and *B*

in which we can call $\mathcal{V}$ the space of *flows* and $\mathcal{V}^*$ the space of *efforts* or the other way around by dualisation.

Depending on the situation, $\mathcal{V}$ van be a scalar, a finite or an infinite dimensional vector space. In the last case, n-forms and Poincaré duality can be used as introduced in [13]. Considering that a port is *the interface* between two "independent" systems, the mathematical formulation describing the port should not be dependent on the states of the two systems and at the same time should be representable at the "input/output" structure of the two system. In multibody dynamics, this is achieved using the structure of Lie groups, in which the port vector space $\mathcal{V}$ is modeled with a Lie algebra, which is not dependent on any element of the group.

A port should also have an orientation indicating the positive direction of power. In bong-graphs, the direction of a port is indicated with a half arrow as shown in Fig. 3.1. Due to the structure of the port, it is then possible in each instant of time to calculate the power flowing in the positive direction as

$$P = e(f). \tag{3.2}$$

Alternatively, by using scattering, the interaction could be represented by wave variables, also known as scattering variables, which can be geometrically defined for finite [9] and infinite dimensional systems [4] in a geometric way. The difference with this formulation is that the power transfer can be then expressed as an algebraic sum of quantities related to the scattering variables rather than a dual-pairing/product of efforts and flows. This approach has some great advantages in certain situations where the energy transport between the two systems is subjected to physical delay. This has brought to novel insight in geometrical telemanipulation [9].

3.3 The Intrinsically Passive Control (IPC) Framework

In [8], the author has introduced a paradigm called Intrinsically Passive Control. The proposed architecture for a controlled robot interacting with the environment is represented in Fig. 3.2. The basic idea is that, as indicated previously, a robot can be modeled as a physical system having two ports, one with the environment and one with the actuators controlled by the control system. The suggested paradigm is that the control should be conceived as a system which will be coupled using the port structure of the actuators to the control robot. The controller, which is implemented in discrete time, is composed of an Intrinsically Passive Controller (IPC) part and a Supervisor part which can inject energy and control the Robot via the IPC controller.

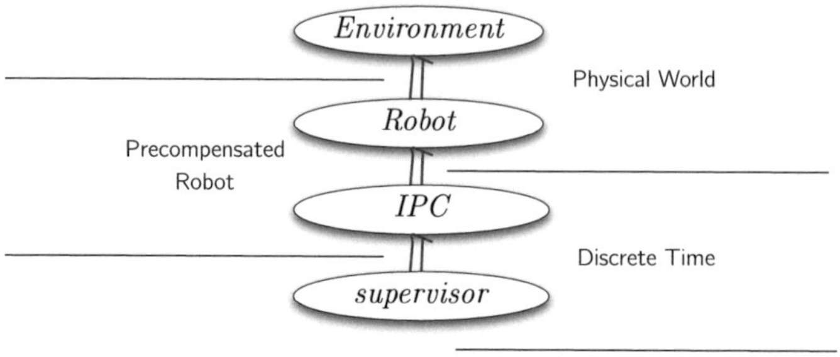

Fig. 3.2 IPC Supervision architecture

This structure has this form because in this way, if the supervisor will not inject energy via the IPC, the energy which can enter the Robot–IPC pair, can only come from the environment. The IPC can be designed on the basis of a model of the Environment, but due to its passive nature, if the Environment will not be as expected, if the supervisor does not inject energy, the interaction will be always passive. This follows the paradigm of what is called *Control by Interconnection* [12].

As it will be formally proved in the next section, if the controlled robot would not be passive seen from the environment side, there exist possible *passive environments* which would destabilise the system when connected to it.

3.4 Passivity as a Must

But why is the concept of port in robotics so important? In robotics the control of robots which interact with an unknown environment should happen stably in interaction with any kind of environment for clear reasons of performance, but more important safety. As said before, once a robot is interconnected with the environment, the stability analysis is only meaningful if the environment is considered as part of the system. Unfortunately, very simplistic and unrealistic models of the environment are used like elastic, purely linear, unilateral or variations of it. The value of such stability proofs is highly discussable considering they only prove stability for a very specific environment. The author argues that in control of systems coupled or interacting with unknown systems, a different paradigm and analysis is necessary as introduced in the previous section.

In this context, the following claims are made:

NP A necessary condition for having stable interaction with an unknown environment is that the controlled robot should result in a passive behaviour seen from the port which interacts with the environment

IPC A necessary condition for achieving the previous point is that, for a physical robot, which is clearly passive as seen composed of a physical system with an interconnection port and a control port, where the controller can supply and drain energy via actuators, the control should be done via interconnection and should be passive by itself following the IPC paradigm.

The previous claims are at this point conjectures which can be specified more clearly in the following problems statements:

Passivity Control Robot (PCR) If a controlled robot is not passive seen from the environment port, there is always a (passive) environment which can destabilise the interconnected system.

Not Passive State FeedBack (NPSF) For any passive robot, a general control which does not specifically address passivity as a port interconnection (IPC), there is always an environment which could result in an unstable interconnected behaviour as described in PCR.

Characterisation of Stable Active Environment (CSAE) Given a Robot controlled passively via interconnection (IPC), we can characterise the active environments which would result in a stable interconnected behaviour.

The argument PCR is important because it proves NP. The argument NPSF could formally prove that the only proper and safe way to control interactive systems should use the IPC methodology for robustness and that any other state feedback cannot ensure stable behaviour under uncertainty of the plant. Last but not least, CSAE would give a method to characterise and relax hypothesis on the passivity of the environment or humans, as often criticised in the haptic literature. In this work PCR will be formally proven. NPSF and CSAE are conjectures at this stage and work is in process to see if they can be formally proved, maybe with extra conditions.

3.4.1 The PCR Problem

The following theorem is a formal proof of PCR.

Theorem 3.1 *Given a non-passive system Σ with input output pair (u, y), there always exist a passive system $\bar{\Sigma}$ which connected to Σ will give rise to an unstable behaviour of the interconnection of Σ and $\bar{\Sigma}$.*

Proof Non-passiveness of Σ implies that $\exists\ \bar{u}(t)$ such that the integral of minus the supply rate is unbounded, which means we can extract infinite energy from the system. Indicate with $\bar{y}(t)$ the output corresponding to the input $\bar{u}(t)$. This means that we can define the extracted energy function $H_o(t)$ as

$$H_o(t) = \int_0^t \langle \bar{u}(s) | \bar{y}(s) \rangle \mathrm{d}s \tag{3.3}$$

By construction $\lim_{t\to\infty} H_o(t) = \infty$. This implies that due to the continuity of $H_o(t)$, $\exists$ a bounded $H_{\min} := \min_t H_o(t)$.

We will now constructively define a passive system $\bar{\Sigma}$ which will generate the input $\bar{u}(t)$.

$$\dot{x} = n(t)\tilde{y} \tag{3.4}$$

$$\tilde{u} = n(t)\frac{\partial H}{\partial x} \tag{3.5}$$

with $H(x) = \frac{1}{2}x^2$ and $n(t) = \frac{\bar{u}(t)}{\frac{\partial H}{\partial x}}$. It is easy to see that the previous system is passive (even conservative) with storage function $H(x)$. By initialising $x(0) = \sqrt{2H_{\min} + \Delta}$ for any $\Delta > 0$, it can be seen that by construction $\frac{\partial H}{\partial x}(t) > 0 \quad \forall t > 0$ and it is therefore always possible to calculate $\bar{u}(t)$. By setting as interconnection $\tilde{u} = \bar{u}$ and $\tilde{y} = \bar{y}$, we by construction have that

$$\lim_{t\to\infty} H_0(x) = \lim_{t\to\infty} H(x) = \infty \Rightarrow x \to \infty$$

which proves instability of the coupled system having a state diverging.

The previous proof is simple and reasonably straightforward, but the theorem's implications are far reaching. First of all, the theorem is general and nonlinear. This means that, if a controlled robot is not passive, it is possible to construct an environment, maybe by a second controlled robot, which would be passive and if connected to the original robot would result in an unstable system. This clearly gives a strong reason to create a passive behaviour for any robot which would potentially interact with an unknown environment in order to ensure stable and safe behaviour.

3.5 Connecting to the Discrete World

Everything done so far is treated in continuous time. One important issue in practical applications is that clearly, the controller will be implemented digitally. In order for this framework to be solid, we therefore need a way to couple the continuous and discrete world which will not violate the energy balance and therefore which will not create or distroy energy in the coupling between the continuous and discrete world. This has been introduced in [11] and will be recalled hereafter.

Consider the port interconnection of a continuous time Hamiltonian system H_C and a discrete Hamiltonian system H_D through a sampler and zero-order hold. Suppose that H_C has an admittance causality (effort in/flow out) and therefore H_D has an impedance causality (flow in/effort out).

During the dynamic evolution of the two systems between time kT and $(k+1)T$, where T is the sampling time and k is a positive integer, the effort supplied to H_C by H_D will be constant due to the zero-order hold assumption. We will indicate this value as $e_d(k)$. If we indicate the power port at the continuous side with $(e(t), f(t))$, we clearly have

$$e(t) = e_d(k) \qquad t \in [kT, (k+1)T]$$

By looking at the energy flow towards the continuous system, we can see that if we indicate with $\Delta H_C^{\text{in}}(k)$ the energy which flows through the input power port from time kT up to time $(k+1)T$, we obtain

$$\begin{aligned}\Delta H_C^{\text{in}}(k) = \int_{kT}^{(k+1)T} e_d^T(k) f(s)\mathrm{d}s \\ = e_d^T(k) \int_{kT}^{(k+1)T} f(s)\mathrm{d}s \\ = e_d^T(k)\,(x((k+1)T) - x(kT))\end{aligned} \tag{3.6}$$

where we indicated with $x()$ the integral of the continuous time flow $f(t)$.

Remark 3.2 It is important to realise that, in most useful mechanical applications like haptics, $e_d(k)$ will correspond to forces/moments that a controller would apply to an inertial element. In this case, $x()$ would be nothing else than a position measurement of the masses the controller pushes on.

It is now straightforward to state the following theorem:

Theorem 3.3 (Sample Data passivity) *If in the situation sketched before, we define for the interconnection port of* H_D

$$f_d(k) := \frac{x(kT) - x((k+1)T)}{T}, \tag{3.7}$$

we obtain an equivalence between the continuous time and discrete time energy flow in the sense that for each n*:*

$$\sum_{i=1}^{n} e_d^T(i) f_d(i) = -\int_0^{nT} e^T(s) f(s)\mathrm{d}s \tag{3.8}$$

Remark 3.4 It is important to notice that the exact equivalence is achieved only by the definition of Eq. 3.7 in which $x()$ is usually the easiest variable to measure in real applications. The negative sign appearing in Eq. 3.8 is consistent with the fact that the power flowing into the continuous system is minus the power flowing into the discrete side.

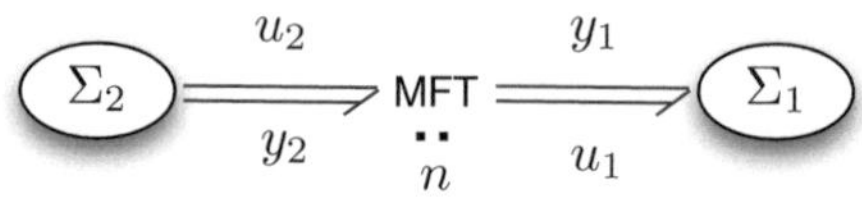

Fig. 3.3 Representation of a power-port to interconnect two systems Σ_1 and Σ_2

3.6 Energy Routing

An important technique which has been originally introduced by Duidam and Stramigioli in [2] and called by Ortega DSER in [7] (Duindam Stramigioli Energy Router) allows to direct energy flows without compromising passivity.

3.6.1 Controlling the Energy Directions and Magnitude Among (Sub)systems

To introduce this, with reference to Fig. 3.3 let us start from the situation in which only two ports are considered connecting two systems Σ_1 and Σ_2 and indicated with $(u_1, y_1) \in \mathcal{V}_1 \times \mathcal{V}_1^*$ and $(u_2, y_2) \in \mathcal{V}_2 \times \mathcal{V}_2^*$ and for which we indicated inputs and outputs of the two systems with u_i and y_i, respectively, for $i = 1, 2$. For simplicity of exposition, let us consider $\mathcal{V}_1 = R^n$ and $\mathcal{V}_2 = R^n$. A power continuous interconnection of the two systems is implemented by using the following relations which correspond in bond graphs to a multidimensional transformer or gyrator

$$u_1 = n \; y_2 \tag{3.9}$$

$$u_2 = n^T \; y_1 \tag{3.10}$$

where n and n^T is any linear map and its dual. Clearly, we have that

$$u_1^T y_1 = y_2^T n^T y_1 = y_2^T u_2 \tag{3.11}$$

which proves energy continuity. In the previous relation, n can be changed continuously or discontinuously and independently of its value the power continuity will hold by construction. We can therefore vary n also as function of the port variables, creating effectively a system which allows energy flow only in a specific direction. Suppose for example we want to force energy flowing from Σ_2 to Σ_1. This can be achieved simply by enforcing the direction of the power. Considering the positive power of Fig. 3.3 goes from Σ_2 to Σ_1, we want to achieve $y_1^T u_1 > 0$ indicating positive power flow toward Σ_1. Using Eqs. (3.9) and (3.10) this can be done by choosing

$$n = \alpha y_1 y_2^T \tag{3.12}$$

for a positive α. It is easy to see that by this construction a negative α will force a flow of energy in the opposite direction and its magnitude will control the amount of energy transfer. At all effects, α can be used to control the amount and direction of energy flow. It is also important to notice that energy will follow in the direction controlled iff energy is available which will result in values of $y_i \neq 0$.

This construction can be easily generalised to the situation in which instead of a two port, we consider a multidimensional Dirac structure connecting n systems Σ_i for $i = 1, \ldots, n$. Suppose that by convention, all positive orientations are chosen towards the systems that the Dirac structure connects. In this case, using the same kind of notation, we would have by power continuity that

$$y_1^T u_1 + \cdots + y_n^T y_n = 0 \tag{3.13}$$

and this will have to be realised by a relation of the form

$$\begin{pmatrix} u_1 \\ \vdots \\ u_n \end{pmatrix} = S \begin{pmatrix} y_1 \\ \vdots \\ y_n \end{pmatrix} \tag{3.14}$$

where S can be a skew symmetric matrix of proper dimension: $S^T = -S$. Suppose it is the goal to control the flow direction and magnitude of energy to the first system $y_1^T u_1$. We have that

$$y_1^T u_1 = y_1^T S_1 \begin{pmatrix} y_2 \\ \vdots \\ y_n \end{pmatrix} \tag{3.15}$$

where $(0\ S_1)$ is the first row of the sknew symmetric matrix S. By clearly choosing

$$S_1 := \alpha_1 y_1 \begin{pmatrix} y_2 & \cdots & y_n \end{pmatrix} \tag{3.16}$$

we can by choosing α_1 choose the direction and magnitude of power flow towards system Σ_1 and this will fix the first row, and for the skew symmetric constraint column, of the matrix S. By proceeding in a similar way, it would be possible to use the extra degrees of freedom still available in the choice of the matrix S in order to select other energy flows to the remaining systems. A similar analysis could also be carried out by using scattering which would directly represent positive and negative energy flows towards the systems and from the systems attached to the Dirac structure.

3.6.2 Energy Tanks and Tracking

Another way to use energy routing is to keep track of the energy which is used to perform a certain operation. This can be used to prevent instability of certain control actions. Suppose for example to control a robot which interacts with an unknown environment, a general control law which would not specifically monitor the amount of energy injected to the system, could potentially destabilise an interaction with an unknown environment as proven previously, if the energy would not be bounded by a passive behaviour. It is therefore useful to have a strategy which is able to allocate a certain *energy budget* to perform a specific operation and take proper actions if this amount of energy has been used. This action may be to adapt the control to prevent instability, or to analyse the situation and possibly adapt the control strategy providing extra energy. This is why the author talks about energy awareness rather than passivity which could seem restricting the applicability of the paradigm.

Consider the energy tank to have an associated positive definite energy function $H(s) = 1/2s^2$ with s a scalar. The energy budget can be initialised by a proper initial value of s. Assume to have n subsystems as in the previous section which need to be controlled and assume to have a control law $u = f(x, y)$ where u represents the column vector of all inputs, x the vector of states of the systems and y the dual outputs of u. To have power continuity we can consider the following interconnection between the energy tank and the systems:

$$\begin{pmatrix} \dot{s} \\ u_1 \\ \vdots \\ u_n \end{pmatrix} = S \begin{pmatrix} \frac{\partial H}{\partial s} \\ y_1 \\ \vdots \\ y_n \end{pmatrix} \tag{3.17}$$

again with a skew symmetric matrix S and also considering that $\dot{H} = \frac{\partial H}{\partial s}\dot{s}$.

It is possible to show that, under the condition that $\frac{\partial H}{\partial s} \neq 0$, it is possible to calculate a skew symmetric matrix S which satisfies the control relation $u = f(x, y)$ and at the same time monitors the energy necessary for that action by the value of the energy function H. By only monitoring the scalar s is therefore possible to see when the available budget of energy has expired. This simple idea, paradoxically can be used to “passivise” any control law, but building a safety mechanism which would prevent to inject indiscriminate energy into a control system leading to instability. In other words, we can implement the control law $u = f(x, y)$ until the energy set in the beginning is finished and then switch to a different control action to prevent loss of passivity and ensure stable interaction with any passive environment as proved in Theorem 3.1.

3.6.3 Projections

In many robotic applications it is useful to use projectors operators in order to implement certain control strategies. For example, in case of under-actuation, it is not possible to servo a complete force which would be desired, but that should be first projected on the subspace of forces which are implementable. If this projection is done naively, it can result in loss of passivity. By using the method shown in the previous section it is possible to monitor the energy consequence of such actions, but for the sake of clarity and with a didactical goal, hereafter, the example of projections will be further constructed.

Consider a control law which would calculate the applied force F to an under-actuated robot. Assume the motion of the robot can be measured and let us indicate with $\dot{x}$ its velocity. From a port point of view, the controlling port of this robot would then be $(\dot{x}, F)$ and $F^T\dot{x}$ would be the power supplied to the robot. Suppose that, for reasons which are not going to be discussed here, we want to apply to the robot an elastic force with some specific geometrical properties. We can create consistently an elastic force by defining an elastic energy function $H(x)$ which, after integrating the velocity of the robot $\dot{x}$ could calculate the force to be applied as $F = \frac{\partial H}{\partial x}$. Unfortunately, due to the under-actuation of the robot, we first need to project the gradient of H to the subspace of applicable forces. If we indicate with P such a projection, we could indicate the control law with:

$$F = P\frac{\partial H}{\partial x}. \tag{3.18}$$

Unfortunately, such an operation alone would break passivity considering that this operator is acting only on the force and not dually on the velocity. The passivity could be recovered by integrating for the state of the spring $P^T\dot{x}$ rather than $\dot{x}$, but this would drastically change the control law because the state of the spring would not be anymore representing the configuration of the robot. This paradox is showing that such a projection on the force only, will inject or extract energy from the system and if we are able to exactly monitor this, we can prevent that the projection action would result in loss of passivity and possibly an unstable behaviour. What we can do is therefore specifically to model the energy which is necessary to recover this passive behaviour. This can be clearly done by framing the control operation of the projection as a general control law $u = f(x, y)$ as explained in Sect. 3.6.2 but we will do it constructively hereafter, in order to give better insight.

Let us indicate with $v = \dot{x}$ the real velocity of the robot and with $\bar{v} := P^T v$. If we want to conserve the integration of v rather than $\bar{v}$ for the state of the controller, we can model this by adding a new power port and using what in bond graphs is called a 0-junction (representing one of Kirchhoff's laws), which is an element whose connected bonds all have the same effort F and for which the flows sum algebraically: $v = \Delta v + \bar{v}$. We can now model the energy used for "balancing" the projection, with a new storage element (energy tank) which we can represent with an energy function $\bar{H}(s) = \frac{1}{2}s^2$. By then setting

$$\Delta v = C \frac{\partial \bar{H}}{s} \tag{3.19}$$

$$\dot{s} = C^T F \tag{3.20}$$

and choosing

$$C = (v - \bar{v}) / \frac{\partial \bar{H}}{\partial s} \tag{3.21}$$

we can easily check that, as long as $s > 0$, and there is energy available in the tank, the projection operator will achieve the original goal without the shortcoming of losing information about the pose of the robot in the elastic control and by having an exact quantification of the energy which such an action requires. If we furthermore slightly modify Eq. (3.21) to be

$$C = \begin{cases} (v - \bar{v}) / \frac{\partial \bar{H}}{\partial s} & s > \varepsilon \text{ or } \Delta v F \geq 0 \\ 0 & s \leq \varepsilon \text{ and } \Delta v F < 0 \end{cases} \tag{3.22}$$

where $\Delta v F > 0$ indicates power flow towards the storage tank $\bar{H}(s)$, we can also handle the singularity. This system will implement the desired compensation as long as energy will be available. Further modifications could for example inject energy to the tank $\bar{H}(s)$ by redirecting energy from possible damping actions as presented in the next section.

3.6.4 What About Damping?

Very often, especially for the control of mechanical systems, damping plays an important role. The effect of damping is clearly to irreversibly extract energy from the system. On the other hand, it may be useful to extract energy without necessarily getting rid of it, but rather store it somewhere else, very much in a similar fashion as introduced in the previous section. From a thermodynamical point of view, dissipation is an irreversible transformation of energy from any domain to the thermal domain leading to an increase of entropy. We can use this metaphor, but from a control point of view, we can buffer this energy and use it for other possible means. This operation does not create energy and it is therefore passive and perfectly consistent with the framework. A small modification of what is presented in Sect. 3.6.2 allows to implement this. If for example we increase the dimension of u and y of 1, we can add a relation:

$$u_{n+1} = B y_{n+1} \tag{3.23}$$

where B could be a varying, but positive damping coefficient. The effect of such an action is that any energy which would be extracted via the port (u_{n+1}, y_{n+1}) would automatically be used to increase the energy buffer $H(s)$ to be used as previously described. The possibility of time varying the damping B allows, from the point of view of control, to shape the dynamics of the system in a desirable way and the presented framework will 'automatically' take care that the energy balance will be accounted for.

3.7 Conclusions

In this chapter some basic concepts of what the author calls energy-aware robotics have been presented. It has been shown that the passive behaviour of a robot which can interact with an unknown environment is a must to ensure stable interaction with any passive environment. Different methodologies have been presented which give an idea on how, thanks to the use of port concepts, it is possible to structure control loops in such a way that all energy flows can be made explicit and passivity ensured. These techniques can also be used in telemanipulation and many other fields of robotics successfully to ensure a stable behaviour.

Acknowledgments I would like to acknowledge a number of people for the many valuable discussions which led to ideas reported in the chapter. Besides Arjan, I would also like to acknowledge Bernhard Maschke, who together with Arjan, infected me for the love of this field and Rob Mahony for great discussions related to geometrical control of quad-copters and under-actuation.

References

1. P.C. Breedveld, Physical systems theory in terms of bond graphs. Ph.D. thesis, Technische Hogeschool Twente, Enschede, The Netherlands, 1984
2. V. Duindam, S. Stramigioli, Port-based asymptotic curve tracking for mechanical systems. Eur. J. Control **10**(5), 411–420 (2004)
3. N. Hogan, Impedance control: an approach to manipulation: part I—theory, part II—implementation, part III—applications. ASME J. Dyn. Syst. Meas. Control **107**(1), 17–24 (1985)
4. A. Macchelli, S. Stramigioli, A.J. van der Schaft, C. Melchiorri, Scattering for Infinite Dimensional Port Hamiltonian Systems, in *Proceedings of the IEEE Conference on Decision and Control*, vol. 4, IEEE, 2002, pp. 4581–4586
5. B.M. Maschke, A.J. van der Schaft, P.C. Breedveld, An intrinsic hamiltonian formulation of network dynamics: non-standard poisson structures and gyrators. J. Franklin Institute **329**(5), 923–966 (1992)
6. H.M. Paynter, *Analysis and Design of Engineering Systems* (M.I.T. Press, Cambridge, 1960)
7. A. Sánchez-squella, R. Ortega, R. Griñó, S. Malo, Dynamic energy router. IEEE Control Syst. 72–80 (2010)
8. S. Stramigioli, *Modeling and IPC control of interactive mechanical systems: a coordinate-free approach*, vol. 266, Lecture Notes in Control and Information Sciences (Springer, London, 2001)

9. S. Stramigioli, A.J. van der Schaft, B.M.J. Maschke, C. Melchiorri, Geometric scattering in robotic telemanipulation. IEEE Trans. Robot. Autom. **18**(4), 588–596 (2002)
10. S. Stramigioli, E. Fasse, J.C. Willems, A rigorous framework for interactive robot control. Int. J. Control **75**(18), 1486–1502 (2002)
11. S. Stramigioli, C. Secchi, A.J. van der Schaft, C. Fantuzzi, Sampled data systems passivity and discrete port-Hamiltonian systems. IEEE Trans. Robot. **21**(4), 574–587 (2005)
12. A.J. van der Schaft, *L2-Gain and Passivity Techniques in Nonlinear Control. Communications and Control Engineering*, 2nd revise edition (Springer, London, 2000)
13. A. van der Schaft, B. Maschke, Hamiltonian formulation of distributed-parameter systems with boundary energy flow. J. Geom. Phys. **42**, 166–194 (2002)
14. J. Won, S. Stramigioli, N. Hogan, Comment on The Equivalence of Second-Order Impedance Control and Proportional Gain Explicit Force Control. June 1996

Chapter 4
Time-Varying Phasors and Their Application to Power Analysis

Dimitri Jeltsema

Abstract The classical complex phasor representation of sinusoidal voltages and currents is generalized to arbitrary waveforms. The method relies on the so-called analytic signal using the Hilbert transform. This naturally leads to the notion of a time-varying power triangle and its associated instantaneous power factor. Additionally, it is shown for linear systems that Budeanu's reactive power can be related to energy oscillations, but only in an average sense. Furthermore, Budeanu's distortion power is decomposed into a part representing a measure of the fluctuation of power around the active power and a part that represents the fluctuation of power around Budeanu's reactive power. The results are presented for single-phase systems.

4.1 Introduction

I first met Arjan when I was a Ph.D. student during his notorious course on nonlinear systems. In the last lecture of the course, Arjan treated a relatively new subject: port-Hamiltonian systems. Port-Hamiltonian systems theory is the result of combining network theory with classical (Hamiltonian) mechanics and nowadays provides the basis for many interesting and novel control methodologies. Port-Hamiltonian systems and related concepts, such as power and energy, remained among my main topics of interest and during the past decade we collaborated on several papers, research projects, national and international courses, and recently we finalized the book "Port-Hamiltonian Systems Theory: An Introductory Overview" [23].

Three years ago, I retrieved my interest in power analysis under nonsinusoidal conditions and during the preparations of our book we had several discussions about this subject and the possibilities to approach the problem from a port-Hamiltonian perspective and Dirac structures in particular. During these discussions, Arjan always came up with the same but very important and fundamental questions: what is this

D. Jeltsema (✉)
Delft Institute of Applied Mathematics, Delft University of Technology,
Mekelweg 4, 2628 Cd Delft, The Netherlands
e-mail: d.jeltsema@tudelft.nl

M.K. Camlibel et al. (eds.), *Mathematical Control Theory I*,
Lecture Notes in Control and Information Sciences 461,
DOI 10.1007/978-3-319-20988-3_4

reactive power, what are its origins, and does it have any physical meaning? With this contribution, I consider it as an honor, on the occasion of the 60th birthday of my scientific collaborator, colleague, and dear friend, to dedicate a chapter to our fruitful quests for interesting and open problems, and to recollect some thoughts and interpretations of reactive power and related concepts.

Happy birthday Arjan and enjoy reading!

4.1.1 Motivation and Background

The usage of alternative sources of power has caused that the problem of energy transfer optimization is increasingly involved with nonsinusoidal signals and nonlinear loads. The power factor (PF) is used as a measure of the effectiveness of the transfer of energy between an electrical source and a load. It is defined as the ratio between the power consumed by a load (real or active power), denoted as P, and the power delivered by a source (apparent power), denoted as S, i.e.,

$$\lambda := \frac{P}{S}. \tag{4.1}$$

The active power is defined as the average of the instantaneous power and apparent power as the product of the root-mean-square norms of the source current and voltage. The standard approach to improve the PF is to place a passive compensator, such as a capacitor or an inductor, parallel to the load. Conceptually, the design of the compensator typically assumes that the source is ideal, i.e., the internal (Thevenin) impedance is negligible, producing a fixed sinusoidal voltage.

If the load is linear and time-invariant (LTI) and the source voltage is sinusoidal, the resulting stationary current generally is a shifted sinusoid, and the PF is the cosine of the phase-shift angle between the source voltage and current. Classically, the remaining part of the power is called reactive power, and is denoted as Q. The relationship between the three types of power is given by

$$S^2 = P^2 + Q^2. \tag{4.2}$$

Thus, any improvement of the PF is accomplished by the reduction of the absolute value of the reactive power, hence reducing the phase shift between the current and the voltage.

For nonsinusoidal voltages and currents, the problem of decomposing the apparent power into active and reactive components is much more involved. Starting from the work of Steinmetz [21], Iliovici [14] and Budeanu [2], many authors have aimed to improve the concept of reactive power in the most general case; see e.g., [1, 8, 11], and the references therein. Every year dozens of articles are published on this subject and most of these contributions aim at decompositions of the load current into physical meaningful orthogonal quantities. The methods and ways of describing the power phenomena and to increase the effectiveness of the energy flow between the source and the load under nonsinusoidal conditions have not been standardized so far and the definition of reactive power has been changed several times [12, 13]. Why is this important? Apart from the economical reasons as electricity is a commodity, one of the main reasons is to reduce the operating costs of the power grid and to protect its reliability.

4.1.2 Contribution and Outline

In this chapter, a different approach is presented that generalizes the classical complex phasor representation of the port voltages and currents. The method relies on the so-called analytic signal using the Hilbert transform. This enables one to translate the power flows proposed in [9, 16] for three-phase systems to single-phase systems and naturally leads to the notion of a time-varying power triangle and its associated instantaneous PF. From an instrumentation and measurement perspective, the introduction of the time-varying power triangle offers interesting properties as it reveals an instantaneous view into the power flows in the system.

A major advantage of the proposed framework is that it is applicable to general loads (e.g., nonlinear and time-varying) as well as to general voltage and current waveforms (e.g., nonsinusoidal, non-periodical, interharmonics, etc.). Additionally, it is shown for LTI systems that Budeanu's reactive power can be related to energy oscillations, but only in an average sense. Furthermore, Budeanu's distortion power is decomposed into a part representing a measure of the fluctuation of power around the active power and a part that represents the fluctuation of power around Budeanu's reactive power. This relaxes some of the assertions in [5].

The remainder of the chapter is organized as follows. In Sect. 4.2, the classical power model for systems operating under sinusoidal conditions is reviewed and an interpretation of the associated active and reactive power is provided from both a time- and frequency-domain perspective. The extension of the classical phasor approach is generalized to time-varying phasors in Sect. 4.3. Section 4.4 revisits the infamous Budeanu power model and provides some novel insights using the time-varying phasor approach. Finally, in Sect. 4.5, some examples are provided to illustrate the theory.

4.1.3 Notation

Given two square integrable T-periodic signals $u(t)$ and $i(t)$, we define the inner product as

$$\langle u, i\rangle := \frac{1}{T}\int_{0}^{T} u(t)i(t)\mathrm{d}t, \tag{4.3}$$

and by $||u|| := \sqrt{\langle u, u \rangle}$ the rms (root-mean-square) value. Time differentiation is denoted by $u'(t) = \frac{du}{dt}(t)$. Voltages are represented in volts [V] and currents are represented in Ampère [A]. However, these units will be omitted in the text.

To simplify the presentation, all voltage and current waveforms are assumed to have zero mean values.

4.2 The Classical Sinusoidal Power Model

Consider the well-known classical case of a single-phase sinusoidal source (power system) transmitting power to a LTI load; see Fig. 4.1. Let the voltage at the load terminals be given by

$$u(t) = U\sqrt{2}\cos(\omega t + \alpha), \tag{4.4}$$

where $\omega = 2\pi/T$. Under the assumption that the voltage at the terminals does not depend on the transmitted current (infinitely strong power system), the associated current reads

$$i(t) = I\sqrt{2}\cos(\omega t + \beta). \tag{4.5}$$

The instantaneous power delivered to the load is given by

$$p(t) = u(t)i(t) = \underbrace{P\big[1 + \cos(2\omega t + 2\alpha)\big]}_{p_a(t)} + \underbrace{Q\sin(2\omega t + 2\alpha)}_{p_r(t)}, \tag{4.6}$$

where P and Q represent the *active* power and the *reactive* power defined by

$$\begin{aligned} P &:= UI\cos(\varphi), \\ Q &:= UI\sin(\varphi), \end{aligned} \tag{4.7}$$

respectively, with $\varphi := \alpha - \beta$ representing the phase shift between $u(t)$ and $i(t)$.

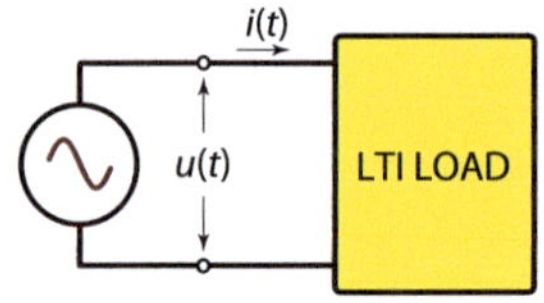

Fig. 4.1 The classical scenario of a single-phase power system transmitting power to a LTI load

4.2.1 On the Meaning of Active and Reactive Power

The term $p_a(t)$ in (4.6) describes the nonnegative component of the instantaneous power, with an average value equal to load's active power P, i.e.,

$$P = \frac{1}{T}\int_0^T p(t)\mathrm{d}t = \frac{1}{T}\int_0^T p_a(t)\mathrm{d}t = UI\cos(\varphi), \tag{4.8}$$

and represents the one-directional flow of energy from the source to the load.

The alternating term $p_r(t)$ in (4.6) is characterized by an amplitude equal to load's reactive power Q and average value equal to zero. This component characterizes the bidirectional flow of the transmitted energy from the source to the load. It is not present if load phase angle is equal to zero. Therefore, in case of a purely resistive load or if the load exhibits phase resonance, bidirectional oscillations in the energy flow between source and load do not exist. For example, if the load in Fig. 4.1 solely consists of a resistor, with resistance R, and is driven by a sinusoidal voltage of the form (4.4), then the associated current reads as in (4.5), with $\beta = \alpha$. Hence, there is no phase shift as $\varphi = 0$, and, according to (4.7), the active power equals $P_R = RI^2$, whereas the reactive power equals $Q_R = 0$.

The alternating component $p_r(t)$ may thus be interpreted as the measure of the backward flow of energy between load's reactance elements and the source. Indeed, if the load in Fig. 4.1 solely consists of an inductor, with inductance L, and is driven by a sinusoidal voltage of the form (4.4), with $\alpha = 0$, then the associated current reads

$$i(t) = \frac{1}{L}\int_0^t u(\tau)\mathrm{d}\tau = \frac{U}{\omega L}\sqrt{2}\sin(\omega t) = I\sqrt{2}\cos\left(\omega t - \tfrac{\pi}{2}\right). \tag{4.9}$$

Hence, the inductor causes a phase shift $\varphi = \frac{\pi}{2}$ and stores a magnetic (co-)energy

$$e_L(t) = \frac{1}{2}Li^2(t) = E_L^{\max}\sin^2(\omega t), \tag{4.10}$$

where $E_L^{\max} = LI^2$. This suggests that the (inductive) reactive power equals

$$Q_L = UI\sin\left(\tfrac{\pi}{2}\right) = UI = \omega LI^2 = \omega E_L^{\max}. \tag{4.11}$$

Alternatively, the (inductive) reactive power can also be expressed in terms of the *average* stored magnetic (co-)energy as

$$Q_L = 2\omega E_L, \tag{4.12}$$

with $E_L = \frac{1}{2}LI^2$. Obviously, the active power of an inductor equals $P_L = 0$.

Similarly, if the load in Fig. 4.1 solely consists of a capacitor, with capacitance C, and is driven by a sinusoidal current of the form (4.5), with $\beta = 0$, then the associated voltage reads

$$u(t) = \frac{1}{C}\int_0^t i(\tau)d\tau = \frac{I}{\omega C}\sqrt{2}\sin(\omega t) = U\sqrt{2}\cos\left(\omega t - \tfrac{\pi}{2}\right). \tag{4.13}$$

Hence, the capacitor causes a phase shift $\varphi = -\frac{\pi}{2}$ and stores an electric (co-)energy

$$e_C(t) = \frac{1}{2}Cu^2(t) = E_C^{\max}\sin^2(\omega t), \tag{4.14}$$

where $E_C^{\max} = CU^2$. This suggests that the (capacitive) reactive power equals

$$Q_C = UI\sin\left(-\tfrac{\pi}{2}\right) = -UI = -\omega CU^2 = -\omega E_C^{\max}. \tag{4.15}$$

Alternatively, the (capacitive) reactive power can also be expressed in terms of the *average* stored electric (co-)energy as

$$Q_C = -2\omega E_C, \tag{4.16}$$

with $E_C = \frac{1}{2}CU^2$. Again, note that $P_C = 0$.

Generally, in case of a load network consisting of LTI resistors, inductors and capacitors, the active power associated to each branch of the network may be expressed as

$$P_b = P_{R_b}, \tag{4.17}$$

where P_{R_b} represents the active power associated to the resistance in the bth branch. Note that $P_{L_b} = P_{C_b} = 0$. The reactive power for each branch is then expressed as

$$Q_b = Q_{L_b} + Q_{C_b} = \omega\left(E_{L_b}^{\max} - E_{C_b}^{\max}\right) = 2\omega\left(E_{L_b} - E_{C_b}\right). \tag{4.18}$$

Then, by Boucherot's theorem [4], the total active power and the total reactive power are obtained by summing over all the branches, i.e.,

$$\begin{aligned} P &= \sum_b P_b, \\ Q &= \sum_b Q_b. \end{aligned} \tag{4.19}$$

Remark 4.1 Note that compensation (reduction) of the reactive power naturally boils down to minimizing the difference between the total (average) magnetic and electric energies stored in the load network. Such perspective on reactive power compensation is known as *energy equalization* [11].

Remark 4.2 It is important to emphasize that the previous interpretations of reactive only apply to LTI systems driven by a purely sinusoidal voltage. If the load is nonlinear and/or time-varying, then it may be proven that reactive power does not uniquely relate to energy accumulation and it may be present in a purely resistive load. This will be exemplified in Sect. 4.5.

4.2.2 The Classical Phasor Representation

Alternatively, a standard method in electrical engineering is to represent the sinusoidal time functions of the voltages and currents by their complex phasor representation [7]

$$\underline{U} = U\sqrt{2}e^{j\alpha},\ \underline{I} = I\sqrt{2}e^{j\beta}, \tag{4.20}$$

where $j := \sqrt{-1}$. This enables one to define the complex power

$$\underline{S} := \frac{1}{2}\underline{U}\,\underline{I}^* = UIe^{j\varphi} = UI\cos(\varphi) + jUI\sin(\varphi) = P + jQ, \tag{4.21}$$

the well-known power triangle (see Fig. 4.2), the PF as $\lambda = \cos(\varphi)$, and the notion of complex impedance [7]

$$\underline{Z} = \frac{\underline{U}}{\underline{I}} = \frac{U}{I}e^{j\varphi} = \frac{P + 2j\omega(E_L - E_C)}{\frac{1}{2}\underline{I}\,\underline{I}^*}, \tag{4.22}$$

where E_L and E_C now represent the *total* mean magnetic and electric energies, respectively. In the same way, the complex admittance reads

$$\underline{Y} = \frac{\underline{I}}{\underline{U}} = \frac{I}{U}e^{-j\varphi} = \frac{P + 2j\omega(E_C - E_L)}{\frac{1}{2}\underline{U}\,\underline{U}^*}. \tag{4.23}$$

The underlying mathematical principle behind the transition from the sinusoidal time functions of the voltages and currents to their complex phasor representation is the so-called analytical signal widely used in telecommunication applications [22].

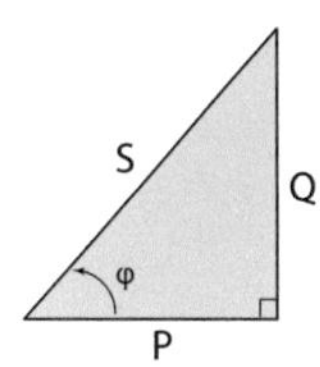

Fig. 4.2 The classical power *triangle* with $S = |\underline{S}|$

The analytic signal corresponding to the voltage (4.4) is given by

$$\underline{u}(t) = U\sqrt{2}\cos(\omega t + \alpha) + jU\sqrt{2}\sin(\omega t + \alpha) = U\sqrt{2}e^{j(\omega t+\alpha)}, \qquad (4.24)$$

and, similarly, the analytic signal corresponding to (4.5) is given by

$$\underline{i}(t) = I\sqrt{2}\cos(\omega t + \beta) + jI\sqrt{2}\sin(\omega t + \beta) = I\sqrt{2}e^{j(\omega t+\beta)}. \qquad (4.25)$$

Hence, the transition from the analytical signal representations (4.24)–(4.25) to the phasors (4.20) is accomplished by multiplying the latter with $e^{-j\omega t}$, which defines a linear (coordinate) transformation.

Furthermore, a straightforward computation shows that

$$\frac{1}{T}\int_0^T \underline{u}(t)\underline{i}^*(t)\mathrm{d}t = \underline{U}\,\underline{I}^*.$$

Thus, the correspondence between sinusoidal signals and their phasor representation is power-preserving once the real voltage and current signals are extended toward their analytic signal representations. This demonstrates that both P and Q are, in fact, *average* quantities.

4.2.3 RL Circuit Example

Consider the uncompensated RL circuit shown in Fig. 4.3 driven by a sinusoidal voltage

$$u(t) = 10\sqrt{2}\cos(t) \Rightarrow \underline{U} = 10\sqrt{2}.$$

The load admittance is given by

$$\underline{Y} = \frac{1}{1+2j} = \frac{1}{5}\sqrt{5}e^{-j\arctan(2)} \Rightarrow \underline{I} = \underline{Y}\,\underline{U} = 2\sqrt{10}e^{-j\arctan(2)}.$$

The complex power is then easily computed as $\underline{S} = 20 + j40$. Hence, the active power is $P = 20$ [W], the reactive power $Q = 40$ [VAr], and the apparent power

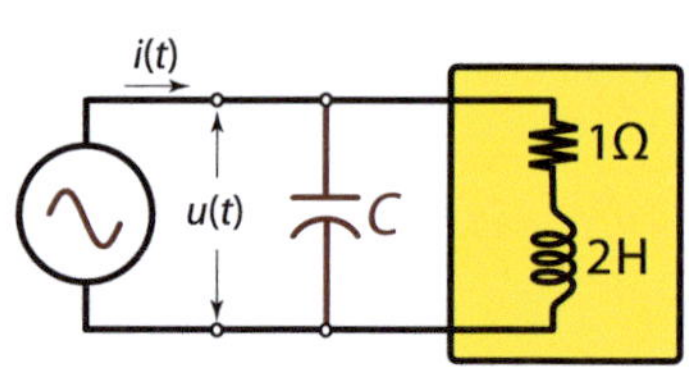

Fig. 4.3 Series RL circuit driven by a sinusoidal voltage: uncompensated ($C = 0$) and compensated ($C > 0$)

$S = |\underline{S}| = 20\sqrt{5}$ [VA]. The results in a PF $\lambda = 0.447$. If a shunt capacitor is placed to compensate Q, then it is clear that a capacitance of $C = 0.4$ [F] is necessary to compensate the effect of the inductance and to drive the PF to unity.

4.3 Time-Varying Phasors

Note that the relationship between the real voltage and current signals and their imaginary counterparts in (4.24)–(4.25) is a 90° backshift operation. For arbitrary waveforms this operation is generalized by the Hilbert transform [22]. Indeed, denoting by $\hat{u}(t) := \mathcal{H}\{u(t)\}$ the Hilbert transform

$$\mathcal{H}\{u(t)\} = \frac{\text{PV}}{\pi}\int_{-\infty}^{\infty}\frac{u(\tau)}{t-\tau}\mathrm{d}\tau, \tag{4.26}$$

with PV the Cauchy principal value, of the real voltage $u(t)$, then from standard complex analysis we know that

$$\underline{u}(t) = U(t)\sqrt{2}e^{j\alpha(t)}, \tag{4.27}$$

where

$$\begin{aligned}
U(t) &= \sqrt{\frac{u^2(t)+\hat{u}^2(t)}{2}} = |\underline{u}(t)|,\\
\alpha(t) &= \arctan\left\{\frac{\hat{u}(t)}{u(t)}\right\} = \operatorname{Im}\{\ln \underline{u}(t)\} = \arg\{\underline{u}(t)\},\\
\omega_\alpha(t) = \alpha'(t) &= \frac{\hat{u}'(t)u(t) - u'(t)\hat{u}(t)}{u^2(t)+\hat{u}^2(t)} = \operatorname{Im}\left\{\frac{\underline{u}'(t)}{\underline{u}(t)}\right\},
\end{aligned}$$

represent the instantaneous amplitude, phase, and frequency, respectively. In a similar fashion, the complex port current can be written as

$$\underline{i}(t) = I(t)\sqrt{2}e^{j\beta(t)}. \tag{4.28}$$

Remark 4.3 It is important to emphasize that in spite of both being measured in radians per second, harmonic and instantaneous radial frequencies are different concepts, which only coincide in the sinusoidal case. Indeed, for a voltage of the form (4.4), we have $\underline{u}(t) = \underline{U}$ and $\alpha(t) = \omega t + \alpha$, and thus $\omega_\alpha(t) = \omega$. See, e.g., [22] for further information. In [16], the representation (4.27) was justified based on Fourier transform. However, as argued in [22], the only way to unambiguously associate $U(t)$ and $\alpha(t)$ with amplitude, phase, and frequency is via the Hilbert transform.

Additionally, note that (4.27) allows to removing the fundamental phase ωt, i.e., $\underline{U}(t) = U(t)\sqrt{2}e^{j\tilde{\alpha}(t)}$, where $\tilde{\alpha}(t) := \alpha(t) - \omega t$.

Remark 4.4 The use of analytic signals, or the voltage and current representations (4.27) and (4.28), is not new in power systems analyses and control. See, for instance, the work of [15]. The Hilbert transform is also successfully used in [19] to derive a single-phase version of the well-known instantaneous p-q theory [8].

4.3.1 Kirchhoff Operators and Tellegen's Theorem

In the time domain, starting from the instantaneous power delivered at the port, i.e., $p(t) = u(t)i(t)$, Tellegen's theorem in generalized form can be stated as [18]

$$\mathcal{A}\{u(t)\}\mathcal{B}\{i(t)\} = \sum_b \mathcal{A}\{u_b(t)\}\mathcal{B}\{i_b(t)\}, \tag{4.29}$$

where $\mathcal{A}$ and $\mathcal{B}$ are so-called Kirchhoff voltage and current operators, respectively. A Kirchhoff voltage (current) operator is defined as an operation that if applied to a set of voltages (currents) that satisfy KVL (KCL) generates a new set of numbers or functions that also satisfy KVL (KCL). These quantities need not have the units of voltages (currents) and may depend on other parameters or variables introduced by the operator. All linear operations that operate in the same way on all branches and ports of the network are Kirchhoff operators. Well-known examples of linear operators are: differentiation, integration, averaging, complex conjugation, and time-shifting.

Since the Hilbert transform is also a linear operator, i.e.,

$$\mathcal{H}\left\{\sum_n c_n f_n\right\} = \sum_n c_n \mathcal{H}\left\{f_n\right\}, \tag{4.30}$$

where c_n are arbitrary numbers and f_n are arbitrary functions for which the Hilbert transform is defined, we may select the Kirchhoff operators in (4.29) as $\mathcal{A} = \mathcal{I} + j\mathcal{H}$ and $\mathcal{B} = \mathcal{I} - j\mathcal{H}$, where $\mathcal{I}$ is the identity operator, i.e., $\mathcal{I}\{f_n\} = f_n$. This yields the complex power balance

$$\big(u(t) + j\hat{u}(t)\big)\big(i(t) - j\hat{i}(t)\big) = \sum_b \big(u_b(t) + j\hat{u}_b(t)\big)\big(i_b(t) - j\hat{i}_b(t)\big). \tag{4.31}$$

This motivates and justifies the developments in the next section.

4.3.2 Time-Varying Complex Power

Starting from the analytical port voltage and current, (4.27) and (4.28), the time-domain nonsinusoidal equivalent of the complex power is defined by the time-varying complex power (compare with (4.31))

$$\underline{S}(t) := \frac{1}{2}\underline{u}(t)\underline{i}^*(t) = U(t)I(t)e^{j\varphi(t)} = P(t) + jQ(t), \tag{4.32}$$

where $\varphi(t) := \alpha(t) - \beta(t)$ denotes the instantaneous phase shift between $\underline{u}(t)$ and $\underline{i}(t)$, and

$$P(t) := \frac{1}{2}\big(u(t)i(t) + \hat{u}(t)\hat{i}(t)\big), \tag{4.33}$$

$$Q(t) := \frac{1}{2}\big(\hat{u}(t)i(t) - u(t)\hat{i}(t)\big), \tag{4.34}$$

or, equivalently,

$$P(t) := \frac{1}{2}\mathrm{Re}\big\{\underline{u}(t)\underline{i}^*(t)\big\}, \quad Q(t) := \frac{1}{2}\mathrm{Im}\big\{\underline{u}(t)\underline{i}^*(t)\big\},$$

represent the time-varying real and imaginary powers, respectively. Furthermore, the time-varying apparent power equals

$$S(t) = |\underline{S}(t)| = U(t)I(t) = \sqrt{P^2(t) + Q^2(t)}, \tag{4.35}$$

which naturally suggests the definition of a time-varying PF

$$\lambda(t) := \frac{P(t)}{S(t)} = \cos\big(\varphi(t)\big), \tag{4.36}$$

and a time-varying power triangle as shown in Fig. 4.4.

Another feature of the analytical representation of the port voltage and current is that the real parts of (4.27) and (4.28) are representing the real port voltage and current, which, in turn, are expressed in a very familiar form:

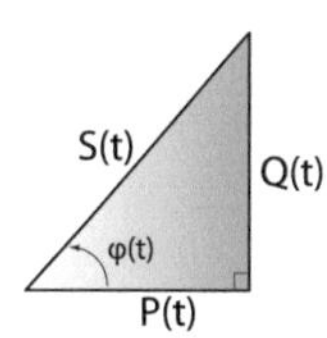

Fig. 4.4 The time-varying power *triangle*

$$u(t) = \mathrm{Re}\{\underline{u}(t)\} = U(t)\sqrt{2}\cos\big(\alpha(t)\big),$$
$$i(t) = \mathrm{Re}\{\underline{i}(t)\} = I(t)\sqrt{2}\cos\big(\beta(t)\big).$$

This means that the instantaneous power (4.6) generalizes to

$$p(t) = P(t)\left[1 + \cos\big(2\alpha(t)\big)\right] + Q(t)\sin\big(2\alpha(t)\big), \tag{4.37}$$

where $P(t)$ and $Q(t)$ are now rather expressed as

$$P(t) = U(t)I(t)\cos\big(\varphi(t)\big),$$
$$Q(t) = U(t)I(t)\sin\big(\varphi(t)\big).$$

Expression (4.37) is extremely general and also holds for non-periodic waveforms (provided (4.26) exists as a principal value). For that reason, we propose to refer to (4.37) as the 'universal power template (UPT).' In Sect. 4.4, one particular application of the UPT is highlighted.

4.3.3 Resistors, Inductors, and Capacitors

Let us next study the time-varying real and imaginary powers associated to the resistor, inductor, and capacitor. Interestingly, these powers bear a marked similarity in form as the powers derived for three-phase systems from the Poynting vector in [9] (see also [16]).

The Resistor

Consider an LTI resistor R driven by a nonsinusoidal voltage $u(t)$. Using Ohm's law $u(t) = Ri(t)$, the associated time-varying real power (4.33) takes the form

$$P_R(t) = \frac{1}{2}\big(Ri^2(t) + R\hat{i}^2(t)\big) = RI^2(t),$$

whereas the imaginary power (4.34) is zero, i.e., $Q_R(t) = 0$, since $\mathcal{H}\{u(t)\} = R\mathcal{H}\{i(t)\}$.

The Inductor

For an LTI inductor $u(t) = Li'(t)$. Using the time stationarity of the Hilbert transform [22], we have that $\mathcal{H}\{i'(t)\} = (\mathcal{H}\{i(t)\})'$. Hence, the real power (4.33) reads

$$P_L(t) = \frac{1}{2}\Big(Li'(t)i(t) + L\hat{i}'(t)\hat{i}(t)\Big) = LI_L'(t)I_L(t) = E_L'(t),$$

where $E_L(t) = \frac{1}{2}LI^2(t)$ represents the envelope of the oscillation of the inductor's magnetic energy storage. The imaginary power (4.34) now takes the form

$$Q_L(t) = \frac{1}{2}\big(L\hat{i}'(t)i(t) - Li'(t)\hat{i}(t)\big),$$

which, after multiplication of the numerator and denominator with $i^2(t) + \hat{i}^2(t)$, yields

$$Q_L(t) = \frac{1}{2}L\frac{\hat{i}'(t)i(t) - i'(t)\hat{i}(t)}{i^2(t) + \hat{i}^2(t)}\big(i^2(t) + \hat{i}^2(t)\big) = \beta'(t)LI^2(t) = 2\omega_\beta(t)E_L(t).$$

Note that if $i(t)$ is of the form (4.5), we have $\omega_\beta(t) = \omega$ and $I(t) = I$, and thus that $Q_L = 2\omega E_L$, as established in (4.12).

The Capacitor

In a similar fashion, for an LTI capacitor $i(t) = Cu'(t)$, the real power (4.33) reads

$$P_C(t) = \frac{1}{2}\big(u(t)Cu'(t) + \hat{u}(t)C\hat{u}'(t)\big) = CU(t)U'(t) = E_C'(t),$$

where $E_C(t) = \frac{1}{2}CU^2(t)$ represents the envelope of the oscillation of the capacitor's electric energy storage. The imaginary power (4.34) now takes the form

$$Q_C(t) = \frac{1}{2}\big(\hat{u}(t)Cu'(t) - u(t)C\hat{u}'(t)\big),$$

which, after multiplication of the numerator and denominator with $u^2(t) + \hat{u}^2(t)$, yields

$$\begin{aligned} Q_C(t) &= \frac{1}{2}C\frac{\hat{u}(t)u'(t) - u(t)\hat{u}'(t)}{u^2(t) + \hat{u}^2(t)}\big(u^2(t) + \hat{u}^2(t)\big) = -\alpha'(t)CU^2(t) \\ &= -2\omega_\alpha(t)E_C(t). \end{aligned}$$

Under sinusoidal conditions, i.e., if $u(t)$ is of the form (4.4), then $\omega_\alpha(t) = \omega$ and $U(t) = U$, and thus $Q_C = -2\omega E_C$, as in (4.16).

4.4 Budeanu's Concept of Reactive and Distortion Power Revisited

Consider a single-phase LTI power system with distorted voltage and current waveforms of the form

$$u(t) = \sum_k U_k\sqrt{2}\cos(k\omega t + \alpha_k), \tag{4.38}$$

$$i(t) = \sum_k I_k\sqrt{2}\cos(k\omega t + \beta_k), \tag{4.39}$$

with $\varphi_k = \alpha_k - \beta_k$. It is readily checked that the active power (from here on denoted by P_A) is straightforwardly obtained from the instantaneous power $p(t) = u(t)i(t)$ after averaging over a period, i.e.,

$$P_A := \frac{1}{T}\int_0^T p(t)\mathrm{d}t = \sum_k U_k I_k \cos(\varphi_k). \tag{4.40}$$

However, how to define and generalize the reactive power?

Inspired by, e.g., Bunet [3] and Boucherot's theorem [4], Budeanu [2] was among the first who tried to find an answer to this question and proposed to define reactive power as

$$Q_B := \sum_k U_k I_k \sin(\varphi_k). \tag{4.41}$$

He also observed that for nonsinusoidal voltages and currents the quadratic sum of the active and reactive power is not equal to the apparent power S as in the sinusoidal case, and ended up with $S^2 = P_A^2 + Q_B^2 + (\text{REST})^2$. To fill in this gap, a new concept

$$D_B := \sqrt{S^2 - P_A^2 - Q_B^2}, \tag{4.42}$$

called distortion (or deformation) power was proposed.

For decades, Budeanu's power model has enjoyed a lot of support and is set down in many publications and academic textbooks on power phenomena in systems with periodical and distorted waveforms, and for a long time has been part of the IEEE Standard [12]. Nevertheless, from the very beginning it has also been criticized by various opponents. Apart from the fact that it took almost 50 years before the first instruments where developed to measure Budeanu's reactive and distortion powers [10], critical questions where raised due to the apparent lack of physical meaning of the distortion power as it does not represent a conserved quantity and the (unauthorized) summing up of amplitudes of oscillating components of different harmonics [20], see also [1] and the references therein. Budeanu's power model was finally vigorously challenged by Czarnecki, and, although the arguments in [5] did not convince adherents of Budeanu's power model instantaneously [8], it was finally abandoned from the latest IEEE Standard [13].

In the following subsections it is shown, using the notion of time-varying phasors and the UPT, that the assertions against Budeanu's power model are either wrong, misinterpreted, or overstressed.

4.4.1 *Budeanu's Reactive Power Represents an Average*

First of all, we note that Budeanu's reactive power (4.41) can be expressed in the time domain using the Hilbert transform as [17]

$$Q_B := \langle \hat{u}, i \rangle = \frac{1}{T} \int_0^T \hat{u}(t) i(t) \mathrm{d}t, \tag{4.43}$$

where we recall that $\hat{u}(t) = \mathcal{H}\{u(t)\}$ denotes the Hilbert transform. Interestingly, using the fact that $\langle \hat{u}, i \rangle = -\langle u, \hat{i} \rangle$, it is readily observed that (4.43) is equivalent to averaging (4.34) over a period, i.e.,

$$Q_B := \frac{1}{T} \int_0^T Q(t) \mathrm{d}t. \tag{4.44}$$

Hence, Budeanu's reactive power Q_B does *not* represent a magnitude or an absolute quantity, but an average; the average of the imaginary power $Q(t)$ in a fashion similar to the active power P_A which represents the average of the real power $P(t)$. Furthermore, this means that Budeanu's reactive power represents the average of the difference between the envelopes of the oscillation of the magnetic and electric energies.

4.4.2 Power Fluctuations

It is correctly observed in [5] that Budeanu's concept of distortion power is not directly related to waveform distortion of the port voltages and currents itself. It may, however, be related to the fluctuations of the real and imaginary powers around their averages, i.e., the active and reactive powers. In this subsection, it is shown that the norms of these fluctuations can be naturally interpreted as distortion powers.

Let $D_P(t) := P(t) - P_A$ and $D_Q(t) := Q(t) - Q_B$ represent the power fluctuations around the active and reactive powers P_A and Q_B, respectively. Furthermore, let $I_P(t) := I(t)\cos(\varphi(t))$ and $I_Q(t) := I(t)\sin(\varphi(t))$, then it is easily shown that $\langle I_P, I_Q \rangle = 0$, i.e., the currents $I_P(t)$ and $I_Q(t)$ are mutually orthogonal. Hence, the 'normed' apparent power can be decomposed into two components:

$$||U||^2||I||^2 = ||U||^2||I_P||^2 + ||U||^2||I_Q||^2,$$

which, in turn, suggest

$$\begin{aligned} ||U||\,||I_P|| &\geq |\langle U, I_P \rangle| \equiv |P_A|, \\ ||U||\,||I_Q|| &\geq |\langle U, I_Q \rangle| \equiv |Q_B|. \end{aligned}$$

If $||U||\,||I_P|| > |\langle U, I_P \rangle|$, the residual is given by

$$D_{P_U}^2 := ||U||^2||I_P||^2 - \langle U, I_P\rangle^2 = \frac{1}{2T^2}\int_0^T\int_0^T \big(U(s)I_P(t) - U(t)I_P(s)\big)^2 \mathrm{d}s\mathrm{d}t.$$

Similarly, if $||U||\,||I_Q|| > |\langle U, I_Q\rangle|$, we have

$$D_{Q_U}^2 := ||U||^2||I_Q||^2 - \langle U, I_Q\rangle^2 = \frac{1}{2T^2}\int_0^T\int_0^T \big(U(s)I_Q(t) - U(t)I_Q(s)\big)^2 \mathrm{d}s\mathrm{d}t.$$

This naturally suggest the decomposition of distortion power into two components:

$$D_B^2 := D_{P_U}^2 + D_{Q_U}^2, \tag{4.45}$$

where D_{P_U} and D_{Q_U} can be considered as a measure of the fluctuation (distortion) around the active power and Budeanu's reactive power, respectively, relative to the voltage amplitude. Hence, we have

$$S^2 = P_A^2 + Q_B^2 + D_B^2 = P_A^2 + D_{P_U}^2 + Q_B^2 + D_{Q_U}^2. \tag{4.46}$$

In the sinusoidal case, $D_{P_U} = D_{Q_U} = 0$, and (4.46) reduces to the well-known standard (static) power triangle.

On the other hand, an equally valid starting point would be by selecting instead of $I_P(t)$ and $I_Q(t)$, the voltages $U_P(t) := U(t)\cos(\varphi(t))$ and $U_Q(t) := U(t)\sin(\varphi(t))$. This suggest to decompose the 'normed' apparent power as

$$||U||^2||I||^2 = ||U_P||^2||I||^2 + ||U_Q||^2||I||^2,$$

and, in a similar fashion as before, gives rise to the distortion powers, D_{P_I} and D_{Q_I}, relative to the current amplitude, and satisfying

$$D_B^2 := D_{P_I}^2 + D_{Q_I}^2. \tag{4.47}$$

Note that, in general, $D_{P_U} \neq D_{P_I}$ and $D_{Q_U} \neq D_{Q_I}$.

4.5 Examples

In this section, two examples are provided to illustrate the previous developments. First, a simple LTI circuit operating under nonsinusoidal conditions is discussed. The second example consists of a periodically switched resistive (triac) circuit.

4.5.1 RL Circuit Example (Cont'd)

Consider again the (uncompensated) series RL circuit as shown in Fig. 4.3, but now supplied by a nonsinusoidal voltage

$$u(t) = 10\sqrt{2}\cos(t) + 5\sqrt{2}\cos(5t). \tag{4.48}$$

In terms of the current amplitude, the complex power reads

$$\underline{S}(t) = P_R(t) + P_L(t) + jQ_L(t) = RI^2(t) + LI'(t)I(t) + j\omega_\beta(t)LI^2(t).$$

The waveforms for $P(t) = P_R(t) + P_L(t)$ and $Q(t) = Q_L(t)$, and their average values $P_A = 20.248$ [W] and $Q_B = 42.475$ [VAr] are depicted in Fig. 4.5. Note that the Budeanu reactive power is clearly related with energy oscillation, but only in an average sense, i.e.,

$$Q_B = \frac{2}{T}\int_0^T \omega_\beta(t)E_L(t)\mathrm{d}t,$$

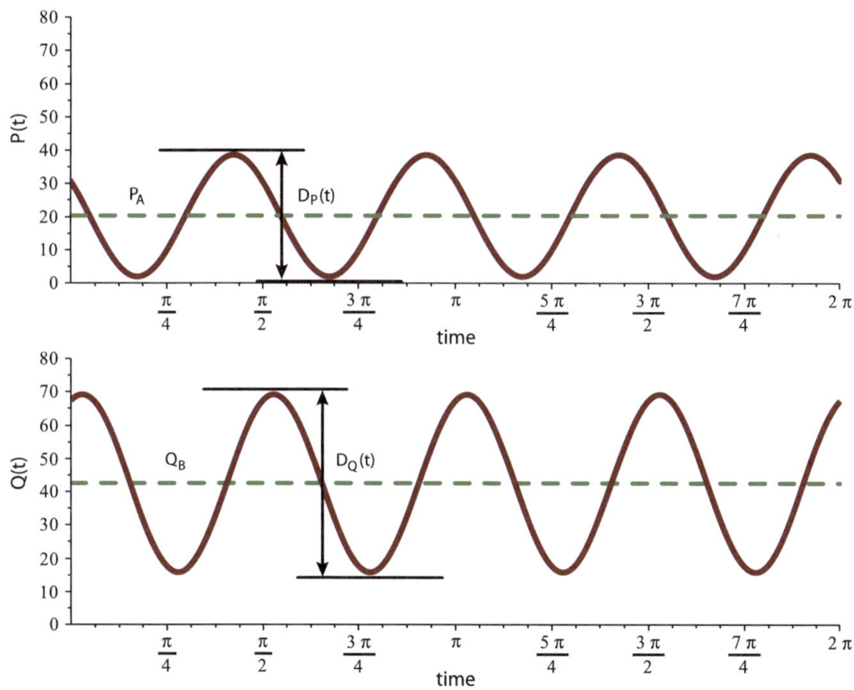

Fig. 4.5 Fluctuation of the time-varying real and imaginary powers around the active power and Budeanu's reactive power for the uncompensated circuit of Fig. 4.3 driven by the nonsinusoidal voltage (4.48)

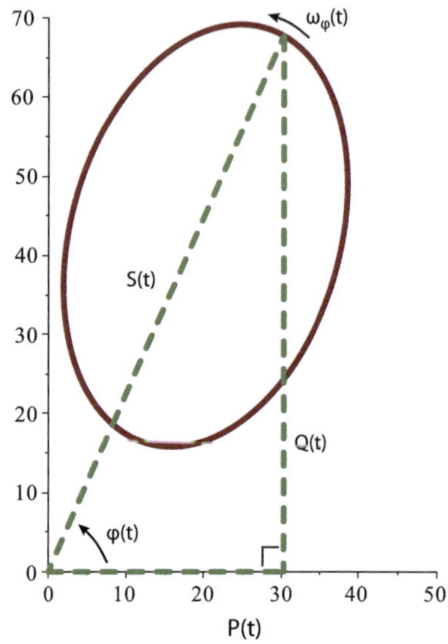

Fig. 4.6 The time-varying power *triangle* for the uncompensated circuit of Fig. 4.3 driven by the nonsinusoidal voltage (4.48)

with $E_L(t) = \frac{1}{2}LI^2(t)$ represents the envelope of the magnetic energy. The power fluctuations $D_P(t)$ and $D_Q(t)$ are also indicated in Fig. 4.5. Furthermore, Fig. 4.6 shows the time-varying power triangle associated to (4.35), which is expanding and contracting at the speed $\omega_\varphi(t) = \varphi'(t)$. Since for this particular example the same current is flowing through both the resistor and the inductor, the 'normed' apparent power can be written as

$$||U||^2||I||^2 = ||RI + LI'||^2||I||^2 + ||\omega_\beta LI||^2||I||^2.$$

It seems therefore most natural to consider the distortion power relative to the port current amplitude. Indeed, the fluctuation around the active power $P_A = R||I||^2$ is caused by the rate of change of $E_L(t)$, i.e., $E_L'(t) = LI'(t)I(t)$. This change of stored energy is due to the variation of the voltage and the current amplitudes and must come from real power. This causes the fluctuation of $D_P(t)$ for which the distortion power $D_{P_I} = ||U_P||\,||I||$ applies, with $||U_P|| = ||LI'||$. The distortion power associated with the fluctuation $D_Q(t)$ equals $D_{Q_I} = ||U_Q||\,||I||$, with $||U_Q|| = ||\omega_\beta LI||$. The values of the distortion power, including the alternative decomposition relative to the voltage amplitude, are computed (in [VAd]) as follows:

D_B	D_{P_U}	D_{Q_U}	D_{P_I}	D_{Q_I}
17.799	13.245	11.891	12.664	12.508

The question that remains is how to improve the PF? It is known [5] that the addition of a shunt capacitor $C = 0.189$ [F] renders $Q_B = 0$. However, the PF then becomes even worse than in the uncompensated case (from $\lambda = 0.403$ to $\lambda = 0.353$) as the distortion power increases to $D_B = 53.654$ [VAd]. Hence, the compensation of Budeanu's reactive power in this way is indeed useless for PF improvement and this was one of the main motivation behind the assertions in [5] against Budeanu's power model.

However, as explained in [24], the main reason why in the above example the compensator current, which renders Budeanu's reactive power to zero, does not reduce the source current—and even increases the distortion power—is that this particular compensator current and the nonactive part of the load current are not mutually orthogonal. The appropriate choice of the current that needs to be compensated is the co-called Budeanu current:

$$i_B(t) := \frac{Q_B}{||\hat{u}||^2}\hat{u}(t). \tag{4.49}$$

Consequently, if the compensator is supplying the Budeanu reactive current to the load, the Budeanu reactive power seen by the source will be zero and the distortion power remains unaltered. As a result, the apparent power decreases. This shows that by choosing the appropriate compensation current the PF increases and that the Budeanu reactive power concept, in general, *does* lead to a compensation scheme that reduces the line losses, except for systems in which $Q_B = 0$ already before compensation.

The compensation results for the RL circuit of Fig. 4.3, supplied with (4.48) and based on compensation of the Budeanu current (4.49), are shown in Fig. 4.7. It should be emphasized that, in general, the compensator supplying the Budeanu current cannot be realized by a single lossless shunt element. In fact, for the given example, it is composed of the same capacitor $C = 0.189$ [F] as before, but in series with a parallel connection of a capacitor $C_x = 0.128$ [F] and an inductor $L_x = 1.805$ [H]. This compensator increases the PF to $\lambda = 0.751$.

Remark 4.5 Although this example demonstrates that, in spite the fact that compensation based on the Budeanu current (if it exists), always leads to an improvement of the PF without altering the distortion power, it may not lead to optimal results as power fluctuations around the average powers may still exist and their compensation using passive filters seems so far not trivial from a time-domain perspective. On the other hand, based on the approach of [19], the power fluctuations can be compensated using an active filter.

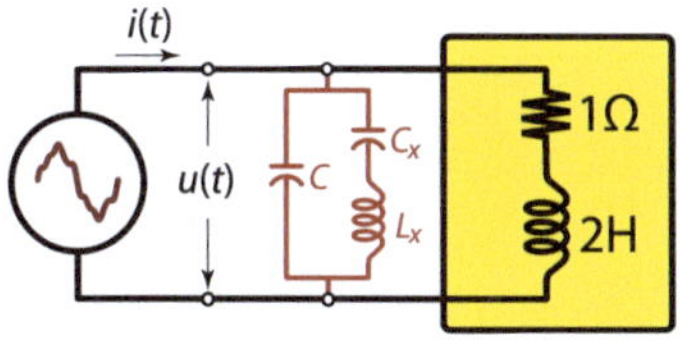

Fig. 4.7 Series RL circuit driven by a nonsinusoidal voltage with compensation network to render $Q_B = 0$ without altering the distortion power

4.5.2 Triac Circuit

Consider the uncompensated (i.e., $C = 0$) triac circuit shown in Fig. 4.8 [6]. Under the assumption that $u(t) = 220\sqrt{2}\sin(t)$, $R = 1\,\Omega$, and a switching angle $\alpha = 135°$, the apparent power equals

$$S = \|u\|\,\|i\| = 14.588\ \text{[kVA]}, \tag{4.50}$$

whereas the active power $P_A = 4.397$ [kW]. This means that the PF is far less than unity, i.e., $\lambda \approx 0.3$. The Budeanu reactive power equals $Q_B = 7.703$ [kVAr], whereas the distortion power and its associated decomposition reads (in [kVAd]):

D_B	D_{P_U}	D_{Q_U}	D_{P_I}	D_{Q_I}
11.582	8.190	8.190	8.671	7.678

It is important to realize that the uncompensated circuit does *not* store any energy. The reason why no energy is accumulated in the circuit becomes apparent when we consider the Lissajous plot of Fig. 4.8. Here it is observed that, although there is a phase shift between the current and voltage caused by the moments that the triac is switching ON, there is no energy accumulation as $i(t) \equiv 0$ whenever $u(t) \equiv 0$, and vice versa. From a frequency-domain perspective, the presence of reactive power can be explained as follows. The fundamental harmonic of the supply current reads

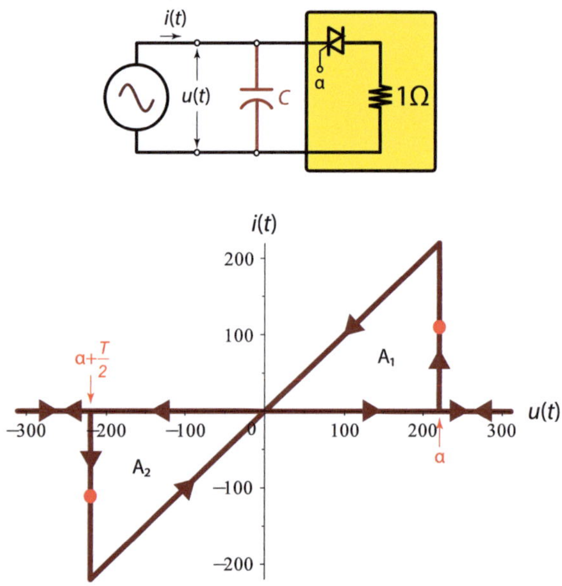

Fig. 4.8 Triac circuit and its associated Lissjous plot. Although there is no energy storage, the circuit exhibits a reactive power that is equal to the sum of the areas $A_1 + A_2$. The reactive power can be fully compensated using a capacitor

$$i_1(t) = 40.317\sqrt{2}\sin(t - 0.334\pi),$$

which is subsequently decomposed into an active component, $i_{1_a}(t)$, that is directly proportional (collinear) with the supply voltage and a quadrature component $i_{1_r}(t)$ as

$$i_1(t) = i_{1_a}(t) + i_{1_r}(t) = 19.987\sqrt{2}\sin(t) - 35.014\sqrt{2}\cos(t).$$

Now, $P_A = \|u\|\,\|i_{1_a}\|$ and $Q_B = \|u\|\,\|i_{1_r}\|$. Thus, the active power is related to the part of the current that is in-phase with the voltage, whereas the reactive power is related to the part that is exactly 90° out-of-phase. The remaining part of the current, $i_d(t) = i(t) - i_1(t)$, represents the harmonics that are due to the triac invoked distortion of the supply voltage and is responsible for the distortion power.

Although there is no energy accumulation, we may conclude that the triac circuit exhibits an inductive-like character since $Q_B > 0$. Hence, it is natural to compensate this behavior by placing a shunt capacitor. The value of the capacitor that fully compensates the reactive power equals $C = 0.159$ [F]. See [6] for more details.

References

1. G. Benysek, M. Pasko (eds.), *Power Theories for Improved Power Quality* (Springer, London, 2012)
2. C.I. Budeanu, *Puissances réactives et fictives* (Inst. Romain de l'Energie, Bucharest, 1927)
3. P. Bunet, Puissance réactives et harmonics. R.G.E., 6 Mars (1926)
4. G. Chateigner, M. Boes, J. Chopin, D. Verkindère, Puissances, facteur de puissance et théorème de Boucherot. Technolgie, 158, Novembre-Décembre 2008
5. L.S. Czarnecki, What is wrong with the Budeanu concept of reactive and distortion power and why it should be abandoned. IEEE Trans. Instr. Meas. **36**(3), 834–837 (1987)
6. L.S. Czarnecki, Physical interpretation of reactive power in terms of the cpc power theory. Electr. Power Qual. Utilisation J. **XIII**(1), 89–95 (2007)
7. C.A. Desoer, E.S. Kuh, *Basic Circuit Theory* (McGraw-Hill Book Company, New York, 1969)
8. A.E. Emanuel, *Power Definitions and the Physical Mechanism of Power Flow* (Wiley-IEEE Press, New York, 2010)
9. A. Fererro, S. Leva, A.P. Morando, An approach to the non-active power concept in terms of the Poynting vector. ETEP **11**(5), 291–299 (2001)
10. P. Filipski, The measurement of distortion current and distortion power. IEEE Trans. Instr. Meas. **IM-33**(1), 36–40 (1984)
11. E. García-Canseco, R. Grino, R. Ortega, M. Salichis, A.M. Stanković, Power-factor compensation of electrical circuits. IEEE Control Syst. Mag. **99**(46), 46–59 (2007)
12. IEEE, Standard 1459–2000. IEEE Power and Energy Society (2000)
13. IEEE, Standard 1459–2010. IEEE Power and Energy Society (2010)
14. M. Iliovici, Définition et mesure de la puissance et de l'energie réactives. Bull. Soc. Franc. Electr. **5**, 931–954 (1925)
15. R.A. Krajewski, A formal aspect of the definition of power. Measurement **8**(2), 77–83 (1990)
16. A.P. Morando, A thermodynamic approach to the instantaneous non-active power. ETEP **11**(6), 357–364 (2001)
17. Z. Nowomiejski, Generalized theory of electric power. Archiv für Electrotechnik **63**, 177–182 (1981)

18. P. Penfield Jr., R. Spence, S. Duinker, *Tellegen's Theorem and Electrical Networks*. Research Monograph No. 58 (MIT Press, Cambridge, 1970)
19. M. Saitou, T. Shimizu, Generalized Theory of Instantaneous Active and Reactive Powers in Single-phase Circuits Based on Hilbert Transform, in *Proceedings of the 33rd Power Electronics Specialists Conference (PESC)*, vol. 3 (2002), pp. 1419–1424
20. W. Shepherd, P. Zand, *Energy Flow and Power Factor in Nonsinusoidal Circuits* (Cambridge University Press, Cambridge, 1979)
21. C.P. Steinmetz, *Theory and Calculation of Alternating Current Phenomena*, 3rd edn. (Electrical World and Engineer, New York, 1900)
22. D. Vakman, *Signals, Oscillations, and Waves. A Modern Approach* (Artech House Inc., Canton, 1998)
23. A.J. van der Schaft, D. Jeltsema, *Port-Hamiltonian Systems Theory: An Introductory Overview, Foundations and Trends in Systems and Control* (Now Publishers Inc., Hanover, 2014)
24. J.L. Willems, Budeanu's reactive power and related concepts revisited. IEEE Trans. Instr. Meas. **60**(4), 1182–1186 (2011)

Chapter 5
Handling Biological Complexity Using Kron Reduction

Bayu Jayawardhana, Shodhan Rao, Ward Sikkema and Barbara M. Bakker

Abstract We revisit a model reduction method for detailed-balanced chemical reaction networks based on Kron reduction on the graph of complexes. The resulting reduced model preserves a number of important properties of the original model, such as, the kinetics law and identity of the chemical species. For determining the set of chemical complexes for the deletion, we propose two alternative methods to the computation of error integral which requires numerical integration of the state equations. The first one is based on the spectral clustering method and the second one is based on the eigenvalue interlacing property of Kron reduction on the graph. The efficacy of the proposed methods is evaluated on two biological models.

5.1 Introduction

Since this chapter is dedicated to Prof. Arjan van der Schaft, we first describe his early work on port-Hamiltonian systems and passivity theory and then describe how his work on chemical reaction network theory which is one of his most recent ventures, is connected with these two concepts. Beginning in the early 1990s, van der Schaft in collaboration with Maschke and Breedveld (see [13–16, 29]), began his work on port-controlled Hamiltonian systems which are commonly referred to as port-Hamiltonian systems. The framework of port-Hamiltonian systems combines the

B. Jayawardhana (✉) · W. Sikkema
Engineering and Technology Institute Groningen, University of Groningen, Nijenborgh 4, 9747AG Groningen, The Netherlands
e-mail: b.jayawardhana@rug.nl, w.sikkema.1@student.rug.nl

S. Rao
Ghent University Global Campus, 119 Songdomunhwa-ro, Yeonsu-gu, Incheon 406-840, South Korea
e-mail: shodhan.rao@ghent.ac.kr

B.M. Bakker
Department of Pediatrics and Systems Biology Centre for Energy Metabolism and Ageing University of Groningen, University Medical Center Groningen, Groningen, The Netherlands
e-mail: b.m.bakker01@umcg.nl

M.K. Camlibel et al. (eds.), *Mathematical Control Theory I*,
Lecture Notes in Control and Information Sciences 461,
DOI 10.1007/978-3-319-20988-3_5

earlier well-known Hamiltonian systems framework in which the system is modeled by using its total stored energy or the Hamiltonian, and the network framework which uses nodes and edges, and is commonly used to model electrical systems. For a detailed explanation of port-Hamiltonian systems, the reader is referred to [26, 27]. The port-Hamiltonian framework allows mainly modeling of *passive* electrical and mechanical systems. By passive systems, we mean systems for which the derivative of the Hamiltonian with respect to time is nonpositive due to dissipation. This derivative is equal to zero for lossless systems and negative for systems with dissipation.

In a first attempt to extend the port-Hamiltonian framework for the modeling of chemical reaction networks, in collaboration with Maschke, van der Schaft published a chapter in Springer lecture notes [28] in 2011. In this work, the Gibbs free energy of a chemical reaction network is considered as the Hamiltonian for its modeling. This work was inspired by the innovative work of Oster, Perelson, and Katchalsky [19, 20] in the area of chemical reaction networks. Later, he refined this work in collaboration with Rao and Jayawardhana who are two of the authors of this manuscript.

Deriving inspiration from the work of Horn, Jackson and Feinberg [5, 8, 9], who can arguably be considered as the founding fathers of chemical reaction network theory, we made a couple of observations. First an easy way of modeling chemical reaction networks is to make use of graphs of complexes of chemical reaction networks. The complexes of a chemical reaction network are the combination of species of the various left- and right-hand sides of the different reactions in the network. The graph of complexes is simply a graph with complexes as nodes and reactions as edges. The complex composition matrix Z, which captures the expression of the various complexes in terms of its constitutive species, and the incidence matrix B corresponding to the graph of complexes can then be used to derive an expression describing the dynamics of a chemical reaction network, given by $\dot{x} = ZBv$, where x denotes the vector of concentrations of the different species and v denotes the vector of the rates of the reactions in the network.

In their seminal papers published in the early 1970s, Horn, Jackson, and Feinberg [6, 8, 9] mainly considered a special class of chemical reaction networks known as complex-balanced networks. A complex-balanced network is one for which there exists a vector of species concentrations at which the combined rate of outgoing reactions from any complex is equal to the combined rate of incoming reactions to the complex, i.e., in some sense each complex of the network is balanced. A detailed-balanced network is a complex-balanced network for which there exists a vector of species concentrations at which the rate of each of the reactions in the network is zero, i.e., in addition to each complex being balanced, each reaction in the network is also balanced. The second observation that we made from [6, 8, 9] is that it is possible to derive a compact mathematical formulation describing the dynamics of complex and detailed-balanced networks in terms of a known equilibrium concentration vector, and a weighted *Laplacian* matrix corresponding to the graph of complexes. This weighted Laplacian matrix is symmetric in the case of detailed-balanced networks, and is balanced meaning that it has zero row and column sums in the case of complex-balanced networks. These properties of the weighted Laplacian matrix allows simple derivation of the previously well-known results from [8, 25] regarding equilibria and asymptotic stability of detailed- and complex-balanced networks (see [23, 30]). It can be shown that our compact mathematical formulation admits a direct port-Hamiltonian interpretation, using the Gibbs free energy of the network as the Hamiltonian and it can be shown that complex-balanced networks are passive systems (see [31] for details).

The graph-theoretic approach for the analysis of detailed- and complex-balanced networks also led to the idea of using the Kron reduction method to reduce models of such networks. Kron reduction method is a well-known method for model reduction of electrical networks (see, for example, [12] and an article written by van der Schaft in [32]) and other complex-networked systems (we refer interested readers to a recent article in [3]). This method exploits the balancedness of the weighted Laplacian matrix which we use in our compact mathematical formulation for the network, in order to perform a meaningful deletion of certain complexes of the network, thereby rewiring the graph of complexes and reducing the number of variables in the corresponding model. In collaboration with two system biologists, Bakker (another author of this chapter) and van Eunen from the Center for Systems Biology, University of Groningen, we generalized this model reduction method so as to be applicable for reaction networks that are governed by a variety of general enzyme kinetic rate laws, involving external inflows and outflows and are not necessarily complex balanced (see [22]). The reader is referred to [21, 22] for the current state of the art in the area of model reduction of biochemical reaction networks. Below, the main features of the model reduction method described in [22] are highlighted.

The method described in [22] reduces the number of reactions, species, and parameters in such a way that the transient behavior of the species concentrations of the reduced model under certain predefined conditions are close to those of the original model. This method proceeds by a simple stepwise reduction in the number of complexes, the effect of which is monitored by an error integral that quantifies how much the transient behavior of the reduced model deviates from that of the original. This method does not rely on prior knowledge about the dynamic behavior or biological function of the network. Consequently, it can be automated. Furthermore, the reduced model largely retains the kinetics and structure of the original model. This enables a direct biochemical interpretation and yields insight into which parts of the network have the highest influence on its behavior. It also accelerates computations and facilitates parameter fitting, especially when we deal with models of huge biochemical reaction networks. One of the drawbacks of this method is that it relies on the computation of error integral which could be time-expensive and depends on a number simulations which increases with the size of the model.

The main contribution of this chapter is to propose two alternative methods to the computation of error integral for determining the best combination of complexes that should be removed from the original network. We restrict ourselves to the class of detailed-balanced chemical reaction networks governed by the law of mass action kinetics. Thermodynamically, the assumption of detailed-balancedness of any reaction network without external fluxes is well-justified as it corresponds to microscopic reversibility.

The first method is based on the *spectral clustering method on a graph* which has been used to solve the ratio-cut and normalized-cut problems [33]. In graph theory, they are related to the problem of clustering the vertices in a graph such that the cost function associated with the weights of the cut-sets[1] is minimized. It has been applied widely for signal and image analyses [24, 33]. In our present context, we adapt the spectral clustering method to cluster complexes, thereby modifying the graph of complexes. Based on the clustering, we can pick complexes for the deletion from weakly coupled clusters since these clusters have minimal influence on the rest of the network.

The second method is based on the interlacing property of eigenvalues of Laplacian matrices associated with undirected graphs. From the classical work of Haemers [7], it is known that Laplacian matrices associated with graphs obey certain eigenvalue interlacing properties. In particular, it is known that the eigenvalues of any principal sub-matrix of a symmetric matrix interlace with the eigenvalues of the original matrix. As a direct consequence, for an undirected graph, the eigenvalues of any Schur complement of the corresponding symmetric Laplacian matrix (which defines the Kron reduction of a graph as will be explained later) interlace with the eigenvalues of the original Laplacian matrix. Based on this property, in our second approach, we look for the best combination of complexes to be deleted by finding a principal sub-matrix that results in a tight eigenvalue interlacing. This approach can be interpreted as finding the set of complexes with fast dynamics and a weak coupling to the rest of the network.

The layout of the chapter is as follows. In Sect. 5.2, we describe the modeling procedure for detailed-balanced mass action kinetics networks using a weighted Laplacian matrix corresponding to the graph of complexes. In Sect. 5.3, we review Kron reduction method for an undirected graph and its application to our chemical reaction network setting as proposed in [22]. The proposed spectral-based approaches are discussed in Sect. 5.4 and the efficacy of our proposed methods are evaluated in Sect. 5.5.

Notation: The space of m-dimensional real vectors is denoted by $\mathbb{R}^m$, the space of m-dimensional real vectors consisting of all strictly positive entries by $\mathbb{R}^m_+$ and the space of m-dimensional real vectors consisting of all nonnegative entries by $\bar{\mathbb{R}}^m_+$. Given $a_1, \ldots, a_n \in \mathbb{R}$, $\text{diag}(a_1, \ldots, a_n)$ denotes the diagonal matrix with diagonal entries $a_1, \ldots, a_n$. The time-derivative $\frac{\mathrm{d}x}{\mathrm{d}t}(t)$ of a vector x depending on time t will be denoted by $\dot{x}(t)$ or $\dot{x}$. The mapping $\text{Ln} : \mathbb{R}^m_+ \to \mathbb{R}^m, x \mapsto \text{Ln}(x)$, is defined as the mapping whose ith component is given as $(\text{Ln}(x))_i := \ln(x_i)$. Similarly, the mapping $\text{Exp} : \mathbb{R}^m \to \mathbb{R}^m_+, \quad x \mapsto \text{Exp}(x)$, is the mapping whose ith component is given as $(\text{Exp}(x))_i := \exp(x_i)$. Also, for any vectors $x, z \in \mathbb{R}^m$ the vector $\frac{x}{z} \in \mathbb{R}^m$ is defined as the elementwise quotient $\left(\frac{x}{z}\right)_i := \frac{x_i}{z_i}$, $i = 1, \ldots, m$.

For $n \in \mathbb{N}$, we define the index set $\mathcal{I}_n := \{1, \ldots, n\}$. For describing sub-matrices, we will use the following notations throughout the paper. Let $a, b \subset \mathcal{I}_n$ be two given subindices of $\mathcal{I}_n$. The sub-matrix of a matrix $\mathcal{L} \in \mathbb{R}^{n \times n}$ whose rows are indexed by

[1]Cut-sets are the edges that connect the vertices of the different clusters.

a and columns are indexed by b is denoted by $\mathcal{L}[a, b]$. Correspondingly, we define the complementary sub-matrices $\mathcal{L}[a, b)$, $\mathcal{L}(a, b]$, $\mathcal{L}(a, b)$ as follows:

$$\mathcal{L}[a, b) := \mathcal{L}[a, \mathcal{I}_n \backslash b], \quad \mathcal{L}(a, b] := \mathcal{L}[\mathcal{I}_n \backslash a, b], \quad \mathcal{L}(a, b) := \mathcal{L}[\mathcal{I}_n \backslash a, \mathcal{I}_n \backslash b].$$

For a symmetric matrix $\mathcal{L} \in \mathbb{R}^{c \times c}$, we arrange the eigenvalues in an increasing order so that

$$\lambda_1(\mathcal{L}) \leq \lambda_2(\mathcal{L}) \leq \cdots \leq \lambda_c(\mathcal{L}).$$

5.2 Detailed-Balanced Chemical Reaction Networks

In this section, we describe the modeling procedure of detailed-balanced mass action networks as in [30]. Consider a reversible reaction networks with r reversible reactions among m chemical species. Assume that the reaction network has c complexes whose expression in terms of the species can be described using the complex composition matrix Z of dimension $m \times c$. The ith column of Z expresses the composition of the ith complex of the network in terms of its m species. As an example, the complex composition matrix for the following reversible network:

$$2X_1 + X_2 \rightleftharpoons X_3 \rightleftharpoons X_1 + 2X_2 \tag{5.1}$$

is given by

$$Z = \begin{bmatrix} 2 & 0 & 1 \\ 1 & 0 & 2 \\ 0 & 1 & 0 \end{bmatrix}$$

The graph of complexes corresponding to a reversible reaction network is a directed graph with complexes as nodes and one edge corresponding to each reversible reaction with direction of the edge given by that of the forward reaction. Note that the modeling and model reduction can be carried out irrespective of the direction that is chosen for the edge corresponding to each of the reversible reactions of the network. One can associate an incidence matrix B of dimension $c \times r$ corresponding to the graph of complexes for which the jth column refers to the jth reaction of the network. If this reaction has the pth complex as the substrate and the qth complex as the product, then the jth column of B has -1 as its pth element, $+1$ as its qth element and all the remaining elements equal to 0. For example, the incidence matrix of the reaction network (5.1) is given by

$$B = \begin{bmatrix} -1 & 0 \\ 1 & -1 \\ 0 & 1 \end{bmatrix}$$

Now if $v \in \mathbb{R}^r$ denotes the vector of reaction rates and x denotes the vector of species concentrations, then the dynamics of the reaction networks can be described using the equation

$$\dot{x} = ZBv$$

Note that v as a function of x depends on the governing law of the reaction network. Here we describe how v can be written as a function of x in case the governing law is mass action kinetics.

The mass action reaction rate of the jth reaction of a chemical reaction network, from a substrate complex $\mathcal{S}_j$ to a product complex $\mathcal{P}_j$, is given as

$$v_j(x) = k_j^{\text{forw}} \prod_{i=1}^{m} x_i^{Z_{i\mathcal{S}_j}} - k_j^{\text{rev}} \prod_{i=1}^{m} x_i^{Z_{i\mathcal{P}_j}}, \tag{5.2}$$

where $Z_{i\rho}$ is the (i, ρ)th element of the complex stoichiometric matrix Z, and $k_j^{\text{forw}}, k_j^{\text{rev}} \geq 0$ are the forward and reverse reaction constants of the jth reaction, respectively.

Equation (5.2) can be rewritten in the following way. Let $Z_{\mathcal{S}_j}$ and $Z_{\mathcal{P}_j}$ denote the columns of the complex stoichiometry matrix Z corresponding to the substrate complex $\mathcal{S}_j$ and the product complex $\mathcal{P}_j$ of the jth reaction. Using the mapping $\text{Ln} : \mathbb{R}_+^c \to \mathbb{R}^c$ as defined at the end of the Introduction, the mass action reaction Eq. (5.2) for the jth reaction takes the form

$$v_j(x) = k_j^{\text{forw}} \exp\left(Z_{\mathcal{S}_j}^T \text{Ln}(x)\right) - k_j^{\text{rev}} \exp\left(Z_{\mathcal{P}_j}^T \text{Ln}(x)\right). \tag{5.3}$$

At this point, we define a detailed-balanced chemical reaction network. A vector of concentrations $x^* \in \mathbb{R}_+^m$ is called a *thermodynamic equilibrium* if $v(x^*) = 0$. Note that at a thermodynamic equilibrium, the rate of each of the reactions in the network is zero. A chemical reaction network $\dot{x} = ZBv(x)$ is called *detailed balanced* if it admits a thermodynamic equilibrium $x^* \in \mathbb{R}_+^m$. It can be shown that a detailed-balanced network is necessarily reversible. Note that $x^* \in \mathbb{R}_+^m$ is a thermodynamic equilibrium, i.e., $v(x^*) = 0$, if and only if

$$k_j^{\text{forw}} \exp\left(Z_{\mathcal{S}_j}^T \text{Ln}(x^*)\right) = k_j^{\text{rev}} \exp\left(Z_{\mathcal{P}_j}^T \text{Ln}(x^*)\right), \quad j = 1, \ldots, r$$

Define the 'conductance' $\kappa_j(x^*) > 0$ of the jth reaction as the common value of the forward and reverse reaction rates at thermodynamic equilibrium x^*, i.e.,

$$\kappa_j(x^*) := k_j^{\text{forw}} \exp\left(Z_{\mathcal{S}_j}^T \text{Ln}(x^*)\right) = k_j^{\text{rev}} \exp\left(Z_{\mathcal{P}_j}^T \text{Ln}(x^*)\right), \quad j = 1, \ldots, r$$

Then the mass action reaction rate (5.3) of the jth reaction can be rewritten as

$$v_j(x) = \kappa_j(x^*) \left[\exp\left(Z_{\mathcal{S}_j}^T \text{Ln}\left(\frac{x}{x^*}\right)\right) - \exp\left(Z_{\mathcal{P}_j}^T \text{Ln}\left(\frac{x}{x^*}\right)\right)\right],$$

where for any vectors $x, z \in \mathbb{R}^m$ the quotient vector $\frac{x}{z} \in \mathbb{R}^m$ is defined elementwise (see the end of the Introduction).

Defining the $r \times r$ diagonal matrix of conductances as

$$\mathcal{K} := \text{diag}\big(\kappa_1(x^*), \ldots, \kappa_r(x^*)\big)$$

it follows that the mass action reaction rate vector of a balanced reaction network can be written as

$$v(x) = -\mathcal{K}B^T \text{Exp}\left(Z^T \text{Ln}\left(\frac{x}{x^*}\right)\right),$$

and thus the dynamics of a balanced reaction network takes the form

$$\dot{x} = -ZB\mathcal{K}B^T \text{Exp}\left(Z^T \text{Ln}\left(\frac{x}{x^*}\right)\right), \quad \mathcal{K} > 0 \tag{5.4}$$

The matrix $\mathcal{L} := B\mathcal{K}B^T$ in (5.4) defines a *weighted Laplacian matrix* for the complex graph, with weights given by the conductances $\kappa_1(x^*), \ldots, \kappa_r(x^*)$. Note that $\mathcal{L}$ is symmetric. Thus Eq. (5.4) can be written as

$$\dot{x} = -Z\mathcal{L}\text{Exp}\left(Z^T \text{Ln}\left(\frac{x}{x^*}\right)\right) \tag{5.5}$$

The above equation is the compact mathematical formulation that was referred to in the Introduction, which is written in terms of a symmetric weighted Laplacian matrix $\mathcal{L}$ and a known equilibrium concentration vector x^* of the network. In addition to the system equation in (5.5), we define the output function y that represents measured or important variables (species concentrations) as follows:

$$y = Cx \tag{5.6}$$

where $y \in \mathbb{R}^p$ is the vector of output variables and $C \in \mathbb{R}^{p \times m}$. Note that y is typically a subset of the set of species, in which case, the matrix C is defined simply by an indicator matrix. We will use this output function to measure the quality of our model reduction method.

5.2.1 Detailed-Balanced CRN with General Kinetics

When enzymatic reactions or allosteric regulation are involved in the network, as commonly found in metabolic pathways, we can generalize (5.5) to take these into account. For describing such reactions, the mass action reaction rate as in (5.2), for every jth reaction, can be generalized to

$$v_j(x) = d_j(x)\left(k_j^{\mathrm{forw}} \prod_{i=1}^{m} x_i^{Z_{i\mathcal{S}_j}} - k_j^{\mathrm{rev}} \prod_{i=1}^{m} x_i^{Z_{i\mathcal{P}_j}} \right), \tag{5.7}$$

where, $d_j : \mathbb{R}^n \to \mathbb{R}_+$ is a positive definite function. In this new formulation, the function d_j models a sigmoidal/Hill function or another nonlinear function that is associated with an enzymatic reaction or allosteric regulation.

Following similar steps as in the mass action case, the detailed-balanced CRN with general kinetics[2] as in (5.7) can be described by

$$\left.\begin{aligned} \dot{x} &= -Z\mathcal{L}(x)\mathrm{Exp}\left(Z^T \mathrm{Ln}\left(\tfrac{x}{x^*}\right)\right) \\ y &= Cx \end{aligned}\right\} \tag{5.8}$$

where the state-dependent weighted balanced Laplacian matrix $\mathcal{L}(x)$ is defined by

$$\mathcal{L}(x) := B\,\mathrm{diag}\big(d_1(x), \ldots, d_r(x)\big)\mathcal{K}\, B^T$$

with $\mathcal{K}$ denoting the conductance matrix as before.

5.3 Kron Reduction

Consider again the graph of complexes of a detailed-balanced CRN with mass action kinetics as discussed in Sect. 5.2 where a weighted Laplacian matrix $\mathcal{L}$ has been defined to describe the interconnecting complexes and their associated reaction rates in (5.5). Similar to the Kron reduction method for electrical circuits, we can potentially reduce the dimension of CRN in (5.5) by applying Kron reduction to the graph of complexes.

The Kron reduction of a detailed-balanced chemical reaction network results in another detailed-balanced chemical reaction network, the vertices of whose complex graph is a subset of the vertices of the complex graph corresponding to the original network. Suppose that $\mathcal{C}_{red} \subset \mathcal{I}_c$ is the set of vertices (i.e., complexes) that we wish to remove from the complex graph corresponding to the original network. Then the Kron reduction of the network results in another detailed-balanced chemical reaction network, whose corresponding Laplacian matrix $\mathcal{L}_{\mathrm{red}}$ is the Schur complement of $\mathcal{L}$ with respect to $\mathcal{L}[\mathcal{C}_{red}, \mathcal{C}_{red}]$, given by

$$\mathcal{L}_{\mathrm{red}} = \mathcal{L}(\mathcal{C}_{\mathrm{red}}, \mathcal{C}_{\mathrm{red}}) - \mathcal{L}(\mathcal{C}_{\mathrm{red}}, \mathcal{C}_{\mathrm{red}}] \left(\mathcal{L}[\mathcal{C}_{\mathrm{red}}, \mathcal{C}_{\mathrm{red}}]\right)^{-1} \mathcal{L}[\mathcal{C}_{\mathrm{red}}, \mathcal{C}_{\mathrm{red}}).$$

The fact that $\mathcal{L}_{red}$ is again a symmetric Laplacian matrix has been shown, for instance, in [3, Lemma 2.1].

[2]For a detailed exposition on detailed-balanced CRNs with general kinetics, we refer interested readers to our work in [10].

The dynamics of the Kron-reduced CRN is then given by

$$\left.\begin{aligned} \dot{x} &= -Z[\mathcal{I}_c, \mathcal{C}_{\text{red}})\mathcal{L}_{\text{red}}\text{Exp}\left(Z[\mathcal{I}_c, \mathcal{C}_{\text{red}})^T \text{Ln}\left(\tfrac{x}{x^*}\right)\right) \\ y_{\text{red}} &= Cx \end{aligned}\right\}. \tag{5.9}$$

Since some of the complexes have been removed from the original state equations, some species in the image of $Z[\mathcal{I}_c, \mathcal{C}_{\text{red}}]$ can be constant, in particular, if they are not in the image of $Z[\mathcal{I}_c, \mathcal{C}_{\text{red}})$. These species can therefore be removed from the state equation, leading to a reduced model. For a detailed-balanced CRN with general kinetics, the Kron reduction method follows the same procedure as above.

The following lemma establishes the spectrum relation of $\mathcal{L}_{\text{red}}$ and its original Laplacian $\mathcal{L}$, which will be useful for our determination of the complex combination for the deletion.

Lemma 5.1 *Consider a weighted symmetric Laplacian matrix $\mathcal{L}$ of a complex graph and its associated Kron-reduced Laplacian $\mathcal{L}_{\text{red}}$ with respect to a set of deleted complexes $\mathcal{C}_{\text{red}}$. Let $k = \dim(\mathcal{C}_{\text{red}})$. Then for every $i = 1, \ldots, c-k$,*

$$\lambda_i(\mathcal{L}) \leq \lambda_i(\mathcal{L}_{\text{red}}) \leq \lambda_i(\mathcal{L}(\mathcal{C}_{\text{red}}, \mathcal{C}_{\text{red}})) \leq \lambda_{i+k}(\mathcal{L}),$$

where $\lambda_i(\mathcal{L})$ (or $\lambda_i(\mathcal{L}_{\text{red}})$) is the ith eigenvalue of $\mathcal{L}$ (or $\mathcal{L}_{\text{red}}$, respectively).

The proof of this lemma follows from [7, Theorem 2.1] or a recent exposition of Kron reduction on graph in [3, Theorem 3.5]. It follows immediately from this lemma that if $\dim(\mathcal{C}_{\text{red}}) = 1$ then

$$\lambda_1(\mathcal{L}) \leq \lambda_1(\mathcal{L}_{\text{red}}) \leq \lambda_2(\mathcal{L}) \leq \lambda_2(\mathcal{L}_{\text{red}}) \leq \cdots \leq \lambda_{c-1}(\mathcal{L}_{\text{red}}) \leq \lambda_c(\mathcal{L}).$$

In other words, the eigenvalues of $\mathcal{L}_{\text{red}}$ interlace those of $\mathcal{L}$.

5.3.1 Error Integral

Although the Kron reduction method as described above involves a fairly straightforward computation, it is not obvious how to determine the set of complexes for removal such that the dynamic behavior of the Kron-reduced CRN remains close to that of the original one.

One approach to do that, which has been proposed in our previous work [22], is to perform an iterative Kron reduction method where at each iteration a removal of a complex that minimizes a cost function is sought for. Since we use the output function to assess the quality of model reduction method, it is assumed that the complexes containing the chemical species in y do not belong to $\mathcal{C}_{\text{red}}$. Based on this assumption, the cost function as given in [22] is a normalized error integral that is defined by

$$J(x) = \sum_{i=1}^{p} \frac{1}{Tp} \int_{t}^{t+T} \left| 1 - \frac{y_{i,\mathrm{red}}(\tau)}{y_i(\tau)} \right| \mathrm{d}\tau, \tag{5.10}$$

where $y_{i,\mathrm{red}}$ and y_i are the ith output of the reduced model (5.9) and of the full model (5.6), respectively. This cost function evaluates the discrepancy of the reduced model's transient behavior compared with that of the full model one on the interval of $[t, t+T]$. It is normalized with respect to the total number of output variables and the length of time interval. Although other type of functions, such as, an L^p-norm-based cost function, can be used instead of (7.10), the normalized error integral as in (7.10) has been found to be effective in our numerical simulations.

One can show that the Kron reduction with respect to a given set of complexes to be deleted $\mathcal{C}_{\mathrm{red}}$ can be done by an iterative Kron reduction with respect to each individual complex in $\mathcal{C}_{\mathrm{red}}$ (see, for example, [3, Lemma 3.3]) and is invariant to the order of complex deletion. This fact supports the aforementioned iterative procedure of finding the combination of complexes for removal.

5.4 Spectral-Based Approaches

In this section, we present two alternative approaches to the iterative procedure of the previous section, for finding the combination of complexes for removal. These approaches are based on the spectral property of $\mathcal{L}$ (or $\mathcal{L}(x)$ for the case of detailed-balanced CRN with general kinetics) so that they do not depend on the numerical integration of the state equations as in (7.10). We show the approach assuming that the complex graph is connected. In case of graphs having more than one connected component, the same approach can be applied for each connected component. Hence, in the following we assume that $\mathcal{L}$ has eigenvalue 0 with multiplicity 1 so that $\lambda_2(\mathcal{L}) > 0$.

5.4.1 Spectral Clustering-Based Approach

For our first approach, we will consider clustering vertices of the complex graph into k clusters such that the combined weight of edges between vertices belonging to different clusters is minimized. More precisely, let us consider the following RatioCut problem [33]

$$\min_{\substack{\mathcal{C}_\ell \\ \ell=1,\dots k}} \sum_{\ell=1}^{k} \frac{W(\mathcal{C}_\ell, \overline{\mathcal{C}}_\ell)}{\dim(\mathcal{C}_\ell)},$$

where $\mathcal{C}_\ell$ denotes the set of vertices (or complexes) in the ℓth cluster, $\overline{\mathcal{C}}_\ell$ is the complement of $\mathcal{C}_\ell$ defined by $\overline{\mathcal{C}}_\ell := \mathcal{I}_c \backslash \mathcal{C}_\ell$ and $W(\mathcal{C}_\ell, \overline{\mathcal{C}}_\ell)$ is the sum of weights in the cut-set of the cut $(\mathcal{C}_\ell, \overline{\mathcal{C}}_\ell)$, i.e.,

$$W(\mathcal{C}_\ell, \overline{\mathcal{C}}_\ell) = \sum_{i \in \mathcal{C}_\ell, j \in \overline{\mathcal{C}}_\ell} -\mathcal{L}_{i,j}.$$

This RatioCut problem can be recasted into another equivalent form by using clustering (or indicator) vectors $u_\ell = \left[u_{1,\ell} \ldots u_{c,\ell}\right]^T$, $\ell = 1, \ldots, k$ defined by

$$u_{j,\ell} = \begin{cases} \frac{1}{\sqrt{\dim(\mathcal{C}_\ell)}} & \text{if } j \in \mathcal{C}_\ell \\ 0 & \text{otherwise,} \end{cases}$$

which are orthonormal vectors. Using u_ℓ, the RatioCut clustering problem can be reformulated as follows:

$$\min_{\substack{\mathcal{C}_\ell \\ \ell=1,\ldots k}} \sum_{\ell=1}^{k} \frac{W(\mathcal{C}_\ell, \overline{\mathcal{C}}_\ell)}{\dim(\mathcal{C}_\ell)} = \min_{\substack{\mathcal{C}_\ell \\ \ell=1,\ldots k}} \sum_{\ell=1}^{k} u_\ell^T \mathcal{L} u_\ell = \min_{\substack{\mathcal{C}_\ell \\ \ell=1,\ldots k}} \mathrm{Tr}(U^T \mathcal{L} U),$$

where Tr is the trace of a matrix and $U = \left[u_1 \; u_2 \; \ldots \; u_k\right]$ satisfies $U^T U = I_{k \times k}$.

Instead of finding minimizing clustering vectors u_ℓ which can be NP-hard, we can look for any orthonormal vectors u_ℓ that minimize the following relaxed RatioCut problem

$$\min_{U \in \mathbb{R}^{c \times k}} \mathrm{Tr}(U^T \mathcal{L} U) \qquad \text{subject to} \quad U^T U = I_{k \times k}.$$

Based on the solution U to this relaxed problem, we cluster the vertices by considering the rows of U as points in the k-dimensional space and by clustering these c points[3] into k clusters using any distance metric. For instance, we can apply the standard k-means algorithm to cluster these points. The resulting clustering result is known to approximate the solution to the original RatioCut problem [33].

Finally, we propose the following algorithm to find the candidate $\mathcal{C}_{\rm red}$ for our Kron reduction:

Spectral clustering-based algorithm:

1. Set $k = 2$ and calculate $\mathcal{L}$ (or $\mathcal{L}(x)$ with x be taken as the species concentration in a given steady state).
2. Obtain k clusters of vertices: $\mathcal{C}_1, \ldots, \mathcal{C}_k$, based on the approximate solution to the aforementioned RatioCut problem.
3. If $y \cap \mathcal{C}_i \neq \emptyset$ for every $i = 1, \ldots, k$ (i.e., every cluster contains some elements of y) then increment k by one (i.e., increase the number of cluster) and return to Step 2. Otherwise define $\overline{\mathcal{C}}_{red}$ as the union of all sets $\mathcal{C}_i$, $i = 1, \ldots, k$, such that $y \cap \mathcal{C}_i = \emptyset$ and we can choose[4] $\mathcal{C}_{\rm red} \subset \overline{\mathcal{C}}_{red}$.

---

[3]For every $i = 1, \ldots, c$, the ith row of U corresponds to the ith vertex.

[4]One can again perform the interative procedure as in Sect. 5.3 to obtain the best combination of complexes $\mathcal{C}_{\rm red}$ from $\overline{\mathcal{C}}_{red}$.

5.4.2 Minimal Eigenvalue Interlacing-based Approach

If one considers that the dynamics of CRN can have multiple timescales since the eigenvalues of $\mathcal{L}$ are related to the rate of decay of complexes, then one can consider removing fast complexes that have minimal influence on the dynamics of the network. This can be done by minimizing the eigenvalue interlacing distance. In this regard, the interlacing property as in Lemma 5.1 can be useful as we demonstrate below.

Suppose that we are looking for a combination of k vertices to be removed for Kron reduction. In order to minimize the influence of the to-be-removed complexes on the rest of the network, we can determine $\mathcal{C}_{\text{red}}$ which solves the following minimization problem:

$$\min_{\mathcal{C}_{\text{red}} \in \binom{\mathcal{I}_c \setminus y}{k}} \sum_{i=2}^{c-k} \lambda_i(\mathcal{L}_{\text{red}}) - \lambda_i(\mathcal{L}), \tag{5.11}$$

where $\binom{\mathcal{I}_c \setminus y}{k}$ is the set of all k-combination from the admissible set of complexes for the deletion $\mathcal{I}_c \setminus y$. Note that the cost function as used above is nonnegative according to the interlacing property in Lemma 5.1.

We summarize our second proposed approach in the following algorithm:

Minimal Eigenvalue Interlacing-based algorithm:

1. Set $k = 1$, set an (averaged and normalized) interlacing distance threshold $\epsilon > 0$ and calculate $\mathcal{L}$ (or $\mathcal{L}(x)$ with x taken as the species concentration in a given steady state).
2. Solve the minimization problem of eigenvalue interlacing as in (5.11). Denote its solution by $\mathcal{C}_{\text{red}}$.
3. If

$$\frac{1}{c-k-1} \sum_{i=2}^{c-k} \frac{\lambda_i(\mathcal{L}_{\text{red}}) - \lambda_i(\mathcal{L})}{\lambda_i(\mathcal{L})} < \epsilon \tag{5.12}$$

then increment k by one and return to Step 2. Otherwise set $\mathcal{C}_{\text{red}}$ from the previous iteration as the desired set of complexes for removal.

Note that in the summation on the left-hand side of (5.12), last term, i.e., the term corresponding to $i = c - k$, contributes much more than the other terms. Therefore, the left-hand side of (5.12) may not be easily interpreted as the normalized deviation of eigenvalues, as will be shown later in the simulation results. One way to overcome this problem is to modify condition (5.12) as

$$\frac{1}{N-1} \sum_{i=2}^{N} \frac{\lambda_i(\mathcal{L}_{\text{red}}) - \lambda_i(\mathcal{L})}{\lambda_i(\mathcal{L})} < \epsilon \tag{5.13}$$

where $N \leq c - k$ is the number of the smallest eigenvalues that are considered to be important.

5.5 Numerical Simulation Results

We evaluate the efficacy of our proposed approaches using two different models. The first one is based on the glycolysis model which has been used in our model reduction paper [22]. The second one is an arbitrary model that is taken from the BioModels database on biological models [1]. This is a model of insulin-dependent glucose metabolism as proposed and studied in [18].

5.5.1 Glycolysis Model

This model describes the glycolysis metabolism based on the work in [4]. The original model in [4] consists of 12 species and 12 reactions and it has successfully been reduced using our Kron reduction approach in [22] to 7 species and 7 reactions. In Table 5.1 below, we reproduce the complexes that are removed at each step of the iterative reduction procedure as described before in Sect. 5.3.

Using the same numerical values as in [22], we apply the spectral clustering-based algorithm to obtain 7 clusters of complexes as shown in Fig. 5.1. It can be seen that four of the complexes in Table 5.1 are in clusters $\mathcal{C}_2$ and $\mathcal{C}_6$ which do not contain any important variables as marked in red color in Fig. 5.1. Since these clusters have minimal cut-set with their neighbors, the corresponding vertices/complexes can be deleted using Kron reduction and it is expected to give us a good reduced model (cf. Table 5.1). Indeed, if we take $\mathcal{C}_{\text{red}} = \{\text{G6P, F6P, P2G, PEP}\}$ then the numerical simulation result gives us an error integral of 0.0701.

We now apply our second proposed approach, i.e., the minimal eigenvalue interlacing-based algorithm to this model and the results are shown in Tables 5.2 and 5.3. From both tables, minimizing the interlacing distance for the first couple of eigenvalues (where we have considered the second and third eigenvalues for the results shown in the lower rows of Tables 5.2 and 5.3) provides a reasonably good combination of complexes for Kron reduction. In particular, if we choose $\epsilon = 0.1$ (i.e., the deviation of eigenvalues of the reduced model should deviate less than 10 %, in average, from those of the full one), then the application of (5.13) leads to F6P

Table 5.1 Order of complex removal using the iterative procedure for the glycolysis model as in [22]

Iteration step	Complex removed	Error integral
1	F6P	0.0002
2	G6P	0.0005
3	P2G	0.0049
4	P3G	0.0147
5	PEP	0.0483

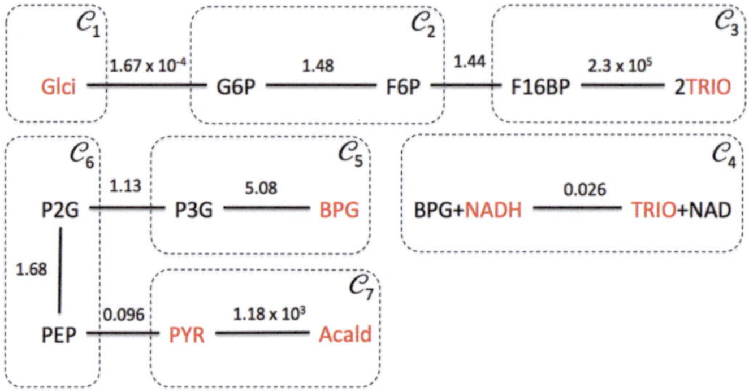

Fig. 5.1 The clustering of complexes in the glycolysis model as used in [22] into 7 clusters using the proposed spectral clustering-based algorithm. The clusters are indicated by the *dashed-line* boxes and the labels $\mathcal{C}_k$, $k = 1, \ldots, 7$ are given on the *top-right corner* of each boxes. The model contains 13 complexes that form a complex graph with three connected components (see [22] for the description and an explanation of all the abbreviations of the complexes). The text in *red* indicates the species defined in output variable y. The number on top of every edge shows the edge weight which are taken from the adjacency matrix in the Laplacian matrix $\mathcal{L}(x)$ using the nominal values of x

Table 5.2 Optimal complexes for removal in the first connected sub-graph containing vertices GLci, G6P, F6P, F16BP and 2TRIO, using the minimal eigenvalue interlacing-based algorithm

Iteration step	Complexes removed	Minimal cost value	λ_2	λ_3	λ_4	λ_5	Error integral
Minimal eigenvalue interlacing-based algorithm with the cost value as in (5.12)							
0	None	0	2.09×10^{-4}	1.04	4.06	4.6×10^5	0
1	F16BP	0.179	2.23×10^{-4}	1.46	4.37		0.6423
2	F6P, F16BP	0.297	2.51×10^{-4}	1.456			0.6417
3	G6P, F6P, F16BP	0.5998	3.348×10^{-4}				0.64
Minimal eigenvalue interlacing-based algorithm with the cost value as in (5.13) and $N = 3$							
1	F6P	0.056	2.23×10^{-4}	1.0927	4.6×10^5		1.83×10^{-4}
2	F16BP, F6P	0.297	2.51×10^{-4}	1.456			0.6417

and P2G as optimal complexes for reduction. In this case, the error integral value associated with the removal of both complexes is 0.025.

However, as shown in Table 5.2, the algorithm still identifies F16BP as a candidate for removal which leads to a large error integral value (which implies that the transient behavior of the reduced model deviates significantly from the full one). On the other hand, our first proposed approach does not identify F16BP as a suitable complex for

Table 5.3 Optimal complexes for removal in the second connected sub-graph containing vertices BPG, P3G, P2G, PEP, PYR and Acald, using the minimal eigenvalue interlacing-based algorithm

Iteration step	Complexes removed	Minimal cost value	λ_2	λ_3	λ_4	λ_5	λ_6	Error integral
Minimal eigenvalue interlacing-based algorithm with the cost value as in (5.12)								
0	None	0	0.068	0.919	4.087	10.856	3.6×10^3	0
1	P3G	83.59	0.077	1.18	4.094	3.6×10^3		0.0046
2	P3G, P2G	296.49	0.092	1.243	3.6×10^3			0.0142
3	P3G, P2G, PEP	1975.5	0.1237	3.6×10^3				0.0478
Minimal eigenvalue interlacing-based algorithm with the cost value as in (5.13) and $N = 3$								
1	P2G	0.1177	0.075	1.045	10.544	3.6×10^3		0.0249
2	P3G, P2G	0.352	0.092	1.243	3.6×10^3			0.0142
3	P3G, P2G, PEP	1978.5	0.1237	3.6×10^3				0.0478

the Kron reduction. This result shows that the spectral-based clustering algorithm outperforms the eigenvalue interlacing-based algorithm.

5.5.2 *Insulin-Signaling-Dependent Glucose Metabolism Model*

The model describes the insulin-signaling-dependent glucose metabolism that includes glycolysis, gluconeogenesis and glycogenesis pathways, all of which are regulated by insulin. The full model consists of 39 reactions (where forward and reverse reactions in a reversible reaction are counted as 2 separate reactions), 23 species, and 23 complexes.[5] Figure 5.2a shows the complex graph of the full model and we refer interested readers to [18] for a description and detailed explanation of the network. The output vector y consists of the concentrations of the species pAkt, GLCex, PEPCK, Glycogen, p1IRS, and F16P.

The iterative reduction procedure as discussed in Sect. 5.3 is performed based on the response to a step increase of external insulin concentration from 0 to 100 nM. For the error integral, we take $t = 0$ and $T = 480$ min. Table 5.4 gives the value of the error integral and the complex deleted at each iterative step. The resulting reduced complex graph is shown in Fig. 5.2b.

[5] Here the complex composition matrix Z is given by an identity matrix.

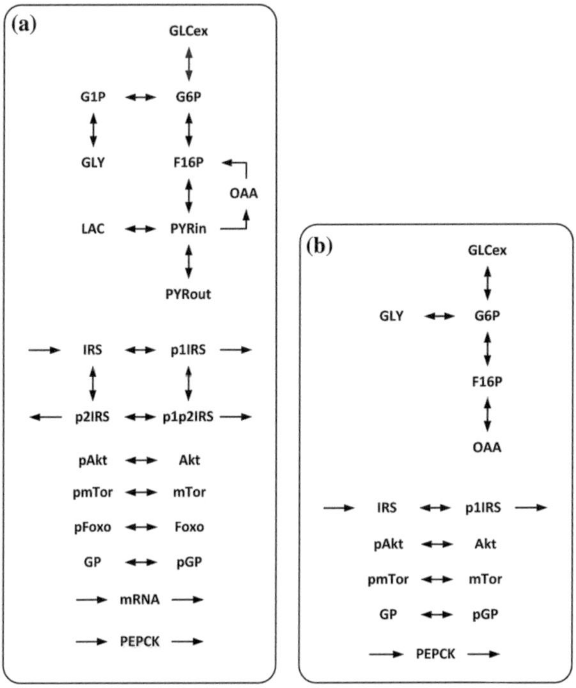

Fig. 5.2 Complex graph of the original and reduced models of the insulin-signaling-dependent glucose metabolism. The *left-hand* panel is a schematic representation of the original model used for model reduction. The full model description and an explanation of all the abbreviations is found in [18]. The *right-hand* panel represents the reduced model after deleting 9 complexes (LAC, PYRin, PYRout, p2IRS, p1p2IRS, G1P, Foxo, pFoxo, mRNA). **a** The complex graph of the full model. **b** The complex graph of the reduced model

Table 5.4 Order of complex removal using the iterative procedure for the insulin-signaling-dependent glucose metabolism model

Iteration step	Complex removed	Error integral
1	LAC	0.0001
2	PYRin	0.0004
3	PYRout	0.0007
4	p2IRS	0.0014
5	p1p2IRS	0.0026
6	G1P	0.0071
7	Foxo	0.0142
8	pFoxo	0.0142
9	mRNA	0.0329
10	G6P	0.1195

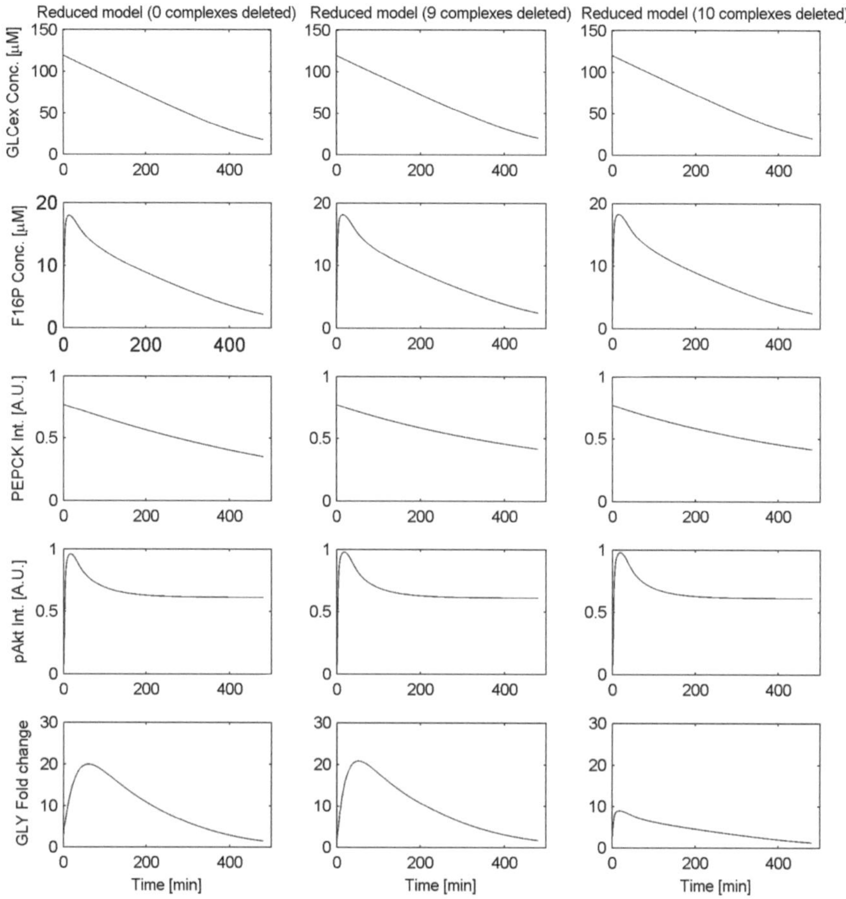

Fig. 5.3 Comparison of transient behavior of species concentrations in the full model and reduced models of insulin-signaling-dependent glucose metabolism model. The figures in the *left column* are the concentration plots of the full model, the figures in the *middle column* are concentration plots of the reduced model with 9 complexes deleted and the figures on the *right column* are the concentration plots of the reduced model with 10 complexes deleted

In Fig. 5.3, we compare the transient behaviors of the species concentrations of the full model with those of the reduced models obtained by deleting 9 and 10 complexes, following the iterative reduction steps as before. It can be observed from these results that the dynamics of the reduced model with 10 complexes deleted, whose error integral value exceeds 0.1, deviates significantly from the full model dynamics. On the other hand, the transient behavior of y of the reduced model with 9 complexes deleted is in close agreement with that of the original model. This reduced model has 14 complexes and 20 reactions and its complex graph is shown in Fig. 5.2b.

Since this network contains a directed sub-graph (the one that interconnects F16P, PYRin and OAA), for evaluating our proposed methods, we replace the cyclic sub-

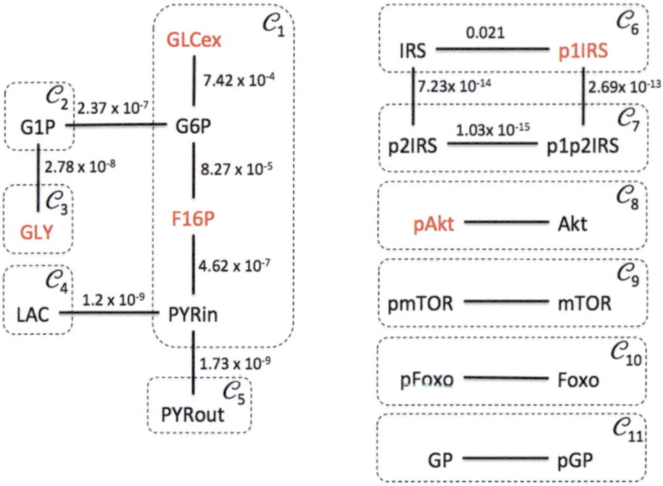

Fig. 5.4 The clustering of complexes in the insulin-signaling-dependent glucose metabolism into 11 clusters using the proposed spectral clustering-based algorithm. The clusters are indicated by the *dashed-line* boxes and the labels $\mathcal{C}_k, k = 1, \ldots, 11$ are given on the *top-right* corner of each boxes. The text in red indicates the species defined in output variable y. The number on top of every edge shows the edge weight which are taken from the adjacency matrix in the Laplacian matrix $\mathcal{L}(x)$ using the nominal values of x

graph F16P↔FYRin→OAA→F16P by a reversible reaction F16P↔PYRin. The equilibrium constant of this reversible reaction is set according to the reaction constants in the original subnetwork. This ad hoc modification gives us an undirected complex graph for which our two proposed methods can be applied.

We apply our spectral clustering-based algorithm to this modified complex graph and the result is shown in Fig. 5.4. The subnetwork containing F16P, GLY, and GLCex is clustered into 5 clusters where three complexes, G1P, LAC, and PYRout are in clusters that do not contain any elements of y. On the other hand, for another subnetwork, there are two complexes, p2IRS and p1p2IRS, which is in a cluster that does not include any element of y. Similar to the result in glycolysis model above, one can observe that these five complexes are also listed in the Table 5.4. Hence removing these complexes via Kron reduction will give us a good reduced model. Indeed, numerical simulation of the Kron-reduced model where $\mathcal{C}_{\text{red}} = \{\text{G1P, LAC, PYRout, p2IRS, p1p2IRS}\}$ gives us an error integral of 0.0071. Since the rest of the network consists of simple sub-graphs with two vertices each, we can delete, for instance the clusters, $\mathcal{C}_9$, $\mathcal{C}_{10}$ and $\mathcal{C}_{11}$.

Using the same modified graph of complexes as above, we apply our second proposed approach to the first connected component of the graph and the result is given in Table 5.5. Our second approach identifies G6P as one of the best candidate for removal, in contrast to the result obtained in Table 5.4 where G6P is shown to

Table 5.5 Optimal complexes for removal in the first connected sub-graph containing vertices F16P, GLY and GLCex, using the minimal eigenvalue interlacing-based algorithm

Iteration step	Complexes removed	Minimal cost value	λ_2 $(\times 10^{-9})$	λ_3 $(\times 10^{-9})$	...	λ_6 $(\times 10^{-5})$	λ_7 $(\times 10^{-4})$	λ_8 $(\times 10^{-3})$	Error integral
Minimal eigenvalue interlacing-based algorithm with the cost value as in (5.12)									
0	none	0	1.32	2.1		0.065	1.2	1.5	0
1	G6P	0.087	1.33	2.2		12	15		0.09916
2	G6P, PYRin	45.486	1.34	2.29		149			0.09923
3	G6P, PYRin, G1P	47.61	2.15	3.52					0.09930
Minimal eigenvalue interlacing-based algorithm with the cost value as in (5.13) and $N = 3$									
1	PYRin	0.0228	1.33	2.17		12.1	15		0.00041
2	G6P, PYRin	0.057	1.34	2.29		149			0.09923
3	G6P, PYRin, G1P	0.115	2.15	3.52					0.09930

be the least preferred complex for removal. Hence for this second model, we can conclude again that the spectral-based clustering approach is a better method than the eigenvalue interlacing-based one to identify complexes for the Kron reduction.

5.6 Conclusion

In this chapter, we propose two approaches for finding the best set of complexes to be deleted for the Kron reduction of a chemical reaction network. The proposed methods are based on the spectral properties of the weighted Laplacian of the complex graph corresponding to the network. The aim of these methods is to provide an alternative to the use of error integral that requires a numerical integration which can be computationally expensive, in particular, if we need to handle a very large dimensional model. We have applied the two approaches on two biological models. For both cases, it has been observed that the spectral-based clustering approach performs better than the eigenvalue interlacing-based approach. The extension of these methods to directed complex graphs, as commonly found in large dimensional biological models, is an interesting topic for further research. The result for directed graphs proposed in [17] can potentially be adapted to modify the spectral-based clustering approach presented in this chapter so as to make it applicable for directed graphs.

Acknowledgments The research is supported by NWO Centres for Systems Biology.

References

1. Biomodels database: an enhanced, curated and annotated resource for published quantitative kinetic models. http://www.ebi.ac.uk/biomodels-main/BIOMD0000000482
2. B. Bollobas, Graduate Texts in Mathematics, *Modern Graph Theory*, vol. 184 (Springer, New York, 1998)
3. F. Döfler, F. Bullo, Kron reduction of graphs with applications to electrical networks. IEEE Trans. Circ. Syst. I **60**(1), 150–163 (2013)
4. K. van Eunen, J.A.L. Kiewiet, H.V. Westerhoff, B.M. Bakker, Testing biochemistry revisited: how in vivo metabolism can be understood from in vitro enzyme kinetics. PLoS Comput. Biol. **8**(4), e1002483 (2012)
5. M. Feinberg, Chemical reaction network structure and the stability of complex isothermal reactors -I. The deficiency zero and deficiency one theorems. Chem. Eng. Sci. **43**(10), 2229–2268 (1987)
6. M. Feinberg, Complex balancing in chemical kinetics. Arch. Ration. Mech. Anal. **49**, 187–194 (1972)
7. W.H. Haemers, Interlacing eigenvalues and graphs. Linear Algebra Appl. **227–228**, 593–616 (1995)
8. F. Horn, R. Jackson, General mass action kinetics. Arch. Ration. Mech. Anal. **47**, 81–116 (1972)
9. F.J.M. Horn, Necessary and sufficient conditions for complex balancing in chemical kinetics. Arch. Ration. Mech. Anal. **49**, 172–186 (1972)

10. B. Jayawardhana, S. Rao, A.J. van der Schaft, Balanced Chemical Reaction Networks Governed by General Kinetics, in *20th International Symposium on Mathematical Theory of Networks and Systems*, Melbourne, July 2012
11. T. Kanungo, D.M. Mount, N.S. Netanyahu, C.D. Piatko, R. Silverman, A.Y. Wu, An efficient k-means Clustering algorithm: analysis and implementation. IEEE Trans. Pattern Anal. Mach. Intell. **24**(7), 881–892 (2002)
12. G. Kron, *Tensor Analysis of Networks* (Wiley, 1939)
13. B.M. Maschke, A.J. van der Schaft, Port-controlled Hamiltonian Systems: Modelling Origins and System-Theoretic Properties, in *Proceedings of the IFAC Symposium on NOLCOS*, Bordeaux, pp. 282–288 (1992)
14. B.M. Maschke, A.J. van der Schaft, System-Theoretic Properties Of Port-controlled Hamiltonian Systems, *Systems and Networks: Mathematical Theory and Applications*, vol. II (Akademie-Verlag, Berlin, 1994), pp. 349–352
15. B.M. Maschke, A.J. van der Schaft, P.C. Breedveld, An intrinsic Hamiltonian formulation of network dynamics: nonstandard poisson structures and gyrators. J. Franklin Inst. **329**, 923–966 (1992)
16. B.M. Maschke, A.J. van der Schaft, P.C. Breedveld, An intrinsic Hamiltonian formulation of the dynamics of LC circuits. IEEE Trans. Circ. Syst. I **42**, 73–82 (1995)
17. M. Meilă, W. Pentney, Clustering by Weighted Cuts in Directed Graphs, in *Proceedings of 2007 SIAM International Conference on Data Mining* (2007)
18. R. Noguchi et al., The selective control of glycolysis, gluconeogenesis and glycogenesis by temporal insulin patterns. Mol. Syst. Biol. **9**, 664 (2013)
19. J.F. Oster, A.S. Perelson, A. Katchalsky, Network dynamics: dynamic modeling of biophysical systems. Q. Rev. Biophys. **6**(1), 1–134 (1973)
20. J.F. Oster, A.S. Perelson, Chemical reaction dynamics, part I: geometrical structure. Arch. Ration. Mech. Anal. **55**, 230–273 (1974)
21. O. Radulescu, A.N Gorban, A. Zinovyev, V. Noel, Reduction of dynamical biochemical reaction networks in computational biology. Front. Genet. **3**, 00131 (2012)
22. S. Rao, A.J. van der Schaft, K. van Eunen, B.M. Bakker, B. Jayawardhana, Model reduction of biochemical reaction networks. BMC Syst. Biol. **8**, 52 (2014)
23. S. Rao, A.J. van der Schaft, B. Jayawardhana, A graph-theoretical approach for the analysis and model reduction of complex-balanced chemical reaction networks. J. Math. Chem **51**(9), 2401–2422 (2013)
24. J. Shi, J. Malik, Normalized cuts and image segmentation. IEEE Trans. Pattern Anal. Mach. Intell. **22**(8), 888–905 (2000)
25. D. Siegel, D. MacLean, Global stability of complex balanced mechanisms. J. Math. Chem. **27**, 89–110 (2000)
26. A.J. van der Schaft, D. Jeltsema, Port-Hamiltonian systems theory: an introductory overview. Found. Trends Syst. Control **1**(2/3), 173–378 (2014)
27. A.J. van der Schaft, L_2-*Gain and Passivity Techniques in Nonlinear Control*, 2nd revised and enlarged edn., (Springer, London, 2000)
28. A.J. van der Schaft, B.M. Maschke, A Port-Hamiltonian Formulation of Open Chemical Reaction Networks, *Advances in the Theory of Control, Signals and Systems*, Lecture Notes in Control and Information Sciences (Springer, New York, 2011), pp. 339–348
29. A.J. van der Schaft, B.M. Maschke, The Hamiltonian formulation of energy conserving physical systems with external ports. Arch. Elektron. Übertragungstech **49**, 362–371 (1995)
30. A. van der Schaft, S. Rao, B. Jayawardhana, On the mathematical structure of balanced chemical reaction networks governed by mass action kinetics. SIAM J. Appl. Math. **73**(2), 953–973 (2013)
31. A.J. van der Schaft, S. Rao, B. Jayawardhana, *A Network Dynamics Approach to Chemical Reaction Networks*. arXiv1502.02247, submitted for publication (2015)
32. A.J. van der Schaft, Characterization and partial synthesis of the behavior of resistive circuits at their terminals. Syst. Control Lett. **59**, 423–428 (2010)
33. U. von Luxburg, A tutorial on spectral clustering. Stat. Comput. **17**(4) (2007)

Chapter 6
Distributed Line Search for Multiagent Convex Optimization

Jorge Cortés and Sonia Martínez

Abstract This note considers multiagent systems seeking to optimize a convex aggregate function. We assume that the gradient of this function is distributed, meaning that each agent can compute its corresponding partial derivative with information about its neighbors and itself only. In such scenarios, the discrete-time implementation of the gradient descent method poses the basic challenge of determining appropriate agent stepsizes that guarantee the monotonic evolution of the objective function. We provide a distributed algorithmic solution to this problem based on the aggregation of agent stepsizes via adaptive convex combinations. Simulations illustrate our results.

6.1 Introduction

This book chapter is related to Arjan's longstanding research interests on nonlinear dynamical systems, systems with interacting continuous and discrete dynamics, and network modeling, analysis, and control of interconnected complex physical and engineering systems. The first author (JC) was a postdoc researcher with Arjan from January 2002 to June 2002 at the University of Twente in Enschede, The Netherlands. Arjan was a member of the Ph.D. committee of the second author (SM) in May 2002 in Madrid, Spain, which ended with a long Spanish meal that took hours to complete. Both authors have always admired the prolific and highly active nature of Arjan, always sharp in identifying important problems and research areas, and his incredible ability to produce impactful work and seminal contributions on which many other researchers, including the authors, have built.

J. Cortés (✉) · S. Martínez
Department of Mechanical and Aerospace Engineering, UC San Diego,
9500 Gilman Dr, La Jolla, CA 92093-0411, USA
e-mail: cortes@ucsd.edu

S. Martínez
e-mail: soniamd@ucsd.edu

M.K. Camlibel et al. (eds.), *Mathematical Control Theory I*,
Lecture Notes in Control and Information Sciences 461,
DOI 10.1007/978-3-319-20988-3_6

A popular approach to the coordination of multiagent systems consists of designing a distributed algorithm that solves an optimization problem encoding the coordination task. This top-bottom method has been very useful in a variety of networked multiagent scenarios, including multi-vehicle coordination, network utility maximization, energy dispatch, and information processing by sensor networks. Due to a lack of centralized authority, the proposed algorithms are to be executed by employing local information only, which allows for greater scalability and robustness to agent failure. In this paper, we consider a particular class of convex optimization problems for which gradient-descent continuous-time algorithms are naturally distributed, meaning that each agent can compute the partial derivative of the function to be optimized with information of its neighbors and itself. While the convergence analysis of these algorithms in continuous time is facilitated by powerful concepts and tools from stability theory, their practical implementation needs to be of discrete-time nature. This requires the determination of an appropriate stepsize along the descent direction. A common approach to solve this problem is the a priori, offline determination of the stepsize using global information. In this manuscript, we instead take the alternative approach of designing distributed procedures that allow agents to coordinate the computation of appropriate stepsizes.

Literature review. This manuscript is a contribution to the recent body of research on distributed optimization by a network of agents subject to intermittent interactions. In these works, the objective function can be expressed as a sum of convex functions and be subject to different inequality and equality constraints; see for example [6, 11, 12, 15, 20]. Building on consensus-based coordination rules [2, 10, 13, 14], the aforementioned efforts lead to discrete-time schemes employing function subgradients. Continuous-time approaches which are robust to errors due to communication and initialization include [16] on undirected networks and [5, 9] on directed networks. With the goal of designing faster algorithms, [17, 18] focus on Newton schemes. Except for [18], which employs a decentralized backtracking line search rule to implement the Armijo rule, and an earlier version [4] of the present work, the aforementioned approaches assume that agents have access to a common (possibly time-varying) stepsize, determined a priori, to implement the distributed algorithm. The recent work [7] instead combines continuous-time computation and discrete-time communication to let individual agents determine autonomously their stepsizes. Our work connects with the literature on algorithms for gradient-descent methods [1]. The classical steepest descent method [3] for unconstrained minimization converges linearly and can show slow performance. However, the understanding of these algorithms is central for theory and design of more sophisticated optimization algorithms [8]. It is within this simple context that we study how a network of agents can determine appropriate stepsizes in a distributed way.

Statement of contributions. We introduce a class of algorithms that allows a group of agents to descend a convex objective function by following an aggregated descent direction. Each agent employs a stepsize that results from a distributed stepsize computation subroutine. This strategy takes as inputs the stepsizes computed by each agent via a line search procedure. By means of a proper initialization, and after only a finite number of rounds, the strategy outputs a vector of stepsizes, one per agent, that agents can readily implement to decrease the function. If let run indefinitely, the strategy converges to a convex combination of stepsizes that guarantees that the function decreases via the steepest descent direction or other alternative aggregated directions of descent.

Organization. Section 6.2 introduces basic preliminaries. Section 6.3 states formally the problem of interest. Section 6.4 introduces several stepsize aggregation models for distributed line search based on convex combinations and Sect. 6.5 presents a provable distributed linear iteration algorithm to compute them. Section 6.6 presents simulations of the resulting algorithms. We gather our conclusions and ideas for future work in Sect. 6.7.

6.2 Preliminaries

This section presents basic notions from graph theory, optimization via gradient descent, and line search.

6.2.1 Notation

We employ $\mathbb{R}^n_{>0}$ (resp. $\mathbb{R}^n_{\geq 0}$) to denote the positive orthant (resp. the nonnegative orthant) of $\mathbb{R}^n$. We use the notation $\mathbf{1}_n \in \mathbb{R}^n_{>0}$ (resp. $\mathbf{1}_{n-1} \in \mathbb{R}^{n-1}_{>0}$) for the vector $(1, \ldots, 1)^T$ (resp. $\mathbf{1}_{n-1} = (1, \ldots, 1)^T$). We denote the eigenvalues of a square matrix $M \in \mathbb{R}^{n\times n}$ as $\lambda_i(M)$, $i \in \{1, \ldots, n\}$. We assume that the eigenvalues are indexed so that $\text{Re}(\lambda_1(M)) \leq \text{Re}(\lambda_2(M)) \leq \cdots \leq \text{Re}(\lambda_n(M))$, where Re denotes the real part of a complex number. We denote by I_n the identity matrix of dimension $n \times n$. The spectral radius of M is $\rho(M) = \max_{i\in\{1,\ldots,n\}} |\lambda_i(M)|$. The essential spectral radius of a matrix M with $\rho(M) = 1$ is $\rho_{\text{ess}}(M) = \max_{i\in\{1,\ldots,n\}}\{|\lambda_i(M)| \mid \lambda_i(M) \neq 1\}$. The notation $M \geq 0$ means that M is positive semidefinite. In particular, $M_1 \geq M_2$ if and only if $M_1 - M_2 \geq 0$. A matrix $M \in \mathbb{R}^{n\times n}$ is Metzler if all its off-diagonal elements are nonnegative. A matrix $M \in \mathbb{R}^{n\times n}_{\geq 0}$ is irreducible if, for any nontrivial partition $J \cup K$ of the index set $\{1, \ldots, n\}$, there exist $j \in J$ and $k \in K$ such that $m_{jk} \neq 0$. We let $\text{span}\{w_1, \ldots, w_l\}$ denote the vector space generated by the vectors $w_1, \ldots, w_l \in \mathbb{R}^n$. Given $g : \mathbb{R} \to \mathbb{R}$ and $h : \mathbb{R} \to \mathbb{R}$, we denote $g(r) \in O(h(r))$ if and only if there is $C > 0$ and r_0 such that $|g(r)| \leq C|h(r)|$, for all $r \geq r_0$.

6.2.2 Graph-Theoretic Notions

We present some basic notions from algebraic graph theory following the exposition in [2]. An *undirected graph*, or simply *graph*, is a pair $G = (V, E)$, where V is a finite set called the vertex set and E is the edge set consisting of unordered pairs of vertices. For $i, j \in V$ and $i \neq j$, the set $\{i, j\}$ denotes an undirected edge, and i and j are *neighbors*. We let $\mathcal{N}_G(i)$ denote the set of neighbors of u_i in G. The graph G is *connected* if for any pair of nodes i, j there exists a sequence of edges $\{i, i_1\}, \{i_1, i_2\}, \dots, \{i_k, j\}$ connecting i with j. The *adjacency matrix* of a graph G is a nonnegative symmetric matrix $A = (a_{ij}) \in \mathbb{R}^{n\times n}_{\geq 0}$ such that $a_{ij} \neq 0$ if and only if $\{i, j\}$ is an edge of the graph. Here, we consider $a_{ij} = 1$, when $\{i, j\} \in E$. Consider the diagonal matrix $D = \operatorname{diag}(A\mathbf{1}_n)$. The *Laplacian matrix* of G is defined as $L = D - A$, which is a symmetric and positive semidefinite matrix. Note that L has an eigenvalue at 0 and $\mathbf{1}_n$ is the corresponding eigenvector. A graph G is connected if and only if L is irreducible and 0 is a simple eigenvalue. Finally, a map $g : \mathbb{R}^n \to \mathbb{R}^n$ is *distributed over* G if, for all $j \in \{1, \dots, n\}$, the component g_j can be expressed as $g_j(x) = g_j(x_{i_1}, \dots, x_{i_{n_j}})$, where $\mathcal{N}_G(j) = \{i_1, \dots, i_{n_j}\}$, for all $x \in \mathbb{R}^n$.

6.2.3 Directions of Descent and Line Search

Given a continuously differentiable function $f : \mathbb{R}^n \to \mathbb{R}$, we let $\nabla f : \mathbb{R}^n \to \mathbb{R}^n$ denote its gradient

$$\nabla f(x) = \Big(\frac{\partial f}{\partial x_1}(x), \dots, \frac{\partial f}{\partial x_n}(x)\Big).$$

Throughout the paper, we use the notation $\nabla_i f$ to refer to the ith component of ∇f. Given a function $f : \mathbb{R}^n \to \mathbb{R}$ and $x \in \mathbb{R}^n$, $v \in \mathbb{R}^n$ is a direction of descent of f at x if there exists $T > 0$ such that

$$f(x + \delta v) < f(x), \quad \delta \in (0, T).$$

If f is continuously differentiable at x, this is equivalent to saying that $\nabla f(x)^T v < 0$. The procedure of calculating the actual step δ to be taken in the direction v is called *line search*. The choice of a stepsize and a direction that guarantees the reduction of the function at each iterate leads to various gradient algorithms. In particular, one could aim to find the best stepsize that optimizes the decrease in the value of f along a direction v, i.e.,

$$\varepsilon_v = \operatorname{argmin}_{\delta\in[0,\infty)} f(x + \delta v). \tag{6.1}$$

Let $h_v(\delta) = f(x + \delta v)$. For a continuously differentiable function, it is not difficult to see that the stepsize (6.1) is characterized by the equation

$$h'_v(\varepsilon_v) = \nabla f(x + \varepsilon_v v)^T v = 0. \tag{6.2}$$

The choice $v = \nabla f$ (which corresponds to the direction that instantaneously descends f the most) leads to the steepest descent method,

$$x^{k+1} = x^k - \alpha_k \nabla f(x^k), \quad k \geq 0,$$

which locally converges to the set of minimizers of f.

6.3 Problem Statement

Consider a network of n agents, indexed by $i \in \{1, \dots, n\}$, with interaction topology described by a graph G. The network state, denoted x, belongs to $\mathbb{R}^n$. Agent i is responsible for the ith component $x_i \in \mathbb{R}$. The results that follow can also be extended to scenarios where each agent supervises several components of the vector $x \in \mathbb{R}^n$, but here we keep the exposition simple. Consider a convex function $f : \mathbb{R}^n \to \mathbb{R}$ whose gradient $\nabla f : \mathbb{R}^n \to \mathbb{R}^n$ distributed over G. Thus, each agent $i \in \{1, \dots, n\}$ can compute

$$v_i(x) = (0, \dots, \nabla_i f(x), \dots, 0), \tag{6.3}$$

with information from its neighbors in G. The next result states that the line search procedure for f and each direction v_i can be carried out in a distributed way.

Lemma 6.1 (Individual agent stepsize computation) *Let $f : \mathbb{R}^n \to \mathbb{R}$ be continuously differentiable and assume $\nabla f : \mathbb{R}^n \to \mathbb{R}^n$ is distributed over G. Let $x \in \mathbb{R}^n$ and $i \in \{1, \dots, n\}$ with $v_i(x)$, as defined in* (6.3), *be nonvanishing. Then, the optimal stepsize $\varepsilon_{v_i(x)}$ along $v_i(x)$ and the associated decrease $\Delta_{v_i(x)}$ in the value of f can be computed with knowledge only of $\{x_i\} \cup \{x_j \mid j \in \mathcal{N}_G(i)\}$.*

Proof For simplicity, we use the shorthand notation h_i, ε_i, and Δ_i to denote $h_{v_i(x)}$, $\varepsilon_{v_i(x)}$, and $\Delta_{v_i(x)}$ respectively. Note that (6.2) in this case reduces to

$$h_i'(\varepsilon_i) = \nabla_i f(x + \varepsilon_i v_i(x))^T \nabla_i f(x) = 0. \tag{6.4}$$

The only difference between $x + \varepsilon_i v_i(x)$ and x is in the ith component, which agent i is responsible for. Since the gradient of f is distributed over G, agent i has all the information required to solve Eq. (6.4). A similar argument holds for the associated decrease in the value of f,

$$\begin{aligned}\Delta_i &= f(x) - f(x + \varepsilon_i v_i(x)) \\ &= h_i(0) - h_i(\varepsilon_i) = -\int_0^{\varepsilon_i} h_i'(\delta)\, d\delta \\ &= -\int_0^{\varepsilon_i} \nabla_i f(x + \delta v_i(x))^T \nabla_i f(x)\, d\delta,\end{aligned} \tag{6.5}$$

which completes the result. □

Note that the line search procedure performed by agent i assumes that all other agents remain fixed. The problem of interest in this paper is the following:

Distributed line search computation problem. Let $x \in \mathbb{R}^n$. Given δ_i such that $f(x + \delta_i v_i(x)) < f(x)$, where $v_i(x)$ is given as in (6.3) for all $i \in \{1, \dots, n\}$, design a distributed algorithm that allows the group of agents to agree on stepsizes $(\varepsilon_1, \dots, \varepsilon_n) \in \mathbb{R}^n_{\geq 0}$ such that

$$f(x + \varepsilon_1 v_1(x) + \cdots + \varepsilon_n v_n(x)) < f(x).$$

In particular, a solution such that $\varepsilon_i = \varepsilon$ for all $i \in \{1, \dots, n\}$, solves the *distributed steepest descent line search computation problem.*

We make the following considerations regarding the above problem. First, note that the choice $\varepsilon_i = \delta_i$, $i \in \{1, \dots, n\}$, is not a solution in general. In principle, there are several ways to approach this problem. For instance, one can resort to parallel algorithms to identify those agents that maximize the function decrease and coordinate their changes in state accordingly via leader election. Instead, here we look for solutions that allow all agents to simultaneously contribute to the decrease of the function.

6.4 Weighted Network-Aggregated Stepsizes

The next result provides guidance as to how the problem stated can be solved. Lemma 6.2 determines how stepsizes based on a convex combination guarantee decrease of cost function.

Lemma 6.2 (Network-aggregated stepsize) *Let $f : \mathbb{R}^n \to \mathbb{R}$ be convex. For $x \in \mathbb{R}^n$, let $w_1, \dots, w_n \in \mathbb{R}^n$ be directions of descent of f from x. Let $\delta_i \in \mathbb{R}_{>0}$ be a stepsize such that $f(x + \delta_i w_i) < f(x)$, for each $i \in \{1, \dots, n\}$. Let $\mu_i \in [0, 1]$, for $i \in \{1, \dots, n\}$, such that $\mu_1 + \cdots + \mu_n = 1$. Then $\mu_1 \delta_1 w_1 + \cdots + \mu_n \delta_n w_n$ is an aggregated direction of descent of f from x, and $f\left(x + \delta \sum_{i=1}^n \mu_i \delta_i w_i\right) < f(x)$.*

Proof The result follows from the following relations:

$$\begin{aligned} f\left(x + \sum_{i=1}^n \mu_i \delta_i w_i\right) &= f\left(\sum_{i=1}^n \mu_i (x + \delta_i w_i)\right) \\ &\leq \sum_{i=1}^n \mu_i f(x + \delta_i w_i) < \sum_{i=1}^n \mu_i f(x) = f(x), \end{aligned}$$

where we have used the fact that f is convex in the first inequality. □

Note that the aggregation procedure in Lemma 6.2 reduces the size of the agent stepsizes, i.e., $\mu_i \delta_i < \delta_i$, $i \in \{1, \ldots, n\}$. This makes sense as the individual agent stepsizes have been computed with the overly optimistic assumption that nobody else would change its state. This reduction in stepsize is the price that the agents have to pay to make sure the aggregate function decreases. The following are particular cases of stepsizes that we consider in the sequel. With the notation of Lemma 6.2, let $w_i = v_i(x)$ be given by (6.3). The *common network-aggregated stepsize vector* is

$$\mu_i = \frac{\frac{1}{\delta_i}}{\frac{1}{\delta_1} + \cdots + \frac{1}{\delta_n}}, \qquad i \in \{1, \ldots, n\}. \tag{6.6}$$

By using this stepsize vector, agents decrease the function along $\nabla f(x)$. The *proportional-to-cost network-aggregated stepsize vector* is

$$\mu_i = \frac{\Delta_i}{\Delta_1 + \cdots + \Delta_n}, \qquad i \in \{1, \ldots, n\}, \tag{6.7}$$

where $\Delta_i = f_i(x) - f_i(x + \delta_i v_i(x))$, for $i \in \{1, \ldots, n\}$. Finally, the *proportional-to-state network-aggregated stepsize vector* is

$$\mu_i = \frac{d_i}{d_1 + \cdots + d_n}, \qquad i \in \{1, \ldots, n\}, \tag{6.8}$$

where $d_i = \delta_i \|v_i(x)\|$, for $i \in \{1, \ldots, n\}$. Note that the weights defined in (6.7) are larger for those agents who offer a larger decrease in the value of the objective function. Thus, they encode a type of "proportional fairness" in the way that each agent can decrease the cost function. A similar consideration applies to (6.8). We call the resulting direction of descent *proportional-to-cost* (resp. *proportional-to-state*) direction of descent.

Lemma 6.2 paves the way for performing line search in a distributed way. Using this result, the agents in the network can collectively fuse their stepsizes in order to guarantee that the resulting network state after updates by all agents decreases the value of the objective function. Remarkably, this is accomplished without the need to share the individual directions of motion of the agents. In particular, the aggregated stepsize models (6.6)–(6.8) take into account the current network state in the determination of the appropriate stepsizes. The challenge is then to perform these stepsize aggregations in a distributed way. We address this in the following section.

6.5 Adaptive Algorithm for Distributed Stepsize Computation

One can implement a number of distributed algorithms to compute the stepsizes (6.6)–(6.8) across the whole network. For instance, average consensus could be employed to compute the corresponding aggregate sums in the denominators of these expressions. This, together with knowledge of the size of the network, would allow each agent to compute the aggregated stepsize. However, the convergence of these algorithms is typically asymptotic, and so it may appear impractical to execute one at each state through the evolution of the network. Instead, we would like to find distributed algorithms that, for each $x \in \mathbb{R}^n$, even if not implementing exactly the models (6.6)–(6.8), (i) can guarantee that the function decreases and (ii) approach asymptotically the directions of descent and stepsizes provided in Lemma 6.2.

6.5.1 Distributed Computation of Convex Combinations

We note that the aggregated stepsize models (6.6)–(6.8) have a similar structure that can be captured as follows: given a vector $y \in \mathbb{R}^n_{>0}$, compute the aggregated vector

$$(y^T \mathbf{1}_n)^{-1} y.$$

Each model corresponds to a different choice of vector y. Specifically, the choice $y = (\frac{1}{\delta_1}, \dots, \frac{1}{\delta_n})$ corresponds to the common network-aggregated stepsize vector, the choice $y = \Delta$ corresponds to the proportional-to-cost network-aggregated stepsize vector, and the choice $y = (\frac{1}{\delta_1}, \dots, \frac{1}{\delta_n})$ corresponds to the proportional-to-state network-aggregated stepsize vector.

In this section, we propose a continuous-time distributed algorithm that allows each agent to compute its component of the aggregated vector $(y^T \mathbf{1}_n)^{-1} y$. Define the matrix $Q(y) \in \mathbb{R}^{n \times n}$ such that

$$Q_{ij}(y) = -y_i y_j, \text{ for } (i, j) \in E$$
$$Q_{ii}(y) = \sum_{j \in \mathcal{N}_G(i)} y_j^2, \text{ for } i \in \{1, \dots, n\}.$$

Three important properties of the matrix $Q(y)$ are that: (i) $Q(y) = Q(y)^T$, (ii) $-Q(y)$ is Metzler, and (iii) $Q(y)$ is irreducible (because G is connected). Consider the function $V : \mathbb{R}^n \to \mathbb{R}$, given by $V(\mu) = \frac{1}{2}\mu^T Q(y)\mu$. Since the network interaction graph G is undirected, it is easy to verify that $V(\mu) = \frac{1}{2}\sum_{i=1}^n \sum_{j \in \mathcal{N}_G(i)} (y_j \mu_i - y_i \mu_j)^2$.

Let us now define the quadratic program

$$\text{minimize} \quad \frac{1}{2}\mu^T Q(y)\mu, \tag{6.9a}$$
$$\text{subject to} \quad \mathbf{1}_n^T \mu = 1. \tag{6.9b}$$

The next result shows that the aggregated vector is the unique solution of this program.

Lemma 6.3 *The unique solution to* (6.9) *is given by* $\mu^\star = (y^T \mathbf{1}_n)^{-1} y$.

Proof It is easy to check that $\mu^\star$ is a solution to the quadratic program (6.9). First, it holds that $(\mu^\star)^T\mathbf{1}_n = (y^T\mathbf{1}_n)^{-1}(y^T\mathbf{1}_n) = 1$. Secondly, note that $(y_j\mu_i^\star - y_i\mu_j^\star)^2 = (y^T\mathbf{1}_n)^{-1}(y_jy_i - y_iy_j)^2 = 0$, thus $\mu^\star$ attains the minimum value of V. To see that $\mu^\star$ is unique, let us study the critical points of V. Any critical point satisfies $\nabla V(\mu) = Q(y)\mu = 0$. Let $\beta = \min_{i\in\{1,\dots,n\}}\sum_{j\in\mathcal{N}_G(i)} y_j^2 + 1$, and consider the matrix $\beta I_n - Q(y)$. Since $-Q(y)$ is Metzler and irreducible, then $\beta I_n - Q(y)$ is a positive and irreducible matrix. By the Perron–Frobenius theorem [19], there exists a unique eigenvector of $\beta I_n - Q(y)$ with positive entries whose corresponding eigenvalue is simple. It is easy to see that $Q(y)y = 0$, and $y \in \mathbb{R}^n_{>0}$, thus y is the Perron eigenvector of $\beta I_n - Q(y)$ and β the corresponding simple eigenvalue. Therefore, 0 is a simple eigenvalue of $Q(y)$ and any critical point of V is of the form ty, with $t \in \mathbb{R}$. The solutions to (6.9) must additionally satisfy the constraint $(ty)^T\mathbf{1}_n = t(y^T\mathbf{1}_n) = 1$. Equivalently, $t = (y^T\mathbf{1}_n)^{-1}$, and, thus, $\mu^\star = ty = (y^T\mathbf{1}_n)^{-1}y$ is the unique solution to (6.9). □

It is easy to see that $Q(y)\mu^\star = 0$. Lemma 6.3 encodes key properties of $Q(y)$ and leads us to design the following distributed algorithm:

$$\dot{\mu} = -LQ(y)\mu, \qquad \mu(0) = \mu_0, \tag{6.10}$$

where $\mu_0 \in \mathbb{R}^n_{>0}$ satisfies $\mu_0^T\mathbf{1}_n = 1$. In coordinates, this can be rewritten as

$$\dot{\mu}_i = -\sum_{j\in\mathcal{N}_G(i)} a_{ij}(\nabla_j V(\mu) - \nabla_i V(\mu)),$$
$$\mu_i(0) = \mu_{0,i},$$

where $\nabla_i V(\mu) = 2\sum_{k\in\mathcal{N}_G(i)}(y_k^2\mu_i - y_iy_k\mu_k)$, leading to a distributed algorithm over G.

Note that the dynamical system (6.10) leaves $\mu(t)^T\mathbf{1}_n = 1$ invariant for all $t \in \mathbb{R}_{\geq 0}$. This can be verified by noting that $\dot{\mu}(t)^T\mathbf{1}_n = (\mu(t))^T QL\mathbf{1}_n = 0$. The following result holds.

Lemma 6.4 *For any $\mu_0 \in \mathbb{R}^n_{>0}$ such that $\mu_0^T\mathbf{1}_n = 1$, the solution of* (6.10) *converges asymptotically to $\mu^\star$, the solution to the quadratic program* (6.9).

Proof The main part of the proof will follow from the application of the LaSalle Invariance Principle with the Lyapunov function V. First, since V is a sum of squares, it is positive semidefinite. Second, $\dot{V} = -(\nabla V)^T LQ\mu = -\mu^T QLQ\mu \leq 0$. Third, we see next that $\lim_{\|\mu\|\to+\infty} V(\mu) = +\infty$ over the line $\mu^T\mathbf{1}_n = 1$. When $\|\mu\| \to +\infty$, note that either $V(\mu) \to 0$ or to $V(\mu) \to +\infty$. But if $V(\mu) \to 0$, then it must be that $y_i\mu_j - y_j\mu_i \to 0$ for all $i \in \{1,\dots,n\}$ and $j \in \mathcal{N}_G(i)$. That is, we converge to solution of (6.9) with arbitrarily large norm. However, this is a contradiction since the solution of (6.9) is unique, and there are no solutions with arbitrarily large norm.

We can now apply the LaSalle Invariance Principle over the space $\mu^T\mathbf{1}_n = 1$. It follows that the trajectories of (6.10) converge to the largest invariant set of $\mu^T\mathbf{1}_n = 1$

contained in $\dot{V}(\mu) = -\mu^T Q(y) L Q(y)\mu = 0$. Since $L = L^T$ and is positive semidefinite, there exists a unique square root $L^{1/2}$ of L such that $L = L^{1/2}L^{1/2}$ and which is positive semidefinite and symmetric [19, Theorem 3.5]. Therefore, we have $\dot{V}(\mu) = -(L^{1/2}Q(y)\mu)^T(L^{1/2}Q(y)\mu) = 0$, which implies $L^{1/2}Q(y)\mu = 0$ and $L^{1/2}L^{1/2}Q(y)\mu = LQ(y)\mu = 0$. From the fact that the graph is connected and undirected, L has a simple eigenvalue at 0 with eigenvector $\mathbf{1}_n$, thus $Q(y)\mu = \alpha\mathbf{1}_n$, for some $\alpha \in \mathbb{R}$.

On the other hand, since y is the Perron eigenvector of $Q(y)$, and $Q(y) = Q(y)^T$, it holds that $0 = y^T Q(y)\mu = \alpha y^T \mathbf{1}_n$. From the fact that $y^T \mathbf{1}_n > 0$, we have $\alpha = 0$. Since 0 is a simple eigenvalue of $Q(y)$, and $Q(y)\mu = 0$, it follows that $\mu = ty$. Finally, the property $\mu^T \mathbf{1}_n = 1$ implies $\mu = \mu^\star$. □

6.5.2 Discrete-Time Implementation and Rate of Convergence

This section focuses on the discrete-time implementation of the dynamics (6.10) and, particularly, on the study of its rate of convergence. This is motivated by two considerations. First, as designed, the algorithm is in continuous time, which requires a continuous flow of information among the agents. Second, the algorithm does not leave $\mathbb{R}^n_{\geq 0}$ invariant because the matrix $-LQ$ is not positive. This means that, even though $\mu^T \mathbf{1}_n$ is conserved, the algorithm cannot be stopped anytime and guarantee that the output is an appropriate convex combination of stepsizes.

Our approach proceeds by using a first-order Euler discretization of (6.10),

$$\mu^{k+1} = (I_n - hLQ(y))\mu^k, \tag{6.11}$$

where $\mu^0 \in \mathbb{R}^n_{>0}$ satisfies $(\mu^0)^T \mathbf{1}_n = 1$. It can be seen that $(\mu^{k+1})^T \mathbf{1}_n = (\mu^k)^T (I_n - hQ(y)L)\mathbf{1}_n = (\mu^k)^T \mathbf{1}_n = 1$. The next result provides a sufficient condition on the stepsize h that guarantees convergence.

Lemma 6.5 *For any $\mu^0 \in \mathbb{R}^n_{>0}$ such that $(\mu^0)^T \mathbf{1}_n = 1$, the solution of* (6.11) *converges asymptotically to $\mu^\star = (y^T \mathbf{1}_n)^{-1} y$ under the assumption that*

$$h < \frac{2}{\lambda_n(L)\lambda_n(Q(y))}.$$

Moreover, the essential spectral radius of $I_n - hLQ(y)$ is upper bounded by $1 - h\lambda_2(L)\lambda_2(Q(y))$.

Proof Recall that 0 is a simple eigenvalue of $Q(y)$ and y is the corresponding eigenvector. Thus, 1 is an eigenvalue of $I_n - hLQ(y)$ with eigenvector y. In order for the discrete-time system to be convergent to $\operatorname{span}\{y\}$, we need to guarantee that $\|\lambda_i(I_n - hLQ(y))\| < 1$, for all $i \geq 2$. Observe that $\lambda_{n-i}(I_n - hLQ(y)) = 1 - h\lambda_i(LQ(y))$, $i \in \{1, \ldots, n\}$.

By [19, Theorem 2.8], $\lambda_i(AB) = \lambda_i(BA)$ for any two pairs of square matrices A, B. Since L is symmetric and positive semidefinite, there exists a unique $L^{1/2}$ such that $L = L^{1/2}L^{1/2}$, where $L^{1/2}$ positive semidefinite and symmetric; see [19, Theorem 3.5]. Thus, $\lambda_i(LQ(y)) = \lambda_i(L^{1/2}L^{1/2}Q(y)) = \lambda_i(L^{1/2}Q(y)L^{1/2})$, $i \in \{1, \ldots, n\}$. Since $L^{1/2}Q(y)L^{1/2}$ is symmetric and positive semidefinite, the eigenvalues of $LQ(y)$ are real and positive.

Let $\lambda_2(Q(y)) > 0$ be the second smallest eigenvalue of $Q(y)$. It is easy to see that $Q(y) - \lambda_2(Q(y))I_n$ is positive semidefinite. Thus,

$$\begin{aligned} L^{1/2}Q(y)L^{1/2} &\geq L^{1/2}Q(y)L^{1/2} \\ &- L^{1/2}(Q(y) - \lambda_2(Q(y))I_n)L^{1/2} = \lambda_2(Q(y))L. \end{aligned}$$

From here we obtain

$$\lambda_i(LQ(y)) = \lambda_i(L^{1/2}Q(y)L^{1/2}) \geq \lambda_2(Q(y))\lambda_i(L),$$

for $i \in \{1, \ldots, n\}$. A similar reasoning leads to the upper bound $\lambda_i(LQ(y)) \leq \lambda_i(L)\lambda_n(Q(y))$, $i \in \{1, \ldots, n\}$.

Thus, for convergence, it is sufficient to show that $2 > h\lambda_i(LQ(y)) > 0$, for $i \geq 2$. From the upper inequality above, the sufficient condition $h\lambda_n(L)\lambda_n(Q(y)) < 2$ follows, leading to the equation stated in the lemma. The above inequalities also guarantees that 0 is a simple eigenvalue of $LQ(y)$, since $\lambda_2(LQ(y)) \geq \lambda_2(L)\lambda_2(Q(y)) > 0$. The dynamic system (6.11) will converge to $\lambda = ty$ for some $t \in \mathbb{R}$ such that $\lambda^T\mathbf{1}_n = ty^T\mathbf{1}_n = 1$. Thus, it must be that $t = (y^T\mathbf{1}_n)^{-1}$, and that $\mu = \mu^\star$. Finally, the essential spectral radius of the matrix $I_n - hLQ(y)$, is its second largest eigenvalue, that is, $\rho_{\text{ess}}(I_n - hLQ(y)) \leq 1 - \lambda_2(L)\lambda_2(Q(y))$. □

Using the bound on the essential spectral radius of $I_n - hLQ(y)$, we next determine a bound on the rate of convergence of the algorithm as follows:

Lemma 6.6 *Let $r > 0$, and let $T_r > 0$ be the time it takes* (6.11) *to reach and remain in the ball of center $\mu^\star$ with radius r. Then*

$$T_r \in O\left(\frac{1}{h\lambda_2(L)\lambda_2(Q(y))}\log\left(\frac{\|\mu^0 - F^\star\mu^0\|_2}{r}\right)\right),$$

where $F^\star = \frac{yz^T}{z^Ty}$ and z is the right eigenvector of $I_n - hLQ(y)$ with eigenvalue 1.

Proof Bounding the eigenvalues of the matrix $I_n - hQ(y)L$ by those of $Q(y)$ and L as in the proof of Lemma 6.5 one can prove that there exists an eigenvector z such that $z^T(I_n - hQ(y)L) = z^T$. Thus, the following holds

$$\lim_{\ell \to +\infty}(I_n - hLQ(y))^\ell = F^\star = \frac{yz^T}{z^Ty}.$$

The rate of convergence of the discrete-time system will be determined by the exponential convergence factor of $(I_n - hLQ(y)) - F^\star$. This factor is equal to the essential spectral radius of $I_n - hLQ(y)$, see [2, Lemma 1.75].

Now, given $\mu^\star$, consider the ball centered at $\mu^\star$ and radius r. Then, the time T_r it takes the discrete-time system to reach and remain this ball from the initial condition μ^0 satisfies

$$T_r \in O\left(\frac{1}{h\lambda_2(L)\lambda_2(Q(y))}\log\left(\frac{\|\mu^0 - F^\star\mu^0\|_2}{r}\right)\right),$$

see [2, Lemma 1.74]. □

Remark 6.7 (Extension to $y \in \mathbb{R}^n_{\geq 0}$) In the previous two subsections we have assumed that $y_i > 0$ for all $i \in \{1, \dots, n\}$. The results can be extended for the case when $y_i = 0$, for some $i \in \{1, \dots, n\}$, by assuming that these nodes act as a relay between any of their neighbors in $G = (V, E)$. To see this, without loss of generality, suppose $y_1 = 0$ only for $i = 1$. In this case, the matrix $Q(y)$ will have an additional eigenvector, $e_1 = (1, 0, \dots, 0)^T \in \mathbb{R}^n$, with zero eigenvalue. Consider the graph $\overline{G} = (\overline{V}, \overline{E})$ where $\overline{V} = \{2, \dots, n\}$, and $\{i, j\} \in \overline{E}$ if and only if $\{i, j\} \in E$ or $\{1, i\}$, $\{1, j\} \in E$. Let $\overline{L}$ be the associated graph Laplacian. Let $\overline{Q}(y)$ be the restriction of $Q(y)$ over $\mathbb{R}^n \setminus \text{span}\{e_1\}$. System (6.11) can be replaced by

$$\begin{aligned}
\mu_1^{k+1} &= \mu_1^k, \\
\overline{\mu}^{k+1} &= (I_n - h\overline{L}\overline{Q}(y))\overline{\mu}^k,
\end{aligned}$$

where a similar bound for h as in Lemma 6.5 can be taken, and $(\mu^0)^T\mathbf{1}_n = 1$. The analysis of the subsystem in $\overline{\mu}$ is similar to the one in μ in (6.11). First, it can be seen that $(\mu^k)^T\mathbf{1}_n = 1$ for all $k \geq 1$. More precisely, $(\mu^{k+1})^T\mathbf{1}_n = \mu_1^{k+1} + (\overline{\mu}^{k+1})^T\mathbf{1}_{n-1} = \mu_1^k + (\overline{\mu}^k)^T(I_n - h\overline{L}\overline{Q}(y))^T\mathbf{1}_{n-1} = \mu_1^k + (\overline{\mu}^k)^T\mathbf{1}_{n-1} = 1$. In particular, we have that $(\overline{\mu}^0)^T\mathbf{1}_{n-1}$ is conserved. The analysis of the system, is similar to the previous discrete-time implementation, and it can be seen that it converges to the convex combination $\mu^\star = (\mu_1^0, \overline{\mu}^\star)$, where

$$\overline{\mu}_i^\star = \frac{(\overline{\mu}^0)^T\mathbf{1}_{n-1}}{y^T\mathbf{1}_n} y_i, \quad i \in \{2, \dots, n\}.$$

•

6.5.3 Distributed Line Search Computation Algorithm

Building on the results from Sects. 6.5.1 and 6.5.2, we describe here a distributed algorithm that allows agents to adapt their stepsizes and solve approximately the distributed (steepest descent) line search computation problem. Agents start from an initial condition μ_0 such that $\mu_0^T\mathbf{1}_n = 1$ (e.g., $\mu_0 = \frac{1}{n}\mathbf{1}_n$). Note that the assumption that agents know n is necessary since it is equal to the dimension of $x \in \mathbb{R}^n$. Then, agents implement (6.11) for an agreed number of rounds N that guarantees $\mu_i^k \geq 0$, for all $i \in \{1, \ldots, n\}$. The algorithm is formally described in Algorithm 1.

The different aggregated stepsize models (6.6)–(6.8) are captured in the algorithm via $\mathcal{R}$. In this way, the choice $y_i = \frac{1}{\delta_i}$ leads to DISTRIBUTED WEIGHTED STEPSIZE (and results in the steepest descent direction). As N grows, this leads to the common network-aggregated stepsize vector (6.6). The choice $y_i = \Delta_i$, $i \in \{1, \ldots, n\}$ as in (6.7) leads to the DISTRIBUTED WEIGHTED STEPSIZE for the proportional-to-cost descent direction. As N grows, this leads to the proportional-to-cost network-aggregated stepsize vector (6.7). Finally, the choice $y_i(f, x) = d_i$, $i \in \{1, \ldots, n\}$ as in (6.8) leads to the DISTRIBUTED WEIGHTED STEPSIZE for the proportional-to-state descent direction. As N grows, this leads to the proportional-to-state network-aggregated stepsize vector (6.7).

Algorithm 1: DISTRIBUTED WEIGHTED STEPSIZE

Executed by: Each agent $i \in \{1, \ldots, n\}$
Data: the function f, the state x, the number of rounds $N \in \mathbb{N} \cup \{0\}$, and aggregated stepsize model $\mathcal{R}$

set $v_i(x) = -(0, \ldots, 0, \nabla_i f(x), 0, \ldots, 0)$
compute stepsize $\delta_i = \varepsilon_i > 0$ satisfying $\nabla_i f(x + \varepsilon_i v_i(x))^T \nabla_i f(x) = 0$
set y_i corresponding to aggregated stepsize model $\mathcal{R}$, send y_i to neighbors, receive $\{y_j \mid j \in \mathcal{N}_G(i)\}$, and compute $Q_{ij}(y)$ for $j \in \mathcal{N}_G(i)$
set $\mu_i^0 = \frac{1}{n}$
for $l \in \{1, \ldots, N\}$ **do**
  $\mu_i^l = ((I_n - hLQ(y))\mu^{l-1})_i$
  send μ_i^l to neighbors, receive $\{\mu_j^l \mid j \in \mathcal{N}_G(i)\}$
end
set $m_i^0 = \mu_i^N$
send m_i^0 to neighbors, receive $\{m_j^0 \mid j \in \mathcal{N}_G(i)\}$
for $l \in \{1, \ldots, N\}$ **do**
  $m_i^l = \min\{m_i^{l-1}, m_j^{l-1} \mid j \in \mathcal{N}_G(i)\}$
  send m_i^l to neighbors, receive $\{m_j^l \mid j \in \mathcal{N}_G(i)\}$
end
while $m_i^n < 0$ **do**
  reassign $\mu_i^N = ((I_n - hLQ(y))\mu^N)_i$
  reset $m_i^0 = \mu_i^N$
  send m_i^0 to neighbors, receive $\{m_j^0 \mid j \in \mathcal{N}_G(i)\}$
  for $l \in \{1, \ldots, N\}$ **do**
    $m_i^l = \min\{m_i^{l-1}, m_j^{l-1} \mid j \in \mathcal{N}_G(i)\}$
    send m_i^l to neighbors, receive $\{m_j^l \mid j \in \mathcal{N}_G(i)\}$
  end
end
change state from x_i to $x_i + \mu_i^N \delta_i \nabla_i f(x)$

The DISTRIBUTED WEIGHTED STEPSIZE algorithm can be informally described as follows. In order to implement a step of the gradient-descent algorithm, each agent outputs first a set of stepsizes μ_i^N, $i \in \{1, \ldots, n\}$. These stepsizes are obtained after applying (6.11) during N iterations. After this, if all of the μ_i^N are positive or zero, which happens when $m_i^n = \min_{j\{1,\ldots,n\}} \mu_j^N \geq 0$, for all $i \in \{1, \ldots, n\}$, then the gradient-descent procedure can be safely implemented. Otherwise, agents iterate (6.11) additional times until the property $\mu_i^N \geq 0, i \in \{1, \ldots, n\}$ holds. The algorithm assumes that $y_i > 0$, for all $i \in \{1, \ldots, n\}$. When $y_i = 0$, agent i should relay information from neighbors to other neighbors at any communication round.

6.6 Simulations

In this section, we include some numerical experiments on a simple mathematical example to illustrate the results. We consider a network of 8 agents subject to a fixed topology corresponding to the graph G depicted in Fig. 6.1a. The function to be optimized $f : \mathbb{R}^8 \to \mathbb{R}^8$ is defined as $f(x) = x^T(I_n + L)x + q^T x$, where $q = (1, -1, 2, 1, 0, -1, 1, 0) \in \mathbb{R}^8$ and L is the graph Laplacian associated with G. It is straightforward to verify that f is convex and distributed over G.

We implement the algorithm in two scenarios. First, we consider DISTRIBUTED WEIGHTED STEPSIZE for the steepest descent direction. Figure 6.1b compares how the function decreases under the centralized steepest descent method (blue), the conservative steepest descent method if all agents had the information to compute the stepsize in (6.6) (red), and the algorithm DISTRIBUTED WEIGHTED STEPSIZE for the steepest descent direction with $N = 4$ (green) and an appropriate h. After $N = 4$, all weights μ_i^N have become positive. As the plot shows, both the conservative steepest descent method and its decentralized version are very close even if the number of rounds ($N = 4$) used to compute the stepsizes in a distributed way is small. The

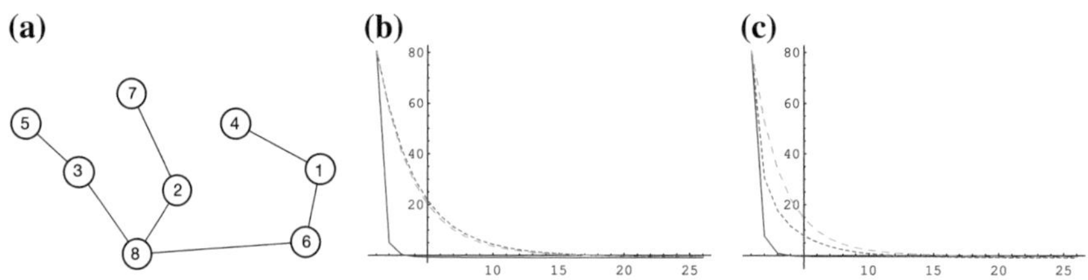

Fig. 6.1 **a** Undirected graph describing the interaction topology of network of 8 agents and evolution of various algorithms along **b** the steepest descent direction and **c** the proportional-to-cost descent direction. Centralized steepest descent is shown in *blue*. In **b**, the conservative steepest descent with stepsize (6.6) is shown in *red*, and the algorithm DISTRIBUTED WEIGHTED STEPSIZE for the steepest descent direction with $N = 4$ is shown in *green*. In **c**, the proportional-to-cost descent method with stepsize (6.7) is shown in *red*, and the algorithm DISTRIBUTED WEIGHTED STEPSIZE for proportional-to-cost descent with $N = 4$ is shown in *green*

differences between the centralized steepest descent method and the other two are to be expected, as the common network-aggregated stepsize vector (6.6) is more conservative in order to guarantee that the function is still decreased.

Second, we consider DISTRIBUTED WEIGHTED STEPSIZE for proportional-to-cost descent. Figure 6.1c compares the evolution of the gradient algorithms following the steepest descent (blue), the decentralized proportional-to-cost descent method if all agents had the information to compute the stepsize as in (6.7) (red), and DISTRIBUTED WEIGHTED STEPSIZE for proportional-to-cost descent with $N = 4$ (green) and an appropriate h. After $N = 4$, all weights μ_i^N are already positive. Similarly as before, and as expected, the function is decreased less rapidly by means of the conservative proportional-to-cost descent direction method and its decentralized version via DISTRIBUTED WEIGHTED STEPSIZE when compared to the centralized steepest descent method. However, the previous two are relatively close, even though the number of rounds $N = 4$ used in DISTRIBUTED WEIGHTED STEPSIZE is low.

6.7 Conclusions

We have considered networked scenarios where a group of agents seeks to optimize a convex aggregate function using gradient information. We have presented a novel distributed algorithm for the computation of aggregated stepsizes that guarantee the decrease of the objective function. We have analyzed the properties of this strategy when implemented both in continuous and discrete time, and characterized its rate of convergence. With a proper initialization, the algorithm gives rise to a convex combination after a finite number of rounds, and can therefore be implemented to fuse the stepsizes of individual agents. Simulations illustrate the results. Future work will be devoted to the analytical characterization of the performance of the proposed strategies, the consideration of scenarios with switching and state-dependent interaction graphs, and the design of distributed line search strategies for higher order (e.g., Newton) schemes.

Acknowledgments Both authors wish to thank Jon Nicolás and Alexandra Cortés-Martínez for constant inspiration and joy. This work was partially supported by grants NSF CMMI-1300272 (JC) and AFOSR-11RSL548 (SM).

References

1. D.P. Bertsekas, J.N. Tsitsiklis, *Parallel and Distributed Computation: Numerical Methods* (Athena Scientific, Belmont, 1997)
2. F. Bullo, J. Cortés, S. Martínez, *Distributed Control of Robotic Networks*, Applied Mathematics Series (Princeton University Press, Princeton, 2009). http://coordinationbook.info
3. A. Cauchy, Méthode générale pour la résolution des systems d'equations simultanées. Comptes rendus de l'Académie des Sciences **25**, 46–89 (1847)

4. J. Cortés, S. Martínez, Distributed line search via dynamic convex combinations, in *IEEE Conference on Decision and Control* (Florence, Italy, 2013), pp. 2346–2351
5. B. Gharesifard, J. Cortés, Distributed continuous-time convex optimization on weight-balanced digraphs. IEEE Trans. Autom. Control **59**(3), 781–786 (2014)
6. B. Johansson, M. Rabi, M. Johansson, A randomized incremental subgradient method for distributed optimization in networked systems. SIAM J. Control Optim. **20**(3), 1157–1170 (2009)
7. S.S. Kia, J. Cortés, S. Martínez, Distributed convex optimization via continuous-time coordination algorithms with discrete-time communication. Automatica **55**, 254–264 (2015)
8. D.G. Luenberger, *Linear and Nonlinear Programming*, 2nd edn. (Addison-Wesley, Reading, 1984)
9. D. Mateos-Núñez, J. Cortés, Noise-to-state exponentially stable distributed convex optimization on weight-balanced digraphs. SIAM J. Control Optim. Submitted (2014)
10. M. Mesbahi, M. Egerstedt, *Graph Theoretic Methods in Multiagent Networks*, Applied Mathematics Series (Princeton University Press, Princeton, 2010)
11. A. Nedic, A. Ozdaglar, Distributed subgradient methods for multi-agent optimization. IEEE Trans. Autom. Control **54**(1), 48–61 (2009)
12. A. Nedic, A. Ozdaglar, P.A. Parrilo, Constrained consensus and optimization in multi-agent networks. IEEE Trans. Autom. Control **55**(4), 922–938 (2010)
13. R. Olfati-Saber, J.A. Fax, R.M. Murray, Consensus and cooperation in networked multi-agent systems. Proc. IEEE **95**(1), 215–233 (2007)
14. W. Ren, R.W. Beard, *Distributed Consensus in Multi-Vehicle Cooperative Control*, Communications and Control Engineering (Springer, London, 2008)
15. P. Wan, M.D. Lemmon, Event-triggered distributed optimization in sensor networks, in *Symposium on Information Processing of Sensor Networks* (San Francisco, CA, 2009), pp. 49–60
16. J. Wang, N. Elia, A control perspective for centralized and distributed convex optimization, in *IEEE Conference on Decision and Control* (Orlando, Florida, 2011), pp. 3800–3805
17. F. Zanella, D. Varagnolo, A. Cenedese, G. Pillonetto, L. Schenato, Newton-Raphson consensus for distributed convex optimization, in *IEEE Conference on Decision and Control* (Orlando, Florida, 2011), pp. 5917–5922
18. M. Zargham, A. Ribeiro, A. Jadbabaie, A distributed line search for network optimization, in *American Control Conference* (Montreal, Canada, 2012), pp. 472–477
19. F. Zhang, *Matrix Theory. Basic Results and Techniques*, Universitext, 2nd edn. (Springer, New York, 2011)
20. M. Zhu, S. Martínez, On distributed convex optimization under inequality and equality constraints. IEEE Trans. Autom. Control **57**(1), 151–164 (2012)

Chapter 7
Optimal Management with Hybrid Dynamics—The Shallow Lake Problem

P.V. Reddy, J.M. Schumacher and J.C. Engwerda

Abstract In this article we analyze an optimal management problem that arises in ecological economics using hybrid systems modeling. First, we introduce a discounted autonomous infinite horizon hybrid optimal control problem and develop few tools to analyze the necessary conditions for optimality. Next, using these tools we study the classical shallow lake problem where the nonlinear lake dynamics is described by hybrid dynamics. We show that our results agree with earlier studies on the problem, that is, variation of system parameters induce bifurcations in the optimal solution.

7.1 Introduction

Most of the optimal decision-making problems studied in economics and ecology are complex in nature. These complexities generally arise while modeling the inherent behavior of the dynamic environment, which includes agents interacting with the system. Modeling with hybrid systems [5, 22] capture some of these complex situations. The behavior of hybrid systems is described by the integration of continuous and discrete dynamics. An abrupt change in the discrete state of the system is called a switch. If a decision-maker influences a switch then it is said to be controlled/external, whereas an internal switch generally results when the continuous state variable satisfies some equality constraints. Some examples in this direction are,

P.V. Reddy
GERAD, HEC Montréal, Montréal, Canada
e-mail: Puduru.Reddy@gerad.ca

J.M. Schumacher (✉) · J.C. Engwerda
Department of Econometrics and Operations Research, Tilburg University, Tilburg, The Netherlands
e-mail: J.M.Schumacher@uvt.nl

J.C. Engwerda
e-mail: J.C.Engwerda@uvt.nl

M.K. Camlibel et al. (eds.), *Mathematical Control Theory I*,
Lecture Notes in Control and Information Sciences 461,
DOI 10.1007/978-3-319-20988-3_7

a firm going bankrupt when its equity is negative and regime shifts in ecology [4, 17] etc. Optimal control of hybrid systems has received considerable interest in control engineering, see for instance [16, 18, 19, 21]. These works include formulation of different versions of the necessary conditions.

In this article, we study an optimal management problem that arises in ecological economics, the so-called shallow lake problem, see [1, 8, 9, 23]. It has been observed that shallow lakes display nonlinearities and hysteresis in their behavior. Economically speaking, these systems offer conflicting services as a resource and a waste sink. As a result, the economic analysis of these systems involves solving a nonstandard optimal control problem, or a nonstandard differential game when seen as a common property resource. The source of nonlinearity in the shallow lake problem is due to convex–concave production function. This implies that the optimal solution associated with the problem displays several interesting qualitative behaviors such as existence of multiple steady states, Skiba points[1] and bifurcations due to parameter variations; see [7] for a detailed analysis. The inflection point of the convex–concave production function acts as a 'threshold,' meaning that the dynamics of the lake differs significantly when the state variable takes values below the inflection point and after crossing it.

In this article, we represent the nonlinear lake dynamics using a simple hybrid system and study the associated optimal management problem. Some literature incorporating threshold effects include [13, 15] and references cited in those papers. In [13], the author uses necessary conditions, in the line of [18], the objective function is quadratic and the state variable admits jumps. In [15], the authors consider an optimal management problem with probabilistic thresholds and use dynamic programming to derive optimal policies.

In this article we do not attempt to solve the optimal controls or equilibrium strategies for a generic class of (hybrid) differential games. Instead, we study the shallow lake problem with hybrid representation and highlight the key differences with the smooth case, i.e., the classical shallow lake problem. This article is organized as follows. In Sect. 7.2, we introduce a class of discounted autonomous infinite horizon optimal control problems with endogenous switching. We review the necessary conditions for this class of problems. Further, for one-dimensional problems we develop some methods to analyze the optimality equations. In Sect. 7.3, we introduce the classical shallow lake problem. Then we represent the nonlinear lake dynamics using simple hybrid dynamics. Next, we study optimal management and open-loop Nash equilibrium policies related to the hybrid version of the shallow lake problem. Finally, Sect. 7.4 concludes.

[1]Starting from such a point the optimal control problem has more than one optimal solution, and as a result the decision-maker is indifferent to a particular solution, see [20].

7.2 Optimal Control of Switched Systems

In this section we review necessary conditions associated with a hybrid optimal control problem. The hybrid system, which is referred to as a switched system, has the following description:

Definition 7.1 (*Switched System*) A switched system is a triple $\mathcal{S} = (\mathcal{I}, \mathcal{F}, \Phi)$ where

- $\mathcal{I}$ is a finite set, called the set of discrete states or modes.
- $\mathcal{F} = \{f_i : \mathbb{R}^n \times \mathbb{R}^m \to \mathbb{R}^n,\ i \in \mathcal{I}\}$ is a collection of vector fields. We denote $\dot{x}(t) = f_i(x(t), u(t))$ to be the vector field associated with the discrete state $i \in \mathcal{I}$.
- A transition from mode i to mode j is triggered by events (internal or external) resulting in an abrupt change in dynamics from f_i to f_j. In this article we consider internal switchings, i.e., transitions from discrete state i to j happen when certain state constraints, say $\phi_{ij}(x(t)) = 0$, are satisfied. Let $\Phi_{ij} := \{x \in \mathbb{R}^n : \phi_{ij}(x) = 0\}$ be the switching surface associated with transition i to j. We denote the joint switching surface by $\Phi := \cup \Phi_{ij}$.

Now, we introduce a class of discounted autonomous infinite horizon optimal control problems with internal switching dynamics, described by $\mathcal{S} = (\mathcal{I}, \mathcal{F}, \Phi)$, as follows.

$$\max J, \quad J = \int_0^{\infty} e^{-rt} g(x(t), u(t))\, \mathrm{d}t \tag{7.1}$$

$$\dot{x}(t) = f_i(x(t), u(t)),\ i \in \mathcal{I},\ f_i \in \mathcal{F} \tag{7.2}$$

$$x(0) = x_0 \in \mathbb{R}^n, u(.) \in \mathcal{U}. \tag{7.3}$$

Assumption 7.2 The real-valued functions $f_i(.), i \in \mathcal{I}$ and $g(.)$ are continuous, $\frac{\partial f_i(.)}{\partial x}$ and $\frac{\partial g(.)}{\partial x}$ exist and are continuous. The control space $\mathcal{U}$ consists of piecewise continuous functions with $u(t) \in U$, where U is a bounded set included in $\mathbb{R}^m$. We assume the left- and right-hand limits for $u(.)$ exist and $x(.)$ is continuous and piecewise continuously differentiable, which satisfies (7.2) for all points t where $u(.)$ is continuous. The initial state satisfies $x_0 \notin \Phi$. Further, we assume that sliding modes do not occur.

We call a pair $(x(.), u(.))$ *admissible* for the problem (7.1)–(7.3) if Assumption 7.2 is satisfied. Let $k(.)$ represent the switching sequence associated with $\mathcal{S}$, i.e., when the system is in mode i at time t we have $k(t) = i$. Since the switchings happen internally we see that $u(.)$ induces a switching sequence $k(.)$. The necessary condition for a pair $(x^*(.), u^*(.))$ to be optimal for the problem (7.1)–(7.3) is given by the following theorem,[2] see Theorem 2.3 of [16] or Theorem 3 of [18] or Theorems 2.2 and 2.3 of [19] for more details.

[2]There exist several variations of the theorem in a more general setting, for instance refer [16, 18, 19, 21]. Here, we consider a specific system $\mathcal{S}$ where switchings happen internally.

Theorem 7.3 (Necessary conditions) *If $(x^*(.),\ u^*(.))$ represent an optimal admissible pair for the problem (7.1)–(7.3), then there exists a piecewise absolutely continuous function $\lambda(.)$ and a constant $\lambda^0 \geq 0$, $\left(\lambda^0, \lambda(t)\right) \neq 0$ on $[0, \infty)$ such that:*

(a) let Hamiltonian be defined as

$$H_k(t, x, u, \lambda, \lambda^0) := \lambda^0 e^{-rt} g(x, u) + \lambda^T f_k(x, u) \tag{7.4}$$

then for a given $(\lambda(t), \lambda^0, x^*(t))$ at a given time t, except at the switching instants, the following maximum condition holds

$$H_{k^*(t)}(x^*(t), u^*(t), \lambda(t), \lambda^0) \geq H_{k^*(t)}(x^*(t), v, \lambda(t), \lambda^0),\ \forall\, v \in U. \tag{7.5}$$

(b) for all $t \geq 0$, except at the switching instants, the costate process $\lambda(t)$ satisfies the following relation:

$$\dot{\lambda}(t) = -\frac{\partial H_{k(t)}}{\partial x}(t, x^*(t), u^*(t), \lambda(t), \lambda^0) \tag{7.6}$$

(c) if τ is a switching instant then the following conditions hold true:

1. $x^*(\tau) \in \Phi$, $\tau \in [0, \infty)$
2. *(costate jump condition)* *there exists a $\beta \in \mathbb{R}$ such that*

$$\lambda(\tau^-) = \lambda(\tau^+) + \beta\left(\phi_{k^*(\tau^-)k^*(\tau^+)}\right)_x (x^*(\tau)) \tag{7.7}$$

3. *(Hamiltonian continuity)*

$$H_{k^*(\tau^-)}(x^*(\tau), u^*(\tau^-), \lambda(\tau^-), \lambda^0) = H_{k^*(\tau^+)}(x^*(\tau), u^*(\tau^+), \lambda(\tau^+), \lambda^0). \tag{7.8}$$

The above necessary conditions, when solved, usually result in multiple candidates as the terminal condition in (7.6) is not specified. The optimal solution is then obtained by comparing the objective evaluated along the candidate trajectories. The non-switched analog of the problem (7.1)–(7.3) is the classical discounted autonomous infinite horizon optimal control problem. For this class of problems an additional necessary condition called as *asymptotic Hamiltonian property* is satisfied. More precisely, if the maximized Hamiltonian associated with the problem is given by $H(t, x^*, u^*, \lambda, \lambda^0)$, then $\lim_{t\to\infty} H(t, x^*(t), u^*(t), \lambda(t), \lambda^0) = 0$, see [11]. Furthermore, when the associated necessary conditions for optimality hold in normal form, i.e., $\lambda^0 = 1$, then the objective value along a candidate[3] trajectory starting at $(x^*(0), u^*(0))$ is given by $\frac{1}{r}H\big(0, x^*(0), u^*(0), \lambda(0), 1\big)$, see [6, Proposition 3.75]. In the next lemma we obtain a hybrid counterpart of this result.

Lemma 7.4 (Objective value given by the Hamiltonian) *Let $(\tilde{x}(.), \tilde{u}(.))$ be an admissible pair that satisfies the necessary conditions (7.4)–(7.8) of Theorem 7.3 with a finite number of switchings. Further, assume $\lambda^0 = 1$. Then the objective value along the trajectory $(\tilde{x}(.), \tilde{u}(.))$ is given by $\frac{1}{r}H_{\tilde{k}(0)}(0, \tilde{x}(0), \tilde{u}(0), \lambda(0), 1)$.*

Proof As the number of switchings is finite, say M, there exists a sequence of switching instants associated with $\tilde{k}(.)$, which we denote as $\tau_1, \tau_2, \ldots, \tau_j, \cdots \tau_M$. Taking the total derivative of the Hamiltonian $H_{\tilde{k}(.)}(.)$ in the interval $t \in \big(\tau_j^+, \tau_{j+1}^-\big)$ we have:

$$\begin{aligned}\frac{dH_{\tilde{k}(t)}}{dt} &= \frac{\partial H_{\tilde{k}(t)}}{\partial t} + \frac{\partial H_{\tilde{k}(t)}}{\partial \tilde{x}}\dot{\tilde{x}}(t) + \frac{\partial H_{\tilde{k}(t)}}{\partial \lambda}\dot{\lambda}(t) + \frac{\partial H_{\tilde{k}(t)}}{\partial \tilde{u}}\dot{\tilde{u}}(t)\\ &= \frac{\partial H_{\tilde{k}(t)}}{\partial t} \quad \text{(last three terms vanish due to necessary conditions)}\\ &= -re^{-rt}g(\tilde{x}(t), \tilde{u}(t))\end{aligned}$$

From the last equality we have $H_{\tilde{k}(\tau_{j+1}^-)} - H_{\tilde{k}(\tau_j^+)} = -r\int_{\tau_j^+}^{\tau_{j+1}^-} e^{-rt}g(\tilde{x}(t), \tilde{u}(t))\,dt$. Again from the necessary conditions we notice that in the last interval, i.e., $t \in [\tau_M^+, \infty)$, $(\tilde{x}(.), \tilde{u}(.))$ maximizes the objective $\int_{\tau_M^+}^{\infty} e^{-rt}g(x(t), u(t))\,dt$. The truncated trajectory $(\tilde{x}(t), \tilde{u}(t))$, $t \in [\tau_M^+, \infty)$ is an optimal admissible pair for the classical discounted infinite horizon optimal control problem

$$\begin{aligned}&\max \int_{\tau_M^+}^{\infty} e^{-rt}g(x(t), u(t))\,dt\\ &\dot{x}(t) = f_{\tilde{k}(\tau_M^+)}(x(t), u(t)),\ x(\tau_M^+) = \tilde{x}(\tau_M^+)\\ &u(t) \in U,\ t \in [\tau_M^+, \infty).\end{aligned}$$

The truncated candidate in the last interval satisfies an additional necessary condition, the asymptotic Hamiltonian property, that is, the maximized Hamiltonian tends to zero when t goes to infinity. As a result, the objective along the truncated trajectory is given by $\frac{1}{r}H_{\tilde{k}(\tau_M^+)}\big(\tau_M^+, \tilde{x}(\tau_M^+), \tilde{u}(\tau_M^+), \lambda(\tau_M^+)\big)$. The objective along $(\tilde{x}(.), \tilde{u}(.))$ is then given by:

[3]Here, the candidate trajectory need not be an optimal solution, but only has to satisfy the necessary conditions.

$$\int_0^{\infty} e^{-rt} g(\tilde{x}(t), \tilde{u}(t))\, \mathrm{d}t = \int_0^{\tau_1^-} e^{-rt} g(\tilde{x}(t), \tilde{u}(t))\, \mathrm{d}t$$
$$+ \sum_{j=1}^{M-1} \int_{\tau_j^+}^{\tau_{j+1}^-} e^{-rt} g(\tilde{x}(t), \tilde{u}(t))\, \mathrm{d}t + \int_{\tau_M^+}^{\infty} e^{-rt} g(\tilde{x}(t), \tilde{u}(t))\, \mathrm{d}t$$
$$= \frac{1}{r}\big(H_{\tilde{k}(0)}(.) - H_{\tilde{k}(\tau_1^-)}(.) + H_{\tilde{k}(\tau_1^+)}(.) \cdots$$
$$- H_{\tilde{k}(\tau_M^-)}(.) + H_{\tilde{k}(\tau_M^+)}(.)\big) = \frac{1}{r} H_{\tilde{k}(0)}(0, \tilde{x}(0), \tilde{u}(0), \lambda(0), 1).$$

The necessary conditions stated in Theorem 7.3 are stated in the present value form. Using the transformation $\mu(t) := e^{rt}\lambda(t)$ the necessary conditions can be reformulated in the current value form; as a result the time dependence of the Hamiltonian in (7.4) through the exponential term can be removed.

Next, when the state and control variables are one-dimensional, we show in the following discussion that it is possible to develop few structural results, in the lines of [23, Sect. 3]. Toward this end, we make the following assumption:

Assumption 7.5 The necessary conditions stated in Theorem 7.3 hold in normal form, i.e., $\mu_0 = 1$. The partial derivative of f_i with respect to u is strictly positive for all $i \in \mathcal{I}$, i.e., $f_{i\,u} > 0$. The current value Hamiltonian defined as $H_i^c(.) := e^{rt} H_i(t, .)$ attains its maximum at u^* in the interior of U for all $i \in \mathcal{I}$. Further, at this point the second derivative of $H_i^c(.)$ is strictly negative, i.e., $H_{i\ uu}^c(x(t), u^*(t), \mu(t), 1) < 0$.

Remark 7.6 Consider a candidate trajectory (x, u). There exists a portion of this trajectory in the interior of a mode $i \in \mathcal{I}$ such that $H_{i\,u}^c(x, u, \mu, 1) = \mu f_{i\,u}(x, u) + g_u(x, u) = 0$. Then the current value costate variable μ can be written as a function of u as $\mu(u; x) = -\frac{g_u(x,u)}{f_{i\,u}(x,u)}$ for every x in the interior of mode i. Using the concavity condition of the current value Hamiltonian, i.e., $H_{i\ uu}^c(x, u, \mu, 1) = \mu f_{i\,uu}(x, u) + g_{uu}(x, u) < 0$, the partial derivative of $\mu(u; x)$ with respect to u for a fixed x is given as:

$$\mu_u(u; x) = \frac{g_u(x, u) f_{i\,uu}(x, u) - g_{uu}(x, u) f_{i\,u}(x, u)}{f_{i\,u}^2(x, u)} = -\frac{H_{i\ uu}^c(x, u, \mu(u; x), 1)}{f_{i\,u}(x, u)}.$$

From Assumption 7.5, the above quantity is nonzero (strictly positive) for all the candidate trajectories in the interior of mode i. The above observation enables to transform, locally in the interior of each mode, the optimality equations formulated in state–costate system to state-input system.

Remark 7.7 Consider a candidate trajectory (x, u) starting at (x_0, u_0). Following Remark 7.6 we define $\Psi(x_0, u_0) := \frac{1}{r} H_{k(0)}^c(x_0, u_0, \mu(u_0; x_0), 1)$. So, if the candidate trajectory undergoes a finite number of switchings then $\Psi(x_0, u_0)$ provides the objective value along the candidate solution. We analyze the variation of the function Ψ in u_0 for a fixed x_0. The partial derivative of Ψ with u_0

is given by $\Psi_{u_0} = \frac{1}{r}(H^c_{k(0)})_{u_0}(x_0, u_0, \mu(u_0; x_0), 1) = \frac{1}{r}(\mu(u_0; x_0) f_{i\,u_0}(x_0, u_0) + \mu_{u_0}(u_0; x_0) f_{k(0)}(x_0, u_0) + g_{u_0}(x_0, u_0)) = \frac{1}{r}\mu_{u_0}(u_0; x_0) f_{k(0)}(x_0, u_0)$. So, for $x_0 \notin \Phi$, the signs of $f_{k(0)}$ and Ψ_{u_0} are the same.

7.3 The Shallow Lake Problem

In this section, we first introduce the classical shallow lake problem and later study the problem where nonlinear lake dynamics is represented using hybrid dynamics. Assume a situation where N economic agents, sharing a natural system, take actions $a_n(t),\ n = 1, 2, \ldots, N$ at time t, and as a result affect the state $x(t)$ of a natural system. The economic agents could be societies, dealing with eutrophication of a lake that they manage. The stock of pollutant in the lake admits a dynamics described by:

$$\dot{x}(t) = \sum_{n=1}^{N} a_n(t) - bx(t) + h(x(t)),\ x(0) = x_0 \geq 0,\ h(x) = \frac{x^2}{1 + x^2}. \tag{7.9}$$

The state variable $x(t)$ could be interpreted as the accumulated phosphorus in a lake. Next, $a(t) := \sum_n a_n(t)$ represents the total input of phosphorus washed into the lake due to farming activities at time t. The rate of loss of phosphorus due to sedimentation is denoted by b. The last term captures internal biological processes for the production of phosphorus. Besides the activity of economic agents, the sources that promote $x(t)$ are the nonlinear internal dynamics captured by the term $h(x(t))$. The second part of the model deals with economic analysis of the agents. An agent n, with action a_n, generates benefits according to a strictly increasing and concave utility function $B(a_n) := \ln(a_n)$. The stock of pollutants $x(t)$ causes damage to the natural system according to a strictly increasing and convex damage function $D(x) := cx^2$, sometimes referred as disutility of agents. Here, the parameter $c > 0$ models the relative cost of pollution. The net profit that an agent n receives at a point of time t is then given by $B(a_n(t)) - D(x(t))$. Each agent uses a strategy $a_n(.)$ to maximize the present value of net benefits over an infinite time horizon, i.e.,

$$\max_{a_n(.)} \int_0^{\infty} e^{-rt} \left(B(a_n(t)) - D(x(t))\right) \mathrm{d}t, \quad n = 1, 2, \ldots, N, \tag{7.10}$$

subject to (7.9), where $r > 0$ is a discount rate. Here, $h(x)$ is assumed to be a convex–concave function; that is, for lower stocks of $x(t)$ there is relatively low marginal return to the system, whereas for the higher stocks this marginal return first increases and then decreases again. When the maximal feedback rate is greater than the decay rate, i.e., $\max(h'(x)) > b$, the system (7.9) exhibits three equilibria, for a certain range of values for a. There will be two stable steady states, one corresponding to lower value of x (clear state) which is highly valued by concerned users of the natural system (could be people using a lake for recreation etc.), but also a relatively higher value of x (polluted state) which is valued by the agents due to economic interest. This nonlinear positive feedback effect is a potential source for complex qualitative behaviors in optimal solutions in the model (7.9) and (7.10). The region of stock near the inflection point of $h(x)$ acts as a soft threshold distinguishing the clear and polluted regions.

7.3.1 Hybrid System Representation of the Shallow Lake Dynamics

In this section we replace the nonlinear production function $h(x)$ in the shallow lake dynamics (7.9) using a Heaviside function, see Fig. 7.1. The resulting description of the hybrid system is given by:

$$\mathcal{S} = \{\mathcal{I}, \mathcal{F}, \Phi\} \text{ where}$$
$$\mathcal{I} = \{1, 2\},\ \mathcal{F} = \{f_i(x, a) = a - bx - \varepsilon_i,\ i = 1, 2,\ \text{with } \varepsilon_1 = 0 \text{ and } \varepsilon_2 = 1\}$$
$$\Phi = x - \Delta$$

$$\dot{x}(t) = \begin{cases} a(t) - bx(t) + \varepsilon_1, & \text{for } x(t) < \Delta \\ a(t) - bx(t) + \varepsilon_2, & \text{for } x(t) > \Delta \end{cases},\ x(0) = x_0 \text{ (given)}. \tag{7.11}$$

The region $x(t) < \Delta$ $(>\Delta)$ constitutes mode 1 (mode 2), which qualitatively describes the lake in clean or oligotrophic (polluted or eutrophic) state. We observe that for certain range of values for a the system (7.11) exhibits two steady states,

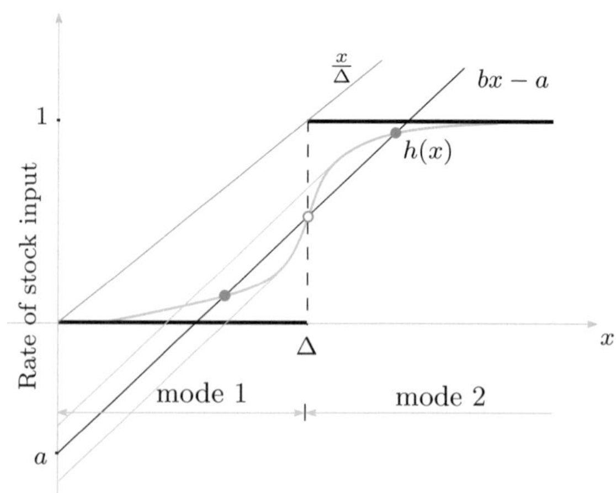

Fig. 7.1 Hybrid representation of $h(x)$ with a Heaviside step function

one each in mode 1 and mode 2. If the decay rate, b, is larger than $\frac{1}{\Delta}$ it is possible to reach the steady state in mode 1 from mode 2 by lowering the external loading a. If $b\Delta < 1$ a steady state in mode 1 cannot be reached from mode 2 even by setting $a = 0$.

7.3.2 Optimal Management

In this section we study the shallow lake problem where the lake dynamics is described by the hybrid system (7.11). First, we study the situation where the economic agents act cooperatively, also called as the optimal management problem. Next, assuming noncooperative behavior, we show that when all the players are identical the resulting noncooperative game can also be solved as an optimization problem. So, we study the optimal management problem in detail. The optimal management problem involves a central authority trying to maximize the joint net profits of economic agents by prescribing a policy for (individual) phosphorus release into the lake. The problem is posed as maximizing the objective

$$\max J, \quad J = \sum_{n=1}^{N} \int_0^{\infty} e^{-rt}(\ln(a_n(t)) - cx^2(t))\,\mathrm{d}t \quad \text{subject to (7.11)}.$$

Following Theorem 7.3, the necessary conditions for $(x^*(t), a_1^*(t), \ldots, a_N^*(t))$ to be optimal for the optimal management problem with simple switching are given as follows:

$$\begin{aligned}
&\text{For each mode } i,\ i = 1, 2\\
&H_i^c(.) = \mu^0\Big(\sum_n \ln a_n(t) - Ncx^2(t)\Big) + \mu(t)\Big(\sum_n a_n(t) - bx(t) + \varepsilon_i\Big)\\
&\frac{\partial H_i^c(.)}{\partial a_n}|_{a_n = a_n^*} = 0 \text{ gives } \frac{\mu^0}{a_n^*(t)} + \mu(t) = 0,\ n = 1, 2, \ldots, N\\
&\dot{x}^*(t) = -\frac{N\mu^0}{\mu(t)} - bx^*(t) + \varepsilon_i,\ x^*(0) = x_0\\
&\dot{\mu}(t) = r\mu(t) - \frac{\partial H_i^c(.)}{\partial x}|_{x=x^*} = (r+b)\mu(t) + 2c\mu^0 Nx^*(t)\\
&(\mu^0, \mu(t)) \neq 0, \forall t \geq 0.
\end{aligned}$$

Here, $x(t) \geq 0$ and $a_n : [0, \infty) \to (0, \infty)$, i.e., interior solutions are considered. Next, we see that if $\mu^0 = 0$ then $\mu(t) = 0$ for all $t \geq 0$. So, necessary conditions hold in normal form, i.e., $\mu^0 = 1$. Notice, in the above state–costate boundary value equations the terminal condition on the costate variable $\mu(t)$ is not specified; as a result we obtain more than one candidate trajectory for an optimal solution. At the switching instant $\tau > 0$, we have $x^*(\tau) = \Delta$ and the costate jump condition given by $\mu(\tau^-) = \mu(\tau^+) + \beta,\ i \neq j,\ i, j = 1, 2,\ \beta \in \mathbb{R}$ holds true. Next, the Hamiltonian continuity property results in the following equations which apply to a switching instant τ:

(switch from mode i to mode j)

$$H_i^c(\tau^-, \Delta, a_1^*(\tau^-), \ldots, a_N^*(\tau^-), \mu(\tau^-)) = H_j^c(\tau^+, \Delta, a_1^*(\tau^+), \ldots, a_N^*(\tau^+), \mu(\tau^+)). \tag{7.12}$$

Following Remark 7.6 the optimal dynamics and switching conditions in $(x^*(t), \mu(t))$ space are transformed, in the interior of each mode, to $(x^*(t), a_1^*(t), \ldots, a_N^*(t))$ space as follows:

$$\dot{x}^*(t) = \begin{cases} \sum_n a_n^*(t) - bx^*(t) + \varepsilon_1, & \text{for } x(t) < \Delta \\ \sum_n a_n^*(t) - bx^*(t) + \varepsilon_2, & \text{for } x(t) > \Delta \end{cases}, \quad x^*(0) = x_0 \text{ (given)}$$

$$\dot{a}_n^*(t) = -(r+b)a_n^*(t) + 2Nca_n^{*2}(t)x^*(t),\ n = 1, \ldots, N.$$

At the switching instant $t = \tau$, we have $x^*(\tau) = \Delta$ and the Hamiltonian continuity conditions given by (7.12) are satisfied. The above necessary conditions lead to an $N+1$ dimensional optimal vector field. In order to analyze the optimal dynamics we consider symmetric strategies, i.e., $a_n^*(t) = a^*(t)/N$. The symmetry assumption brings on that the optimal vector field reduces to a two-dimensional one which can be analyzed using the phase plane diagram. The necessary conditions with symmetry assumption are now given as:

$$\dot{x}^*(t) = \begin{cases} a^*(t) - bx^*(t) + \varepsilon_1, & \text{for } x(t) < \Delta \\ a^*(t) - bx^*(t) + \varepsilon_2, & \text{for } x(t) > \Delta \end{cases}, \quad x^*(0) = x_0 \text{ (given)} \tag{7.13}$$

$$\dot{a}^*(t) = -(r+b)a^*(t) + 2ca^{*2}(t)x^*(t) \text{ (except at the switching instants).} \tag{7.14}$$

Again, at the switching instant $t = \tau$ the Hamiltonian continuity condition leads to the following equation:

$$\ln a^*(\tau^-) + \frac{b\Delta - \varepsilon_{k(\tau^-)}}{a^*(\tau^-)} = \ln a^*(\tau^+) + \frac{b\Delta - \varepsilon_{k(\tau^+)}}{a^*(\tau^+)}, \tag{7.15}$$

we recall the variable $\varepsilon_{k(.)}$ is defined by $\varepsilon_1 = 0$ and $\varepsilon_2 = 1$. Next, we consider the situation where the economic agents act noncooperatively in realizing their profits. We have the following lemma.

Lemma 7.8 ((Symmetric) open-loop Nash equilibrium) *Let us assume that players are identical (symmetric), i.e., $a_i(.) = a_j(.)$. The open-loop Nash equilibrium problem can be solved as an optimal management problem with c replaced by $\frac{c}{N}$.*

The proof of the lemma follows directly from the usual Nash equilibrium inequalities. We know that $(\bar{a}_1, \ldots, \bar{a}_n, \ldots, \bar{a}_N)$ constitutes the Nash equilibrium strategy if $\bar{a}_n$ satisfies $J_n(\bar{a}_1, \ldots, a_n, \ldots, \bar{a}_N) \leq J_n(\bar{a}_1, \ldots, \bar{a}_n, \ldots, \bar{a}_N)$, $\forall a_n$ for player n. Upon writing the necessary conditions for this maximization problem and with symmetry assumption the statement of the lemma follows immediately. So, the open-loop Nash equilibrium problem with symmetry assumption is a potential game.[4] Next, using Lemma 7.4 the benefit player n receives in cooperation along a candidate trajectory starting at (x_0, a_0) can be easily computed as:

$$J_n^{\text{opt}}(x_0, a_0) = \frac{1}{r}\left(\ln\left(\frac{a_0}{N}\right) + \frac{bx_0 - \varepsilon_{k(0)} + 1}{a_0} - cx_0^2 - 1\right). \tag{7.16}$$

Similarly, the benefit player i receives, in noncooperation, along a candidate trajectory starting at (x_0, a_0) can be shown as:

$$J_n^{\text{olne}}(x_0, a_0) = \frac{1}{r}\left(\ln\left(\frac{a_0}{N}\right) + \frac{bx_0 - \varepsilon_{k(0)} + 1}{\frac{a_0}{N}} - cx_0^2 - N\right). \tag{7.17}$$

In the following discussion we study the optimal management problem in detail.

7.3.2.1 Phase Plane Analysis

The equilibrium points of the optimal vector field (7.13) and (7.14) are:

$$\text{mode 1:}\ (x_{eq}, a_{eq}) = \left\{(0,0), \left(\sqrt{\frac{r+b}{2cb}}, \sqrt{\frac{b(r+b)}{2c}}\right), \left(-\sqrt{\frac{r+b}{2cb}}, -\sqrt{\frac{b(r+b)}{2c}}\right)\right\}$$

$$\text{mode 2:}\ (x_{eq}, a_{eq}) = \left\{\left(\frac{1}{b}, 0\right), \left(\frac{1}{2b} + \sqrt{\frac{1}{4b^2} + \frac{r+b}{2cb}}, \sqrt{\frac{1}{4} + \frac{b(r+b)}{2c}} - \frac{1}{2}\right), \right.$$
$$\left. \left(\frac{1}{2b} - \sqrt{\frac{1}{4b^2} + \frac{r+b}{2cb}}, -\sqrt{\frac{1}{4} + \frac{b(r+b)}{2c}} - \frac{1}{2}\right)\right\}.$$

Since $a(t) > 0$, only the second equilibrium point is chosen for each of the modes. Let x_{eq}^1 and x_{eq}^2 denote these equilibrium points. Then we have, $0 < x_{eq}^1 < x_{eq}^2$ and $x_{eq}^2 > \frac{1}{b}$. The eigenvalues of the Jacobian matrix, for the linearized dynamics near the equilibrium points, are:

$$\text{mode 1:}\quad \frac{r}{2} \pm \frac{\sqrt{8b^2 + 8br + r^2}}{2}$$

$$\text{mode 2:}\quad \frac{r}{2} \pm \frac{\sqrt{r^2 + 8b^2 + 8br + 4c - 4\sqrt{c^2 + 2brc + 2b^2c}}}{2}.$$

[4] A potential game [12] facilitates to compute Nash equilibria as an optimization problem instead of a fixed point problem.

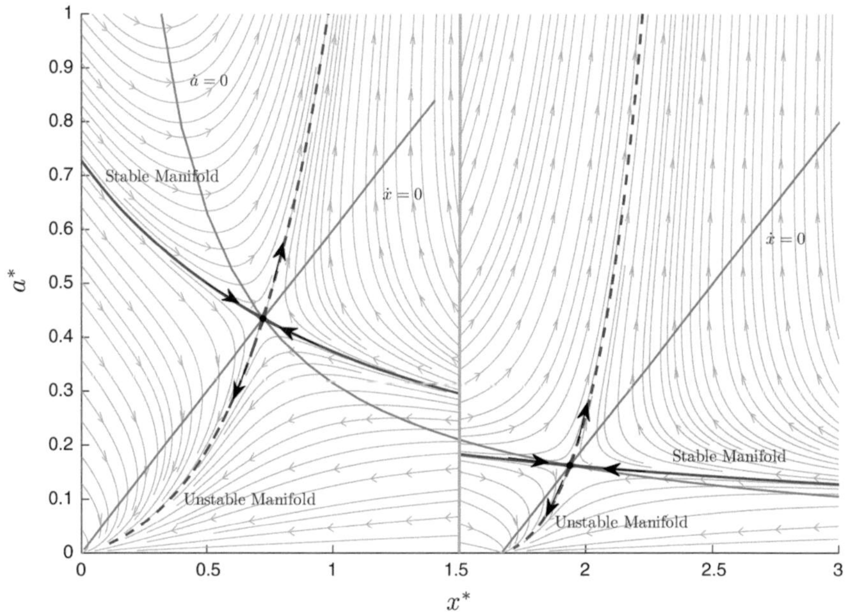

Fig. 7.2 Switching dynamics (7.13)–(7.15) with $b = 0.6$, $c = 0.5$, $r = 0.03$ and $\Delta = 1.5$

The equilibrium point in mode 1 is clearly a saddle point. Next, we have $8b^2 + 8br + 4c - 4\sqrt{c^2 + 2bcr + 2b^2c} = 4\sqrt{c + 2b(r+b)}(\sqrt{c + 2b(r+b)} - \sqrt{c}) > 0$. So, the equilibrium point in mode 2 is also a saddle point. Figure 7.2 illustrates the phase portrait of the optimal dynamics (7.13)–(7.15). Any trajectory approaching the surface at $x = \Delta$ undergoes a switching according to the rule (7.15). Next, we analyze these switching rules in detail.

7.3.2.2 Switching Rules

Before proceeding with the actual switching analysis we discuss solvability of the equation

$$s(q, m) = \ln(q) + \frac{m}{q} = p, \; m, p \in \mathbb{R}, \; q > 0. \tag{7.18}$$

If $m = 0$ then $q = e^p$. We consider the case $m \neq 0$. After rearranging terms the above equation can be written as $ye^y = l$, $y = -\frac{m}{q}$, $l = -me^{-p}$. The solution of the reformulated equation is given by $y = \mathcal{W}(l)$, where $\mathcal{W}(.)$ is the Lambert W function [3]. Here, $\mathcal{W}(z)$ is single valued for $\{z \geq 0\} \cup \{-\frac{1}{e}\}$, multiple valued

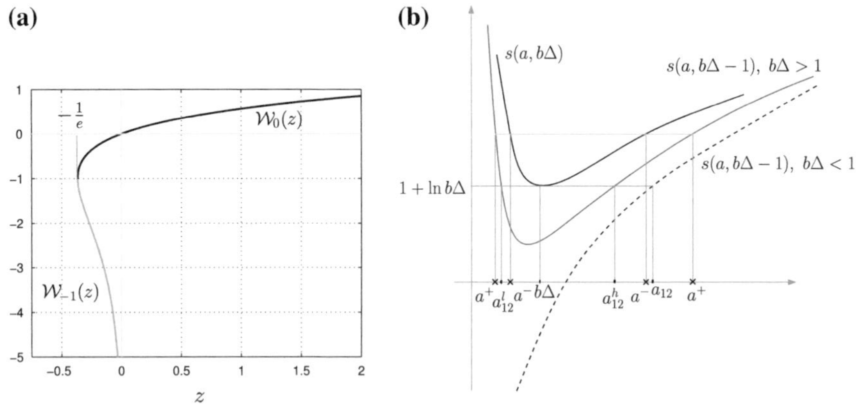

Fig. 7.3 Switching analysis. **a** Two branches of Lambert W function. **b** Analysis of jumps in control action at the switching instant τ

for $-\frac{1}{e} < z < 0$, and not defined for $z < -\frac{1}{e}$. Figure 7.3a shows two branches of $\mathcal{W}(z)$ denoted as $\mathcal{W}_0(z)$ and $\mathcal{W}_{-1}(z)$. Thus, the solution of (7.18) is given by $q = -\frac{m}{\mathcal{W}(-me^{-p})}$.

(1) Switch from mode 1 to mode 2
When the optimal system switches from mode 1 to mode 2, following (7.15), the jump in the control satisfies

$$s(a^-, b\Delta) = s(a^+, b\Delta - 1).$$

Here, $s(., b\Delta) : (0,\ \infty) \rightarrow [\ln b\Delta + 1,\ \infty)$. If $b\Delta = 1$, then $a^+ = e^{-s(a^-, b\Delta)}$. If $b\Delta \neq 1$, then $a^+ = -\frac{b\Delta - 1}{\mathcal{W}\left(-(b\Delta-1)e^{-s(a^-, b\Delta)}\right)}$. For $b\Delta < 1$, we have $-(b\Delta - 1)e^{-s(a^-, b\Delta)} > 0$. So, a jump results in a^+ in the interval $[a_{12},\ \infty)$, $a_{12} = \frac{1 - b\Delta}{\mathcal{W}_0\left(\left(\frac{1}{b\Delta} - 1\right)\frac{1}{e}\right)}$. Here, $\mathcal{W}(z)$ increases for $z > 0$. For $b\Delta > 1$, we have $-(b\Delta - 1)e^{-s(a^-, b\Delta)} < 0$. So, a jump occurs at a_l^+ in the interval $(0,\ a_{12}^l]$ and at a_h^+ in the interval $[a_{12}^h,\ \infty)$ where $a_{12}^l = \frac{1 - b\Delta}{\mathcal{W}_0\left(\left(\frac{1}{b\Delta} - 1\right)\frac{1}{e}\right)}$ and $a_{12}^h = \frac{1 - b\Delta}{\mathcal{W}_{-1}\left(\left(\frac{1}{b\Delta} - 1\right)\frac{1}{e}\right)}$. Here, $\mathcal{W}(z)$ decreases for $z < 0$. Superscripts l and h denote the lower and higher values which are computed at different branches of $\mathcal{W}(z)$ for $-\frac{1}{e} < z < 0$.

(2) Switch from mode 2 to mode 1
When the optimal system switches from mode 2 to mode 1, following (7.15), the jump in the control satisfies

$$s(a^-, b\Delta - 1) = s(a^+, b\Delta).$$

Then, $a^+ = \frac{b\Delta}{\mathcal{W}\left(-b\Delta e^{-s(a^-,b\Delta-1)}\right)}$ which is well defined only if the condition $0 < b\Delta e^{-s(a^-,b\Delta-1)} \le \frac{1}{e}$ holds true, and this implies $s(a^-, b\Delta - 1) \ge \ln b\Delta + 1$. Further, for $b\Delta e^{s(a^-,b\Delta-1)} = \frac{1}{e}$ which implies $s(a^-, b\Delta - 1) = \ln b\Delta + 1$. So, a^- should satisfy $s(a^-, b\Delta - 1) \ge \ln b\Delta + 1$ for a jump to happen from mode 2 to mode 1. In such a case, a jump results in two points, namely $a_l^+ \in (0, b\Delta]$ and $a_h^+ \in [b\Delta, \infty)$.

A graphical illustration of the switchings is given in Fig. 7.3b. Here, a^- is called a predecessor of a^+ (a^+ is a successor of a^-). Notice, there always exists a successor during transitions from mode 1 to mode 2, whereas some points on the switching surface may not have predecessors in mode 1. Further, in some cases there exist more than one successor or predecessor. These characteristics of the optimal vector field should be considered while analyzing the candidates for optimal solution. We discuss these issues in the next section.

7.3.2.3 Analysis of Optimal Vector Field (7.13)–(7.15)

In this subsection we use the results from Sects. 7.3.2.1 and 7.3.2.2 to analyze the optimal system (7.13)–(7.15) and arrive at conclusions regarding the optimal solution and control actions. First, we notice that a solution, denoted by $\gamma(t)$, of the optimal system (7.13)–(7.15) starting at a point $(x_0, a_0) \in \mathbb{R}_+^2$ either

1. converges to one of the equilibrium points as $t \to \infty$, or
2. leads to a control $a^*(t)$ that goes to infinity in a finite time, or
3. converges to a closed orbit.

In the following discussion we analyze the above three scenarios in detail.

(1) Solutions approaching stable equilibrium points
First, we notice that a trajectory $\gamma(t)$ approaching any equilibrium point admits a finite number of switchings. As a result, the truncated trajectory in the last interval satisfies necessary conditions similar to a classical problem. So, the transversality condition, given by $\lim_{t\to\infty} -\frac{Ne^{-rt}}{a^*(t)} = 0$, holds true.[5] Next, we show that the trajectory $\gamma(t)$ approaching the stable equilibrium points $(0, 0)$ and $\left(\frac{1}{b}, 0\right)$ fails to satisfy the transversality condition. First, consider a linearization around stable equilibrium point $(0, 0)$. The eigenvectors associated with the eigenvalues are $\left[1\ 0\right]^T$ and $\left[1\ -r\right]^T$. Thus, the trajectory $\gamma(t)$ approaching the stable equilibrium points can be approximated as $\gamma(t) = \left[x^*(t)\ a^*(t)\right]^T = c_1 e^{-bt}\left[1\ 0\right]^T + c_2 e^{-(r+b)t}\left[1\ -r\right]^T + \left[o(e^{-bt})\ o(e^{-(r+b)t})\right]^T$. We see that the transversality condition is violated for trajectories approaching stable equilibrium points, i.e., $\lim_{t\to\infty} -\frac{e^{-rt}}{-c_2 r e^{-(r+b)t} + o(e^{-(r+b)t})} =$

[5] The transversality condition $\lim_{t\to\infty} e^{-rt}\lambda(t) = 0$ is satisfied if $\liminf_{t\to\infty} x(t) > 0$, see [7].

$\lim_{t\to\infty} \frac{e^{bt}}{c_2 r - o(1)} \neq 0$. Following the same reasoning it can be shown that trajectories approaching the other stable equilibrium point also fail to satisfy the transversality condition.

(2) Solutions going to infinity
In the following discussion, given in the lines of [23, Sects. A.2 and A.3], we show two properties for the solutions that grow without bound. First, we show that it is not possible for solutions going to infinity that $a(t)$ remains bounded. For solutions going to infinity we have $x(t) \geq \delta > 0$ for all t. As a result, we have $\dot{a}(t) = -(r+b)a(t) + 2cx(t)a^2(t) \geq -(r+b)a(t) + 2c\delta a^2(t)$. Since the solution grows unbounded there exists a t_* such that $a(t_*) = \frac{r+b}{c\delta}$. So, we have $\dot{a}(t) \geq c\delta a^2(t)$ for $t \geq t_*$. Now, setting $v(t_*) = a_*$, the equation $\dot{v}(t) = c\delta v^2(t)$ has a solution $v(t) = \frac{u_*}{1 - a_* c\delta(t - t_*)}$ for all $t \geq t_*$ which goes to infinity in finite time. By Gronwall's inequality we have $a(t) \geq v(t)$ for all $t \geq t_*$. So, $a(t)$ goes to infinity in finite time as well. Next, we show that it is not possible for a trajectory $\gamma(t)$ to grow unbounded while $x^*(t)$ remains bounded. If the latter condition holds, then we have $x^*(t) < M$ for $t > 0$, which implies $\dot{x}^* = a^* - bx^* > a - bM$ for dynamics in mode 1 and $\dot{x}^* = a^* - bx^* + 1 > a^* - bM + 1$ for dynamics in mode 2. However, since $(x^*(t), a^*(t)) \to \infty$, there exists $T_0 > 0$ such that $a^*(t) > bM + 2$ for all $t > T_0$. So, for $T = T_0 + M$, $x^*(T) \geq M$, which contradicts the assumption $x^*(t) < M$ for all t. The solutions with finite escape time are not admissible, see Assumption 7.2. So, we are left with candidates that approach saddle equilibria and closed orbits.

(3) Solutions converging to a closed orbit
In this section we recall from the appendix an extension of Bendixson criterion for hybrid systems, see Theorem 7.11. The state–costate dynamics associated with the optimal management problem is given by $\left(\dot{x}^* = \frac{\partial H_i^c}{\partial \mu},\ \dot{\mu} = r\mu - \frac{\partial H_i^c}{\partial x^*}\right)$ for mode i, $i = 1, 2$. At the switching instant τ there may be a jump in the costate variable which is described by $\mu(\tau^+) = R(\mu(\tau^-))$. The mapping R should be such that the following Hamiltonian continuity property is satisfied:

$$H_i^c(x^*(\tau^-), a^*(\tau^-), \mu(\tau^-)) = H_j^c(x^*(\tau^+), a^*(\tau^+), \mu(\tau^+)) \tag{7.19}$$

with $x^*(\tau^-) = x^*(\tau^+) = \Delta$ and $\mu(\tau^+) = R(\mu(\tau^-))$. Next we consider the jump at the entry point $(\Delta, \mu(\tau^-))$. Differentiating the relation (7.19) with respect to $\mu(\tau^-)$ we obtain the relation

$$\frac{\partial H_i^c}{\partial \mu(\tau^-)} = \frac{\partial H_j^c}{\partial \mu(\tau^+)} R'(\mu(\tau^-)).$$

Here, the dependence through the control is ignored because we have $\frac{\partial H_i^c}{\partial a} = 0$ at the optimum $a = a^*$ for $i = 1, 2$. The growth factor at the left hand side of (7.24) can now be computed as

$$\left| \frac{\frac{\partial H_j^c}{\partial \mu(\tau^+)}}{\frac{\partial H_i^c}{\partial \mu(\tau^-)}} R'(\mu(\tau^-)) \right| = 1.$$

Together with the fact that the divergence is positive (equal to r) away from the discontinuity line, this proves on the basis of the Theorem 7.11 that there can be no closed orbits in the hybrid system described in state–costate coordinates. The dynamics (7.13)–(7.15) is topologically equivalent to the optimal dynamics described in state–costate coordinates; the former is obtained by the transformation $a = -\frac{1}{\mu}$. So, there are no closed orbits in the planar dynamical system (7.13)–(7.15).

7.3.2.4 Candidates and Objective

Let W_i^u and W_i^s denote the unstable and stable manifolds in the mode i. If $x_0 \in \mathbb{R}_+$ is the initial state of the lake then the candidates are obtained by first tracing the trajectories backwards starting at the equilibrium points. Let $\gamma(t) \in \mathbb{R}_+^2$ be one such candidate, then the initial nutrient loading, i.e., $a(0) = a_0$, is obtained as the intersection of $\gamma(t)$ with the line $x = x_0$, and as a result multiple candidates, starting at x_0, are possible. Since we only consider candidate trajectories that eventually reach saddle points. Notice, these candidates undergo only a finite number of switchings, and as a result Lemma 7.4 can be used to compare the objectives along the candidate trajectories. The optimal vector field of the classical shallow lake problem with smooth nonlinearities admits complex qualitative behaviors such as multiple steady states, existence of indifference or Skiba points and bifurcations due to variations in the parameters b, c and r, refer [7] for a complete analysis. In the present model with hybrid approximation, we consider bifurcations due to variations in the switching surface. Further, we make the following assumption:

Assumption 7.9 The switching surface does not coincide with the equilibrium points, i.e., $\Delta \notin \left\{0, \frac{1}{b}, x_{eq}^1, x_{eq}^2\right\}$.

(D) Bifurcations due to switching surface

The qualitative behavior of the optimal dynamics depends upon the position of the switching surface. Let S_i and U_i be points where the stable and unstable manifolds in mode i touch the switching surface. We consider the following situations:

$\underline{b\Delta < 1}$: First, we notice that a trajectory approaching the switching surface from mode 1, after entering mode 2 satisfies $\dot{x} = a - bx + 1$. Near the switching surface in mode 2 we have $\dot{x}(\tau^+) = a - (b\Delta - 1) > 0$. So, the trajectory never returns to mode 1, i.e., if the lake switches to turbid state it can never return to clear state.

(a) Consider the case with $\Delta < x_{eq}^1$ as illustrated in Fig. 7.4a. For any $x_0 < \Delta$, the optimal candidate is the one that switches to the point S_2 from mode 1. If S_2 does not have a predecessor in mode 1 then there is no optimal solution.[6] If $x_0 > \Delta$, then the optimal candidate is the trajectory starting at $(x_0, W_2^s(x_0))$. So, the admissible candidates converge to the steady state in mode 2.
(b) Consider the case $\Delta > x_{eq}^1$ as illustrated in Fig. 7.4b. Following the discussion in Sect. 7.3.2.3.(2) there is a nonempty closed region Ω such that solutions

[6] Transversality condition allows for trajectories with $a(t)$ going to ∞.

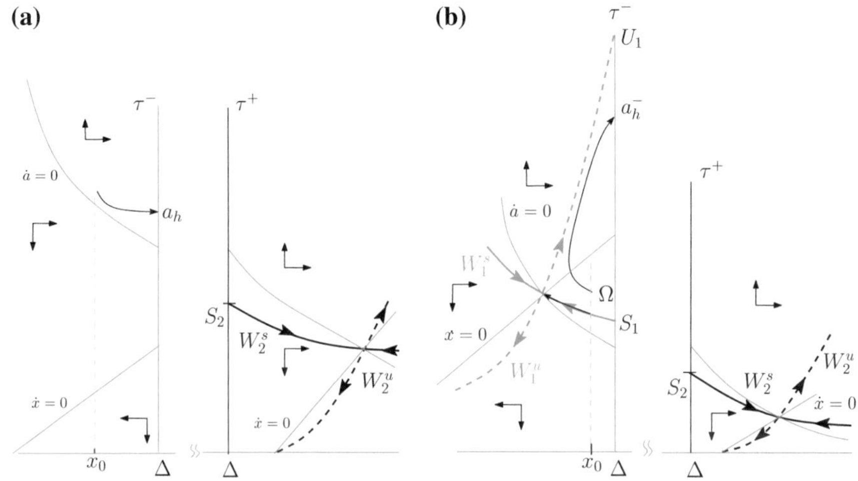

Fig. 7.4 Bifurcation analysis for $b\Delta < 1$. **a** $\Delta < x_{eq}^1$. **b** $\Delta > x_{eq}^1$

originating in Ω escape to infinity in a finite time. The trajectories can reach the steady states either in mode 1 or in mode 2. However, starting in mode 2 the steady state in mode 1 cannot be reached, whereas the steady state in mode 2 can be reached starting in mode 1.

$b\Delta > 1$: In this case, the economic agents can reverse the lake to mode 1 from mode 2 by lowering the nutrient loading, i.e., $\dot{x}(\tau^+) = a - (b\Delta - 1) < 0$. We have the following three cases:

(c) Consider the case with $\Delta < x_{eq}^1$ as illustrated in Fig. 7.5a. A trajectory starting in mode 1 either switches to mode 2 or approaches the origin. Trajectories reaching the equilibrium point in mode 2 are optimal. The existence of the closed region Ω_2 follows from the discussion in Sect. 7.3.2.3.(1). A detailed analysis includes tracing the predecessors for the point S_2 on the surface $x = \Delta$ at τ^-.
(d) Consider the case with $x_{eq}^1 < \Delta < x_{eq}^2$ as illustrated in Fig. 7.5b. The trajectories can reach either of the steady states in mode 1 and mode 2 by first reaching the points S_1 and S_2. So, the candidates are obtained by finding the predecessors of these points using the switching rules devised in Sect. 7.3.2.2.
(e) Consider the case with $\Delta > x_{eq}^2$ as illustrated in Fig. 7.5c. Trajectories reaching the steady state in mode 1 are obtained by using the switching rules for finding the predecessors.

Next, we demonstrate the subtleties in finding the candidates for a specific choice where the parameters satisfy $b\Delta > 1$ and $\Delta > x_{eq}^2$. Then we explain in detail about the candidates for all initial states x_0. Toward that end, we have the following procedure to generate points on the switching surface which eventually reach the steady state in mode 1.

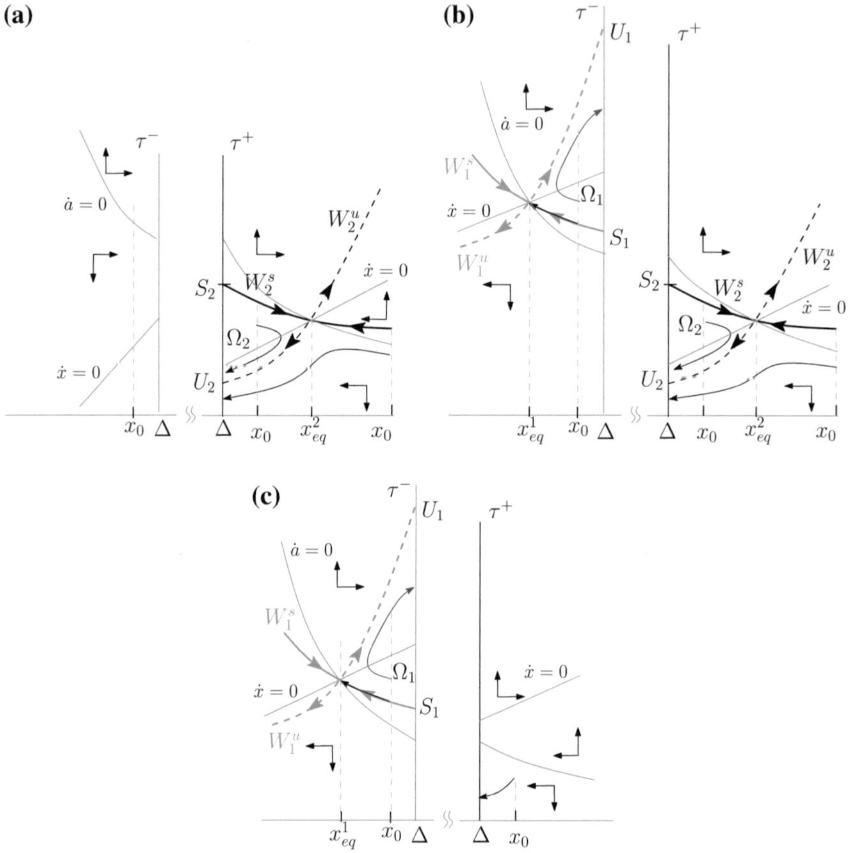

Fig. 7.5 Bifurcation analysis for $b\Delta > 1$. **a** $\Delta < x_{eq}^1$. **b** $x_{eq}^1 < \Delta < x_{eq}^2$. **c** $x_{eq}^2 < \Delta$

We notice that the above procedure generates sequences that satisfy the following condition:

$$g_0 < g_1 < \cdots < g_k < \cdots < a_{12}^l < b\Delta - 1 < a_{12}^h < \cdots < h_k < \cdots < h_1 < h_0$$
$$\text{and} \quad d_0 < d_1 < \cdots < d_k < \cdots < b\Delta < \cdots < e_k < \cdots < e_1 < e_0.$$

A graphical illustration of the algorithm is given by Fig. 7.6. Any trajectory starting at (Δ, d_k), (Δ, e_k), and (Δ, g_k) will spiral out and eventually reach the steady state in mode 1. Here, in the discrete part of the trajectory, the transition from e_k to d_k is a switch (with jump) from mode 1 to mode 1. Notice, also that the point $(\Delta, b\Delta)$ is an equilibrium point for the discrete dynamics described by the jump e_k to d_k. At this point the vector field in the x direction changes its sign in mode 1. Next, we consider three situations with variation of x_0 and find the candidates for each one of them.

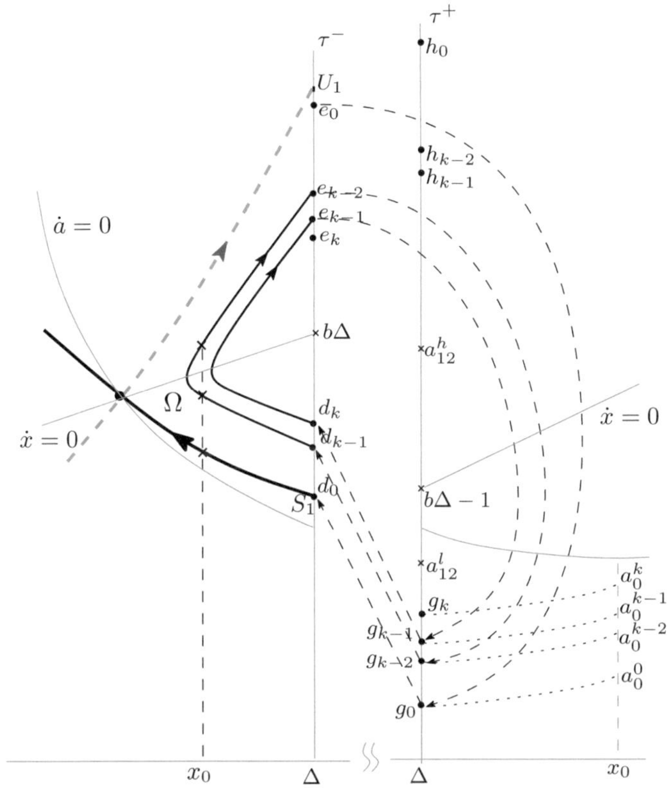

Fig. 7.6 Graphical illustration of Algorithm 2

Algorithm 2: Algorithm for generating candidates with $b\Delta > 1$ and $\Delta > x_{eq}^2$

---

1 Construct sequences $d_k, e_k, g_k, h_k, k = 1, 2, \ldots$ using the following steps:

1. For $k = 0$, set $d_0 = S_1$ and obtain $e_0,\ g_0,\ h_0$ by solving the equation (follows from Sect. 7.3.2.2)

$$s(d_0, b\Delta) = s(e_0, b\Delta) = s(g_0, b\Delta - 1) = s(h_0, b\Delta - 1)$$
$$\text{such that } d_0 < b\Delta < e_0,\ g_0 < a_{12}^l < b\Delta - 1 < a_{12}^h < h_0.$$

2. If $e_0 < U_1$ go to step 3 else STOP.
3. For, $k \geq 1$, solve the boundary value problem to obtain $a(0)$ (solution exists due to the property of the region Ω)

$$\dot{x} = a - bx,\ \dot{a} = -(r + b)a + 2ca^2x,\ x(0) = \Delta,\ x(\tau) = \Delta,\ a(\tau) = e_{k-1}.$$

   Set $d_k = a(0)$.
4. Solve $s(d_k, b\Delta) = s(e_k, b\Delta) = s(g_k, b\Delta - 1) = s(h_k, b\Delta - 1),\ d_k < b\Delta < e_k,\ g_k < a_{12}^l < b\Delta - 1 < a_{12}^h < h_k$.
5. Set $k = k + 1$ and go to step 3.

- Consider the case with $x_0 > \Delta$, i.e., starting in mode 2. Let (x_0, a_0^k) denote the initial state of the trajectory which reaches the point (Δ, g_k) (shown as dotted lines in Fig. 7.6). Then a candidate trajectory starting at (x_0, a_0^k) will undergo k cycles, that spirals out, before reaching the stable manifold W_1^s starting at d_0. So, we have countably infinite candidates that satisfy the necessary conditions.
- Consider the case with $x_{eq}^1 < x_0 < \Delta$, i.e., starting in mode 1. If $e_0 > U_1$, then there may exist two candidates that approach steady state in mode 1. The first one is the stable manifold. The other one may lie above the unstable manifold W_1^u which approaches e_0, then switches to mode 2 at g_0 and switches back to mode 1 at d_0. If $e_0 < U_1$, then there always exists one candidate that lies on the stable manifold W_1^s. Further, we observe that trajectories starting at d_{k-1} and ending at e_k intersect the line $x = x_0$ at two points or at one point (tangential intersection). So, depending upon the location of x_0 we have either $2L$ or $2L + 1$ candidates, where L represents the number of paths that intersect the section $x = x_0$.
- Consider the case with $x_0 < x_{eq}^1$. If $e_0 < U_1$, then there exists one candidate that lies on the stable manifold W_1^s. If $e_0 > U_1$, then there may exist an additional candidate that reaches e_0 switches to mode 2 at g_0 and returns to mode 1 at d_0.

Remark 7.10 The objective along each candidate trajectory is calculated using Lemma 7.4, and the optimal solution is obtained using Remark 7.7. The qualitative behavior of the optimal vector field alters when $b\Delta$ crosses a certain level. The scenarios discussed above consider all the possibilities that can arise due to parameter variations. These situations can be easily checked once the parameter values are known. However, for the classical shallow lake problem one has to resort to numerical simulations to analyze the bifurcations, see Fig. 4 of [23]. The situations, where there exists only one equilibrium point in the phase plane, illustrated in Figs. 7.4a and 7.5a, c are similar to plots (i), (vii), and (viii) in Fig. 4 of [23]. Similarly, the situations, where there exist two equilibrium points in the phase plane, illustrated in Figs. 7.4b and 7.5b are similar to plots (ii)–(vii) in Fig. 4 of [23].

7.3.2.5 Numerical Illustration

Consider the shallow lake system with parameter values $b = 1$, $c = 0.6$, $r = 0.03$, $\Delta = 1.6$, and $N = 4$. For this choice of parameters, we have $b\Delta = 1.6 > 1$, $U_1 = 5.03493$, and $e_0 = 4.9259$. The candidates for optimal solution are obtained by following the previous discussion. Optimal candidates starting at various initial states are illustrated, in small roman letters, and the benefit of each player along these trajectories is calculated according to Eq. (7.16), in cooperation, and (7.17), in noncooperation. First, we consider the optimal management case and the results are illustrated in Fig. 7.7a. For initial state $x_0 = 1.06$, we obtain three candidates labeled as (ii), (iii), and (iv). When following the paths (iii) and (iv), the agents increase the level of nutrient loading till the lake switches to mode 2 and instantaneously drop the levels to be able to switch back to mode 1 along the stable manifold W_1^s. We observe that $f_1(x_0, a_0) < 0$ for candidates (ii) and (iii). So, from Remark 7.7, we

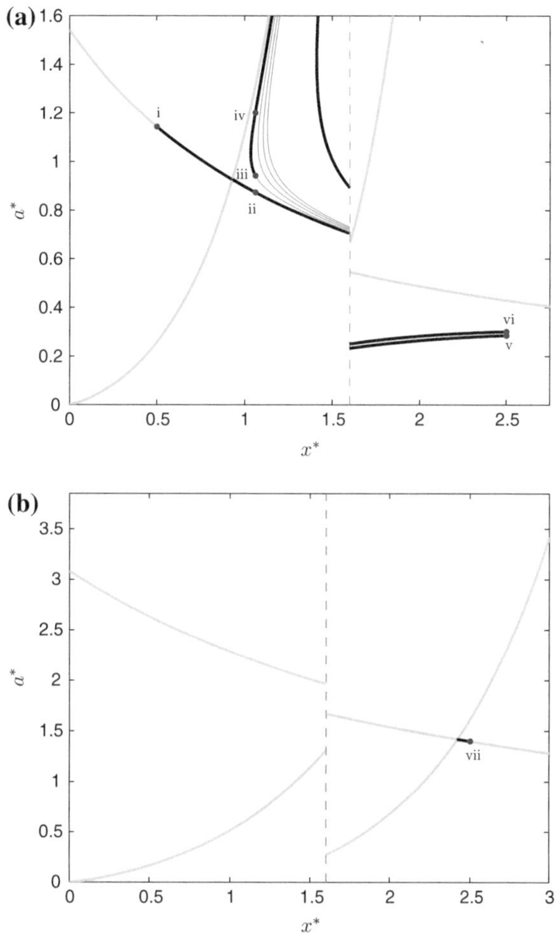

Fig. 7.7 Candidates for various initial states. *Thick gray lines* indicate stable and unstable manifolds, *dashed line* indicates the switching surface. *Thick dark lines* indicate the candidates. **a** Phase portrait with $b = 1, c = 0.6, r = 0.03$ and $\Delta = 1.6$. **b** Phase portrait with $b = 1$, $c = 0.15, r = 0.03$ and $\Delta = 1.6$

see that following the trajectory (ii) results in higher benefits, see Table 7.1. When the agents start in mode 2, i.e., $x_0 > 1.6$, there exist countably infinite candidates each undergoing a finite number of cycles before reaching the steady state in mode 1. For instance, on path (vi) the agents can alter the nutrient levels, i.e., a decrease and increase cycle, 39 times before reaching the steady state in mode 1. We observe that $f_2(x_0, a_0) < 0$ for the candidates (v) and (vi). Again, from Remark 7.7 it can be inferred that following trajectory (v) results in higher benefits, see Table 7.1a.

For the (symmetric) open-loop Nash equilibrium, following Lemma 7.8, we analyze the optimal vector field with $c = 0.6$ replaced with $\frac{c}{N} = 0.15$. The phase plane diagram for the optimal vector field with $c = 0.15$, ceteris paribus, is illustrated in Fig. 7.7b. We notice that only steady state in mode 2 can be achieved by the agents. So, the trajectory (vii) corresponds to the open-loop Nash equilibrium path with $x_0 = 2.5$ and the welfare parameter set to $c = 0.6$. We notice that the choice of

Table 7.1 Performance of the candidate trajectories

(a) With $b = 1, c = 0.6, r = 0.03$ and $\Delta = 1.6$, see Fig. 7.7a

Initial state	Candidate			Objective
	Starting at $(x_0,\ a_0)$	# of cycles	Reaches steady state in mode #	$N = 4$ optimal management
0.50	(0.50, 1.1425) (i)	0	1	−65.5150
1.06	(1.06, 0.8733) (ii)	0	1	−66.0710
	(1.06, 0.9420) (iii)	1	1	−66.4980
	(1.06, 1.1998) (iv)	1	1	−66.4940
2.50	(2.50, 0.2844) (v)	0	1	−70.6235
	(2.50, 0.2851)	1	1	−71.0171
	(2.50, 0.2859)	2	1	−71.3826
	⋮	⋮	⋮	⋮
	(2.50, 0.3003)	38	1	−78.1329
	(2.50, 0.3005) (vi)	39	1	−78.2385
	⋮	⋮	⋮	⋮

(b) With $b = 1, c = 0.15, r = 0.03$ and $\Delta = 1.6$, see Fig. 7.7b

Initial state	Candidate			Objective	
	Starting at $(x_0,\ a_0)$	# of cycles	Reaches steady state in mode #	$N = 4$ optimal management	$N = 4$ open-loop Nash equilibrium ($c = 0.6$)
0.50	(0.50, 3.500000)	–	–	−11.1676*	
1.70	(1.70, 1.635331)	0	2	−63.3300	
2.50	(2.50, 1.398072) (vii)	0	2	−63.8598	−150.3192

*The candidate starting at (0.5, 3.5) switches to mode 2 at time $t = 0.4732$ and the system (7.13)–(7.15) has a finite escape time at $t = 0.7844$. So, the objective is computed in the interval $[0, 0.7844)$. We discard this candidate as Assumption 7.2 is violated

institutional arrangement also affects the qualitative behavior of the optimal solution; more clearly, when players play cooperatively steady state in mode 1 is attained, whereas a noncooperative behavior leads to the steady state in mode 2. Further, it is clear from Table 7.1a, b that each player receives greater benefits in cooperation. Now, consider the effect of reducing the c from 0.6 to 0.15 on optimal management. Since, in the latter case the players incur less costs, toward cleaning activities, there is an incentive for increasing the nutrients and as a result the optimal vector field results in the steady state in mode 2.

7.4 Conclusions

In this article we introduce a class of discounted autonomous infinite horizon hybrid optimal control problems and provide the necessary conditions for optimality. Restricting the state and control variables to dimension one we obtain additional insights on the necessary conditions. Using these tools we study the shallow lake problem where nonlinear lake dynamics are approximated using a simple hybrid system. Assuming symmetry in agents' actions, we solve the associated optimal management problem using relevant necessary conditions. The hybrid approximation leads to simple dynamics within each mode and a complex jump rule near the switching surface. The dynamic behavior of the switched vector field is similar, qualitatively, to the smooth version. However, there are some differences. The bifurcations in the present analysis are governed by simple rules, a set of inequalities, which can be checked/verified once the parameters are given. In the previous works these bifurcation scenarios were established numerically, though their existence was proved using continuation methods. Further, we provide a bifurcation analysis of the vector field. We observe that the variation of switching surface induces bifurcations in this vector field.

There are several open issues that require considerable attention. It was shown in [23] that existence of Skiba points is closely related to heteroclinic connections[7] in the optimal vector field. A piecewise linear approximation of the convex–concave production function [20] also results in multiple steady states, see Figs. 7.4b and 7.5b, then to see whether Skiba points exist in the optimal switching dynamics would be interesting. Feedback policies are generally preferred over open-loop policies as they can quickly adapt to changes in the state variable. However, designing the feedback policies is difficult due to computational burden; see [2] for some preliminary work.

Appendix

Bendixson Criterion for a System with Jumps

Consider a planar dynamical system that has a discontinuity across the line $x = \Delta$, and that moreover exhibits jumps when it reaches the line of discontinuity. The system may be described as follows:

$$\dot{x}(t) = P(x(t), y(t)), \quad x(t) \neq \Delta \tag{7.20}$$

$$\dot{y}(t) = Q(x(t), y(t)), \quad x(t) \neq \Delta \tag{7.21}$$

$$(x(t^+), y(t^+)) = (x(t^-), R(y(t^-))), \quad \text{for } t \text{ such that } x(t^-) = \Delta, \tag{7.22}$$

[7]This happens when a branch of an unstable manifold of an equilibrium point coincides with a branch of a stable manifold of a different equilibrium point.

where

$$(P(x, y), Q(x, y)) = \begin{cases} (P_1(x, y), Q_1(x, y)), & x < \Delta \\ (P_2(x, y), Q_2(x, y)), & x > \Delta \end{cases}$$

and where all the functions P_1, P_2, Q_1, Q_2, and R are smooth. The vector fields are supposed to be such that solutions are well defined and lead off the discontinuity line. The following extension of the classical Bendixson criterion is proposed, see also [10, 14] for different formulations.

Theorem 7.11 (Bendixson criterion) *Suppose there is a simply connected region in the plane in which the following conditions are satisfied:*

$$\frac{\partial P}{\partial x}(x, y) + \frac{\partial Q}{\partial x}(x, y) > 0, \; x \neq \Delta \tag{7.23}$$

$$\left| \frac{P_{k^+}(\Delta, R(y))}{P_{k^-}(\Delta, y)} R'(y) \right| \geq 1 \; \textit{with} \; \begin{cases} k^+ = \textit{mode after jump} \\ k^- = \textit{mode before jump.} \end{cases} \tag{7.24}$$

Then in this region there are no closed orbits. A similar statement holds with both inequalities reversed.

Sketch of the proof Let us assume that there exists a closed orbit in a system described by (7.20)–(7.22). Then this implies there exists a bounded area that is invariant under the flow (including jumps) (7.20)–(7.22). Next, we consider volume elements within this area. The divergence of the vector field at a point (x, y) is the rate of change of surface area of a volume element that is carried along the flow starting from (x, y). Since this rate is positive, from (7.23), the volume elements expand during the continuous part of the flow.

Next, we analyze what happens to the volume element when the system reaches the discontinuity line, experiences a jump, and leaves the discontinuity line again. Imagine a small rectangle with width δx and height δy that is taken by the flow in mode k^- to meet the discontinuity line at the point (Δ, y). We can approximate the vector field near this point as a constant field with components $(P_{k^-}(\Delta, y), Q_{k^-}(\Delta, y))$; likewise, the vector field near the exit point $(\Delta, R(y))$ can be approximated as a constant field with components $(P_{k^+}(\Delta, R(y)), Q_{k^+}(\Delta, R(y)))$. The volume element that enters in a neighborhood of (Δ, y) emerges in a neighborhood of $(\Delta, R(y))$ as approximately a rectangle with width $|P_{k^+}(\Delta, R(y))|\delta t$ and height $R(y+\delta y)-R(y)$, where δt is the time it takes for the flow near the entry point (Δ, y) to move the far side of the volume element to the discontinuity line. This time δt is approximately equal to $\frac{\delta x}{|P_{k^-}(\Delta, y)|}$. The height of the volume element that emerges after the jump is approximately equal to $R'(y)\delta y$. The ratio of change of surface of area of the volume element before and after experiencing the jump is then given by $\left| \frac{P_{k^+}(\Delta, R(y))}{P_{k^-}(\Delta, y)} R'(y) \right|$. Next, from (7.24) this ratio is greater than or equal to 1; this implies, along with (7.23), that all volume elements within the region enclosed by the closed orbit are expanding during the continuous part of the flow and not contracting during the discrete (jump) part. Since the area of the region as a whole cannot expand as it is enclosed by a closed orbit, we have a contradiction.

References

1. W.A. Brock, D. Starrett, Managing systems with non-convex positive feedback. Env. Resour. Econ. **26**, 575–602 (2003)
2. P.E. Caines, M.S. Shaikh, Optimality zone algorithms for hybrid systems: efficient algorithms for optimal location and control computation, in *Hybrid Systems: Computation and Control '06*, pp. 123–137 (2006)
3. R. Corless, G. Gonnet, D. Hare, D. Jeffrey, D. Knuth, On the Lambert W function. Adv. Comput. Math. **5**, 329–359 (1996)
4. R. Cross, M. Grinfeld, H. Lamba, Hysteresis and economics. IEEE Control Syst. **29**(1), 30–43 (2009)
5. R. Goebel, R. Sanfelice, A. Teel, Hybrid dynamical systems. IEEE Control Syst. **29**(2), 28–93 (2009)
6. D. Grass, J.P. Caulkins, G. Feichtinger, G. Tragler, *Optimal Control of Nonlinear Processes: With Applications in Drugs, Corruption, and Terror* (Springer, Heidelberg, 2008)
7. T. Kiseleva, F.O.O. Wagener, Bifurcations of optimal vector fields in the shallow lake model. J. Econ. Dyn. Control **34**(5), 825–843 (2010)
8. G. Kossioris, M. Plexousakis, A. Xepapadeas, A. de Zeeuw, K.-G. Mäler, Feedback Nash equilibria for non-linear differential games in pollution control. J. Econ. Dyn. Control **32**(4), 1312–1331 (2008)
9. K.-G. Mäler, A. Xepapadeas, A. de Zeeuw, The economics of shallow lakes. Environ. Resour. Econ. **26**, 603–624 (2003)
10. J. Melin, Does distribution theory contain means for extending Poincaré-Bendixson theory? J. Math. Anal. Appl. **303**(1), 81–89 (2005)
11. P. Michel, On the transversality conditions in infinite horizon optimal problems. Econometrica **50**(4), 975–985 (1982)
12. D. Monderer, L.S. Shapley, Potential games. Games Econ. Behav. **14**(1), 124–143 (1996)
13. E. Nævdal, Dynamic optimisation in the presence of threshold effects when the location of the threshold is uncertain—with an application to a possible disintegration of the western antarctic ice sheet. J. Econ. Dyn. Control **30**(7), 1131–1158 (2006)
14. A. Pogromsky, H. Nijmeijer, J. Rooda, A negative Bendixson-like criterion for a class of hybrid systems. IEEE Trans. Autom. Control **52**(4), 586–595 (2007)
15. S. Polasky, A. de Zeeuw, F. Wagener, Optimal management with potential regime shifts. J. Env. Econ. Manag. **62**(2), 229–240 (2011)
16. P. Riedinger, C. Iung, F. Kratz, An optimal control approach for hybrid systems. Eur. J. Control **9**(5), 449–458 (2003)
17. M. Scheffer, S. Carpenter, J.A. Foley, C. Folke, B. Walker, Catastrophic shifts in ecosystems. Nature **413**(6856), 591–596 (2001)
18. A. Seierstad, S.D. Stabrun, Control problems with surfaces triggering jumps in the state variables. Optimal Control Appl. Methods **31**(2), 117–136 (2010)
19. M.S. Shaikh, P.E. Caines, On the hybrid optimal control problem: theory and algorithms. IEEE Trans. Autom. Control **52**(9), 1587–1603 (2007)
20. A.K. Skiba, Optimal growth with a convex-concave production function. Econometrica **46**(3), 527–39 (1978)
21. H.J. Sussmann, A Maximum principle for hybrid optimal control problems, in *Proceedings of the 38th IEEE Conference on Decision and Control*, vol. 1, pp. 425–40 (1999)

22. A.J. van der Schaft, J.M. Schumacher, *An Introduction to Hybrid Dynamical Systems*, Lecture Notes in Control and Information Sciences (Springer, London, 2000)
23. F.O.O. Wagener, Skiba points and heteroclinic bifurcations, with applications to the shallow lake system. J. Econ. Dyn. Control **27**(9), 1533–1561 (2003)

Chapter 8
Modeling Perspectives of Hybrid Systems and Network Systems

Jun-ichi Imura and Takayuki Ishizaki

Abstract This article presents two topics, i.e., well-posedness of piecewise affine systems, and model reduction of network systems. The well-posedness problem, i.e., the problem of existence and uniqueness of solutions, of hybrid systems is one of the fundamental research topics, which the first author has collaborated with Prof. Arjan van der Schaft in 1998. Some results are revisited by focusing on the class of bimodal piecewise affine systems. The latter discusses the most recent topic that both Arjan and the first author have common interest in. In particular, the clustering-based $H_\infty-$ and H_2-model reduction approaches of large-scale network systems, which have been independently developed by the authors, are represented in a unified way.

8.1 Introduction

I, the first author, has started with research topics on hybrid systems since I stayed in Twente University for one year from May 1998 as a visiting researcher under Professor Arjan van der Schaft. In those days, Arjan tried to publish a book entitled "An Introduction to Hybrid Dynamical Systems" with van der Schaft and Schumacher [1]. I had a lucky opportunity to read this first draft with great interest. In particular, the concept of complementarity systems and its well-posedness problem were very impressive for me, and started with the well-posedness problem of bimodal piecewise linear systems together with Arjan [2, 3]. Since then, this topic brought me various kinds of results on modeling, analysis, and control synthesis of hybrid systems including feedback well-posedness and stabilizability of piecewise affine systems [4, 5], controllability analysis of piecewise affine systems [6, 7], discrete abstraction of nonlinear systems [8], and so on. The first part of this article revisits

J. Imura (✉) · T. Ishizaki
Graduate School of Information Science and Engineering, Tokyo Institute of Technology, 2-12-1 Oh-Okayama, Meguro, Tokyo 152-8552, Japan
e-mail: imura@mei.titech.ac.jp

T. Ishizaki
e-mail: ishizaki@mei.titech.ac.jp

M.K. Camlibel et al. (eds.), *Mathematical Control Theory I*,
Lecture Notes in Control and Information Sciences 461,
DOI 10.1007/978-3-319-20988-3_8

the well-posedness issue of hybrid systems, which I look back with valuable collaboration with Arjan.

The second part focuses on more recent topic of model reduction of large-scale network systems, which recently gave common interest to Arjan and myself. Throughout the study of discrete abstraction of nonlinear systems, which produces a kind of graph structure for approximately expressing complex system behavior based on bisimilarity notation, I and my colleagues also had great interest in model reduction of large-scale network systems. We thus have developed a clustering-based approach in the framework of projective model reduction [9–12], which we call clustered model reduction. This is also a kind of structure-preserving model reduction methods. On the other hand, for the concept of the port-controlled Hamiltonian systems preserving the essential property of physical structure, proposed by Arjan and B.M. Maschke (e.g., [13, 14]), the problem of model reduction preserving such physical structure is naturally and relevantly induced. Most recently, Arian and his colleagues have solved this problem by a clustering framework, where the strict H_2 norm-approximation-error evaluation and an extension to the case of second-order systems are provided [15, 16]. This result is also based on the research works by H. Trentleman and his colleagues [17]. The second part of this article provides a summary on our previous results including H_2/H_∞-norm-error evaluation and extensions to the case of second-order subsystems in a unified way. We hope this unified approach will provide any further common framework with the works by Arjan, Harry, and their colleagues. In addition, as an application of clustered model reduction, we present our recent result on the design of a projective state observer, which estimates the average state behavior of large-scale network systems according to the above clustered model reduction [18]. Numerical simulations on power systems show that the method is effective.

Notation We denote the set of real numbers by $\mathbb{R}$, the n-dimensional identity matrix by I_n, the ith column of I_n by e_i, the cardinality of a set $\mathcal{I}$ by $|\mathcal{I}|$, the l_p-norm of a vector x by $\|x\|_{l_p}$, the Frobenius norm of a matrix M by $\|M\|_{\mathrm{F}}$, the l_2-induced norm of a matrix M by $\|M\|$, and the l_∞-induced norm of a matrix $M \in \mathbb{R}^{n\times m}$ is defined by

$$\|M\|_{l_\infty} := \max_{i\in\{1,\dots,n\}} \sum_{j=1}^{m} |M_{i,j}|$$

where $M_{i,j}$ denotes the (i, j)-element of M. The positive (negative) definiteness of a matrix $M = M^T$ is denoted by $M \succ 0 (M \prec 0)$. Furthermore, we denote the block diagonal matrix having matrices $M_1, \dots, M_n$ on its block diagonal by $\mathrm{diag}(M_1, \dots, M_N)$. Finally, the $\mathcal{H}_\infty$-norm and $\mathcal{H}_2$-norm of a stable transfer matrix G are denoted by $\|G(s)\|_{\mathcal{H}_\infty}$ and $\|G(s)\|_{\mathcal{H}_2}$, respectively.

8.2 Revisit: Well-Posedness of Piecewise Affine Systems

8.2.1 Motivating Example

Consider a 2-tank system in Fig. 8.1, where x_i is the deviation of the water level from the equilibrium state x_{ie}, and u_i is the volume of water discharged from the tap i. We assume that u_i is an input, i.e., $u_i = u_{ie}$, where u_{ie} is constant, and the valve at the tap is open or closed according to the rule shown in Fig. 8.1. Equations of motion of this system are given by

$$\dot{x} = \begin{cases} \begin{bmatrix} -1 & 0 \\ 1 & -1 \end{bmatrix} x & \text{if } x_2 \leq 1 \\ \begin{bmatrix} 0 & 0 \\ 0 & -1 \end{bmatrix} x + \begin{bmatrix} u_{1e} \\ -u_{1e} \end{bmatrix} & \text{if } x_2 > 1 \end{cases} \tag{8.1}$$

where $x = [x_1, x_2]^T$. The coefficients are normalized to 1 for brevity, and the equilibrium state and input satisfy $-x_{1e} + u_{1e} = 0$, $-x_{2e} + x_{1e} + u_{2e} = 0$, $x_{1e} > 0$, and $0 < x_{2e} < 1$. Although this tank system is nonlinear, we here consider the linearization of the system at the equilibrium since the solution behavior will be essentially similar to that of the original system.

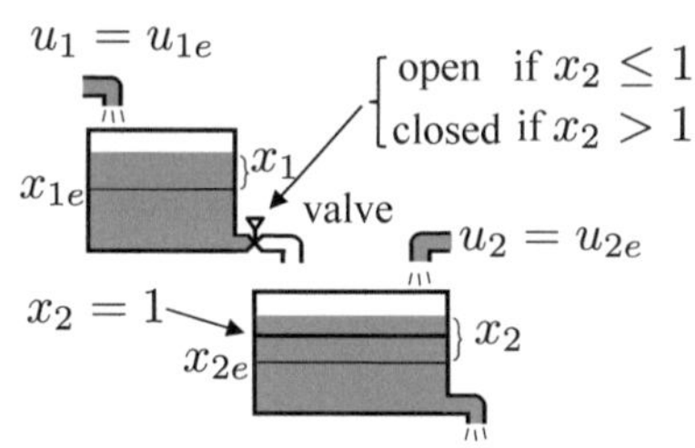

Fig. 8.1 2-tank system with a valve

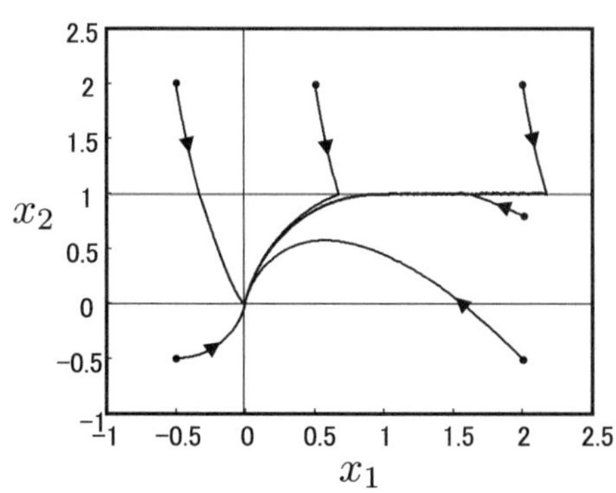

Fig. 8.2 Trajectories for the system in Fig. 8.1

Figure 8.2 shows trajectories of the system from six different initial states. There exists a sliding motion when $x(0) = [2\ \ 2]^{\mathsf{T}}$ and $[2\ \ 0.8]^{\mathsf{T}}$, where, in fact, chattering phenomena happen due to numerical simulation. Since we consider the Open/Closed motion of the valve in this case, such phenomena is not desirable. It is important to specify a condition on discontinuity of the vector field to avoid such phenomena. The next section gives a solution to this question.

8.2.2 Well-Posedness Condition

Consider the discontinuous system

$$\dot{x} = f_I(x) \text{ if } x \in \mathcal{X}_I, \quad I \in \mathcal{I}, \tag{8.2}$$

where $x \in \mathbb{R}^n$, $\mathcal{I} = \{1, 2, \ldots, M\}$, and $\mathcal{X}_I$ is a closed subset of $\mathbb{R}^n$ satisfying

$$\mathrm{int}\,\mathcal{X}_I \neq \emptyset, \quad \bigcup_{I=1}^{M} \mathcal{X}_I = \mathbb{R}^n, \quad \mathrm{int}\,\mathcal{X}_I \bigcap \mathrm{int}\,\mathcal{X}_J = \emptyset.$$

Then a solution of this system is defined as follows:

Definition 8.1 (*Extended Carathéodory solution*) Suppose that an initial state $x(t_0) = x_0 \in \mathbb{R}^n$ is given. Then if on $[t_0, t_1)$ for some $t_1 > t_0$, $x(t)$ satisfies

$$x(t) = x_0 + \int_{t_0}^{t} f_{I(\tau)}(x(\tau))d\tau, \tag{8.3}$$

and there is no left accumulation point of event times, $x(t)$ is said to be a solution of (8.2) on $[t_0, t_1)$ in the sense of Carathéodory for $x(t_0) = x_0$.

Note that this notion of solutions does not admit sliding motions and left-Zeno behavior, although the right-Zeno behavior is regarded as a solution. The system (8.2) is said to be well-posed if for every initial state $x(t_0) \in \mathbb{R}^n$, there exists a right unique extended Caratéodory solution of (8.2) on $[t_0, \infty)$.

The notion of smooth continuation is very important for characterizing the well-posedness property [19]. Consider a solution of $\dot{x} = f_I(x)$ in (8.2) with a fixed I. If for an initial state $x(t_0) = x_0$ there exists an $\varepsilon > 0$ such that $x(t) \in \mathcal{X}_I$ for all $t \in [0, \varepsilon]$, we say that smooth continuation is possible from x_0 in $\mathcal{X}_I$. Furthermore, we call the *smooth continuation set*, denoted by $\mathcal{S}_I$, the set of all x_0 from which smooth continuation is possible in $\mathcal{X}_I$.

Obviously $\mathcal{S}_I \subseteq \mathcal{X}_I$ holds, and the smooth continuation set $\mathcal{S}_I$ expresses the region of existence of solutions $x(t)$ of the system, while the difference set $\mathcal{X}_I - \mathcal{S}_I$ expresses all the state from which there exists no solution $x(t)$.

Then we have the following theorem [3]:

Theorem 8.2 *The system (8.2) is well-posed if and only if the following two conditions:*

(a) $\bigcup_{I\in\mathcal{I}} \mathcal{S}_I = \mathbb{R}^n$.
(b) For every $I_1,\ I_2 \in \mathcal{I}$*, there exists an* $\varepsilon > 0$ *such that both solutions* $x(t)$ *of* $\dot{x} = f_I(x)$, $I = I_1, I_2$ *are the same on* $[t_0, t_0+\varepsilon)$ *for every* $x_0 \in \mathcal{S}_{I_1} \bigcap \mathcal{S}_{I_2}$.

To derive an explicit representation of the above conditions, consider

$$\dot{x} = \begin{cases} A_1 x \text{ if } Cx \geq 0, \\ A_2 x \text{ if } Cx \leq 0. \end{cases} \tag{8.4}$$

Denote by T_1 and T_2 the observability matrices of (C, A_1) and (C, A_2), respectively, and by m_1 and m_2 their observability indexes. We also let $\mathcal{L}_+$ be the set of $n \times n$ lower triangular matrices with all diagonal elements positive. Then, the conditions (a) and (b) in Theorem 8.2 are reduced into the following conditions [3]:

Theorem 8.3 *The system (8.4) is well-posed if and only if the following conditions hold:*

(a) $m_1 = m_2$,
(b) $T_2 = MT_1$ *for some* $M \in \mathcal{L}_+$,
(c) $(A_1 - A_2)x = 0$ *for all* $x \in KerT_1$.

The smooth continuation set for $\mathcal{X}_1 := \{x \in \mathbb{R}^n \mid Cx \geq 0\}$ is given by $\mathcal{S}_1 = \{x \in \mathbb{R}^n \mid T_1 x \succeq 0\}$, where $x \succeq 0$ expresses the lexicographic inequality, i.e., for each i, $x_j = 0\ (j = 1, 2, \ldots, i-1)$ and $x_i > 0$, or $x = 0$. This comes from the fact that for sufficiently $\varepsilon > 0$, $y(t)(:= Cx(t)) = y(t_0) + \dot{y}(t_0)(t-t_0) + \ddot{y}(t_0)(t-t_0)^2 + \cdots \geq 0$ holds for all $t \in [t_0, t_0+\varepsilon)$. Thus $\mathcal{S}_1 \bigcup \mathcal{S}_2 = \mathbb{R}^n$, which corresponds to condition (a) in Theorem 8.2, implies condition (b) in Theorem 8.3. Conditions (a) and (c) in Theorem 8.3 come from condition (b) in Theorem 8.2.

Note that the above conditions can be easily checked. A similar but rather complicated necessary and sufficient condition can be obtained for bimodal piecewise affine systems [4], and a sufficient condition for a multimodel piecewise affine system with external inputs to be well-posed can be also obtained [5]. In addition, the feedback well-posedness condition, which implies that the system can be made well-posed by a feedback controller, can be characterized for bimodal piecewise affine systems [4].

8.3 Clustered Model Reduction of Network Systems

In this section, we briefly summarize our clustered model reduction method for linear network systems, which belongs to a type of structured model reduction methods. In this model reduction, toward the preservation of network structure of systems, clustering of subsystems is performed according to a notion of uncontrollability of local states, called cluster reducibility. All mathematical proofs of theoretical results are omitted due to page limitation; see [9–12] for details.

8.3.1 Clustered Model Reduction Problem

We first deal with a stable linear network system denoted by

$$\Sigma : \dot{x} = Ax + Bu, \quad A \in \mathbb{R}^{n\times n}, \quad B \in \mathbb{R}^{n}, \tag{8.5}$$

whose network structure is represented by the Boolean structure of A. For simplicity, we consider only single-input systems while a similar result can be obtained also for multi-input systems. To formulate a clustered model reduction problem for Σ in (8.5), we introduce the following notion of network clustering:

Definition 8.4 For $\mathbb{L} := \{1, \ldots, L\}$, the family of an index set, $\{\mathcal{I}_{[l]}\}_{l\in\mathbb{L}}$, is called a *cluster set*, each of whose elements is referred to as a cluster, if each element $\mathcal{I}_{[l]} \subseteq \{1, \ldots, n\}$ satisfies

$$\bigcup_{l\in\mathbb{L}} \mathcal{I}_{[l]} = \{1, \ldots, n\}, \quad \mathcal{I}_{[l]} \cap \mathcal{I}_{[l']} = \emptyset, \quad l \neq l'.$$

Furthermore, an *aggregation matrix* compatible with $\{\mathcal{I}_{[l]}\}_{l\in\mathbb{L}}$ is defined by

$$P := \Pi \operatorname{diag}(p_{[1]}, \ldots, p_{[L]}) \in \mathbb{R}^{n\times L}, \tag{8.6}$$

where $p_{[l]} \in \mathbb{R}^{|\mathcal{I}_{[l]}|}$ such that $\|p_{[l]}\| = 1$, and the permutation matrix Π is defined as

$$\Pi := [e_{\mathcal{I}_{[1]}}, \ldots, e_{\mathcal{I}_{[L]}}] \in \mathbb{R}^{n\times n}, \quad e_{\mathcal{I}_{[l]}} \in \mathbb{R}^{n\times|\mathcal{I}_{[l]}|}.$$

In this definition, the aggregation matrix P clearly satisfies $P^{\mathsf{T}}P = I_L$, i.e., all column vectors of P are orthonormal. Using the aggregation matrix P in (8.6), we define the aggregated model of Σ in (8.5) by

$$\hat{\Sigma} : \begin{cases} \dot{\xi} = P^{\mathsf{T}}AP\xi + P^{\mathsf{T}}Bu \\ \hat{x} = P\xi. \end{cases} \tag{8.7}$$

Note that each state of the aggregated model $\hat{\Sigma}$ represents an approximant of the clustered states, given by $e_{\mathcal{I}_{[l]}}^{\mathsf{T}} x \in \mathbb{R}^{|\mathcal{I}_{[l]}|}$. The trajectory of each state of $\hat{\Sigma}$ aims at tracing the trajectory of a kind of centroid compatible with the clustered states of Σ. In this notation, we formulate a clustered model reduction problem as follows:

Probelm 8.5 Let a stable linear system Σ in (8.5) be given. Given a constant $\varepsilon \geq 0$, find a stable aggregated model $\hat{\Sigma}$ in (8.7) such that

$$\|G(s) - \hat{G}(s)\|_{\mathcal{H}_2} \leq \varepsilon \quad \text{or} \quad \|G(s) - \hat{G}(s)\|_{\mathcal{H}_\infty} \leq \varepsilon, \tag{8.8}$$

where

$$G(s) := (sI_n - A)^{-1}B, \quad \hat{G}(s) := P(sI_L - P^{\mathsf{T}}AP)^{-1}P^{\mathsf{T}}B \tag{8.9}$$

denote the transfer matrices of Σ and $\hat{\Sigma}$, respectively.

In traditional model reduction methods, each state of the reduced model is usually obtained as a linear combination of *all* states of the original system [20]. This can be rephrased as that the projection matrix has no specific sparse structure. Note that the aggregation matrix P in (8.6) is *block-diagonally structured*. In this sense, our problem formulation clearly contrasts with the traditional model reduction problems.

8.3.2 Controllability Characterizations for Clustered Model Reduction

In systems and control theory, Σ in (8.5) is said to be *controllable* if there exists an input function u such that the state x is moved from any initial state to any other final state in a finite time interval. One best-known characterization of controllability is the Kalman rank condition, i.e., Σ is controllable if and only if $[B, AB, \dots, A^{n-1}B]$ has full row rank [20]. However, the Kalman rank condition is not necessarily useful for model reduction because it cannot capture the controllability of systems quantitatively. Such a quantitative characterization of controllability plays an important role in performing an approximation error analysis in model reduction.

In view of this, let us seek some other characterizations of controllability that have good compatibility with model reduction. One of useful controllability characterizations is given by the controllability Gramian, related to the $\mathcal{H}_2$-norm of linear systems. It is known that a stable linear system Σ in (8.5) is controllable if and only if the controllability Gramian, defined as

$$M := \int_0^\infty e^{At}B(e^{At}B)^{\mathsf{T}}dt \in \mathbb{R}^{n\times n}, \tag{8.10}$$

is nonsingular. It will turn out below that this characterization based on the controllability Gramian can be used to evaluate the approximation error of clustered model reduction in terms of the $\mathcal{H}_2$-norm.

To devise a controllability characterization compatible with the $\mathcal{H}_\infty$-norm, we provide the following lemma that gives a particular realization of Σ, called the controller-Hessenberg form:

Lemma 8.6 *For any linear system Σ in (8.5), there exists a unitary matrix $H \in \mathbb{R}^{n\times n}$ such that $\mathfrak{A} := H^{\mathsf{T}}AH \in \mathbb{R}^{n\times n}$ and $\mathfrak{B} := H^{\mathsf{T}}B \in \mathbb{R}^n$ are in the form of*

$$\mathfrak{A} = \begin{bmatrix} \alpha_{1,1} & \alpha_{1,2} & \cdots & \cdots & \alpha_{1,n} \\ \alpha_{2,1} & \alpha_{2,2} & \alpha_{2,3} & \cdots & \alpha_{2,n} \\ 0 & \ddots & \ddots & \ddots & \vdots \\ \vdots & \ddots & \ddots & \ddots & \vdots \\ 0 & \cdots & 0 & \alpha_{n,n-1} & \alpha_{n,n} \end{bmatrix}, \quad \mathfrak{B} = \begin{bmatrix} \beta_1 \\ 0 \\ 0 \\ \vdots \\ 0 \end{bmatrix}. \tag{8.11}$$

Furthermore, the dimension of the controllable subspace of Σ is given by

$$\nu := \begin{cases} \min_{i \in \{1,\dots,n-1\}} \{i : \alpha_{i+1,i} = 0\}, & \text{if } \prod_{i=1}^{n-1} \alpha_{i+1,i} = 0, \\ n, & \text{otherwise.} \end{cases} \tag{8.12}$$

Note that the controller-Hessenberg form of Σ in Lemma 8.6 has the serially cascaded structure as shown in (8.11). From this particular structure, it follows that Σ is controllable if and only if $\alpha_{i+1,i} \neq 0$ for all $i \in \{1, \dots, n-1\}$. Controllability characterizations in the following lemma will be used to give a solution to Problem 8.5:

Lemma 8.7 *Let a stable linear system Σ in (8.5) be given. For the controllability Gramian M in (8.10), define $\Phi_{\mathcal{H}_2} \in \mathbb{R}^{n \times n}$ such that $M = \Phi_{\mathcal{H}_2} \Phi_{\mathcal{H}_2}^{\mathsf{T}}$. Furthermore, for $\mathfrak{A}$ and $\mathfrak{B}$ with H in Lemma 8.6, define*

$$\Phi_{\mathcal{H}_\infty} := H \mathrm{diag}(\gamma_1, \dots, \gamma_n) \in \mathbb{R}^{n \times n}, \quad \gamma_i := \left\| e_i^{\mathsf{T}} (s I_n - \mathfrak{A})^{-1} \mathfrak{B} \right\|_{\mathcal{H}_\infty}. \tag{8.13}$$

Then, Σ is controllable if and only if $\Phi_{\mathcal{H}_p}$ is nonsingular, where $p = 2$ or $p = \infty$.

8.3.3 Clustered Model Reduction Theory

8.3.3.1 Exact Clustered Model Reduction

In this subsection, we first consider the case where no approximation error is caused by the cluster aggregation. To do this, we introduce the following notion of the reducibility of clusters:

Definition 8.8 Let a linear system Σ in (8.5) be given. A cluster $\mathcal{I}_{[l]}$ is said to be *reducible* if there exist a scalar rational function $G_{[l]}^{\star}$ and a vector $\eta_{[l]} \in \mathbb{R}^{|\mathcal{I}_{[l]}|}$ such that

$$e_{\mathcal{I}_{[l]}}^{\mathsf{T}} G(s) = \eta_{[l]} G_{[l]}^{\star}(s), \tag{8.14}$$

where G is defined as in (8.9).

This definition of cluster reducibility represents that the states corresponding to $\mathcal{I}_{[l]}$ have the same trajectories for all input signals. The following theorem shows that the cluster reducibility can be characterized by a kind of local singularity of $\Phi_{\mathcal{H}_p}$ defined in Lemma 8.7:

Theorem 8.9 *Let a stable linear system Σ in (8.5) be given. With the same notation as that in Lemma 8.7, a cluster $\mathcal{I}_{[l]}$ is reducible if and only if there exist $\phi^{\star}_{[l]} \in \mathbb{R}^{1\times n}$ and $\eta_{[l]} \in \mathbb{R}^{|\mathcal{I}_{[l]}|}$ such that*

$$e^{\mathsf{T}}_{\mathcal{I}_{[l]}} \Phi_{\mathcal{H}_p} = \eta_{[l]} \phi^{\star}_{[l]}, \tag{8.15}$$

where $p = 2$ or $p = \infty$. In addition, if $\mathcal{I}_{[l]}$ is reducible, then $\eta_{[l]}$ coincides with a multiple of $-e^{\mathsf{T}}_{\mathcal{I}_{[l]}} A^{-1} B$. Moreover, if all clusters are reducible, then the aggregated model $\hat{\Sigma}$ in (8.7) given by $p_{[l]} = \|\eta_{[l]}\|^{-1} \eta_{[l]}$ is stable and satisfies

$$G(s) = \hat{G}(s), \tag{8.16}$$

where G and $\hat{G}$ are defined as in (8.9).

Theorem 8.9 shows that the cluster reducibility is characterized by linear dependence among the row vectors of $\Phi_{\mathcal{H}_p}$. However, the cluster reducibility is generally restrictive for the reduction of dimensions. This is because it represents a kind of structured uncontrollability representing that the controllable subspace of $e^{\mathsf{T}}_{\mathcal{I}_{[l]}} x$ is one-dimensional.

8.3.3.2 Approximation Error Evaluation for Clustered Model Reduction

In what follows, aiming at more significant dimension reduction, we consider the case where a degree of approximation errors is caused by cluster aggregation. In this situation, even if the original system Σ in (8.5) is stable, the aggregated model $\hat{\Sigma}$ in (8.7) is not necessarily stable. In clustered model reduction, the stability preservation is to be guaranteed on the basis of the following two facts:

Lemma 8.10 *Let a stable linear system Σ in (8.5) be given. If*

$$A + A^{\mathsf{T}} \prec 0, \tag{8.17}$$

then the aggregated model $\hat{\Sigma}$ in (8.7) is stable for any cluster set $\{\mathcal{I}_{[l]}\}_{l\in\mathbb{L}}$.

Lemma 8.11 *Let $A \in \mathbb{R}^{n\times n}$ be such that*

$$DA + A^{\mathsf{T}} D \prec 0 \tag{8.18}$$

for a diagonal matrix $D \succ 0$. Then

$$\tilde{A} + \tilde{A}^{\mathsf{T}} \prec 0, \quad \tilde{A} := D^{\frac{1}{2}} A D^{-\frac{1}{2}}, \tag{8.19}$$

where $D^{\frac{1}{2}} \succ 0$ is a diagonal matrix whose diagonal elements are the square roots of those of D.

Lemma 8.10 shows that, if A is negative definite as in (8.17), the stability of aggregated models can be guaranteed for any choice of cluster sets. Furthermore, Lemma 8.11 shows that any stable system having a diagonal Lyapunov function as in (8.18) is diagonally similar to a system having a negative definite system matrix as in (8.19). Note that a similarity transformation (coordinate transformation) by a diagonal matrix does not break the network structure, i.e., the Boolean structure, of the original system. Thus, by combining Lemmas 8.10 and 8.11, we can theoretically guarantee the stability preservation in clustered model reduction for the class of systems having diagonal Lyapunov functions.

In the following, focusing especially on this class of stable network systems, we analyze the approximation error in clustered model reduction. To this end, we introduce a weaker notion of cluster reducibility as follows:

Definition 8.12 Let a stable linear system Σ in (8.5) be given. With the same notation as that in Lemma 8.7, a cluster $\mathcal{I}_{[l]}$ is said to be *θ-reducible* with respect to the $\mathcal{H}_p$-norm if there exists $\phi^{\star}_{[l]} \in \mathbb{R}^{1\times n}$ such that

$$\begin{cases} \left\| e^{\mathsf{T}}_{\mathcal{I}_{[l]}} \Phi_{\mathcal{H}_2} - \eta_{[l]} \phi^{\star}_{[l]} \right\|_{\mathrm{F}} \le \sqrt{|\mathcal{I}_{[l]}|}\, \theta, & p = 2, \\ \left\| e^{\mathsf{T}}_{\mathcal{I}_{[l]}} \Phi_{\mathcal{H}_\infty} - \eta_{[l]} \phi^{\star}_{[l]} \right\|_{l_\infty} \le \theta, & p = \infty \end{cases} \tag{8.20}$$

for $\eta_{[l]} = -e^{\mathsf{T}}_{\mathcal{I}_{[l]}} A^{-1} B$.

In Definition 8.12, the constant $\theta \ge 0$ represents the degree of cluster reducibility. In the case of $p = 2$, the scaling by $\sqrt{|\mathcal{I}_{[l]}|}$ is introduced for technical reasons. It can be easily verified by Theorem 8.9 that the θ-reducibility with $\theta = 0$ is equivalent to the cluster reducibility in Definition 8.8. In the following theorem, on the basis of the θ-reducibility, we perform approximation error evaluation in clustered model reduction:

Theorem 8.13 *Let a stable linear system Σ in (8.5) be given and assume that (8.17) holds. Furthermore, let $\gamma > 0$ be such that*

$$A + A^{\mathsf{T}} + \gamma^{-1}(AA^{\mathsf{T}} + I_n) \prec 0, \tag{8.21}$$

and either $p = 2$ or $p = \infty$. If all clusters are θ-reducible with respect to the $\mathcal{H}_p$-norm, then the aggregated model $\hat{\Sigma}$ in (8.7) given by $p_{[l]} = \|\eta_{[l]}\|^{-1}\eta_{[l]}$ is stable and satisfies

$$G(0) = \hat{G}(0), \quad \|G(s) - \hat{G}(s)\|_{\mathcal{H}_p} \leq \gamma \sqrt{\sum_{l=1}^{L} |\mathcal{I}_{[l]}|(|\mathcal{I}_{[l]}| - 1)}\, \theta, \tag{8.22}$$

where G and $\hat{G}$ are defined as in (8.9).

Theorem 8.13 shows a linear relation between the approximation error caused by cluster aggregation and the parameter θ expressing the degree of cluster reducibility. Thus, we can use θ as a criterion to regulate the approximation error of the resultant aggregated model. In this sense, Theorem 8.13 gives a strategy for reasonable cluster construction.

On the basis of the premise that $\theta \geq 0$ is given and $\Phi_{\mathcal{H}_p}$ is calculated in advance, we propose an algorithm to construct a set of θ-reducible clusters. Assuming that a set of θ-reducible clusters $\mathcal{I}_{[1]}, \ldots, \mathcal{I}_{[l-1]}$ are already formed, we consider determining a new cluster $\mathcal{I}_{[l]}$. Let

$$\mathcal{N} := \{1, \ldots, n\} \backslash \bigcup_{i=1}^{l-1} \mathcal{I}_{[i]}.$$

When constructing $\mathcal{I}_{[l]}$, we first select an index $j \in \mathcal{N}$. Then, letting either $p = 2$ or $p = \infty$, we find all indices $i \in \mathcal{N}$ such that

$$\left\| \phi_i - \eta_i \eta_j^{-1} \phi_j \right\|_{l_p} \leq \theta, \tag{8.23}$$

where $\phi_i \in \mathbb{R}^{1 \times n}$ denotes the ith row vector of $\Phi_{\mathcal{H}_p}$ and $\eta_i \in \mathbb{R}$ denotes the ith entry of $\eta = -A^{-1}B$. We notice that (8.23) is a sufficient condition for (8.20) with $\phi_{[l]}^{\star} = \eta_j^{-1}\phi_j$; thereby verifying that the new cluster $\mathcal{I}_{[l]}$ is θ-reducible.

8.3.3.3 Generalization to Second-Order Networks

As giving an advanced result on clustered model reduction, we generalize the results in Sect. 8.3.3.2 to those in the case of interconnected second-order systems. More specifically, we deal with a class of interconnected second-order systems denoted by

$$\Sigma : \ddot{x} + D\dot{x} + Kx = Fu, \tag{8.24}$$

where $D = D^{\mathsf{T}} \in \mathbb{R}^{n \times n}$ and $K = K^{\mathsf{T}} \in \mathbb{R}^{n \times n}$ are assumed to be positive definite, and $F \in \mathbb{R}^n$. The network structure of Σ can be represented as the Boolean structure of K. Using the aggregation matrix P in (8.6), we define the aggregated model of Σ in (8.24) by

$$\hat{\Sigma} : \begin{cases} \ddot{\xi} + P^{\mathsf{T}} D P \dot{\xi} + P^{\mathsf{T}} K P \xi = P^{\mathsf{T}} F u, \\ \hat{x} = P\xi. \end{cases} \tag{8.25}$$

Note that the aggregated model $\hat{\Sigma}$ is stable for any P because $P^{\mathsf{T}}DP$ and $P^{\mathsf{T}}KP$ are also positive definite. In this notation, similarly to Problem 8.5, we address the following clustered model reduction problem for interconnected second-order systems:

Probelm 8.14 Let a stable second-order system Σ in (8.24) be given. Given a constant $\varepsilon \geq 0$, find a stable aggregated model $\hat{\Sigma}$ in (8.25) such that (8.8) for

$$G(s) := (s^2 I_n + sD + K)^{-1}F, \quad \hat{G}(s) := P(s^2 I_L + sP^{\mathsf{T}}DP + P^{\mathsf{T}}KP)^{-1}P^{\mathsf{T}}F \tag{8.26}$$

denoting the transfer matrices of Σ and $\hat{\Sigma}$, respectively.

To give a solution to this problem, let us represent Σ in (8.24) by the first order form as

$$\Sigma : \begin{cases} \dot{X} = AX + Bu, \\ x = CX, \end{cases} \tag{8.27}$$

where $X := [x^{\mathsf{T}}, \dot{x}^{\mathsf{T}}]^{\mathsf{T}} \in \mathbb{R}^{2n}$, and

$$A := \begin{bmatrix} 0 & I_n \\ -K & -D \end{bmatrix} \in \mathbb{R}^{2n \times 2n}, \quad B := \begin{bmatrix} 0 \\ F \end{bmatrix} \in \mathbb{R}^{2n}, \quad C := \begin{bmatrix} I_n & 0 \end{bmatrix} \in \mathbb{R}^{n \times 2n}.$$

On the basis of this representation, as a generalization of Definition 8.12, we define the notion of θ-reducibility for second-order systems as follows:

Definition 8.15 Let a stable second-order system Σ in (8.24) be given. For $p = 2$ or $p = \infty$, define $\Phi_{\mathcal{H}_p} \in \mathbb{R}^{2n \times 2n}$ similarly to those in Lemma 8.7. A cluster $\mathcal{I}_{[l]}$ is said to be θ-*reducible* with respect to the $\mathcal{H}_p$-norm if there exist $\phi^{\star}_{[l]} \in \mathbb{R}^{1 \times 2n}$ and $\psi^{\star}_{[l]} \in \mathbb{R}^{1 \times 2n}$ such that

$$\begin{cases} \max\left\{ \left\| e^{\mathsf{T}}_{\mathcal{I}_{[l]}} \Phi^{(1)}_{\mathcal{H}_2} - \eta_{[l]} \phi^{\star}_{[l]} \right\|_{\mathrm{F}}, \left\| e^{\mathsf{T}}_{\mathcal{I}_{[l]}} \Phi^{(2)}_{\mathcal{H}_2} - \eta_{[l]} \psi^{\star}_{[l]} \right\|_{\mathrm{F}} \right\} \leq \sqrt{|\mathcal{I}_{[l]}|}\theta, & p = 2 \\ \max\left\{ \left\| e^{\mathsf{T}}_{\mathcal{I}_{[l]}} \Phi^{(1)}_{\mathcal{H}_\infty} - \eta_{[l]} \phi^{\star}_{[l]} \right\|_{l_\infty}, \left\| e^{\mathsf{T}}_{\mathcal{I}_{[l]}} \Phi^{(2)}_{\mathcal{H}_\infty} - \eta_{[l]} \psi^{\star}_{[l]} \right\|_{l_\infty} \right\} \leq \theta, & p = \infty \end{cases}$$

for $\eta_{[l]} = -e^{\mathsf{T}}_{\mathcal{I}_{[l]}} K^{-1} F$, where $\Phi^{(1)}_{\mathcal{H}_p} \in \mathbb{R}^{n \times 2n}$ and $\Phi^{(2)}_{\mathcal{H}_p} \in \mathbb{R}^{n \times 2n}$ denote the upper and lower half components of $\Phi_{\mathcal{H}_p}$, respectively.

In Definition 8.15, $\Phi^{(1)}_{\mathcal{H}_p}$ and $\Phi^{(2)}_{\mathcal{H}_p}$ correspond to the controllability Gramians with respect to the position and velocity of states. Then, Theorem 8.13 can be generalized to second-order systems as follows:

Theorem 8.16 *Let a stable second-order system Σ in (8.24) be given. If all clusters are θ-reducible for the $\mathcal{H}_p$-norm, then the aggregated model $\hat{\Sigma}$ in (8.25) given by $p_{[l]} = \|\eta_{[l]}\|^{-1}\eta_{[l]}$ is stable and satisfies (8.22) for (8.26) with*

$$\gamma := \sqrt{2} \left\| P^{\mathsf{T}} \left(s^2 I_L + s P^{\mathsf{T}} D P + P^{\mathsf{T}} K P \right)^{-1} \left[P^{\mathsf{T}} K \;\; P^{\mathsf{T}} D \right] - \left[I_n \;\; 0 \right] \right\|_{\mathcal{H}_\infty}.$$

Similarly to Theorem 8.13, we can derive an approximation error bound for clustered model reduction of second-order systems. In Theorem 8.16, even though the value of γ is not computable a priori, i.e., before determining the aggregation matrix P, the parameter θ can be used to regulate the approximation error of the resultant aggregated model.

8.3.4 Application to Average State Observer

Based on the results of Sect. 8.3.3, this section presents a design method of reduced-order observers for average state estimation of large-scale network systems, which we called here a projective state observer. This has been developed by the authors and their colleagues [18]. It is remarked that the physical meaning of the average state variable of the original systems can be preserved in the obtained reduced-order observer by using a block-diagonal structured projection matrix.

Consider also a stable linear system Σ of (8.5) as a large-scale network system, where the measurement output $y \in \mathbb{R}^{m_y}$ is given by $y = Cx$, and $u \in \mathbb{R}^{m_u}$ (i.e., $B \in \mathbb{R}^{n \times m_u}$). Motivated by a reduced-order model of Σ given by $\hat{\Sigma}$ of (8.7), we consider the following observer, called here a projective state observer:

$$O : \begin{cases} \dot{\hat{x}} = P^{\mathsf{T}} A P \hat{x} + P^{\mathsf{T}} B u + H(y - C P \hat{x}) \\ z = \hat{x}. \end{cases} \tag{8.28}$$

Then the projective state observer problem is to find P and H such that the estimation error $Px - \hat{x} (=: e)$ is within the specified precision. The dynamics of the error system is given by

$$\begin{bmatrix} \dot{e} \\ \dot{x} \end{bmatrix} = \begin{bmatrix} P^{\mathsf{T}} A P - H C P & (P^{\mathsf{T}} A - H C)(I_n - P P^{\mathsf{T}}) \\ 0 & A \end{bmatrix} \begin{bmatrix} e \\ x \end{bmatrix} + \begin{bmatrix} 0 \\ B \end{bmatrix} u. \tag{8.29}$$

Thus the estimation error depends on the external input u and the initial state x_0 as well as the initial estimation error e_0, which is denoted by $e(t) = e(t; e_0, x_0, u)$. Since the dynamics of the error system is linear, we can independently consider $e(t; e_0, 0, 0)$, $e(t; 0, x_0, 0)$, and $e(t; 0, 0, u)$. For simplicity of explanation, we only consider here the case of $e(t; 0, 0, u)$. See [18] for further details. In this case, owing to the cascaded structure of the error dynamics, the estimation error with for an impulse input u is characterized by

$$\|e(t; 0, 0, u)\|_{\mathcal{L}_2} \leq \|\Gamma(s)\|_{\mathcal{H}_\infty} \|(I - P P^{\mathsf{T}})(sI - A)^{-1} B\|_{\mathcal{H}_2} \tag{8.30}$$

where $\Gamma(s)$ is given by a certain system that includes P and H. Thus for a given $\varepsilon > 0$, we first consider to determine P such that $\|(I - PP^{\mathsf{T}})(sI - A)^{-1}B\|_{\mathcal{H}_2} \leq \rho$, which can be solved by the clustered model reduction, and then for a given P, determine H in solving a kind of $\mathcal{H}_\infty$ state feedback control problem with $\|\Gamma(s)\|_{\mathcal{H}_\infty} \leq \varepsilon/\rho$.

Figure 8.3 shows a network of 54 power generators based on the IEEE 118 bus system, where each generator has two-dimensional system, and its reduced-order network model of 9 dimension obtained according to the above model reduction procedure. We also show the average behavior of the state variable (i.e., angular velocity ω_i) of the original system with solid lines and the corresponding state behavior of the projective state observer with dotted lines in Fig. 8.4. We can see that both trajectories are almost the same and the proposed observer works effectively.

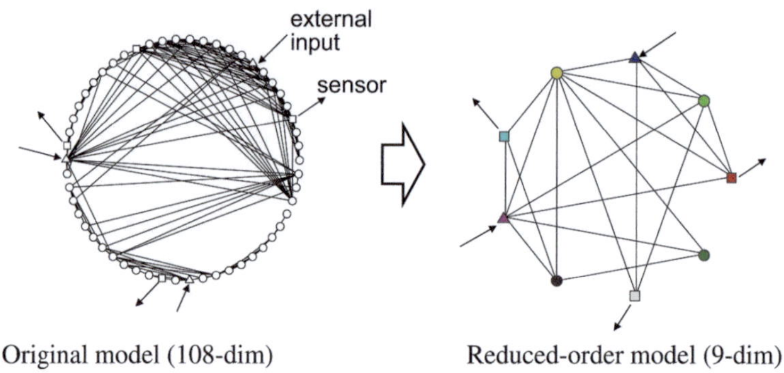

Fig. 8.3 Power network based on IEEE 118 bus system

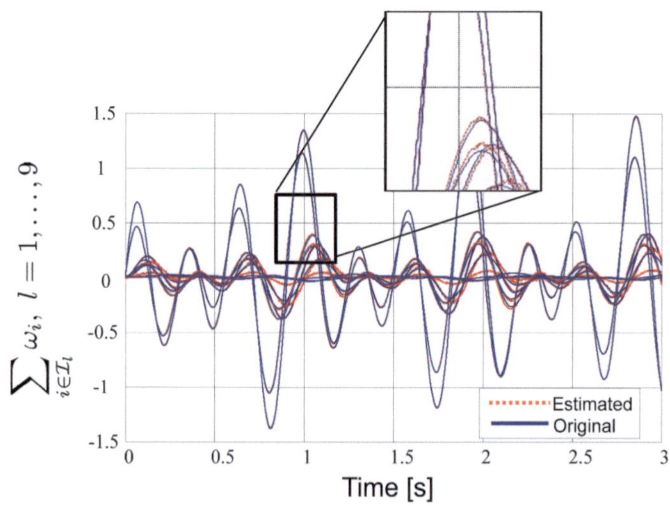

Fig. 8.4 Simulation result of a projective state observer

8.4 Conclusion

This article discussed two topics, i.e., one the revisit of the well-posedness issue of hybrid systems, and the other is the clustered model reduction of large-scale network systems. The former treats how to characterize an essential structure of mathematical models including switching phenomena in a proper way, while the latter deals with how to extract a strongly controllable network system from a large-scale network system. In this sense, their research directions are very similar, and the beginning of them was the first author's stay at Twente University under Arjan's host.

Acknowledgments The authors would like to thank Dr. T. Sadamoto and Mr. F. Watanabe for presenting Figs. 8.3 and 8.4. This work is supported by JSPS KAKENHI Grand 26249062.

References

1. A.J. van der Schaft, H. Schumacher, An Introduction to Hybrid Dynamical Systems. Lecture Notes in Control and Information Sciences, vol. 251 (Springer, London, 2000)
2. J. Imura, A.J. van der Schaft, Well-posedness of a class of piecewise linear systems with no jumps, in *Hybrid systems: Computation and Control*, Lecture Notes in Computer Sciences 1569, ed. by F.W. Vaandrager, J.H. van Schuppen (Springer, London, 1999), pp. 123–136
3. J. Imura, A.J. van der Schaft, Characterization of well-posedness of piecewise linear systems. IEEE Trans. Autom. Control **45**(9), 1600–1619 (2000)
4. J. Imura, Classification and stabilizability analysis of bimodal piecewise affine systems. Int. J. Robust Nonlinear Control **12**(10), 897–926 (2002)
5. J. Imura, Well-posedness analysis of switch-driven hybrid systems. IEEE Trans. Autom. Control **48**(11), 1926–1935 (2003)
6. S. Azuma, J. Imura, Polynomial-time probabilistic controllability analysis of discrete-time piecewise affine systems. IEEE Trans. Autom. Control **52**(11), 2029–2046 (2007)
7. S. Azuma, E. Yangisawa, J. Imura, Controllability analysis of biosystems based on piecewise affine systems approach. IEEE Trans. Autom. Control and IEEE Trans. Circuits Syst. I, **53**(1), 139–152 (2008)
8. Y. Tazaki, J. Imura, Discrete abstractions of nonlinear systems based on error propagation analysis. IEEE Trans. Autom. Control **57**(3), 550–564 (2012)
9. T. Ishizaki, K. Kashima, J. Imura, K. Aihara, Reaction-Diffusion Clustering of Single-Input Dynamical Networks, in *Proceedings of the 50th IEEE Conference on Decision and Control*, pp. 7837–7842 (2011)
10. T. Ishizaki, K. Kashima, A. Girard, J. Imura, L. Chen, K. Aihara, Clustering-based $\mathcal{H}_2$-state aggregation of positive networks and its application to reduction of chemical master equations, in Proceedings of the 51st IEEE Conference on Decision and Control, 4175–4180 (2012)
11. T. Ishizaki, K. Kashima, J. Imura, K. Aihara, Model reduction and clusterization of large-scale bidirectional networks. IEEE Trans. Autom. Control **59**(1), 48–63 (2014)
12. T. Ishizaki, J. Imura, Clustered model reduction of interconnected second-order systems. Nonlinear Theory Appl. Inst. Electron. Inf. Commun. Eng. **6**(1), 26–37 (2015)
13. A.J. van der Schaft, B.M. Maschke, The Hamiltonian formulation of energy conserving physical systems with external ports. Archiv für Elektronik und Übertragungstechnik **49**, 362–371 (1995)
14. A.J. van der Schaft, B.M. Maschke, Port-Hamiltonian systems on graphs. SIAM J. Control Optim. **51**(2), 906–937 (2013)

15. A.J. van der Schaft, On Model Reduction of Physical Network Systems, in *Proceedings of the 21st International Symposium on Mathematical Theory of Networks and Systems*, pp. 1419–1425 (2014)
16. N. Monshizadeh, A.J. van der Schaft, Structure-Preserving Model Reduction of Physical Network Systems by Clustering, in *Proceedings of the 53rd IEEE Conference on Decision and Control*, pp. 4434–4440 (2014)
17. N. Monshizadeh, M.K. Camlibel, H.L. Trentelman, Projection based model reduction of multi-agent systems using graph partitions. IEEE Trans. Control Netw. Syst. **1**(2), 145–154 (2014)
18. T. Sadamoto, T. Ishizaki, J. Imura, Projective state observers for large-scale linear systems, in *Proceedings of the 13th European Control Conference*, pp. 2969–2974 (2014)
19. A.J. van der Shaft, H. Schumacher, The complementary-slackness class of hybrid systems. Math. Control Signals Syst. **9**, 266–301 (1996)
20. A.C. Antoulas, *Approximation of Large-scale Dynamical Systems* (Society for Industrial and Applied Mathematics, Philadelphia, 2005)

Chapter 9
Control of HVDC Transmission Systems: From Theory to Practice and Back

Daniele Zonetti and Romeo Ortega

Abstract The problem of modeling and control of multi-terminal high-voltage direct-current transmission systems is addressed in this chapter, which contains three main contributions. First, to propose a unified, physically motivated, modeling framework—based on port-Hamiltonian systems representations—of the various network topologies used in this application. Second, to prove that the system can be globally asymptotically stabilized with a decentralized PI control that exploits its passivity properties. Close connections between the proposed PI and the popular Akagi's PQ instantaneous power method are also established. Third, to reveal the transient performance limitations of the proposed controller that, interestingly, is shown to be intrinsic to PI passivity-based control. The performances of the controller are verified via simulations on a three-terminal benchmark example.

9.1 Introduction

We have witnessed in the last few years an ever widespread utilization of renewable energy utilities, mainly based on wind and solar power [6, 17]. Because of its intermittent nature the integration of this generating units to the existing alternating-current (AC) distribution network poses a challenging problem [5, 23]. For this, and other reasons related to reduced losses and problems with reactive power and voltage stability in AC systems, the option of high-voltage direct-current (HVDC)

This work is dedicated to Arjan van der Schaft, excellent teacher and dearest friend.

D. Zonetti (✉) · R. Ortega
Laboratoire des Signaux et Systèmes, 3, rue Joliot Curie,
91192 Gif-sur-Yvette, France
e-mail: daniele.zonetti@lss.supelec.fr

R. Ortega
e-mail: romeo.ortega@lss.supelec.fr

M.K. Camlibel et al. (eds.), *Mathematical Control Theory I*,
Lecture Notes in Control and Information Sciences 461,
DOI 10.1007/978-3-319-20988-3_9

transmission systems is gaining wide popularity, see [6, 19, 21] for additional motivations and details.

The main components of an HVDC system are AC to DC power converters, transmission lines and voltage bus capacitors. The power converters connect the AC sources—that are associated to renewable generating units or to AC grids—to an HVDC grid through voltage bus capacitors. Two notable features distinguish HVDC systems from standard AC ones: the absence of loads and the central role played by the power converters, whose dynamics is highly *nonlinear*.

For its correct operation HVDC systems—like all electrical power systems—must satisfy a large set of different regulation objectives that are, typically, associated to the multiple time-scale behavior of the system. One way to deal with this issue, that prevails in practice, is the use of hierarchical architectures. These are nested control loops, at different time scales, each one providing references for an inner controller [20, 35]. In this chapter we focus on the "innermost" control loop for HVDC transmission systems, that is, the control at the power converter level—in the sequel we will refer to this level as *inner-loop* control. The objective of the inner-loop control is to asymptotically drive the HVDC system towards a desired steady-state regime determined by the user. Regulation should be achieved selecting a suitable switching policy for the converters. A major practical constraint is that the control should be *decentralized*. That is, the controller of each power converter has only available for measurement its corresponding coordinates, with no exchange of information between them.

An essential step in the development of suitable analysis and synthesis tools for physical systems is the proper selection of the dynamic model describing its behaviour. In this respect, port-Hamiltonian models are gaining widespread popularity in many different engineering applications. This is particularly true for electrical systems, where they are now the dominating description. The development of these models is essentially due to the work of Arjan van der Schaft—mainly in collaboration with Bernard Maschke—and credit should be given to them for this fundamental contribution that has helped to bridge the gap between theory and applications.

The chapter is structured as follows. In Sect. 9.2, the mathematical model of the system is established. Then, to determine the achievable behaviors, a study of the assignable equilibria is necessary. This analysis is done in Sect. 9.3. Main results are next developed in Sect. 9.4, with the design of the decentralized passivity-based PI controller. Slow transients exhibited in simulations motivate the subsequent performance analysis, that is carried-out in Sect. 9.5.

Notation All vectors are column vectors. Given positive integers n, m we use $\underline{0}_n \in \mathbb{R}^n$ to denote the vector of all zeros, $\mathbb{1}_n \in \mathbb{R}^n$ the vector with all ones, $\mathbb{I}_n$ the $n \times n$ identity matrix, $\underline{0}_{n\times m}$ the $n \times m$ column matrix of all zeros. $x := \mathrm{col}(x_1, \ldots, x_n) \in \mathbb{R}^n$ denotes a vector with entries $x_i \in \mathbb{R}$, when clear from the context we simply write $x := \mathrm{col}(x_i)$. $\mathrm{diag}\{a_i\}$ is a diagonal matrix with entries $a_i \in \mathbb{R}$ and $\mathrm{bdiag}\{A_i\}$ denotes a block diagonal matrix with entries the matrices A_i. For a function $f : \mathbb{R}^n \to \mathbb{R}$, ∇f denotes the transpose of its gradient. The subindex i, preceded by a comma when necessary, denotes elements corresponding to the ith subsystem.

9.2 Energy-based Port-Hamiltonian Modeling

In [10] it was shown that electrical power systems can be represented by a directed graph[1] where the relevant electrical components correspond to edges and the buses correspond to nodes. Moreover, to underscore the physical structure of the components, they are modeled as pH systems. In this section we apply the same procedure to describe the dynamics of HVDC transmission systems.

9.2.1 Assumptions

As indicated in the Introduction, in HVDC transmission systems no loads exist and the relevant components are: VSRs, RL transmission lines and voltage bus capacitors. Throughout the chapter we make the following assumptions, which are widely accepted in practice.

(A1) Balanced operation of the three phase line voltages.
(A2) Synchronized operation of the VSRs.[2]
(A3) Ideal four quadrant operation of the VSRs.

Assumptions A1 and A2 considerably simplify the modeling and control problems, as they allow the description of the three-phase dynamics of the VSRs in suitably oriented $dq0$ reference frames, where the value of the *zero*-component is always zero, thus reducing the three AC quantities to two DC quantities. This allows us to express the regulation objective as a standard *equilibrium stabilization* problem of the nonlinear dynamical system describing the behavior of the HVDC system. Assumption A3 directly follows by assuming HVDC transmission systems based on VSRs instead of current source rectifiers, which is an alternative converter topology used in HVDC systems. As a matter of fact, since the VSRs do not depend on line-commutations, all the four quadrants of the operating plane are possible, hence Assumption A3 is automatically satisfied for the system under consideration [1].

[1]A directed graph is an ordered 3-tuple, $\mathcal{G} = \{\mathcal{V}, \mathcal{E}, \Pi\}$, consisting of a finite set of nodes $\mathcal{V}$, a finite set of directed edges $\mathcal{E}$ and a mapping Π from $\mathcal{E}$ to the set of ordered pairs of $\mathcal{V}$, where no self-loops are allowed.

[2]Synchronized operation of the VSRs is usually achieved via robust phase-locked-loop detection of the latching frequencies [35].

9.2.2 Network Topologies: A Graph Description

We can mainly distinguish two kinds of topologies used in HVDC transmission systems: *radial* and *meshed* topology [4, 11, 13], which are illustrated in Fig. 9.1. The radial topology is widely used for systems in which a certain number of off-shore stations feeds on-shore stations with no connection between them. This is the case for example of on-shore stations situated on opposite seacoasts while the off-shore stations are placed in their middle [4, 21]. However, in a more general setting we have to consider the situation in which the stations are directly connected with lines, that corresponds to a meshed topology. In the interest of brevity, we present here a systematic way to build global pH models only for the meshed topology. For a radial topology, analogous results can be obtained, for which the interested reader is referred to [36].

In order to give a formal representation of a topology we adopt the following definitions. We call a bus a *VSR-bus* if a VSR is connected to it and we call a bus a *capacitor-bus* when only a capacitor is connected to it. Furthermore, we call a bus a *reference-bus* when all the voltages of the buses in the network are measured with respect to it. As the reference-bus is assumed to be at ground potential, it is also denoted as *ground*. A general topology is then described by the incidence matrix $\mathcal{M}$ associated to the graph, where the nodes represent the ground, the VSR and the capacitor-buses; the edges represent the VSRs, the lines and the single capacitors that are interconnected to the ground or to the voltage buses.

In a meshed topology each VSR is connected to the ground and to a VSR-bus, while the lines directly connect VSR-buses, according to a determined meshed structure. The number n of VSRs is the same of voltage buses, ground excluded, and is lower or equal to the number ℓ of lines. Formally this can be represented by a graph $\mathcal{G} := \{\mathcal{V}, \mathcal{E}, \Pi\}$ constituted by: $n+1$ ordered nodes, where n nodes are associated to the VSR-buses and one node to the ground; $n+\ell$ ordered edges, where n edges are associated to the VSRs and ℓ edges to the lines. The incidence matrix then, following the mentioned order, takes the form

$$\mathcal{M} = \begin{bmatrix} \mathbb{I}_n & M \\ -\mathbb{1}_n^\top & \underline{0}_\ell^\top \end{bmatrix} \in \mathbb{R}^{(n+1)\times(n+\ell)}, \tag{9.1}$$

where M is the incidence matrix of the subgraph obtained eliminating the VSRs edges and the ground node.

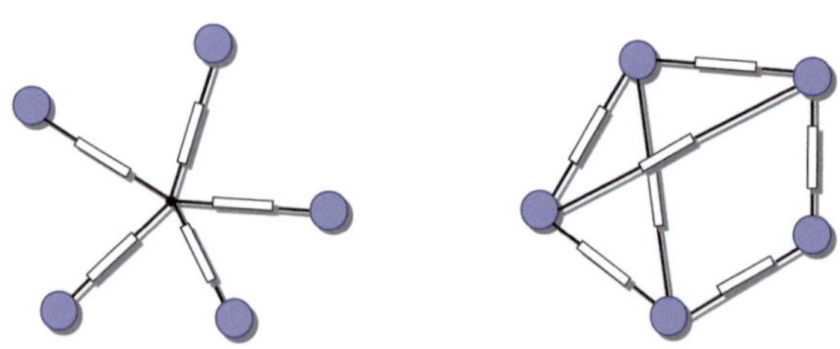

Fig. 9.1 Nodal representation of HVDC transmission systems with radial and meshed topologies

Remark 9.1 In a meshed topology the only relevant components are the VSRs and the RL transmission lines. As a matter of fact, because a VSR is associated to each node, the voltage bus capacitors can be represented by an equivalent VSR output capacitor, that results to be the parallel interconnection of all capacitors attached to the node.

9.2.3 Port-Hamiltonian Models of the Elements

As explained above the edges of the graph $\mathcal{G}$ contain the electrical components of the HVDC system, namely n VSRs and ℓ RL transmission lines, while the nodes are the buses. In this section we derive a pH representation of these elements, which are then interconnected—through power preserving interconnections—via the graph. Besides its physically appealing nature, the choice of a pH model is motivated by the fact that—similarly to [15]—this structure is instrumental to derive the passivity property exploited in the controller design.

9.2.3.1 Voltage Source Rectifiers

In [9, 15, 36] the well-known average model of a single VSR shown in Fig. 9.2, expressed in dq-coordinates and written in (perturbed) pH form is given.

Similarly, a *set* of n VSRs can also be represented in pH form as

$$\begin{aligned} \dot{x}_R &= [\mathcal{J}_R(u) - \mathcal{R}_R]\nabla\mathcal{H}_R + E_1 V - E_3 i_R \\ v_R &= E_3^\top \nabla\mathcal{H}_R, \end{aligned} \tag{9.2}$$

where we use the following definitions.

- State space variables the collection of inductors fluxes $(\phi_{d,i}, \phi_{q,i})$ and capacitor charges $q_{c,i}$ of every VSR, that is, $x_R := \text{col}(\text{col}(\phi_{d,i}), \text{col}(\phi_{q,i}), \text{col}(q_{c,i})) \in \mathbb{R}^{3n}$.

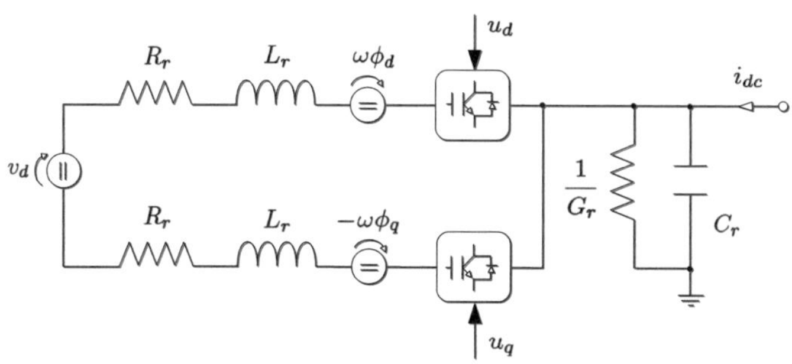

Fig. 9.2 Schematic diagram of the equivalent circuit of a VSR in dq frame

- Energy function $\mathcal{H}_R(x_R) := \frac{1}{2} x_R^\top Q_R x_R$, with[3]

$$Q_R := \text{bdiag}\{L_R^{-1}, L_R^{-1}, C_R^{-1}\}, \quad L_R := \text{diag}\{L_{r,i}\}, \quad C_R := \text{diag}\{C_{r,i}\}$$

  where $L_{r,i}$, $C_{r,i}$ are the inductance and capacitance of each VSR, respectively.
- Duty cycles $u := \text{col}(u_{Rd}, u_{Rq}) \in \mathbb{R}^{2n}$, where $u_{Rd} := \text{col}(u_{d,i})$ and $u_{Rq} := \text{col}(u_{q,i})$.
- External voltage sources $V := \text{col}(v_{d,i}) \in \mathbb{R}^n$, where $v_{d,i}$ is the d component of the AC sources. These voltages are assumed constant and positive.
- Port variables the out-going currents $i_R := \text{col}(i_{dc,i}) \in \mathbb{R}^n$ and the voltages $v_R := \text{col}(v_{dc,i}) \in \mathbb{R}^n$.
- Interconnection matrix

$$\mathcal{J}_R(u) := \sum_{i=1}^{n} (\mathcal{J}_{R0,i} L_{r,i} \omega_i + \mathcal{J}_{Rd,i} u_{d,i} + \mathcal{J}_{Rq,i} u_{q,i}) \tag{9.3}$$

  where ω_i are the AC sides frequencies and

$$\mathcal{J}_{R0,i} := \begin{cases} -1 \text{ in } (i,n+i) \\ 1 \text{ in } (n+i,i) \\ 0 \text{ elsewhere} \end{cases} \quad \mathcal{J}_{Rd,i} := \begin{cases} 1 \text{ in } (i,2n+i) \\ -1 \text{ in } (2n+i,i) \\ 0 \text{ elsewhere} \end{cases} \quad \mathcal{J}_{Rq,i} := \begin{cases} -1 \text{ in } (n+i,2n+i) \\ 1 \text{ in } (2n+i,n+i) \\ 0 \text{ elsewhere} \end{cases}$$

- Dissipation matrix $\mathcal{R}_R := \text{bdiag}\{R_R, R_R, G_R\}$, where $R_R := \text{diag}\{R_{r,i}\}$ and $G_R := \text{diag}\{G_{r,i}\}$, with $R_{r,i}$, $G_{r,i}$ the resistance and conductance of each VSR.
- Port matrices $E_1 : \left[\mathbb{I}_n \; 0 \; 0\right]^\top$, $E_3 := \left[0 \; 0 \; \mathbb{I}_n\right]^\top \in \mathbb{R}^{3n \times n}$.

Remark 9.2 Notice that, in view of the skew-symmetry of $\mathcal{J}_R(u)$, the VSRs satisfy the power balance equation

$$\underbrace{\dot{\mathcal{H}}_R}_{\text{stored power}} = \underbrace{-x_R^\top Q_R \mathcal{R}_R Q_R x_R}_{\text{dissipated power}} + \underbrace{x_R^\top Q_R E_1 V - x_R^\top Q_R E_3 i_R}_{\text{supplied power}} \tag{9.4}$$

9.2.3.2 Transmission Lines

A set of ℓ RL transmission lines can be represented by the pH system

$$\begin{aligned} \dot{x}_L &= -\mathcal{R}_L \nabla \mathcal{H}_L + v_L \\ i_L &= -\nabla \mathcal{H}_L, \end{aligned} \tag{9.5}$$

with the following definitions.

[3]Unless indicated otherwise all physical parameters of the system are positive constants.

- State space variables the collection of inductor fluxes $x_L := \mathrm{col}(\phi_{\ell,i}) \in \mathbb{R}^\ell$ of every line.
- Energy function

$$\mathcal{H}_L(x_L) := \frac{1}{2} x_L^\top Q_L x_L, \qquad Q_L := \mathrm{diag}\{\frac{1}{L_{\ell,i}}\},$$

  where $L_{\ell,i}$ is the inductance of the line.
- Port variables the voltages at the terminals $v_L := col(v_{L,i}) \in \mathbb{R}^\ell$ and the inductors currents $i_L := \mathrm{col}(i_{\ell,i}) \in \mathbb{R}^\ell$.
- Dissipation $\mathcal{R}_L = \mathrm{diag}\{R_{\ell,i}\}$, with $R_{\ell,i}$ the resistance of the line.

9.2.4 Overall Interconnected System

The interconnection laws can be obtained following the approach used in [33], where the Kirchhoff's current and voltage laws (KCL and KVL, respectively) are expressed in relation to the incidence matrix. For a *meshed* topology we have then

$$\begin{aligned} &[\mathrm{KCL}]\ \mathcal{M}\mathcal{I}_e = \underline{0}_{n+1} \\ &[\mathrm{KVL}]\ \mathcal{M}^\top \mathcal{V} = \mathcal{V}_e, \end{aligned} \tag{9.6}$$

where $\mathcal{I}_e := \mathrm{col}(i_R, i_L)$, $\mathcal{V}_e := \mathrm{col}(v_R, v_L)$ and $\mathcal{V} := \mathrm{col}(v_1, \ldots, v_n)$, v_0 are the edge currents, the edge voltages, the nodes potentials and the ground potential, respectively. The ground potential $v_0 = 0$ by definition. From (9.6) and (9.1) then follows

$$\begin{aligned} i_R + M i_L = \underline{0}_n, \quad -\mathbb{1}_n^\top i_R = 0, \\ v = v_R, \quad M^\top v = v_L. \end{aligned} \tag{9.7}$$

Recalling the expression for i_L from (9.2) and v_R from (9.5) we have

$$i_R = M \nabla \mathcal{H}_L, \qquad v_L = M^\top E_3^\top \nabla \mathcal{H}_R. \tag{9.8}$$

To obtain the overall pH representation it is sufficient to combine (9.2), (9.5) and (9.8), thus leading to:

$$\dot{x} = [\mathcal{J}(u) - \mathcal{R}]\nabla \mathcal{H} + EV, \tag{9.9}$$

with the following definitions.

- State space variables $x := \mathrm{col}(x_R, x_L) \in \mathbb{R}^{3n+\ell}$.
- Energy function $\mathcal{H}(x) := \mathcal{H}_R(x) + \mathcal{H}_L(x)$.
- Duty cycles (controls) $u := \mathrm{col}(u_{Rd}, u_{Rq}) \in \mathbb{R}^{2n}$.
- Interconnection matrix

$$\mathcal{J}(u) := \begin{bmatrix} J_R(u) & -E_3 M \\ M^\top E_3^\top & \underline{0}_{\ell\times\ell} \end{bmatrix}, \tag{9.10}$$

- Dissipation matrix

$$\mathcal{R} := \text{bdiag}\{\mathcal{R}_R, \mathcal{R}_L\} > 0. \tag{9.11}$$

- Input matrix $E := \left[E_1^\top \underline{0}_{\ell\times n}^\top\right]^\top$.

Remark 9.3 To simplify the notation in the pH representation we have selected a state representation of the system using energy variables, that is, inductor fluxes and capacitor charges, instead of the more commonly used co-energy variables, *i.e.*, inductor currents and capacitor voltages. See (9.22) and [24] for the derivation of the pH model in the latter coordinates. We recall that they are related by

$$i_d = \frac{\phi_d}{L}, \quad i_q = \frac{\phi_q}{L}, \quad v_C = \frac{q_C}{C}, \quad i_L = \frac{\phi_L}{L_\ell}. \tag{9.12}$$

Remark 9.4 For ease of presentation we have assumed that the state of the system lives in $\mathbb{R}^{3n+\ell}$. Due to physical and technological constraints it is actually only defined in a subset of $\mathbb{R}^{3n+\ell}$. In particular, the voltage of the DC link v_C is strictly bounded away from zero.

9.3 Assignable Equilibria

A first step towards the development of a control strategy for the system (9.9) is the definition of its achievable, steady-state behavior, which is determined by the assignable equilibria. That is, the (constant) vectors $x^\star \in \mathbb{R}^{3n+\ell}$ such that

$$[\mathcal{J}(u^\star) - \mathcal{R}]\nabla\mathcal{H}(x^\star) + EV = \underline{0}_{3n+\ell}$$

for some (constant) vector $u^\star \in \mathbb{R}^{2n}$. To identify this set we establish the following lemmata.

Lemma 9.5 *The equilibria of the transmission line coordinates are given by*

$$x_L^\star = (\mathcal{R}_L Q_L)^{-1} M^\top E_3^\top Q_R x_R^\star. \tag{9.13}$$

Proof Setting to zero the left-hand side of (9.5), calculated at $x_L^\star$, gives

$$\underline{0}_\ell = -\mathcal{R}_L Q_L x_L^\star + v_L^\star \quad \Rightarrow \quad x_L^\star = (\mathcal{R}_L Q_L)^{-1} v_L^\star.$$

Moreover, from (9.8) we have $v_L^\star = M^\top E_3^\top Q_R x_R^\star$, that replaced in the equation above completes the proof.

Lemma 9.6 *The equilibria of the VSRs coordinates are the solution of the n quadratic equations, $i = 1 \dots n$*

$$-\frac{R_i}{L_{r,i}^2}\left[(\phi_{d,i}^\star)^2 + (\phi_{q,i}^\star)^2\right] - \frac{G_i}{C_{r,i}^2}(q_{C,i}^\star)^2 + \frac{v_{d,i}}{L_{r,i}}\phi_{d,i}^\star - \frac{1}{C_{r,i}}q_{C,i}^\star i_{dc,i}^\star = 0, \quad (9.14)$$

with $\mathrm{col}(i_{dc,i}^\star) = M\mathcal{R}_L^{-1}M^\top \mathrm{col}(q_{C,i}^\star)$.

Proof In [27] it is shown that the set of admissible equilibria of a VSR is obtained by setting equal to zero the power balance of the VSR, that for n VSRs is equivalent to (9.14). It is now sufficient to recall definitions

$$\mathrm{col}(i_{dc,i}^\star) = i_R^\star, \qquad E_3^\top Q_R x_R^\star = \mathrm{col}(q_{C,i}^\star),$$

together with (9.8), (9.13) to complete the proof.

We are now ready to present the main result of the section, whose proof follows immediately from the lemmata above.

Proposition 9.7 *The set of assignable equilibria of the system* (9.9) *is given by*

$$\mathcal{E} := \{x^\star \in \mathbb{R}^{3n+\ell} \mid (9.13) \textit{ and } (9.14) \textit{ hold}\}. \quad (9.15)$$

From the derivations above it is clear that the equilibria of the network are univocally determined by the equilibria of the VSRs. Moreover, the latter should satisfy the quadratic equations (9.14), which are the well-known *power flow steady-state equations* (PFSSE) of the individual VSR subsystems. A question of interest is how to select from this set the equilibrium points that correspond to some *desired behavior*. In the latter definition there are many practical considerations to be taken into account, see for example [14, 29].

Remark 9.8 It is well-known that for affine systems of the form $\dot{x} = f(x) + g(x)u$ the assignable equilibrium set is given by

$$\{x^\star \in \mathbb{R}^n \mid g^\perp(x^\star)f(x^\star) = 0\}$$

where $g^\perp(x)$ is a full-rank left annihilator of $g(x)$. Moreover, given $x^\star$, the corresponding equilibrium control $u^\star$ is *univocally* determined by

$$u^\star = -\left[(g^\top g)^{-1}g^\top f\right](x^\star).$$

Since (9.9) is clearly of this form this relations hold true for the HVDC system under study. See [27] for additional details on this issue.

Remark 9.9 Differently from the single VSR case, the set of assignable equilibria does not coincide, but is strictly contained, in the set where the power of the system is balanced, that is

$$\mathcal{E} \subset \mathcal{P}, \qquad \mathcal{P} := \{x^\star \in \mathbb{R}^{3n+\ell} \mid \dot{\mathcal{H}}_R = 0\}.$$

This fact is clearly explained in [27], where it is proved that a necessary condition for $\mathcal{E} \equiv \mathcal{P}$, is the system to be of co-dimension one.

9.4 Main Result: Inner Loop Control

As indicated in the Introduction we are mainly interested in the inner-loop control of HVDC transmission systems, that is, the control at the VSR level. For this problem we present in this section a decentralized, globally asymptotically stabilizing, PI–PBC for the HVDC transmission system (9.9). The construction of the controller is inspired by our previous works on PI–PBC reported in [15, 18], which exploit the property of passivity of the *incremental model*. The interested reader is referred to these references for additional details.

To place the proposed PI–PBC in context, in the last part of this section we briefly review the most commonly used inner-loop controls for HVDC transmission systems and establish the connection with the widely popular Akagi's PQ method.

9.4.1 Passivity of the Incremental Model

Along the lines of Proposition 1 in [15], it is possible to establish passivity of the incremental model of the overall HVDC transmission system (9.9) with respect to a suitable defined output. As is well-known, global regulation of a passive output can be achieved with a simple PI controller. Regulation of the state to the desired equilibrium then follows provided a suitable detectability assumption is satisfied [11].

Proposition 9.10 *Consider the HVDC transmission system* (9.9). *Let* $x^\star \in \mathcal{E}$ *be the desired equilibrium with corresponding (univocally defined) control* $u^\star \in \mathbb{R}^{2n}$. *Define the error signals*

$$\tilde{x} = x - x^\star, \quad \tilde{u} = u - u^\star \tag{9.16}$$

and the output signal

$$y := \begin{bmatrix} \mathrm{col}(y_{d,i}) \\ \mathrm{col}(y_{q,i}) \end{bmatrix} \in \mathbb{R}^{2n}, \tag{9.17}$$

with

$$y_{d,i} := x_R^{*\top} Q_R \mathcal{J}_{Rd,i} Q_R x_R, \qquad y_{q,i} := x_R^{*\top} Q_R \mathcal{J}_{Rq,i} Q_R x_R.$$

The mapping $\tilde{u} \rightarrow y$ *is* passive. *More precisely, the system verifies the dissipation inequality*

$$\dot{\mathcal{H}}_d \leq y^\top \tilde{u}, \tag{9.18}$$

with storage function $\mathcal{H}_d(\tilde{x}) = \frac{1}{2}\tilde{x}^\top Q\tilde{x}$.

Proof The proof mimics the proof of Proposition 1 in [15]. We first notice that

$$\mathcal{J}(u)Qx = \mathcal{J}_0 Qx + g(x)u,$$

where we defined

$$\mathcal{J}_0 := \begin{bmatrix} \sum_{i=1}^n (\mathcal{J}_{R0,i} L_{r,i}\omega_i) & -E_3 M \\ M^\top E_3^\top & \underline{0}_{\ell\times\ell} \end{bmatrix}, \quad g(x) := \begin{bmatrix} g_{Rd}(x_R) & g_{Rq}(x_R) \\ \underline{0}_{\ell\times n} & \underline{0}_{\ell\times n} \end{bmatrix},$$

with

$$\begin{aligned} g_{Rd}(x_R) &:= \left[\mathcal{J}_{Rd,1} Q_R x_R \ \dots \ \mathcal{J}_{Rd,n} Q_R x_R\right] \\ g_{Rq}(x_R) &:= \left[\mathcal{J}_{Rq,1} Q_R x_R \ \dots \ \mathcal{J}_{Rq,n} Q_R x_R\right]. \end{aligned}$$

Hence, it is possible to write (9.9) in the alternative form

$$\begin{aligned} \dot{x} &= (\mathcal{J}_0 - \mathcal{R})Qx + EV + g(x)u \\ &= (\mathcal{J}_0 - \mathcal{R})Q(\tilde{x} + x^\star) + EV + g(x)(\tilde{u} + u^\star) \\ &= (\mathcal{J}_0 - \mathcal{R})Q\tilde{x} + g(x)\tilde{u} + g(\tilde{x})u^\star \end{aligned} \tag{9.19}$$

where we have used (9.16) to get the second equation and the fact that the assignable equilibria $x^\star$ and corresponding (constant) control $u^\star$ satisfy

$$(\mathcal{J}_0 - \mathcal{R})Qx^\star + EV + g(x^\star)u^\star = 0,$$

to obtain the third equation.

The derivative of $\mathcal{H}_d$ along the trajectories of the incremental model (9.19) yields

$$\dot{\mathcal{H}}_d = -\tilde{x}^\top Q\mathcal{R}Q\tilde{x} + \tilde{x}^\top Qg(x)\tilde{u} = -\tilde{x}^\top Q\mathcal{R}Q\tilde{x} + y^\top\tilde{u}$$

where skew-symmetry of $\mathcal{J}_0$, $\mathcal{J}_{Rd,i}$ and $\mathcal{J}_{Rq,i}$ is used in the first equation, and the fact that the output signal can be rewritten as

$$y = g^\top(x^\star)Qx = g^\top(x^\star)Q\tilde{x}$$

to obtain the second identity. The proof is completed recalling that the dissipation matrix verifies $\mathcal{R} > 0$ to get the bound (9.18).

9.4.2 PI Passivity-based Control

We are in position to present the first main result.

Proposition 9.11 *Consider the HVDC transmission system* (9.9), *with a desired steady-state* $x^\star \in \mathcal{E}$, *in closed-loop with the decentralized PI control*

$$u = -K_P y - K_I \zeta, \quad \dot{\zeta} = y, \tag{9.20}$$

with y *given in* (9.17) *and gain matrices*

$$K_P = \text{bdiag}\{k_{P,i}\} \in \mathbb{R}^{2n \times 2n}, \quad K_I = \text{bdiag}\{k_{I,i}\} \in \mathbb{R}^{2n \times 2n}, \tag{9.21}$$

with $k_{P,i}, k_{I,i} \in \mathbb{R}^{2\times 2}$ arbitrary *positive definite matrices. The equilibrium point* $(x^\star, K_I^{-1} u^\star)$ *is globally asymptotically stable (GAS).*

Proof Define the Lyapunov function candidate

$$W(\tilde{x}, \tilde{\zeta}) := \mathcal{H}_d(\tilde{x}) + \frac{1}{2}\tilde{\zeta}^\top K_I \tilde{\zeta},$$

where $\tilde{\zeta} := \zeta - K_I^{-1} u^\star$. The derivative of $W(x, \zeta)$ along the trajectories of the closed-loop system (9.19), (9.20) is given by

$$\begin{aligned} \dot{W} &= -\tilde{x}^\top \mathcal{Q}\mathcal{R}\mathcal{Q}\tilde{x} + y^\top \tilde{u} + \tilde{\zeta}^\top K_I y \\ &= -\tilde{x}^\top \mathcal{Q}\mathcal{R}\mathcal{Q}\tilde{x} + y^\top \tilde{u} - y^\top (K_P y + \tilde{u}) \\ &= -\tilde{x}^\top \mathcal{Q}\mathcal{R}\mathcal{Q}\tilde{x} - y^\top K_P y \le 0 \end{aligned}$$

that proves global stability. Asymptotic stability follows, as done in [15], using LaSalle's arguments. Indeed, from the inequality above and the definition of $\mathcal{R}$ in (9.11) it is clear that all components of the error state vector $\tilde{x}$ tend to zero.

Remark 9.12 The proposed PI–PBC is decentralized in the sense that, for its implementation, each VSR control requires only the measurement of its corresponding inductor currents and capacitor voltage. Guaranteeing this property motivates our choice of block diagonal gain matrices (9.21).

Remark 9.13 The PI–PBC requires the selection of the desired values for the inductor currents and capacitor voltages that, clearly, cannot all be chosen arbitrarily. Instead, they have to be selected from the set of assignable equilibrium points $\mathcal{E}$, that is determined by the PFSSE. This set has a rather simple structure: the quadratic equation (9.14) defines the VSRs variables from which we *univocally* determine the transmission lines coordinates via (9.13).

9.4.3 Other Inner-loop Controllers Reported in the Literature

In this section we review some of the inner-loop controllers for VSRs reported in the literature. The vast majority of the papers reported on this topic—and, in general, of control of power converters [20, 25]—uses the description of the dynamics in co-energy variables. To facilitate the reference to these works, to some of which we refer here, we give the following model that is immediately obtained from (9.2) and (9.12) as[4]

$$\begin{aligned} L\dot{i}_d &= -Ri_d + L\omega i_q - v_C u_d + v_d \\ L\dot{i}_q &= -L\omega i_d - Ri_q - v_C u_q \\ C\dot{v}_C &= i_d u_d + i_q u_q - Gv_C - i_{dc}. \end{aligned} \tag{9.22}$$

The total energy of the VSR is

$$\mathcal{H}(i_d, i_q, v_C) := \frac{1}{2}\left(Li_d^2 + Li_q^2 + Cv_C^2\right),$$

and the power balance is

$$\dot{\mathcal{H}} = -R(i_d^2 + i_q^2) - Gv_C^2 + P - P_{dc}, \tag{9.23}$$

where we have defined the active and DC powers

$$P = v_d i_d, \qquad P_{dc} = v_C i_{dc}. \tag{9.24}$$

It is also common to define the reactive power as $Q = v_d i_q$.

A caveat regarding the subsequent analysis is, however, necessary. When the VSRs are connected to the transmission lines the currents i_{dc} are linked to the currents on the line via (9.7), which are clearly nonconstant. However, to simplify the analysis, we exploit the fact that their rate of change is slow (with respect to the VSR dynamics) and assume that they are *constant*. Under this assumption the assignable equilibrium set of (9.22) is given as

$$\mathcal{E} = \{x \in \mathbb{R}^3 \mid R(i_d^2 + i_q^2) - v_d i_d + Gv_C^2 + i_{dc}v_C = 0\}. \tag{9.25}$$

---

[4]For ease of presentation we restrict the discussion here to a *single* VSR. The extension to multiple VSRs being straightforward.

Since v_d and i_{dc} are constant, it is then clear that the regulation of P, Q and P_{dc} are equivalent to the regulation of i_d, i_q and v_C, respectively. In practice, because of the small losses of the VSR, the value of P slightly differs from P_{dc}, and consequently there is no interest in regulating the pair i_d and v_C at the same time.

In the literature it is common to distinguish two modes of operation for a VSR:

- *PQ control mode*, when the VSR is required to control the active and reactive power. This is achieved regulating to zero the output

$$y_I = \begin{bmatrix} i_d - i_d^{\text{ref}} \\ i_q - i_q^{\text{ref}} \end{bmatrix}, \tag{9.26}$$

  where the superscript $(\cdot)^{\text{ref}}$ is used to denote reference values—that *do not* necessarily belong to the assignable equilibrium set. These kind of schemes are also called *direct current control* [30].
- *DC voltage control mode*, when the VSR is required to control reactive power and DC voltage. In this case, the regulated output is

$$y_V = \begin{bmatrix} v_C - v_C^{\text{ref}} \\ i_q - i_q^{\text{ref}} \end{bmatrix}. \tag{9.27}$$

  These kind of schemes are also called *direct output voltage control* [30].

To regulate the outputs (9.26) and (9.27) different controllers have been proposed in the literature, ranging from simple PI control [22, 25] to feedback linearization [7, 8, 31]. Some of these papers include an (invariably local) stability analysis. In Sect. 9.5 we prove that y_I and y_V, used for the PI's or with respect to which feedback linearization is performed, have *unstable zero dynamics*. Consequently, applying high gains in the PIs will induce instability and the internal behavior of the feedback linearizing schemes will be unstable.[5] Simulations in Sect. 9.5.4 show that instability indeed arises for these schemes.

For the sake of comparison we write now the passive output (9.17) in co-energy variables for a single VSR as

$$y = \begin{bmatrix} v_C^\star i_d - i_d^\star v_C \\ v_C^\star i_q - i_q^\star v_C \end{bmatrix}, \tag{9.28}$$

where we recall that $(i_d^\star, i_q^\star, v_C^\star) \in \mathcal{E}$, that is, they belong to the assignable equilibrium set.

Remark 9.14 The PI–PBC is *universal*, in the sense that it can operate either in *PQ* or *DC voltage control mode*, depending on which equilibria are assigned as desired references, and which one is consequently determined via the PFSSE. One important

[5] This well-known phenomenon of nonlinear systems [16] is akin to cancellation of unstable zeros of the plant with the unstable poles of the controller in linear systems.

advantage of this universal feature is that there is *no need to switch* between different controllers when the VSRs are requested to change their mode of operation—this is in contrast with other inner-loop schemes that require switchings between controllers, which is clearly undesirable in practice.

9.4.4 Relation of PI–PBC with Akagi's PQ Method

A dominant approach for the design of controllers for reactive power compensation using active filters (for three-phase circuits) is the PQ instantaneous power method proposed by Akagi et al. [2]. It consists in an outer-loop that generates references for the inner PI loops. The references are selected in order to satisfy a very simple heuristic: the AC active power P has to be equal to the DC power P_{dc}, thus ensuring the maximal power transfer from AC to DC side, and the reactive power should take a desired value. Now, using (9.24) define the active AC and DC powers at the equilibrium as

$$P^\star = v_d i_d^\star, \quad P_{dc}^\star = v_C^\star i_{dc}.$$

Consider then the following equivalences

$$P^\star P_{dc} = P_{dc}^\star P \Leftrightarrow v_C^\star i_d = i_d^\star v_C \Leftrightarrow y_1 = 0,$$

with y_1 the first component of the passive output (9.28). Similarly, for the reactive power

$$Q^\star P_{dc} = P_{dc}^\star Q \Leftrightarrow v_C^\star i_q = i_q^\star v_C \Leftrightarrow y_2 = 0,$$

with y_2 the second component of the passive output (9.28). In other words, the objective of the PI–PBC to drive the passive output y to zero can be reinterpreted as a power equalization objective identical to the one used in Akagi's PQ method.

9.5 Performance Limitations of Inner-Loop PIs

Quality assessment of control algorithms is a difficult task—epitomized by the classical performance versus robustness tradeoff, neatly captured by the stability margins in linear designs. The situation for nonlinear systems, where the notions of (dominant) poles and frequency response are specious, is far more complicated. In any case, it is well-known that the achievable performance in control systems is limited by the presence of minimum phase zeros [12, 26, 28].

In this section an attempt is made to evaluate the performance limitations of the inner-loop PI controllers discussed in the previous section. Towards this end, we compute the zero dynamics of the VSR system (9.22) for the outputs y (9.17), y_I (9.26) and y_V (9.27). All three outputs have relative degrees $\{1, 1\}$, hence their zero dynamics is of order one but, while it is exponentially stable for the passive output y it turns out that—for normal operating regimes of the VSR—it is *unstable* for y_I and y_V. If the zero dynamics is *unstable* cranking up the controller gains yields an unstable behavior. This should be contrasted with the passive output y that, as shown in Proposition 16.3 yields an asymptotically stable closed-loop system for all positive gains.[6]

To simplify the derivations we consider only the case of $i_q^\star = 0$. This assumption is justified since it corresponds to fixing to zero the desired value of the reactive power, which is a common operating mode of VSRs. Moreover, this is done without loss of generality because it is possible to show—alas, with messier calculations—that the stability of the zero dynamics is the same for the case of $i_q^\star \neq 0$. This case may arise when the VSR is associated to an AC grid and not to a renewable energy source. In this section we also prove that the (first order) zero dynamics associated to (9.17), is "extremely slow"—with respect to the overall bandwidth of the VSR. Since this zero "attracts" one of the poles of the closed-loop system it stymies the achievement of fast transient responses.

9.5.1 Zero Dynamics Analysis of the Passive Output y

Before presenting the main result of this subsection we make the important observation that the zero dynamics of the VSR model (9.22) and of its corresponding incremental version are the same. Indeed, the zero dynamics describes the behavior of the dynamical system restricted to the set where the output is zero. Since the incremental model dynamics is the *same* as the original model dynamics—simply adding and substracting a constant—their zero dynamics coincide.

Proposition 9.15 *Fix* $(i_d^\star, i_q^\star, v_C^\star) \in \mathcal{E}$ *with* $i_q^\star = 0$. *The zero dynamics*[7] *of the VSR* (9.22) *with respect to the output* (9.28) *is* exponentially stable *and is given by*

$$\dot{v}_C = -\lambda v_C + \lambda v_C^\star, \qquad \lambda := \frac{R(i_d^\star)^2 + G(v_C^\star)^2}{L(i_d^\star)^2 + C(v_C^\star)^2}. \tag{9.29}$$

Proof Setting the output (9.28) identically to zero and using the fact that $i_q^\star = 0$ we get

$$i_d = \frac{i_d^\star}{v_C^\star} v_C, \qquad i_q = \frac{i_q^\star}{v_C^\star} v_C = 0. \tag{9.30}$$

[6]This discussion pertains only to the behavior of the adopted mathematical model of the VSR. In practice, other dynamical phenomena and unmodeled effects may trigger instability even for the PI–PBC.

[7]With some abuse of notation, the zero dynamics is represented using the same symbols of the system dynamics.

Replacing (9.30) into (9.22) gives

$$L\frac{i_d^\star}{v_C^\star}\dot{v}_C = -R\frac{i_d^\star}{v_C^\star}v_C - v_C u_1 + v_d \tag{9.31}$$

$$0 = -L\omega\frac{i_d^\star}{v_C^\star}v_C - v_C u_2 \tag{9.32}$$

$$C\dot{v}_C = \frac{i_d^\star}{v_C^\star}v_C u_1 - Gv_C - i_{dc}. \tag{9.33}$$

To eliminate u_1 we multiply (9.33) by $\frac{v_C^\star}{i_d^\star}$ and add it to (9.31) yielding

$$\left(\frac{Cv_C^\star}{i_d^\star} + \frac{Li_d^\star}{v_C^\star}\right)\dot{v}_C = -\left(\frac{Ri_d^\star}{v_C^\star} + \frac{Gv_C^\star}{i_d^\star}\right)v_C + v_d - \frac{v_C^\star}{i_d^\star}i_{dc}.$$

The proof is completed noting from (9.25) that, for $(i_d^\star, i_q^\star, v_C^\star) \in \mathcal{E}$ with $i_q^\star = 0$, we have

$$v_d - \frac{v_C^\star}{i_d^\star}i_{dc} = \frac{R(i_d^\star)^2 + G(v_C^\star)^2}{i_d^\star}$$

and pulling out the common factor $\frac{1}{i_d^\star v_C^\star}$.

Remark 9.16 The parameters R and G, that represent the losses in the VSR, are usually small—compared to L and C. Consequently, λ will also be a small value, placing the pole of the zero dynamics very close to the origin and inducing slow convergence.

Remark 9.17 It is interesting to note that the rate of exponential convergence of the zero dynamics can be rewritten

$$\lambda = \frac{1}{2}\frac{P^\star - P_{dc}^\star}{\mathcal{H}(i_d^\star, i_q^\star, v_C^\star)}$$

that is half the ratio between the steady-state dissipated power and the steady-state energy of the system. This relationship holds true also for the case $i_q^\star \neq 0$.

Remark 9.18 In [37] it has been observed that the addition of an outer-loop controller, in the form of a voltage droop—that is the *de facto* standard in the power systems community [3, 14]—allows to overcome the performance limitations of the PI–PBC. The droop controller has been originally proposed to robustify inner-loop controllers with respect to unexpected (large) perturbations. The ability to overcome the performance limitations of the PI–PBC, further substantiate the interest in this control scheme. Unfortunately, the addition of the outer loop destroys the passivity property instrumental for the stability analysis of the PI–PBC. Current research is under way to establish some stability properties of the PI–PBC plus droop control.

9.5.2 Zero Dynamics Analysis of y_I

Before analyzing the zero dynamics of the PQ and DC voltage control outputs, (9.26) and (9.27), respectively, we recall that the references do not necessarily belong to the assignable equilibrium set. However, we make the reasonable assumption that, for the chosen reference values, the zero dynamics admits an *equilibrium*—if this is not the case the zero dynamics is unstable. Moreover, similarly to the case of the passive output, we will take $i_q^{\text{ref}} = 0$.

Proposition 9.19 *Fix* $i_d^{\text{ref}} \in \mathbb{R}$, $i_q^{\text{ref}} = 0$. *The zero dynamics of the VSR* (9.22) *with respect to the output* (9.26) *is given by*

$$C\dot{v}_C = -Gv_C + \frac{\alpha_I}{v_C} - i_{dc}^{\text{ref}}, \qquad \alpha_I := v_d i_d^{\text{ref}} - R(i_d^{\text{ref}})^2 \tag{9.34}$$

where i_{dc}^{ref} *is a constant value for* i_{dc} *satisfying*

$$(i_{dc}^{\text{ref}})^2 > 4G\alpha_I. \tag{9.35}$$

- *If* $\alpha_I > 0$ *the zero dynamics has one equilibrium and it is* stable.
- *If* $\alpha_I < 0$ *the zero dynamics has two equilibria one stable and one unstable.*
- *If* $\alpha_I = 0$ *the zero dynamics is a linear asymptotically* stable *system.*

Proof Setting the output (9.26) equal to zero with $i_q^* = 0$ and replacing into (9.22) gives

$$0 = -Ri_d^{\text{ref}} - v_C u_1 + v_d \tag{9.36}$$

$$0 = -L\omega i_d^{\text{ref}} - v_C u_2 \tag{9.37}$$

$$C\dot{v}_C = i_d^{\text{ref}} u_1 - Gv_C - i_{dc}^{\text{ref}}, \tag{9.38}$$

where we have added the superscript $(\cdot)^{\text{ref}}$ to i_{dc}. Replacing u_1 obtained from (9.36) into (9.38) yields directly (9.34). Condition (9.35) is then necessary and sufficient for the existence of a (real) equilibrium of (9.34). If $\alpha_I = 0$ the dynamics reduces to

$$C\dot{v}_C = -Gv_C - i_{dc}^{\text{ref}}.$$

The proof is completed, recalling that $v_C > 0$ and looking at the plots of the right hand side of (9.34) for α_I positive and negative in Fig. 9.3.

Remark 9.20 From Fig. 9.3, if $\alpha_I < 0$, it is easy to see that the stable equilibrium point is the largest one. For standard values of the system parameters it turns out that this equilibrium is located beyond the physical operating regime of the system, hence it is of no practical interest.

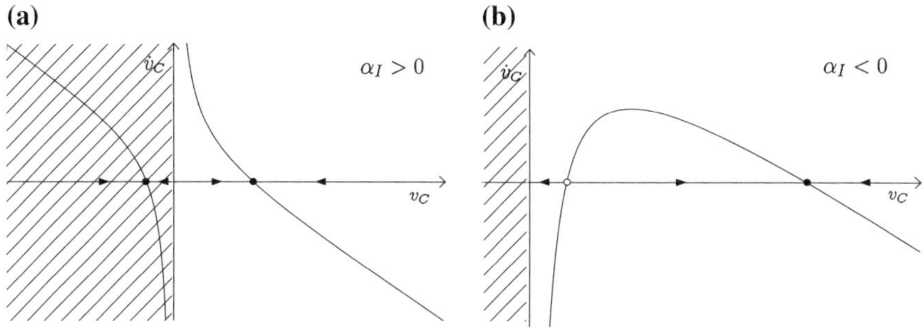

Fig. 9.3 Plot of $\dot{v}_C$ versus v_C for the cases of **a** $\alpha_I > 0$ and **b** $\alpha_I < 0$. The *arrows* in the *horizontal* axis indicate the direction of the flow of the zero dynamics

Remark 9.21 The parameters R and G are usually very small and i_{dc}^{ref} can take positive or negative values in standard operations. Then, condition (9.35) is always verified while α_I can take positive or negative values.

Remark 9.22 The situation $\alpha_I = 0$, when the zero dynamics is linear and asymptotically stable, is unattainable in applications. Indeed, assuming that in steady-state all signals converge to their reference values, it can be shown that $\alpha_I = 0$ if and only if $Gv_C + i_{dc} = 0$ that, given the small values of G is not realistic in practice.

9.5.3 Zero Dynamics Analysis of y_V

Proposition 9.23 *Fix $v_C^{\text{ref}} \in \mathbb{R}$, $i_q^{\text{ref}} = 0$. The zero dynamics of the VSR (9.22) with respect to the output (9.27) is given by*

$$L\frac{di_d}{dt} = -Ri_d - \frac{\alpha_V}{i_d} + v_d, \qquad \alpha_V := i_{dc}^{\text{ref}} v_C^{\text{ref}} + G(v_C^{\text{ref}})^2 \tag{9.39}$$

where i_{dc}^{ref} is a constant value for i_{dc} satisfying

$$v_d^2 > -4R\alpha_V. \tag{9.40}$$

- *If $\alpha_V < 0$ the zero dynamics has two equilibria and they are both* stable.
- *If $\alpha_V > 0$ the zero dynamics has two equilibria one stable and one unstable.*
- *If $\alpha_V = 0$ the zero dynamics is a linear asymptotically* stable *system.*

Proof Setting the output (9.27) equal to zero with $i_q^* = 0$ and replacing into (9.22) gives

$$L\frac{di_d}{dt} = -Ri_d - v_C^{\text{ref}} u_1 + v_d \tag{9.41}$$

$$0 = -L\omega i_d - v_C^{\text{ref}} u_2 \tag{9.42}$$

$$0 = i_d u_1 - G v_C^{\text{ref}} - i_{dc}. \tag{9.43}$$

Replacing u_1 obtained from (9.43) into (9.41) yields directly (9.39). Condition (9.40) is necessary and sufficient for the existence of a (real) equilibrium of (9.39). The proof is completed invoking the same arguments used in the proof of Proposition 9.19 and are omitted for brevity.

Remark 9.24 Remarks 9.21 and 9.22 apply *verbatim* to (9.39) and α_V of Proposition 9.23.

9.5.4 Simulated Evidence of the Performance Limitations

Although Proposition 9.15 proves that the zero dynamics for the passive output y is exponentially stable it turns out that, for the components used in standard HVDC transmission system, the convergence rate is $\lambda \approx 0.04$, which is extremely slow. As indicated above this dominating dynamics stymies the achievement of fast transient responses—situation that is shown in the following simulations. Also, we present simulated evidence of unstable behavior of PI inner-loops using the outputs (9.26) and (9.27).

We consider a three-terminals HVDC transmission system with a simple *meshed* topology, that is illustrated in Fig. 9.4, where the corresponding graph is also given. The model of the system is given by (9.9), that is a system of dimension $3n + \ell = 11$ with $2n = 6$ inputs. Parameters of the VSRs and of the transmission lines are given in Table 9.1.

We define the following *control objectives*: all the stations are required to regulate the reactive power to zero; the stations associated to the wind farms (WF1, WF2)

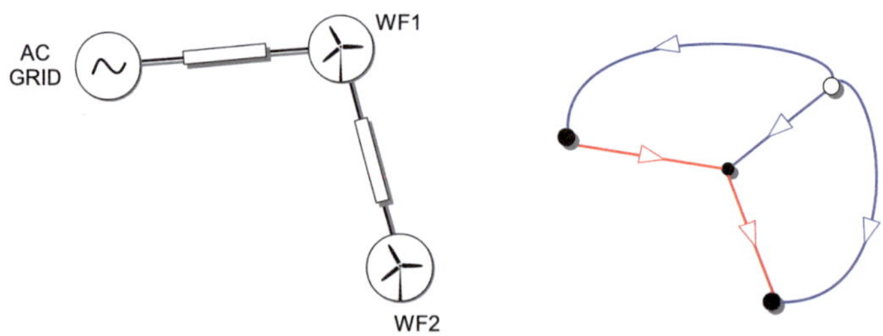

Fig. 9.4 Schematic representation of an HVDC transmission system constituted by three stations, associated to two wind farms (WFs) and an AC grid, with associated graph. The graph is represented by *filled circles* for the *VSRs-buses* and the *unfilled circle* for the *ground* node. *Blue* and *red* edges characterize VSRs and lines, respectively

Table 9.1 System parameters

$R_{r,i}$	$G_{r,i}$	$L_{r,i}$	$C_{r,i}$	V_i	$R_{\ell,12}$	$L_{\ell,12}$	$R_{\ell,23}$	$L_{\ell,23}$	ω_i
0.01 Ω	0 Ω^{-1}	40 mH	20 μF	130 kV	26 Ω	3.8 mH	20 Ω	2.5 mH	50 Hz

Table 9.2 System references $i^{\star}_{d,i}$(A), $v^{\star}_{C,i}$(kV) associated to the slack bus (SB) and wind farms (WFs)

	SB (A)	WF_1 (A)	WF_2 (A)	SB (kV)	WF_1 (kV)	WF_2 (kV)
0	−1260	900	1000	100	142.595	158.951
T	−1588	900	1800	100	153.650	179.691
$2T$	−266	500	−200	100	109.004	104.004
$3T$	905	−400	−200	100	69.419	60.877
$4T$	−849	1300	−200	100	128.708	124.532

are required to regulate the active power to desired (constant) values; the remaining station, called *slack bus* (SB), must regulate the voltage around its nominal value. In Table 9.2, the corresponding references of direct current and DC voltages are furnished, together with the corresponding assignable equilibria, that are calculated via the PFSSE defined by (9.7). Changes in references occur every T s over a time interval of $5T$ s. It should be noticed that from 0 to $2T$ the power flow is uniquely directed from both wind farms stations to the AC grid, while at $2T$, and next $3T$ the wind farms stations start demanding power to the AC grid, thus reversing the direction of the power flow. This situation can arise when the power produced by the wind farms is insufficient to supply local loads.

9.5.4.1 PI–PBC

In this section we present simulations on the three-terminals benchmark example of the decentralized PI–PBC defined in Sect. 9.4.2, which illustrate the stability properties and performance limitations discussed in the previous sections. Setting $T = 2000$ s the controllers (9.20) are designed with identical parameters and diagonal matrices $k_{P,i} = \text{diag}\{1, 1\}$, $k_{I,i} = \text{diag}\{10, 10\}$. The behavior of the VSRs are depicted in Fig. 9.5.

As expected, the direct currents of each station attain the assignable equilibria defined in Table 9.2, while the quadrature currents are always kept to zero after a very short transient. Moreover, the DC voltage at the slack bus is maintained near the nominal value of 100 kV, as required, while the DC voltage variation at the wind farms stations, balances the fluctuation of power demand. Even though the desired steady-state is attained, for all practical purposes, the convergence time of direct currents and DC voltages is extremely slow. This poor transient performance

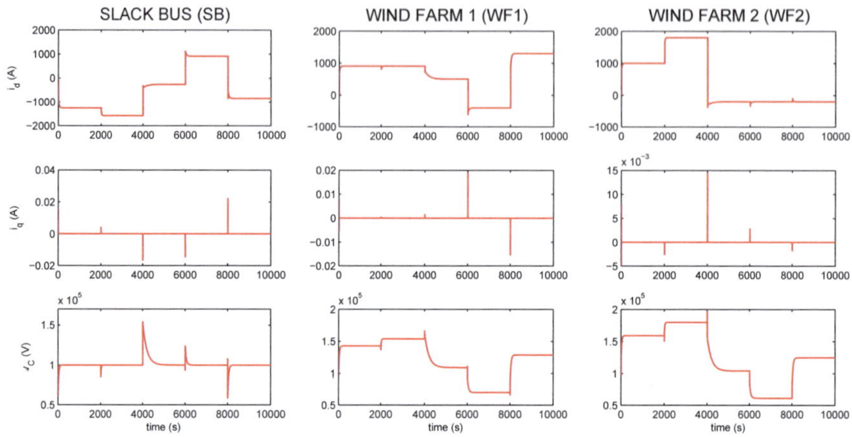

Fig. 9.5 Responses of VSRs variables under the decentralized PI–PBC

behavior is independent of the controller gains. Indeed, extensive simulations show that the system maintains the same slow convergence time even with larger gains, thus validating the performance limitations analysis realized in Sect. 9.5.1.

9.5.4.2 PQ and DC Voltage Controllers

We next analyze the behavior of the system under the standard PQ and DC voltage controllers of Sect. 9.4.3. In agreement with the control requirements described above, two PQ controllers are designed to regulate direct and quadrature currents of the wind farms stations, and one DC voltage controller is designed to regulate DC voltage and quadrature current of the slack bus. We consider simple PI controllers defined over the outputs (9.26), (9.27), designed with identical gains $k_{P,i}$, $k_{I,i}$, that are tuned via simulations. The behavior of the VSRs are depicted in Fig. 9.6, with $T = 4$ s. This value should be contrasted with the value ($T = 2000$ s) used for the PI–PBC. It is easy to see that the PQ and DC voltage controllers correctly (and rapidly) regulate the station at the desired references between 0 and 8 s. This good behavior is not surprising, because PQ controllers applied to VSRs that are injecting power and a DC voltage controller applied to VSRs that is absorbing power, have associated globally asymptotically stable zero dynamics, as proved in Sects. 9.5.2, 9.5.3. On the other hand, as shown in the figures, when at stations WF1 and WF2 the power flow is reversed (respectively at $t = 12$ s and $t = 8$ s), the correspondent DC voltages go unstable, because in these cases the zero dynamics is unstable. Similar unstable behavior appears also at the slack bus station.

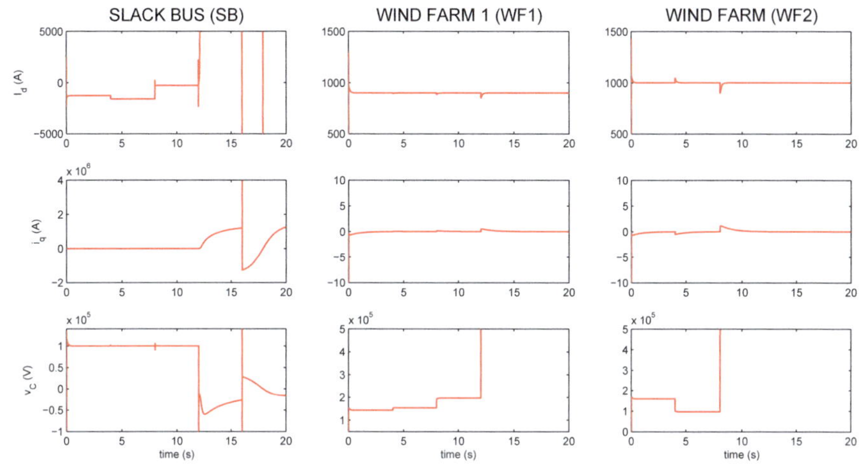

Fig. 9.6 Responses of VSRs variables under the decentralized PQ and DC voltage controllers

9.6 Conclusions and Future Perspectives

The work covers different aspects of modeling, analysis and control of multi-terminal HVDC transmission system. The main contribution is a decentralized, globally asymptotically stable, PI control for a very general class of multi-terminal HVDC transmission systems. For this purpose, starting from a graph description of the network, a pH representation has been obtained, thus revealing the intrinsic passivity properties of the system. The main result is a direct extension of the previous works on PI control of VSRs, to a sufficiently general interconnected system, with the important property that the control is decentralized, a fundamental requirement for large-scale systems. To provide some connections between the proposed controller and standard techniques, widely used in literature, a comparative analysis of stability and performances is provided, shedding some light on limitations and benefits of different approaches. In particular it is proved—and validated via simulations—that the popular current and voltage control techniques possibly lead to unstable behaviors of the controlled system, while the proposed PI–PBC, although ensuring convergence, has clear performance limitations. The theoretical analysis that substantiates these claims is based on a detailed, nonlinear, zero dynamics analysis of a single VSR with respect to the outputs used for all these controllers.

A future research line pertains to the use of more accurate models for the description of the system, that may improve the control quality. For instance, the behavior of long transmission lines is best described by means of the Telegrapher's equations, thus leading to an infinite dimensional pH representation, which can still be handled with existing theory [34]. Because of the addition of standard outer controllers destroys the passivity property—that is instrumental for the stability analysis of the PI–PBC—current research is under way to establish some stability properties of the

PI–PBC plus the commonly employed outer controllers, like for example the ubiquitous droop control. A further possibility is the development of a new provably stable outer-loop controller that would replace the droop. It would also be of interest to investigate new strategies for power flow optimization, moving away from the PFSSE. A final, long term, objective is the experimental validation of the proposed PI–PBC plus droop scheme.

Acknowledgments This work was supported by the Ministry of Education and Science of Russian Federation (Project 14.Z50.31.0031), Alstom Grid and partially supported by the iCODE institute, research project of the Idex Paris-Saclay.

References

1. A.M. Abbas, P.W. Lehn, PWM based VSC-HVDC systems—a review, in *Power Energy Society General Meeting*, 2009. PES '09. IEEE, pp. 1–9, July 2009
2. H. Akagi, *Instantaneous Power Theory and Applications to Power Conditioning* (Wiley, Newark, 2007)
3. M. Andreasson, M. Nazari, D.V. Dimarogonas, H. Sandberg, K.H. Johansson, M. Ghandhari, in *Distributed Voltage and Current Control of Multi-Terminal High-Voltage Direct Current Transmission Systems*. ArXiv e-prints, Nov. 2013
4. M.K. Bucher, R. Wiget, G. Andersson, C.M. Franck, Multiterminal HVDC Networks—what is the preferred topology? Power Delivery, IEEE Trans. **29**(1), 406–413 (2014)
5. J.M. Carrasco, L.G. Franquelo, J.T. Bialasiewicz, E. Galvan, R.C.P. Guisado, M.A.M. Prats, J.I. Leon, N. Moreno-Alfonso, Power-electronic systems for the grid integration of renewable energy sources: a survey. Ind. Electron., IEEE Trans. **53**(4), 1002–1016 (2006)
6. S. Chatzivasileiadis, D. Ernst, G. Andersson, The global grid. CoRR abs/1207.4096 (2012)
7. H. Chen, Z. Xu, F. Zhang, Nonlinear control for VSC based HVDC system. in *Power Engineering Society General Meeting*. IEEE, p. 5, 2006
8. Y. Chen, J. Dai, G. Damm, F. Lamnabhi-Lagarrigue, Nonlinear control design for a multi-terminal VSC-HVDC system. in *Control Conference (ECC), 2013 European*, pp. 3536–3541, July 2013
9. G. Escobar, A.J. Van Der Schaft, R. Ortega, A Hamiltonian viewpoint in the modeling of switching power converters. Automatica **35**(3), 445–452 (1999)
10. S. Fiaz, D. Zonetti, R. Ortega, J.M.A. Scherpen, A.J. van der Schaft, A port-Hamiltonian approach to power network modeling and analysis. Europ. J. Control **19**(6), 477–485 (2013)
11. N. Flourentzou, V.G. Agelidis, G.D. Demetriades, VSC-based HVDC power transmission systems: an overview. Power Electron., IEEE Trans. **24**(3), 592–602 (2009)
12. B.A. Francis, G. Zames, On H^{∞}-optimal sensitivity theory for siso feedback systems. Autom. Control, IEEE Trans. **29**(1), 9–16 (1984)
13. O. Gomis-Bellmunt, J. Liang, J. Ekanayake, R. King, N. Jenkins, Topologies of multiterminal HVDC-VSC transmission for large offshore wind farms. Electr. Power Syst. Res. **81**(2), 271–281 (2011)
14. T.M. Haileselassie, T. Undeland, K. Uhlen, *Multiterminal HVDC for offshore windfarms control strategy*. European Power Electronics and Drives Association, 2009
15. M. Hernandez-Gomez, R. Ortega, F. Lamnabhi-Lagarrigue, G. Escobar, Adaptive PI stabilization of switched power converters. Control Syst. Technol., IEEE Trans. **18**(3), 688–698 (2010)
16. A. Isidori, *Nonlinear Control Systems*, 3rd edn. (Springer, New York, Secaucus, 1995)
17. A. Jager-Waldau, Photovoltaics and renewable energies in europe. Renew. Sustain. Energy Rev. **11**(7), 1414–1437 (2007)

18. B. Jayawardhana, R. Ortega, E. Garcia-Canseco, F.F. Castaños, *Passivity of Nonlinear Incremental Systems: Application to PI Stabilization of Nonlinear RLC Circuits*. In Decision and Control, 2006 45th IEEE Conference on. pp. 3808–3812 (2006)
19. S.G. Johansson, G. Asplund, E. Jansson, R. Rudervall, *Power system stability benefits with VSC DC-transmission systems*. In CIGRE Conference, Paris, France (2004)
20. M.P. Kazmierkowski, R. Krishnan, F. Blaabjerg, J.D. Irwin, *Control in Power Electronics: Selected Problems* (Academic Press Series in Engineering, Elsevier Science, 2002)
21. N.M. Kirby, M.J. Luckett, L. Xu, W. Siepmann, HVDC transmission for large offshore windfarms. in *AC-DC Power Transmission, 2001*. Seventh International Conference on (Conf. Publ. No. 485), pp. 162–168. IET, (2001)
22. T. Lee, Input-output linearization and zero-dynamics control of three-phase AC/DC voltage-source converters. Power Electron., IEEE Trans. **18**(1), 11–22 (2003)
23. H. Lund, Large-scale integration of wind power into different energy systems. Energy **30**(13), 2402–2412 (2005)
24. M. Perez, R. Ortega, J.R. Espinoza, Passivity-based PI control of switched power converters. Control Syst. Technol., IEEE Trans. **12**(6), 881–890 (2004)
25. R.T. Pinto, S.F. Rodrigues, P. Bauer, J. Pierik, Comparison of direct voltage control methods of multi-terminal dc (MTDC) networks through modular dynamic models. in *Power Electronics and Applications (EPE 2011), Proceedings of the 2011–14th European Conference on*, pp. 1–10 (Aug 2011)
26. L. Qiu, E.J. Davison, *Performance Limitations of Non-minimum Phase Systems in the Servomechanism Problem*, pp. 337–349 (1993)
27. S. Sanchez, R. Ortega, R. Gri no, G. Bergna, M. Molinas-Cabrera, Conditions for Existence of equilibrium points of systems with constant power loads. In Decision and Control, 2013 52nd IEEE Conference on, Firenze, Italy, 2013
28. M.M. Seron, J.H. Braslavsky, G.C. Goodwin, *Fundamental Limitations in Filtering and Control*, 1st edn. (Springer Publishing Company, Incorporated, 2011)
29. S. Shah, R. Hassan, J. Sun, HVDC transmission system architectures and control—a review. In Control and Modeling for Power Electronics (COMPEL), 2013 IEEE 14th Workshop on, pp. 1–8, June 2013
30. D. Shuai, X. Zhang, Input-output linearization and stabilization analysis of internal dynamics of three-phase AC/DC voltage-source converters. In Electrical Machines and Systems (ICEMS), 2010 International Conference on, pp. 329–333, Oct. 2010
31. J.-L. Thomas, S. Poullain, A. Benchaib, Analysis of a robust DC-bus voltage control system for a VSC transmission scheme. In AC-DC Power Transmission, 2001. Seventh International Conference on (Conf. Publ. No. 485), pp. 119–124, Nov. 2001
32. A.J. van der Schaft, *$\mathcal{L}_2$-gain and Passivity Techniques in Nonlinear Control*. Communications and Control Engineering (Springer, Berlin, 2000)
33. A.J. van der Schaft, Characterization and partial synthesis of the behavior of resistive circuits at their terminals. Syst. Control Lett. **59**(7), 423–428 (2010)
34. A.J. van der Schaft, D. Jeltsema, *Port-Hamiltonian Systems Theory: An Introductory Overview*. (Now publishers Inc, 2014)
35. A. Yazdani, R. Iravani, *Voltage-Sourced Controlled Power Converters - Modeling* (Control and Applications, Wiley IEEE, 2010)
36. D. Zonetti, R. Ortega, A. Benchaib, A globally asymptotically stable dencentralized PI controller for multi-terminal high-voltage DC transmission systems. in *Proceedings of the 13th European Control Conference on*, June 2014
37. D. Zonetti, R. Ortega, A. Benchaib, Modeling and control of high-voltage direct-current transmission systems: from theory to practice and back. CoRR abs/1406.4392 (2014)

Chapter 10
A Complement on Elimination and Realization in Rational Representations

Harry L. Trentelman, Tjerk W. Stegink and Sasanka V. Gottimukkala

Abstract In this paper we study a number of problems in the context of rational representations of behaviors. In that context, a given proper real rational matrix can represent three behaviors. In the first place it can represent an input–output behavior. Second, it can represent the kernel behavior of the rational 'differential operator' associated with the rational matrix. Third, it can represent the image behavior asociated with the rational matrix. On the other hand, every proper real rational matrix admits a realization as a finite-dimensional linear state-space system. Such realization can represent three system behaviors: an input-state-output behavior, an output nulling behavior, or a driving variable behavior. In this paper we will study the relation between the three external behaviors of these state representations, and the behaviors given by the three rational representations associated with the underlying rational matrix. Preliminary results from [5] will be complemented to obtain necessary and sufficient conditions such that the respective external behaviors are equal.

10.1 Introduction

Arjan and I (the first author) met around 35 years ago. Arjan had already been a Ph.D. student with Jan Willems in Groningen for a longer time, and I had just started. In those days it was still allowed to smoke inside university buildings, and I recall a particular instant that I entered Jan's office, while Arjan was undergoing his supervision by Jan. Apart from the fact that the blackboard was filled with Hamiltonian systems,

---

H.L. Trentelman (✉) · T.W. Stegink
Johann Bernoulli Institute for Mathematics and Computer Science, University of Groningen, P. O Box 800, 9700 AV, Groningen, The Netherlands
e-mail: h.l.trentelman@rug.nl

T.W. Stegink
e-mail: t.w.stegink@rug.nl

S.V. Gottimukkala
Hightech Solutions B.V. Apeldoorn, Apeldoorn, The Netherlands
e-mail: svg1729@gmail.com

M.K. Camlibel et al. (eds.), *Mathematical Control Theory I*,
Lecture Notes in Control and Information Sciences 461,
DOI 10.1007/978-3-319-20988-3_10

he and Jan were staring in the direction of this blackboard, with thick clouds of cigar smoke circling around. I instantly realized that I witnessed real science here: staring at the board, trying to get somewhere, cigars firmly held in the hand. At that time, Arjan developed scientifically into one of the leaders in the area of nonlinear control, in particular working on the differential geometric approach to nonlinear systems. Together with other Dutch experts in that area he had founded an unofficial society, called the Dutch Nonlinear Systems Group. I myself worked on the almost version of the linear geometric approach. One of the major achievements in my research career is that I was declared an 'almost' member of that illustrious society as a present after my Ph.D. defense.

After a long period at different universities, Arjan and I finally ended up together in Groningen again, as colleagues in the same research group. While writing this article on representations of behaviors, a typically Groningen subject indeed, I would like to congratulate him on the occasion of his 60th birthday.

As is well known, behaviors of linear differential systems admit different kinds of representations. Although the key idea of the behavioral approach to systems and control is that a mathematical model of a dynamical system is essentially formed by its set of trajectories, called the behavior of the system, it also offers the flexibility that this behavior can be represented in many different ways, as, for example, the kernel or image of a polynomial differential operator, or as the external behavior of a polynomial latent variable representation, see [11].

Another class of representations of behaviors consist of several forms of state representations, which involve a latent variable having the *property of state*, see [11]. State representations of behaviors are, for example, the classical input/state/output (ISO) representations, driving variable representations, and output nulling representations, see [6, 16].

A more classical concept in modeling linear input–output systems is that of transfer matrix. In the context of causal, linear, finite dimensional, time-invariant systems, transfer matrices are proper real rational matrices. In an attempt to further bridge the gap between the behavioral time domain approach and the transfer matrix, frequency domain, approach, in [17], the concepts of rational kernel and image representation were elaborated. One of the highlights of this work is that, for the proper real rational transfer matrix of any given linear input–output system, it gives a sound and simple time domain interpretation of the representation $y(t) = G(\frac{\mathrm{d}}{\mathrm{d}t})u(t)$ as an alternative for the mathematically unsatisfactory frequency domain representation $\hat{y}(s) = G(s)\hat{u}(s)$.

From classical realization theory, see [2], it is well known that every proper real rational matrix $G(s)$ admits a *realization*. In particular, a quadruple of matrices (A, B, C, D) is called a realization of $G(s)$ if $G(s) = C(sI - A)^{-1}B + D$. This raises a number of questions on the relationship between rational kernel and image representations on the one hand, and input-state-output, driving variable, and output nulling representations on the other. In particular, given a proper real rational matrix $G(s)$ with realization (A, B, C, D), the natural question arises: what is the relation between the behavior represented by $y = G(\frac{\mathrm{d}}{\mathrm{d}t})u$ and the input–output behavior of the representation $\dot{x} = Ax + Bu,\ y = Cx + Du$? Under what conditions are these behaviors equal? A similar question arises for output nulling representations: what is the relation between the behavior represented by the rational kernel representation $G(\frac{\mathrm{d}}{\mathrm{d}t})w = 0$ and the external behavior of the output nulling representation $\dot{x} = Ax + Bw,\ 0 = Cx + Dw$? For driving variable representations: under what conditions is the external behavior represented by the rational image representation $w = \text{im } G(\frac{\mathrm{d}}{\mathrm{d}t})l$ equal to external behavior of the driving variable representation $\dot{x} = Ax + Bv,\ w = Cx + Dv$? Partial answers to these questions have been obtained in [5]. In the present paper we will present new results that complement these earlier ones, and a complete answer to the questions posed will be given. In order to address the above questions, we will first review results from [5] on the problem of elimination of the state variable from input-state-output, driving variable, and output nulling representations. This problem consists of finding polynomial kernel representations of the external behaviors associated with these state representations.

The outline of this paper is as follows. In the next section, Sect. 10.2, we will give some preliminaries and introduce the notation used in this paper. In Sect. 10.3, we will review various representations of behaviors of linear differential systems. In Sect. 10.4, we will review results from [5] on the problem of eliminating the state from a given input-state-output representation, and, likewise, from a given output nulling representation. In the case of driving variable representations we will discuss how to eliminate both the state and the driving variable. Section 10.5 contains the main results of this paper. Here, in three subsequent subsections, we will give complete answers to the problems posed above. We will conclude this paper with some final remarks in Sect. 10.6.

10.2 Preliminaries and Notation

In this paper we will use the standard notation $\mathbb{R}$ and $\mathbb{C}$ for the fields of real and complex numbers. We use $\mathbb{R}^{\mathrm{n}},\ \mathbb{R}^{\mathrm{n}\times\mathrm{m}}$, etc., for the real linear spaces of vectors and matrices with components in $\mathbb{R}$. The space of all infinitely often differentiable functions from $\mathbb{R}$ to $\mathbb{R}^{\mathrm{w}}$ will be denoted by $\mathfrak{C}^{\infty}(\mathbb{R}, \mathbb{R}^{\mathrm{w}})$. The field of real rational functions in the indeterminate s will be denoted by $\mathbb{R}(s)$. The ring of real polynomials in the indeterminate s will be denoted by $\mathbb{R}[s]$. $\mathbb{R}(s)_p$ will denote the ring of proper rational functions in the indeterminate s with real coefficients. We will use $\mathbb{R}(s)^{\mathrm{n}}$, $\mathbb{R}(s)^{\mathrm{n}\times\mathrm{m}}$, $\mathbb{R}[s]^{\mathrm{n}}$, $\mathbb{R}[s]^{\mathrm{n}\times\mathrm{m}}$, etc. for the spaces of vectors and matrices with components in $\mathbb{R}(s)$ and $\mathbb{R}[s]$, respectively. If the dimensions are clear from the context, we will use the notation $\mathbb{R}(s)^{\bullet\times\mathrm{m}}$, $\mathbb{R}(s)^{\mathrm{n}\times\bullet}$, $\mathbb{R}[s]^{\bullet\times\bullet}$ or $\mathbb{R}(s)^{\bullet\times\bullet}$, etc. A square non-singular polynomial matrix U is called *unimodular* if it has an inverse which is again a polynomial matrix. $F \in \mathbb{R}(s)^{\mathrm{n}\times\mathrm{n}}$ is *biproper* if $\det(F) \neq 0$ and F, F^{-1} are both proper. A polynomial matrix $P \in \mathbb{R}[s]^{\bullet\times\bullet}$ is called left prime over $\mathbb{R}[s]$ if it has a polynomial right inverse, i.e., if there exists a polynomial matrix Q such that $PQ = I$, the identity matrix. Left primeness is equivalent with the condition that $P(\lambda)$ has full row rank for all $\lambda \in \mathbb{C}$.

For a given finite-dimensional linear system $\dot{x} = Ax + Bu,\ y = Cx + Du$, we will denote its reachable subspace by $\mathcal{R}$. This is the smallest A-invariant subspace containing im B. The unobservable subspace will be denotesd by $\mathcal{N}$, and is the largest A-invariant subspace contained in ker C.

10.3 Representations of Behaviors

In this section we will review the basic material on representations of behaviors of linear differential systems. A linear differential system is defined as a system $\Sigma = (\mathbb{R}, \mathbb{R}^{\mathtt{w}}, \mathfrak{B})$ whose behavior $\mathfrak{B}$ is the solution space of given finite set of higher order constant coefficient linear differential equations. By representing the given set of linear differential equations in terms of a polynomial matrix, for any linear differential system $\Sigma = (\mathbb{R}, \mathbb{R}^{\mathtt{w}}, \mathfrak{B})$ there obviously exists a real polynomial matrix R with $\mathtt{w}$ columns, i.e., $R \in \mathbb{R}[s]^{\bullet \times \mathtt{w}}$, such that

$$\mathfrak{B} = \{w \in \mathfrak{C}^{\infty}(\mathbb{R}, \mathbb{R}^{\mathtt{w}}) \mid R(\frac{\mathrm{d}}{\mathrm{d}t})w = 0\}. \tag{10.1}$$

The representation (10.1) is called a *polynomial kernel representation* of $\mathfrak{B}$, and we often write $\mathfrak{B} = \ker\ R(\frac{\mathrm{d}}{\mathrm{d}t})$. For a detailed exposition on polynomial representations of behaviors we refer to [11].

In addition to polynomial representations, behaviors admit rational representations. Rational representations were formally introduced in [17]. Earlier work on rational representations for $\mathcal{L}_2$-behaviors can be found in [15], and was later extended in [8, 9]. In order to introduce rational representations, we need the concept of left coprime factorization of a rational matrix over $\mathbb{R}[s]$. A factorization of $G \in \mathbb{R}(s)^{\bullet \times \bullet}$ as $G = P^{-1}Q$ with $P, Q \in \mathbb{R}[s]^{\bullet \times \bullet}$ is called a left coprime factorization if $\begin{bmatrix} P & Q \end{bmatrix}$ is left prime over $\mathbb{R}[s]$ and $\det(P) \neq 0$. Following [17], if $G = P^{-1}Q$ is a left coprime factorization over $\mathbb{R}[s]$ then we *define* w to be a solution of $G(\frac{\mathrm{d}}{\mathrm{d}t})w = 0$ if it is a solution of the differential equation $Q(\frac{\mathrm{d}}{\mathrm{d}t})w = 0$. Likewise, we define

$$\ker\ G(\frac{\mathrm{d}}{\mathrm{d}t}) := \ker\ Q(\frac{\mathrm{d}}{\mathrm{d}t}). \tag{10.2}$$

If G is a rational matrix, we call a representation of $\mathfrak{B}$ as $G(\frac{\mathrm{d}}{\mathrm{d}t})w = 0$ a *rational kernel representation* of $\mathfrak{B}$ and sometimes write $\mathfrak{B} = \ker\ G(\frac{\mathrm{d}}{\mathrm{d}t})$. For additional material on rational representations we refer to [4] and [5].

In addition to kernel representations, which are in terms of the variable w to be modeled only (called the *manifest variable* or *external variable*), behaviors admit representations in which auxiliary variables may appear. The most general form of such representation of a behavior $\mathfrak{B}$ is the so-called *latent variable representation*, which has the following form:

$$R(\frac{\mathrm{d}}{\mathrm{d}t})w = M(\frac{\mathrm{d}}{\mathrm{d}t})l. \tag{10.3}$$

In the above differential equation, w is the manifest variable and l is called the *latent variable*. The (w, l)'s satisfying (10.3) are given by ker $\left[R(\frac{\mathrm{d}}{\mathrm{d}t}) \; -M(\frac{\mathrm{d}}{\mathrm{d}t})\right] =: \mathfrak{B}_{\mathrm{full}}$, and $\mathfrak{B}_{\mathrm{full}}$ is called the full behavior. The manifest behavior is given by $(\mathfrak{B}_{\mathrm{full}})_w$, which is the projection of the full behavior onto the behavior of the external variable w. The matrices R and M can be polynomial matrices or rational matrices. If (10.3) involves polynomial matrices only then we call it a *polynomial latent variable representation*. In general, we call it a *rational latent variable representation*.

A special class of latent variable representations consists of the so-called *state representations*. These are latent variable representations in which the latent variable has the *property of state*, see [6, 7, 14, 16]. In the following, we will briefly review the basics of the three most important kinds of state representations, namely, input-state-output representations, driving variable representations, and output nulling representations.

Let $A \in \mathbb{R}^{\mathtt{n}\times\mathtt{n}}$, $B \in \mathbb{R}^{\mathtt{n}\times\mathtt{m}}$, $C \in \mathbb{R}^{\mathtt{p}\times\mathtt{n}}$, $D \in \mathbb{R}^{\mathtt{p}\times\mathtt{m}}$, and consider the equations

$$\frac{\mathrm{d}}{\mathrm{d}t}x = Ax + Bu, \quad y = Cx + Du. \tag{10.4}$$

The full behavior represented by these equations is given by

$$\mathfrak{B}_{ISO}(A, B, C, D) := \{(u, y, x) \in \mathfrak{C}^\infty(\mathbb{R}, \mathbb{R}^{\mathtt{m}}) \times \mathfrak{C}^\infty(\mathbb{R}, \mathbb{R}^{\mathtt{p}}) \times \mathfrak{C}^\infty(\mathbb{R}, \mathbb{R}^{\mathtt{n}}) \mid (10.4) \text{ holds}\}.$$

In (10.4), we interpret (u, y) as manifest variable and x as latent variable. Thus, $\mathfrak{B}_{ISO}$ is a latent variable representation of its external behavior given by

$$\mathfrak{B}_{ISO}(A, B, C, D)_{\mathrm{ext}} = \{(u, y) \mid \exists x \text{ such that } (u, y, x) \in \mathfrak{B}_{ISO}(A, B, C, D)\}.$$

In fact, in (10.4), x is a state variable, u has the property of input, and y the property of output, see [11]. Further, if $\mathfrak{B} = \mathfrak{B}_{ISO}(A, B, C, D)_{\mathrm{ext}}$ then we call $\mathfrak{B}_{ISO}$ an *input-state-output representation* of $\mathfrak{B}$.

Next, let $A \in \mathbb{R}^{\mathtt{n}\times\mathtt{n}}$, $B \in \mathbb{R}^{\mathtt{n}\times\mathtt{v}}$, $C \in \mathbb{R}^{\mathtt{w}\times\mathtt{n}}$, $D \in \mathbb{R}^{\mathtt{w}\times\mathtt{v}}$, and consider the equations

$$\frac{\mathrm{d}}{\mathrm{d}t}x = Ax + Bv, \quad w = Cx + Dv. \tag{10.5}$$

The full behavior represented by these equations is given by

$$\mathfrak{B}_{DV}(A, B, C, D) := \{(w, x, v) \in \mathfrak{C}^\infty(\mathbb{R}, \mathbb{R}^{\mathtt{w}}) \times \mathfrak{C}^\infty(\mathbb{R}, \mathbb{R}^{\mathtt{n}}) \times \mathfrak{C}^\infty(\mathbb{R}, \mathbb{R}^{\mathtt{v}}) \mid (10.5) \text{ holds}\}.$$

In (10.5), we interpret w as manifest variable and (x, v) as latent variables. Thus, $\mathfrak{B}_{DV}$ is a latent variable representation of its external behavior given by

$$\mathfrak{B}_{DV}(A, B, C, D)_{\mathrm{ext}} = \{w \mid \exists (x, v) \text{ such that } (w, x, v) \in \mathfrak{B}_{DV}(A, B, C, D)\}.$$

In fact, in (10.5), x is a state variable and v is an auxiliary variable, called the *driving variable*. Further, if $\mathfrak{B} = \mathfrak{B}_{DV}(A, B, C, D)_{\text{ext}}$ then we call $\mathfrak{B}_{DV}$ a *driving variable representation* of $\mathfrak{B}$.

Finally, let $A \in \mathbb{R}^{\mathtt{n}\times\mathtt{n}}$, $B \in \mathbb{R}^{\mathtt{n}\times\mathtt{w}}$, $C \in \mathbb{R}^{\mathtt{p}\times\mathtt{n}}$, $D \in \mathbb{R}^{\mathtt{p}\times\mathtt{w}}$ and consider the equations

$$\frac{\mathrm{d}}{\mathrm{d}t}x = Ax + Bw, \quad 0 = Cx + Dw. \tag{10.6}$$

The full behavior represented by these equations is given by

$$\mathfrak{B}_{ON}(A, B, C, D) := \{(w, x) \in \mathfrak{C}^{\infty}(\mathbb{R}, \mathbb{R}^{\mathtt{w}}) \times \mathfrak{C}^{\infty}(\mathbb{R}, \mathbb{R}^{\mathtt{n}}) \mid (10.6) \text{ holds}\}.$$

In (10.6), we interpret w as manifest variable and x as a latent variable. Thus, $\mathfrak{B}_{ON}$ is a latent variable representation of its external behavior given by

$$\mathfrak{B}_{ON}(A, B, C, D)_{\text{ext}} = \{w \mid \exists x \text{ such that } (w, x) \in \mathfrak{B}_{ON}(A, B, C, D)\}.$$

Also in (10.6), x is a state variable. Further, if $\mathfrak{B} = \mathfrak{B}_{ON}(A, B, C, D)_{\text{ext}}$ then we call $\mathfrak{B}_{ON}$ an *output nulling representation* of $\mathfrak{B}$.

An important concept in the behavioral approach is the property of *controllability*. By now, the definition of controllability of behaviors of linear differential systems is well known, and for its definition we refer to [11]. It was shown in [11] that controllable behaviors admit, yet another special kind of latent variable representations called *image representations*. Consider the differential equation

$$w = M(\frac{\mathrm{d}}{\mathrm{d}t})\ell, \tag{10.7}$$

where $M \in \mathbb{R}[s]^{\mathtt{w}\times\mathtt{l}}$. The w's that satisfy (10.7) are given by

$$\mathfrak{B} = \{w \in \mathfrak{C}^{\infty}(\mathbb{R}, \mathbb{R}^{\mathtt{w}}) \mid \exists \ell \in \mathfrak{C}^{\infty}(\mathbb{R}, \mathbb{R}^{\mathtt{l}}) \text{ such that } w = M(\frac{\mathrm{d}}{\mathrm{d}t})\ell\}. \tag{10.8}$$

A representation of $\mathfrak{B}$ as in (10.8) is called a *polynomial image representation*. Like for polynomial kernel representations previously in this paper, this can be extended to rational image representations as well, see [17]. Let $M \in \mathbb{R}(s)^{\mathtt{w}\times\mathtt{l}}$. Then a meaning to the equation $w = M(\frac{\mathrm{d}}{\mathrm{d}t})\ell$ can be given by interpreting it as

$$\begin{bmatrix} I & -M(\frac{\mathrm{d}}{\mathrm{d}t}) \end{bmatrix} \begin{bmatrix} w \\ \ell \end{bmatrix} = 0. \tag{10.9}$$

Then, if $M = P^{-1}Q$ is a left coprime factorization over $\mathbb{R}[s]$, $\begin{bmatrix} I & -M \end{bmatrix} = P^{-1}\begin{bmatrix} P & -Q \end{bmatrix}$ clearly is a left coprime factorization of $\begin{bmatrix} I & -M \end{bmatrix}$, and therefore, by the definition in (10.2), Eq. (10.9) holds if and only if $P(\frac{\mathrm{d}}{\mathrm{d}t})w = Q(\frac{\mathrm{d}}{\mathrm{d}t})\ell$. Thus $w = M(\frac{\mathrm{d}}{\mathrm{d}t})\ell$ should be interpreted as $P(\frac{\mathrm{d}}{\mathrm{d}t})w = Q(\frac{\mathrm{d}}{\mathrm{d}t})\ell$ and

$$\mathfrak{B} := \{w \mid \exists \ell \text{ such that } w = M(\frac{\mathrm{d}}{\mathrm{d}t})\ell\} = \{w \mid \exists \ell \text{ such that } P(\frac{\mathrm{d}}{\mathrm{d}t})w = Q(\frac{\mathrm{d}}{\mathrm{d}t})\ell\}. \tag{10.10}$$

This representation of $\mathfrak{B}$ is called a *rational image representation*.

10.4 Elimination from State Representations

In this section we will review results on the elimination problem for state representations from [5]. For a given state representation, the elimination problem is to obtain a polynomial kernel representation of its external behavior, thus 'eliminating' the state variable from the original representation. Before addressing this problem, we will first review some basic material on the general elimination problem in the polynomial representation context. First, recall the notion of minimal left annihilator of a polynomial matrix (see [19]).

Definition 10.1 Let $M \in \mathbb{R}[s]^{\mathtt{m}\times\bullet}$. Then $X \in \mathbb{R}[s]^{\bullet\times\mathtt{m}}$ is called a minimal left annihilator of M if :

1. X has full row rank,
2. X is left annihilator of M, i.e. $XM = 0$, and
3. any left annihilator of M is a multiple of X, i.e. $X_1M = 0$ implies $X_1 = RX$ for some polynomial matrix R.

The following well known result gives a rank characterization of the minimal left annihilator of a given polynomial matrix. For a proof we refer to [5].

Proposition 10.2 *Let $M \in \mathbb{R}[s]^{\mathtt{p}\times\mathtt{q}}$. Then $X \in \mathbb{R}[s]^{\mathtt{n}\times\mathtt{p}}$ is a minimal left annihilator of M if and only if X is left prime over $\mathbb{R}[s]$, $XM = 0$ and* $\mathrm{rank}(X) = \mathtt{p} - \mathrm{rank}(M)$.

Using the above characterization of a minimal left annihilator, we have the following proposition that plays a crucial role in the rest of this section. The result is well known and has appeared in various forms in the literature before (see [11], Theorem 6.2.6, [3, 10]).

Proposition 10.3 *Let $\mathfrak{B}_{\mathrm{full}} \in \mathfrak{L}^{\mathtt{w}+\mathtt{l}}$ be represented by the polynomial latent variable representation $R(\frac{\mathrm{d}}{\mathrm{d}t})w = M(\frac{\mathrm{d}}{\mathrm{d}t})\ell$, where $R \in \mathbb{R}[s]^{\mathtt{p}\times\mathtt{w}}$, $M \in \mathbb{R}[s]^{\mathtt{p}\times\mathtt{l}}$. Then the manifest behavior $(\mathfrak{B}_{\mathrm{full}})_w$ has kernel representation $(XR)(\frac{\mathrm{d}}{\mathrm{d}t})w = 0$, where X is a minimal left annihilator of M.*

An immediate consequence of Proposition 10.2 is the following:

Lemma 10.4 *Let $G \in \mathbb{R}(s)^{\mathtt{p}\times\mathtt{w}}$. Let $G = AB^{-1}$ be a factorization such that $A \in \mathbb{R}[s]^{\mathtt{p}\times\mathtt{w}}$ and $B \in \mathbb{R}[s]^{\mathtt{w}\times\mathtt{w}}$. Let $L_2^{-1}L_1$ be a left coprime factorization of G over $\mathbb{R}[s]$. Then $\begin{bmatrix} L_1 & -L_2 \end{bmatrix}$ is a minimal left annihilator of $\begin{bmatrix} B \\ A \end{bmatrix}$.*

Also the following easy result will be instrumental in the sequel:

Lemma 10.5 *Let $M \in \mathbb{R}[s]^{\mathtt{p}\times\mathtt{q}}$ be partitioned as $M = \begin{bmatrix} M_1 & M_2 \end{bmatrix}$. If X_1 is a minimal left annihilator of M_1, and X_2 is a minimal left annihilator of X_1M_2 then X_2X_1 is a minimal left annihilator of M.*

We will now address the problem of eliminating the state in an input-state-output representation of a given behavior. This problem was considered before in [11] for the single input, single output case. Consider the input-state-output representation $\frac{\mathrm{d}}{\mathrm{d}t}x = Ax + Bu$, $y = Cx + Du$, and as before denote its full behavior by $\mathfrak{B}_{ISO}(A, B, C, D)$. Our aim is to find a polynomial kernel representation of the external behavior $\mathfrak{B}_{ISO}(A, B, C, D)_{\mathrm{ext}}$. Obviously, by choosing a basis of the state space $\mathbb{R}^{\mathtt{n}}$ adapted to the decomposition $\mathbb{R}^{\mathtt{n}} = \mathcal{X}_1 \oplus \mathcal{X}_2$, with $\mathcal{X}_1 := \mathcal{R}$, we may assume that the matrices A, B, C are in the form

$$A = \begin{bmatrix} A_{11} & A_{12} \\ 0 & A_{22} \end{bmatrix}, \; B = \begin{bmatrix} B_1 \\ 0 \end{bmatrix}, \text{ and } C = \begin{bmatrix} C_1 & C_2 \end{bmatrix}, \tag{10.11}$$

such that (A_{11}, B_1) is a controllable pair. The dimensions of A_{11} and A_{22} are assumed to be $\mathtt{n}_1 \times \mathtt{n}_1$ and $\mathtt{n}_2 \times \mathtt{n}_2$ respectively.

Theorem 10.6 *Let $\mathfrak{B}_{ISO}(A, B, C, D)$ be the full behavior induced by the input-state-output representation $\frac{\mathrm{d}}{\mathrm{d}t}x = Ax + Bu$, $y = Cx + Du$. Assume that A, B, C have the form* (10.11). *Let $L_2^{-1}L_1 = C_1(sI - A_{11})^{-1}$ and $K_2^{-1}K_1 = (L_1A_{12} + L_2C_2)(sI - A_{22})^{-1}$ be left coprime factorizations over $\mathbb{R}[s]$. Then*

$$\mathfrak{B}_{ISO}(A, B, C, D)_{\mathrm{exr}} = \ker[\; K_2(L_1B_1 + L_2D)(\frac{\mathrm{d}}{\mathrm{d}t}) \quad - (K_2L_2)(\frac{\mathrm{d}}{\mathrm{d}t})].$$

Proof Clearly, the full behavior is represented by $R(\frac{\mathrm{d}}{\mathrm{d}t})w = M(\frac{\mathrm{d}}{\mathrm{d}t})x$, where

$$M = \begin{bmatrix} M_1 & M_2 \end{bmatrix} = \begin{bmatrix} sI - A_{11} & -A_{12} \\ 0 & sI - A_{22} \\ C_1 & C_2 \end{bmatrix} \text{ and } R = \begin{bmatrix} B_1 & 0 \\ 0 & 0 \\ -D & I \end{bmatrix}. \tag{10.12}$$

Note that $X_1M_1 = 0$ and $X_2X_1M_2 = 0$, where $X_1 = \begin{bmatrix} L_1 & 0 & -L_2 \\ 0 & I & 0 \end{bmatrix}$ and $X_2 = \begin{bmatrix} K_2 & K_1 \end{bmatrix}$. It follows from Lemma 10.4 that X_1 and X_2 are minimal left annihilators of M_1 and X_1M_2, respectively. Using Lemma 10.5, we then have that X_2X_1 is a minimal left annihilator of M. Then, by Proposition 10.3 it is evident that the external behavior, i.e., $\mathfrak{B}_{ISO}(A, B, C, D)_{\mathrm{ext}}$, is given by

$$\ker(X_2X_1R)(\frac{\mathrm{d}}{\mathrm{d}t}) = \ker\;[K_2(L_1B_1 + L_2D)(\frac{\mathrm{d}}{\mathrm{d}t}) \quad - (L_2K_2)(\frac{\mathrm{d}}{\mathrm{d}t})].$$

The case of output nulling representations was treated in detail in [5], see also [13]. We will confine ourselves here to formulating the result. The proof follows the same limes as the proof of Theorem 10.6.

Theorem 10.7 *Let* $\mathfrak{B}_{ON}(A, B, C, D)$ *be the full behavior induced by the output nulling representation* $\frac{\mathrm{d}}{\mathrm{d}t}x = Ax + Bw,\ 0 = Cx + Dw$. *Assume that* A, B, C *have the form* (10.11). *Let* $L_2^{-1}L_1 = C_1(sI - A_{11})^{-1}$ *and* $K_2^{-1}K_1 = (L_1A_{12} + L_2C_2)(sI - A_{22})^{-1}$ *be left coprime factorizations over* $\mathbb{R}[s]$. *Then*

$$\mathfrak{B}_{ON}(A, B, C, D)_{\text{ext}} = \ker\ K_2(L_1B_1 + L_2D)(\frac{\mathrm{d}}{\mathrm{d}t}).$$

Finally, we turn to elimination in driving variable representations. Here the problem is to eliminate both the state variable as well as the driving variable, and obtain a polynomial kernel representation of the external behavior. This problem is dealt within the following theorem, which was proven in [5].

Theorem 10.8 *Let* $\mathfrak{B}_{DV}(A, B, C, D)$ *be the full behavior induced by the driving variable representation* $\frac{\mathrm{d}}{\mathrm{d}t}x = Ax + Bv,\ w = Cx + Dv$. *Assume that* A, B, C *are as in* (10.11). *Let* $L_2^{-1}L_1 = C_1(sI - A_{11})^{-1}$ *and* $K_2^{-1}K_1 = (L_1A_{12} + L_2C_2)(sI - A_{22})^{-1}$ *be left coprime factorizations over* $\mathbb{R}[s]$. *Then*

$$\mathfrak{B}_{DV}(A, B, C, D)_{\text{ext}} = \ker\ (QK_2L_2)(\frac{\mathrm{d}}{\mathrm{d}t}),$$

where Q *is any minimal left annihilator of* $K_2(L_1B_1 + L_2D)$.

10.5 Rational Representations and Realizations

In this section we will formally state and address the problems announced in the introduction to this paper. The following problems will be considered:

1. Let $G(s)$ be a proper real rational matrix and let (A, B, C, D) be a realization of G, i.e. $G(s) = C(sI - A)^{-1}B + D$. Consider the input–output behavior $\{(u, y) \mid y = G(\frac{\mathrm{d}}{\mathrm{d}t})u\}$ associated with G, where $y = G(\frac{\mathrm{d}}{\mathrm{d}t})u$ should be interpreted as the rational kernel representation $[G(\frac{\mathrm{d}}{\mathrm{d}t})\ \ -I]\begin{bmatrix} u \\ y \end{bmatrix} = 0$. The problem is to obtain necessary and sufficient conditions such that

$$\mathfrak{B}_{ISO}(A, B, C, D)_{\text{ext}} = \{(u, y) \mid y = G(\frac{\mathrm{d}}{\mathrm{d}t})u\}. \tag{10.13}$$

2. Let $G(s)$ be a proper real rational matrix and let (A, B, C, D) be a realization of G. The problem is to find necessary and sufficient conditions such that

$$\mathfrak{B}_{ON}(A, B, C, D)_{\text{ext}} = \ker\ G(\frac{\mathrm{d}}{\mathrm{d}t}).$$

3. Let $G(s)$ be a proper real rational matrix and let (A, B, C, D) be a realization of G. The problem is to obtain necessary and sufficient conditions such that

$$\mathfrak{B}_{DV}(A, B, C, D)_{\mathrm{ext}} = \mathrm{im}\ G(\frac{\mathrm{d}}{\mathrm{d}t}).$$

In the remainder of this section we will subsequently address all three problems stated here.

We start off with a lemma that will be instrumental in solving the first of these problems. The result states that condition (10.13) is equivalent with the condition that a particular rational matrix obtained from the triple (A, B, C) is polynomial.

Lemma 10.9 *Let $G \in \mathbb{R}(s)_p^{\bullet\times\bullet}$. Let (A, B, C, D) be a realization of G such that (10.11) holds. Let $L_2^{-1}L_1 = C_1(sI - A_{11})^{-1}$ be a left coprime factorizations over $\mathbb{R}[s]$. Then the following statements are equivalent:*

1. $\mathfrak{B}_{ISO}(A, B, C, D)_{\mathrm{ext}} = \{(u, y) \mid y = G(\frac{\mathrm{d}}{\mathrm{d}t})u\}$,
2. $(L_1A_{12} + L_2C_2)(sI - A_{22})^{-1}$ *is a polynomial matrix.*

Proof (1 ⇒ 2) Assume that condition 1. holds. We know that

$$\begin{aligned}&\ker\,[G(\frac{\mathrm{d}}{\mathrm{d}t}) - I]\\ &= \ker\,[(C_1(sI - A_{11}))B_1 + D)(\frac{\mathrm{d}}{\mathrm{d}t}) - I]\\ &= \ker L_2^{-1}[L_1B_1 + L_2D - L_2](\frac{\mathrm{d}}{\mathrm{d}t}).\end{aligned}$$

It is easily verified that $L_2^{-1}[L_1B_1 + L_2D - L_2])$ is also a left coprime factorization over $\mathbb{R}[s]$. From the definition in (10.2) it then follows that

$$\ker\,[G(\frac{\mathrm{d}}{\mathrm{d}t}) - I] = \ker\,[L_1B_1 + L_2D - L_2](\frac{\mathrm{d}}{\mathrm{d}t}). \tag{10.14}$$

Let $K_2^{-1}K_1 = (L_1A_{12}+L_2C_2)(sI-A_{22})^{-1}$ be a left coprime factorization over $\mathbb{R}[s]$. Then, from Theorem 10.6 we have

$$\mathfrak{B}_{ISO}(A, B, C, D)_{\mathrm{exr}} = \ker K_2[L_1B_1 + L_2D - L_2](\frac{\mathrm{d}}{\mathrm{d}t}). \tag{10.15}$$

Since L_2 is nonsingular $[L_1B_1 + L_2D - L_2]$ has full row rank. Also, since K_2 is nonsingular, $K_2[L_1B_1 + L_2D - L_2]$ has full row rank. Since

$$\ker K_2[L_1B_1 + L_2D - L_2](\frac{\mathrm{d}}{\mathrm{d}t}) = \ker\,[L_1B_1 + L_2D - L_2](\frac{\mathrm{d}}{\mathrm{d}t}),$$

it follows from Theorem 3.6.2 in [11] that there exists a unimodular matrix U such that

$$[L_1B_1 + L_2D \;\; -L_2] = UK_2[L_1B_1 + L_2D \;\; -L_2] \tag{10.16}$$

We prove now that K_2 is unimodular. Rewriting Eq. (10.16) we have

$$(I - UK_2)[L_1B_1 + L_2D \;\; -L_2] = 0. \tag{10.17}$$

Since the second factor in this equation has full row rank, it follows that $UK_2 = I$ and hence K_2 is unimodular. It follows that $K_2^{-1}K_1 = (L_1A_{12} + L_2C_2)(sI - A_{22})^{-1}$ is a polynomial matrix.
(2 $\Rightarrow$ 1) Assume $(L_1A_{12} + L_2C_2)(sI - A_{22})^{-1}$ is a polynomial matrix. Let $K_2 = I$ and $K_1 = (L_1A_{12}+L_2C_2)(sI-A_{22})^{-1}$. Then $K_2^{-1}K_1 = (L_1A_{12}+L_2C_2)(sI-A_{22})^{-1}$ is a left coprime factorization over $\mathbb{R}[s]$. Then the result follows from Theorem 10.6 and the fact that (10.14) holds.

A similar result as Lemma 10.9 will be instrumental to treat the case of output nulling representations:

Lemma 10.10 *Let $G \in \mathbb{R}(s)_p^{\bullet\times\bullet}$ have full row rank. Let (A, B, C, D) be a realization of G in the form (10.11). Let $L_2^{-1}L_1 = C_1(sI-A_{11})^{-1}$. Then the following statements are equivalent:*

1. $\mathfrak{B}_{ON}(A, B, C, D)_{\text{ext}} = \ker G(\frac{\mathrm{d}}{\mathrm{d}t})$.
2. $(L_1A_{12} + L_2C_2)(sI - A_{22})^{-1}$ *is a polynomial matrix.*

Proof (1 $\Rightarrow$ 2) Assume that condition 1. holds. We have

$$\ker G(\frac{\mathrm{d}}{\mathrm{d}t}) = \ker\,(C_1(sI - A_{11}))B_1 + D)(\frac{\mathrm{d}}{\mathrm{d}t}) = \ker L_2^{-1}(L_1B_1 + L_2D)(\frac{\mathrm{d}}{\mathrm{d}t}).$$

Obviously, $L_2^{-1}(L_1B_1 + L_2D)$ is also a left coprime factorization over $\mathbb{R}[s]$. By definition we then have

$$\ker G(\frac{\mathrm{d}}{\mathrm{d}t}) = \ker L_2^{-1}(L_1B_1 + L_2D)(\frac{\mathrm{d}}{\mathrm{d}t}) = \ker\,(L_1B_1 + L_2D)(\frac{\mathrm{d}}{\mathrm{d}t}). \tag{10.18}$$

On the other hand, after taking left coprime factorisation

$$K_2^{-1}K_1 = (L_1A_{12} + L_2C_2)(sI - A_{22})^{-1},$$

from Theorem 10.7 we obtain

$$\mathfrak{B}_{ON}(A, B, C, D)_{\text{ext}} = \ker K_2(L_1B_1 + L_2D)(\frac{\mathrm{d}}{\mathrm{d}t}). \tag{10.19}$$

Since $G = L_2^{-1}(L_1B_1+L_2D)$ and G has full row rank, we must have that $L_1B_1+L_2D$ has full row rank. Since $\ker K_2(L_1B_1 + L_2D)(\frac{\mathrm{d}}{\mathrm{d}t}) = \ker\,(L_1B_1 + L_2D)(\frac{\mathrm{d}}{\mathrm{d}t})$ it then follows from Theorem 3.6.2 in [11] that

$$L_1B_1 + L_2D = UK_2(L_1B_1 + L_2D) \tag{10.20}$$

for some unimodular matrix U. Again we prove now that K_2 is unimodular. Rewriting equation (10.20) we get

$$(I - UK_2)(L_1B_1 + L_2D) = 0. \tag{10.21}$$

Since $L_1B_1 + L_2D$ has full row rank we must have $UK_2 = I$, so K_2 is unimodular. Hence $K_2^{-1}K_1 = (L_1A_{12} + L_2C_2)(sI - A_{22})^{-1}$ is a polynomial matrix.
($2 \Rightarrow 1$) Assume $(L_1A_{12} + L_2C_2)(sI - A_{22})^{-1}$ is a polynomial matrix. Let $K_2 = I$ and $K_1 = (L_1A_{12}+L_2C_2)(sI-A_{22})^{-1}$. Then $K_2^{-1}K_1 = (L_1A_{12}+L_2C_2)(sI-A_{22})^{-1}$ is a left coprime factorization over $\mathbb{R}[s]$. Then the result follows from Theorem 10.7 together with (10.18).

In order to proceed now, we will need a finer decomposition of the state space than the one used in (10.11), namely the classical Kalman decomposition of linear state space systems. This will be reviewed now. For a given triple (A, B, C) with $A \in \mathbb{R}^{n\times n}$, $B \in \mathbb{R}^{n\times m}$, and $C \in \mathbb{R}^{p\times n}$, let $\mathcal{R}$ be the reachable subspace, and $\mathcal{N}$ the unobservable subspace. Now define $\mathcal{X}_1 := \mathcal{R} \cap \mathcal{N}$, and let $\mathcal{X}_2$ be a subspace such that $\mathcal{X}_1 \oplus \mathcal{X}_2 = \mathcal{R}$. Let $\mathcal{X}_3$ be such that $\mathcal{X}_1 \oplus \mathcal{X}_3 = \mathcal{N}$. Finally, let $\mathcal{X}_4$ be such that $\mathcal{X}_1 \oplus \mathcal{X}_2 \oplus \mathcal{X}_3 \oplus \mathcal{X}_4 = \mathbb{R}^n$.

Then, with respect to a basis adapted to this decomposition, A, B and C have the form

$$A = \left[\begin{array}{cc|cc} A_{11} & A_{12} & A_{13} & A_{14} \\ 0 & A_{22} & 0 & A_{24} \\ \hline 0 & 0 & A_{33} & A_{34} \\ 0 & 0 & 0 & A_{44} \end{array}\right], \quad B = \left[\begin{array}{c} B_1 \\ B_2 \\ \hline 0 \\ 0 \end{array}\right], \text{ and } C = \left[\begin{array}{cc|cc} 0 & C_2 & 0 & C_4 \end{array}\right]. \tag{10.22}$$

The lines in the matrices above indicate that, in fact, the system can be interpreted to be in the form (10.11). Note that $\mathcal{R} + \mathcal{N} = \mathcal{X}_1 \oplus \mathcal{X}_2 \oplus \mathcal{X}_3$. It is well known that the following properties hold:

1. (A_{22}, B_2, C_2) is controllable and observable.
2. $\left(\begin{bmatrix} A_{11} & A_{12} \\ 0 & A_{22} \end{bmatrix}, \begin{bmatrix} B_1 \\ B_2 \end{bmatrix}, \begin{bmatrix} 0 & C_2 \end{bmatrix}\right)$ is controllable.
3. $\left(\begin{bmatrix} A_{22} & A_{24} \\ 0 & A_{44} \end{bmatrix}, \begin{bmatrix} B_2 \\ 0 \end{bmatrix}, \begin{bmatrix} C_2 & C_4. \end{bmatrix}, \right)$ is observable.

In the following three subsections, we will now resolve the three problems posed in the introduction to this section.

10.5.1 Realization and Input-State-Output Representation

The following theorem is one of main results of this paper. It states that the external behavior of an input-state-output representation is equal to the input–output behavior associated with its associated transfer matrix if and only if the sum of the reachable subspace and the unobservable subspace is equal to the entire state space.

Theorem 10.11 *Let $G \in \mathbb{R}(s)_p^{\mathtt{p}\times\mathtt{m}}$. Let (A, B, C, D) be a realization of G with $A \in \mathbb{R}^{\mathtt{n}\times\mathtt{n}}$, $B \in \mathbb{R}^{\mathtt{n}\times\mathtt{m}}$, $C \in \mathbb{R}^{\mathtt{p}\times\mathtt{n}}$, $D \in \mathbb{R}^{\mathtt{p}\times\mathtt{m}}$. Then*

$$\mathfrak{B}_{ISO}(A, B, C, D)_{\mathrm{ext}} = \{(u, y) \mid y = G(\frac{\mathrm{d}}{\mathrm{d}t})u\}$$

if and only if $\mathcal{R} + \mathcal{N} = \mathbb{R}^{\mathtt{n}}$.

Proof (⇐) Assume $\mathcal{R} + \mathcal{N} = \mathbb{R}^n$. Without loss of generality we can assume that (A, B, C, D) is of the form

$$A = \left[\begin{array}{cc|c} A_{11} & A_{12} & A_{13} \\ 0 & A_{22} & 0 \\ \hline 0 & 0 & A_{33} \end{array}\right], B = \left[\begin{array}{c} B_1 \\ B_2 \\ \hline 0 \end{array}\right], C = \left[\begin{array}{cc|c} 0 & C_2 & 0 \end{array}\right]. \tag{10.23}$$

The idea is now to apply Lemma 10.9 and check that condition 2. in that lemma holds. As stated in Lemma 10.9 take a left coprime factorization

$$L_2^{-1}L_1 = \begin{bmatrix} 0 & C_2 \end{bmatrix} \begin{bmatrix} Is - A_{11} & -A_{12} \\ 0 & sI - A_{22} \end{bmatrix}^{-1} = \begin{bmatrix} 0 & C_2(sI - A_{22})^{-1} \end{bmatrix}. \tag{10.24}$$

Since L_2 is nonsingular we can split up L_1 into blocks of appropriate sizes: $L_1 = \begin{bmatrix} 0 & L_{12} \end{bmatrix}$. We now check that

$$(\begin{bmatrix} 0 & L_{12} \end{bmatrix} \begin{bmatrix} A_{13} \\ 0 \end{bmatrix} + L_2 \cdot 0)(sI - A_{33})^{-1} = \begin{bmatrix} 0 & 0 \end{bmatrix}. \tag{10.25}$$

Since this is a polynomial matrix it follows that condition 2. in Lemma 10.9 holds. From this the claim follows.

(⇒) Assume $\mathcal{R} + \mathcal{N} \neq \mathbb{R}^n$. Without loss of generality assume that the system is in the form (10.22), with $A_{11} \in \mathbb{R}^{\mathtt{n}_1\times\mathtt{n}_1}$, $A_{22} \in \mathbb{R}^{\mathtt{n}_2\times\mathtt{n}_2}$, $A_{33} \in \mathbb{R}^{\mathtt{n}_3\times\mathtt{n}_3}$, $A_{44} \in \mathbb{R}^{\mathtt{n}_4\times\mathtt{n}_4}$ and B, C with appropriate dimensions. Note that by our assumption the matrix A_{44} is non void, i.e. $\mathtt{n}_4 > 0$. Likewise, the corresponding 0-matrix in B and the matrix C_4 are really present. Recall that the subsystem

$$\left(\begin{bmatrix} A_{22} & A_{24} \\ 0 & A_{44} \end{bmatrix}, \begin{bmatrix} B_2 \\ 0 \end{bmatrix}, \begin{bmatrix} C_2 & C_4 \end{bmatrix}, D\right) \tag{10.26}$$

is observable. Define

$$\mathcal{O}(s) := \begin{bmatrix} A_{22} - sI & A_{24} \\ 0 & A_{44} - sI \\ C_2 & C_4 \end{bmatrix}$$

By observability we have $\text{rank}(\mathcal{O}(\lambda)) = \text{n}_2 + \text{n}_4$ for all $\lambda \in \mathbb{C}$.

Note again that the system (10.22) is obviously of the form (10.11). Let

$$L_2^{-1} L_1 = \begin{bmatrix} 0 & C_2 \end{bmatrix} \begin{bmatrix} sI - A_{11} & -A_{12} \\ 0 & sI - A_{22} \end{bmatrix}^{-1} = \begin{bmatrix} 0 & C_2(sI - A_{22})^{-1} \end{bmatrix} \tag{10.27}$$

be a left coprime factorization over $\mathbb{R}[s]$. Let $L_1 = \begin{bmatrix} 0 & L_{12} \end{bmatrix}$ be split up into appropriate sized blocks. Then $L_2^{-1} L_{12} = C_2(sI - A_{22})^{-1}$ is also a left coprime factorization over $\mathbb{R}[s]$.

Since $\begin{bmatrix} L_{12} & L_2 \end{bmatrix}$ has full row rank for all $\lambda \in \mathbb{C}$, there exist polynomial matrices $M_1 \in \mathbb{R}[s]^{\text{n}_2 \times \text{n}_2}$, $M_2 \in \mathbb{R}[s]^{\text{n}_2 \times \text{p}}$ such that the matrix

$$U_1 = \begin{bmatrix} L_{12} & 0 & L_2 \\ 0 & I & 0 \\ M_1 & 0 & M_2 \end{bmatrix} \tag{10.28}$$

is unimodular. By premultiplying $\mathcal{O}(\lambda)$ by $U_1(\lambda)$, with $\lambda \in \mathbb{C}$, we obtain

$$U_1(\lambda)\mathcal{O}(\lambda) = \begin{bmatrix} L_{12} & 0 & L_2 \\ 0 & I & 0 \\ M_1 & 0 & M_2 \end{bmatrix} \begin{bmatrix} A_{22} - \lambda I & A_{24} \\ 0 & A_{44} - \lambda I \\ C_2 & C_4 \end{bmatrix} = \begin{bmatrix} 0 & L_{12}A_{24} + L_2 C_4 \\ 0 & A_{44} - \lambda I \\ M_1 A_{24} + M_2 C_2 & M_1 A_{24} + M_2 C_4 \end{bmatrix} \tag{10.29}$$

and $\text{rank}(U_1(\lambda)\mathcal{O}(\lambda)) = \text{n}_2 + \text{n}_4$ for all $\lambda \in \mathbb{C}$. In this formula we have suppressed some of the lambda's. Now consider the rational matrix

$$\begin{aligned} R(s) &:= \left(\begin{bmatrix} 0 & L_{12} \end{bmatrix} \begin{bmatrix} A_{13} & A_{14} \\ 0 & A_{24} \end{bmatrix} + L_2 \begin{bmatrix} 0 & C_4 \end{bmatrix} \right) \begin{bmatrix} sI - A_{33} & -A_{34} \\ 0 & sI - A_{44} \end{bmatrix}^{-1} \\ &= \begin{bmatrix} 0 & L_{12}A_{24} + L_2 C_4 \end{bmatrix} \begin{bmatrix} sI - A_{33} & -A_{34} \\ 0 & sI - A_{44} \end{bmatrix}^{-1} \\ &= \begin{bmatrix} 0 & (L_{12}A_{24} + L_2 C_4)(sI - A_{44})^{-1} \end{bmatrix}. \end{aligned} \tag{10.30}$$

Since, $\mathfrak{B}_{ISO}(A, B, C, D)_{\text{ext}} = \{(u, y) \mid y = G(\frac{\text{d}}{\text{d}t})u\}$, Lemma 10.9 says that $R(s)$ given by (10.30) is a polynomial matrix. Define a polynomial matrix K by

$$K(s) := (L_{12}A_{24} + L_2 C_4)(sI - A_{44})^{-1}.$$

Since $\begin{bmatrix} -I & K(\lambda) \end{bmatrix}$ also has full row rank for all $\lambda \in \mathbb{C}$ there exist polynomial matrices $N_1 \in \mathbb{R}[s]^{\text{n}_4 \times \text{p}}$, $N_2 \in \mathbb{R}[s]^{\text{n}_4 \times \text{n}_4}$ such that the matrix

$$U_2 = \begin{bmatrix} -I & K & 0 \\ N_1 & N_2 & 0 \\ 0 & 0 & I \end{bmatrix} \tag{10.31}$$

is unimodular. Now define $\mathcal{O}_1(s) := U_2(s)U_1(s)\mathcal{O}(s)$. Then

$$\mathcal{O}_1(s) = \begin{bmatrix} 0 & 0 \\ 0 & (N_1K + N_2)(A_{44} - sI) \\ M_1A_{24} + M_2C_2 & M_1A_{24} + M_2C_4 \end{bmatrix}. \tag{10.32}$$

We know that $\mathrm{rank}(\mathcal{O}_1(\lambda)) = \mathrm{rank}(\mathcal{O}(\lambda)) = \mathrm{n}_2 + \mathrm{n}_4$ for all λ and hence

$$\mathrm{rank} \begin{bmatrix} 0 & (N_1K + N_2)(A_{44} - \lambda I) \\ M_1A_{24} + M_2C_2 & M_1A_{24} + M_2C_4 \end{bmatrix} = \mathrm{n}_2 + \mathrm{n}_4$$

for all λ. The latter is, however, equivalent with

$$\det(M_1A_{24} + M_2C_2)\det(N_1K + N_2)\det(A_{44} - \lambda I) \neq 0 \text{ for all } \lambda \in \mathbb{C}.$$

This leads to a contradiction, so $\mathcal{X}_4$ can not be present in the direct sum decomposition, and hence $\mathcal{R} + \mathcal{N} = \mathbb{R}^n$.

The above theorem settles the first problem posed in this section. We will now turn to the second one on driving variable representations.

10.5.2 Realization and Output Nulling Representation

In [5] it was shown that if G is a proper rational matrix and (A, B, C, D) is a realization, then controllability of the pair (A, B) is a *sufficient* condition for the equality $\mathfrak{B}_{ON}(A, B, C, D)_{\mathrm{ext}} = \ker\ G(\frac{\mathrm{d}}{\mathrm{d}t})$ to hold. It was also shown in [5] that the controllability condition is *not necessary*. The following theorem is the second main result of this paper. It sharpens the result from [5] and gives necessary *and* sufficient conditions:

Theorem 10.12 *Let $G \in \mathbb{R}(s)_p^{\mathrm{p}\times\mathrm{m}}$. Let (A, B, C, D) be a realization of G. If $\mathcal{R}+\mathcal{N} = \mathbb{R}^{\mathrm{n}}$ then $\mathfrak{B}_{ON}(A, B, C, D)_{\mathrm{ext}} = \ker G(\frac{\mathrm{d}}{\mathrm{d}t})$. Moreover, if G has full row rank then $\mathfrak{B}_{ON}(A, B, C, D)_{\mathrm{ext}} = \ker G(\frac{\mathrm{d}}{\mathrm{d}t})$ if and only if $\mathcal{R} + \mathcal{N} = \mathbb{R}^{\mathrm{n}}$.*

Proof ($\Leftarrow$) Assume $\mathcal{R} + \mathcal{N} = \mathbb{R}^{\mathrm{n}}$. Again we may assume that (A, B, C) has the form (10.23). As stated in Theorem 10.7, take a left coprime factorization

$$L_2^{-1}L_1 = \begin{bmatrix} 0 & C_2 \end{bmatrix} \begin{bmatrix} sI - A_{11} & -A_{12} \\ 0 & sI - A_{22} \end{bmatrix}^{-1} = \begin{bmatrix} 0 & C_2(sI - A_{22})^{-1} \end{bmatrix}. \tag{10.33}$$

Since L_2 is nonsingular we can split up L_1 into blocks of appropriate sizes: $L_1 = \begin{bmatrix} 0 \ L_{12} \end{bmatrix}$. We now check that

$$(\begin{bmatrix} 0 \ L_{12} \end{bmatrix} \begin{bmatrix} A_{13} \\ 0 \end{bmatrix} + L_2 \cdot 0)(sI - A_{33})^{-1} = \begin{bmatrix} 0 \ 0 \end{bmatrix}. \tag{10.34}$$

Note that

$$G(s) = L_2^{-1}(L_1 \begin{bmatrix} B_1 \\ B_2 \end{bmatrix} + L_2 D).$$

Since this is also a left coprime factorization, by definition we have

$$\ker \ G(\frac{\mathrm{d}}{\mathrm{d}t}) = \ker \ (L_1 \begin{bmatrix} B_1 \\ B_2 \end{bmatrix} + L_2 D)(\frac{\mathrm{d}}{\mathrm{d}t}). \tag{10.35}$$

On the other hand, since (10.34) admits the trivial coprime factorization $K_2^{-1} K_1$ with $K_2 = I$ and $K_1 = 0$, Theorem 10.7 confirms that (10.35) equals $\mathfrak{B}_{ON}(A, B, C, D)_{\text{ext}}$.
($\Rightarrow$) the proof of the second statement follows exacty the same reasoning as the corresponding part of Theorem 10.11 and uses Lemma 10.10.

10.5.3 Realization and Controllability of Driving Variable Representations

In this subsection we will resolve the third problem posed in this section, and establish necessary and sufficient conditions on the matrices A, B and C such that $\mathfrak{B}_{DV}(A, B, C, D)_{\text{ext}} = \text{im } G(\frac{\mathrm{d}}{\mathrm{d}t})$ for a given realization (A, B, C, D) of G.

It is obvious that if $\mathfrak{B}_{DV}(A, B, C, D)_{\text{ext}} = \text{im } G(\frac{\mathrm{d}}{\mathrm{d}t})$, then $\mathfrak{B}_{DV}(A, B, C, D)_{\text{ext}}$ must be controllable, as it is then represented in terms of a (rational) image representation, which always represents a controllable system. We will now prove a lemma that states that also the converse holds:

Lemma 10.13 *Let $G \in \mathbb{R}(s)_p^{\bullet \times \bullet}$. Let (A, B, C, D) be a realization of G. Then the following statements are equivalent.*

1. $\mathfrak{B}_{DV}(A, B, C, D)_{\text{ext}}$ *is controllable,*
2. $\mathfrak{B}_{DV}(A, B, C, D)_{\text{ext}} = \text{im } G(\frac{\mathrm{d}}{\mathrm{d}t})$.

Proof (1. $\Rightarrow$ 2.) Assume $\mathfrak{B}_{DV}(A, B, C, D)_{\text{ext}}$ is controllable. From Theorem 10.8 it follows that $\mathfrak{B}_{DV}(A, B, C, D)_{\text{ext}} = \ker (Q_2 K_2 L_2)(\frac{\mathrm{d}}{\mathrm{d}t})$, where Q_2 is a minimal left annihilator of $K_2(L_1 B_1 + L_2 D)$. Since the external behavior $\mathfrak{B}_{DV}(A, B, C, D)_{\text{ext}}$ is controllable it follows that $Q_2(\lambda) K_2(\lambda) L_2(\lambda)$ has full row rank for all $\lambda \in \mathbb{C}$. Therefore $Q_2(\lambda) K_2(\lambda)$ has full row rank for all $\lambda \in \mathbb{C}$. Let Q_1 be a minimal left annihilator of $L_1 B_1 + L_2 D$. Since obviously $Q_2 K_2$ is an annihilator as well, we must have $Q_2K_2 = XQ_1$ for some polynomial matrix X. As $Q_2(\lambda)K_2(\lambda)$ has full row rank for all $\lambda \in \mathbb{C}$ also $X(\lambda)$ has full row rank for all λ. From the fact that $\text{rank}(K_2(L_1B_1 + L_2D)) = \text{rank}(L_1B_1 + L_2D)$ it follows that the number of rows of Q_2 and Q_1 are the same. Hence X is square, and therefore unimodular. This now implies that $\mathfrak{B}_{DV}(A, B, C, D)_{\text{ext}} = \ker\,(Q_2K_2L_2)(\frac{d}{dt}) = \ker\,(Q_1L_2)(\frac{d}{dt})$.

Finally, we will prove that $\ker\,(Q_1L_2)(\frac{d}{dt}) = \text{im}\,G(\frac{d}{dt})$. Indeed, we have $G = L_2^{-1}(L_1B_1 + L_2D)$ is a left coprime factorization, so

$$\text{im}\,G(\frac{d}{dt}) = \{w \mid \text{ there exists } l \text{ such that } L_2(\frac{d}{dt})w = (L_1B_1 + L_2D)(\frac{d}{dt})l\}.$$

Clearly the full (w, l)-behavior of this is represented by the polynomial latent variable representation $L_2(\frac{d}{dt})w = (L_1B_1 + L_2D)(\frac{d}{dt})l$. Since Q_1 is a minimal left annihilator of $L_1B_1 + L_2D$, Proposition 10.3 then tells us that the result of eliminating l here is represented by $(Q_1L_2)(\frac{d}{dt})w = 0$. This proves our claim. The converse implication (2. $\Rightarrow$ 1.) is trivially true.

Thus, our third problem will be resolved if we can find necessary and sufficient conditions for controllability of the external behavior of a driving variable representation. We will investigate this now.

It will turn out that this involves the notion of *weakly unobservable subspace* associated with the system (A, B, C, D). This notion was studied in detail in [12]. We will review its definition and properties here. Originally, the weakly unobservable subspace was studied in the context of the *disturbance decoupling problem* for input-state-output systems. It consists of all initial states for which there exists an input function that makes the corresponding output function identically equal to zero:

Definition 10.14 Consider the system $\dot{x} = Ax + Bu,\ y = Cx + Du$, with state space $\mathbb{R}^n$. For given initial state $x_0 \in \mathbb{R}^n$ and input function u, denote the corresponding output by $y_u(t, x_0)$. Then the weakly unobservable subspace is defined as

$$\mathcal{V}^*(A, B, C, D) := \{x_0 \in \mathbb{R}^n \mid \text{ there exists } u \text{ such that } y_u(t, x_0) = 0 \text{ for all } t \geq 0\}$$

This subspace also plays a prominent role in a generalization of the classical Kalman decomposition, the so-called *ninefold canonical decomposition*, see [1]. It has the following feedback characterisation: $\mathcal{V}^*(A, B, C, D)$ is the largest subspace $\mathcal{V}$ of $\mathbb{R}^n$ for which there exists an F such that $(A + BF)\mathcal{V} \subset \mathcal{V}$ and $(C + DF)\mathcal{V} = \{0\}$. It can even be computed recursively from the data (A, B, C, D) in finitely many steps, see [12], Chap. 7.

An important property that we will need is that the sum of the reachable subspace and the weakly unobservable subspace is the so-called *output null controllable subspace* (see Exercise 4.1 in [12]):

Lemma 10.15 *Consider the system* $\dot{x} = Ax + Bu,\ y = Cx + Du$. *let* $\mathcal{R}$ *be the reachable subspace. Then we have:*

$$\mathcal{R} + \mathcal{V}^*(A, B, C, D) =$$
$$\{x_0 \in \mathbb{R}^n \mid \textit{there exists } u \textit{ and } T > 0 \textit{ such that } y_u(t, x_0) = 0 \textit{ for all } t \geq T\}.$$

For this reason, if $\mathcal{R} + \mathcal{V}^*(A, B, C, D) = \mathbb{R}^n$, then the system is sometimes called *output null controllable*.

We will now proceed with stating the third main result of this paper, giving necessary and sufficient conditions for controllability of the external behavior of a driving variable representation. The result states that controllability is equivalent to output null controllability, with the driving variable interpreted as input, and the manifest variable as output.

Theorem 10.16 *Let $G \in \mathbb{R}(\xi)_p^{p\times m}$. Let (A, B, C, D) be a realization of G. Then the following two statements are equivalent:*

1. *$\mathfrak{B}_{DV}(A, B, C, D)_{\text{ext}}$ is controllable,*
2. *$\mathcal{R} + \mathcal{V}^*(A, B, C, D) = \mathbb{R}^n$.*

Proof (1. $\Rightarrow$ 2.) For a given initial state x_0 and driving variable trajectory v, the resulting external trajectory will be denoted by $w_v(t, x_0)$. Now let $x_0 \in \mathbb{R}^n$. Consider the two external trajectories $w_1(t) := w_0(t, x_0)$ and $w_2(t) := 0$. By controllability there exists a third external trajectory, say $w_v(t, \bar{x}_0)$, and $T > 0$ such that $w_v(t, \bar{x}_0) = w_1(t)$ for $t \leq 0$ and $w_v(t, \bar{x}_0) = w_2(t) = 0$ for $t \geq T$. We will now first prove that, in fact, $\bar{x}_0 - x_0 \in \mathcal{V}^*(A, B, C, D)$.

Indeed, by linearity we have $w_v(t, \bar{x}_0 - x_0) = 0$ for all $t \leq 0$. Now consider the time-reversed system $\dot{x} = -Ax - Bv,\ w = Cx + Dv$. Let $w_{v_R}(t, \bar{x}_0 - x_0)$ denote its external trajectory corresponding to the time-reversed driving variable $v_R(t) := v(-t)$ and initial state $\bar{x}_0 - x_0$. It is easily seen that for all $t \geq 0$ we have $w_{v_R}(t, \bar{x}_0 - x_0) = w_v(-t, \bar{x}_0 - x_0) = 0$. This implies that $\bar{x}_0 - x_0 \in \mathcal{V}^*(-A, -B, C, D)$, the weakly unobservable subspace associated with the time-reversed system. By their feedback characterization, we have $\mathcal{V}^*(-A, -B, C, D) = \mathcal{V}^*(A, B, C, D)$, which proves the claim that $\bar{x}_0 - x_0 \in \mathcal{V}^*(A, B, C, D)$.

Next, it follows from the characterisation in Lemma 10.15 that $\bar{x}_0 \in \mathcal{R} + \mathcal{V}^*(A, B, C, D)$. Wrapping things up then leads to $x_0 = \bar{x}_0 - (\bar{x}_0 - x_0) \in \mathcal{R} + \mathcal{V}^*(A, B, C, D)$, which proves condition 2.

(2. $\Rightarrow$ 1.) By linearity it suffices to prove that for every external trajectory $w(t)$ there exists $T > 0$ and an external trajectory $w'(t)$ such that $w'(t) = w(t)$ for $t \leq 0$ and $w'(t) = 0$ for $t \geq T$. let $w(t) = w_v(t, x_0)$. Since $x_0 \in \mathcal{R} + \mathcal{V}^*(A, B, C, D)$, by Lemma 10.15 there exists a driving variable trajectory $v_1(t)$ such that $w_{v_1}(t, x_0) = 0$ for $t \geq T$. Define now $\bar{v}(t) := v(t)$ for $t \leq 0$ and $\bar{v}(t) := v_1(t)$ for $t \geq 0$. Define $w'(t) := w_{\bar{v}}(t, x_0)$. Then $w'(t) = w(t)$ for $t \leq 0$ and $w'(t) = 0$ for $t \geq T$. Hence $\mathfrak{B}_{DV}(A, B, C, D)_{\text{ext}}$ is controllable.

Thus we immediately obtain the following corollary:

Corollary 10.17 *Let $G \in \mathbb{R}(\xi)_p^{p\times m}$. Let (A, B, C, D) be a realization of G. Then the following two statements are equivalent:*

1. $\mathfrak{B}_{DV}(A, B, C, D)_{\text{ext}} = \text{im}\, G(\frac{\mathrm{d}}{\mathrm{d}t})$,
2. $\mathcal{R} + \mathcal{V}^*(A, B, C, D) = \mathbb{R}^{\mathrm{n}}$.

This result generalizes Theorem 4.5 in [5] that states that controllability of the pair (A, B) is a sufficient condition for condition 1. to hold.

We note that the necessary and sufficient condition for our third problem differs from the one that we established for the first two problems. It turns out, however, that the condition $\mathcal{R} + \mathcal{N} = \mathbb{R}^{\mathrm{n}}$ does provide a necessary and sufficient condition for controllability of the (w, v)-behavior associated with a driving variable representation, so of the external behavior with only the state x eliminated. Denote this behavior by

$$\mathfrak{B}_{DV}(A, B, C, D)_{(v,w)} := \{(v, w) \mid \text{there exists}\, x \,\text{such that}\, (x, v, w) \in \mathfrak{B}_{DV}(A, B, C, D)\}$$

Then the final theorem of this paper states the following:

Theorem 10.18 *Let* $\mathfrak{B}_{DV}(A, B, C, D)$ *be the full behavior of the driving variable representation* (10.5). *Then the following statements hold:*

1. $\mathfrak{B}_{DV}(A, B, C, D)$ *is controllable if and only if* $\mathcal{R} = \mathbb{R}^{\mathrm{n}}$.
2. $\mathfrak{B}_{DV}(A, B, C, D)_{(v,w)}$ *is controllable if and only if* $\mathcal{R} + \mathcal{N} = \mathbb{R}^{\mathrm{n}}$.
3. $\mathfrak{B}_{DV}(A, B, C, D)_{\text{ext}}$ *is controllable if and only if* $\mathcal{R} + \mathcal{V}^*(A, B, C, D) = \mathbb{R}^{\mathrm{n}}$.

Proof 1. The full behavior of the driving variable representation is equal to $\ker R(\frac{\mathrm{d}}{\mathrm{d}t})$, with

$$R(s) := \begin{bmatrix} sI - A & -B & 0 \\ -C & -D & I \end{bmatrix}$$

Obviously $\mathfrak{B}_{DV}(A, B, C, D)$ is controllable if and only if $R(\lambda)$ has full row rank for all $\lambda \in \mathbb{C}$. This is, however, equivalent with the condition that $\begin{bmatrix} \lambda I - A & -B \end{bmatrix}$ has full row rank for all $\lambda \in \mathbb{C}$. This holds if and only if the pair (A, B) is controllable, equivalently $\mathcal{R} = \mathbb{R}^{\mathrm{n}}$.

2. This follows immediately from Theorem 10.11 by interpreting the (w, v)- behavior as input output behavior of the representation $\dot{x} = Ax + Bv,\ w = Cx + Dv$ with input v and output w.
3. This is a restatement of Theorem 10.16.

10.6 Concluding Remarks

In this paper we have considered three open problems on the relation between the external behaviors of state representations associated with realizations of proper real rational matrices, and the behaviors represented by these rational matrices. These problems had been studied before in [5], where however only sufficient conditions were obtained. In the present paper we have complemented the results in [5] by establishing necessary and sufficient conditions in terms of the reachability subspace, the unobservable subspace and the weakly unobservable subspace associated with the realization.

References

1. H. Aling, J.M. Schumacher, A nine-fold canonical decomposition for linear systems. Int. J. Control **39**(4), 779–805 (1984)
2. P.J. Antsaklis, A.N. Michel, *A Linear Systems Primer.* (Birkhäuser Boston, 2007)
3. L. Ehrenpreis, *Fourier Analysis In Several Complex Variables, Pure and Applied Mathematices*, vol. XVII (Wiley-Interscience Publishers A Division of John Wiley and Sons, New York, 1970)
4. S.V. Gottimukkala, S. Fiaz, H.L. Trentelman, Equivalence of rational representations of behaviors. Syst. Control Lett. **60**, 119–127 (2011)
5. S.V. Gottimukkala, H.L. Trentelman, S. Fiaz, Realization and elimination in rational representations of behaviors. Syst. Control Lett. **62**, 708–714 (2012)
6. M. Kuijper, *First-Order Representations of Linear Systems* (Birkhäuser, 1994)
7. H.B. Minh, *Model Reduction in a Behavioral Framework, Doctoral Dissertation* (Rijksuniversiteit Groningen, 2009). http://dissertations.ub.rug.nl/faculties/science/2009/h.b.minh
8. M.E.C. Mutsaers, S. Weiland, Rational representations and controller synthesis of $\mathcal{L}$ behaviors. Automatica **48**, 1–14 (2012)
9. M.E.C. Mutsaers, *Control Relevant Model Reduction and Controller Synthesis for Complex Dynamical Systems*, Doctoral dissertation, Technische Universiteit Eindhoven, 2012. http://repository.tue.nl/734624
10. V.P. Palamodov, *Linear Differential Operators With Constant Coefficients*, Translated from the Russian by A.A. Brown. Die Grundlehren der mathematischen Wissenschaften, Band 169, (Springer, New York, 1970)
11. J.W. Polderman, J.C. Willems, *Introduction to Mathematical Systems Theory: a Behavioral Approach* (Springer, Berlin, 1997)
12. H.L. Trentelman, A.A. Stoorvogel, M.L.J. Hautus, *Control Theory Linear System* (Springer, London, 2001)
13. M.E. Valcher, A Note on the Driving Variable Realization of Behaviors, in *Proceedings of 42nd IEEE Conference on Decision and Control*, pp. 1627–1632, 9–12, Maui, Hawai (USA), Dec. (2003)
14. S. Weiland, *Theory of Approximation and Disturbance Attenuation for Linear Systems* (Doctoral Dissertation, Rijksuniversiteit Groningen, 1991)
15. S. Weiland, A.A. Stoorvogel, Rational representations of behaviors: interconnectability and stabilizability. Math. Control Signal Syst. **10**, 125–164 (1997)
16. J.C. Willems, Input-output and state-space representations of finite-dimensional linear time-invariant systems. Linear Algebra Appl. **50**, 581–608 (1983)
17. J.C. Willems, Y. Yamamoto, Behaviors defined by rational functions. Linear Algebra Appl. **425**, 226–241 (2007)
18. K. Zhou, J.C. Doyle, *Essentials of Robust Control* (Prentice Hall, 1998)
19. E. Zerz, *Topics in Multidimensional Linear Systems Theory*, Springer Lecture Notes in Control and Information Sciences 2564 (Springer, London, 2000)

Chapter 11
Modeling and Analysis of Energy Distribution Networks Using Switched Differential Systems

Jonathan C. Mayo-Maldonado and Paolo Rapisarda

Abstract It is a pleasure to dedicate this contribution to Prof. Arjan van der Schaft on the occasion of his 60th birthday. We study the dynamics of energy distribution networks consisting of switching power converters and multiple (dis-)connectable modules. We use parsimonious models that deal effectively with the variant complexity of the network and the inherent switching phenomena induced by power converters. We also present the solution to instability problems caused by devices with negative impedance characteristics such as constant power loads. Elements of the behavioral system theory such as linear differential behaviors and quadratic differential forms are crucial in our analysis.

11.1 Introduction

In recent years, the development of a new paradigm of energy generation and distribution systems has become a pressing research question. Issues such as the urge to reduce CO_2 emissions, the compelling advantages of renewable energy generation, and the undesirable power losses in complex transmission lines, have motivated the development of distributed energy generation systems based on renewable energies [31]. However, the intermittent nature of renewable energies is reflected in the characteristics of the voltages/currents (e.g., amplitude and frequency) provided by transducers, prompting to regulate such variables to satisfy the nominal requirements of the the loads.

In order to achieve voltage/current/frequency regulation and distribution of electricity, interconnections of power converters are implemented; however, their interaction can display unstable behaviors (see [3, 30, 32]). A common example of this

J.C. Mayo-Maldonado (✉) · P. Rapisarda
Vision, Learning and Control Group, School of Electronics and Computer Science, University of Southampton, Southampton SO17 1BJ, UK
e-mail: jcmm1g11@ecs.soton.ac.uk

P. Rapisarda
e-mail: pr3@ecs.soton.ac.uk

M.K. Camlibel et al. (eds.), *Mathematical Control Theory I*,
Lecture Notes in Control and Information Sciences 461,
DOI 10.1007/978-3-319-20988-3_11

issue is the *negative impedance instability* produced by current/voltage controlled converters behaving as *constant power loads* (see [17]). In order to address instability problems, we first need to choose a modeling framework that is suitable to describe the network characteristics. We consider the network as a complex switched system whose dynamic modes with variant state space dimension are induced by switching power converters and the arbitrary (dis-)connection of loads.

There exist traditional approaches to switched systems based on *state space-* (see e.g., [7]) and *descriptor form-* (see e.g., [23]) representations, where the dynamic modes share a global state space. However, the fact that the dynamic modes of energy distribution networks do not necessarily share the same state space engenders three main disadvantages in current approaches:

(1) *Loss of parsimony*. The complexity of "lower order modes" needs to be increased by adding fictitious variables and equations, only to satisfy a predefined global structure, see [11, 12].
(2) *State representations are not given a priori*. The modeling of elements of the network as impedances offer considerable computational advantages when dealing with complex scenarios (see e.g., [6]). Such approach leads directly to higher order descriptions and not state space representations. Consequently, additional computations must be performed to derive state space models.
(3) *Loss of modularity*. The incremental modeling of the dynamic modes in the bank is not permitted, i.e., new dynamic modes cannot be added to the underlying bank without altering the existing ones. The need to allow for incremental modeling arises naturally in an energy distribution network when new loads are connected to the network, see [11].

These issues motivated the development of the *switched linear differential systems framework (SLDS)* in [9–12, 18, 19], which is not representation-oriented, and thus permits the use of the type of models that are most natural for each application (e.g., the modeling of impedances). This approach is based on the concepts of behavioral system theory, and allows the modeling of dynamic modes expressed by sets of linear differential equations that do not necessarily share the same state space, as well as the introduction to new dynamic modes to the bank without altering the existing ones. In this chapter, we study the notion of *passivity* in the SLDS framework, using *quadratic differential forms* (see [27]) as a tool to model energy functions of the network. We also derive a systematic procedure to design passive stabilizing filters in terms of standard *bilinear-* and *linear matrix inequalities*, that can be easily constructed from the higher order models.

11.2 Notation

We use the following notation. The space of $\mathtt{n}$ dimensional real vectors is denoted by $\mathbb{R}^{\mathtt{n}}$, and that of $\mathtt{m} \times \mathtt{n}$ real matrices by $\mathbb{R}^{\mathtt{m}\times\mathtt{n}}$. $\mathbb{R}^{\bullet\times\mathtt{m}}$ denotes the space of real matrices with $\mathtt{m}$ columns and an unspecified finite number of rows. Given matrices

$A, B \in \mathbb{R}^{\bullet \times \mathrm{m}}$, $\mathrm{col}(A, B)$ denotes the matrix obtained by stacking A over B. The ring of polynomials with real coefficients in the indeterminate s is denoted by $\mathbb{R}[s]$; the ring of two-variable polynomials with real coefficients in the indeterminates ζ and η is denoted by $\mathbb{R}[\zeta, \eta]$. $\mathbb{R}^{\mathrm{r} \times \mathrm{w}}[s]$ denotes the set of all $\mathrm{r} \times \mathrm{w}$ matrices with entries in s, and $\mathbb{R}^{\mathrm{n} \times \mathrm{m}}[\zeta, \eta]$ that of $\mathrm{n} \times \mathrm{m}$ polynomial matrices in ζ and η. The set of rational $\mathrm{m} \times \mathrm{n}$ matrices is denoted by $\mathbb{R}^{\mathrm{m} \times \mathrm{n}}(s)$. The set of infinitely differentiable functions from $\mathbb{R}$ to $\mathbb{R}^{\mathrm{w}}$ is denoted by $\mathfrak{C}^{\infty}(\mathbb{R}, \mathbb{R}^{\mathrm{w}})$. $\mathfrak{D}(\mathbb{R}, \mathbb{R}^{\mathrm{w}})$ is the subset of $\mathfrak{C}^{\infty}(\mathbb{R}, \mathbb{R}^{\mathrm{w}})$ consisting of compact support functions. For a function $f : [t - \varepsilon, t) \to \mathbb{R}^{\bullet}$ we set the notation $f(t^-) := \lim_{\tau \nearrow t} f(\tau)$; and similarly for $f : (t, t + \varepsilon] \to \mathbb{R}^{\bullet}$ we set $f(t^+) := \lim_{\tau \searrow t} f(\tau)$, provided that these limits exist.

We also use standard concepts and notation of the behavioral setting, in particular those of *linear differential behaviors*, *state maps* and *quadratic differential forms*. A simplified collection of the theory that is relevant for the presented results can be found in Appendix A, p. 2046 of [12].

11.3 Modeling of Energy Distribution Networks

Consider the energy distribution network in Fig. 11.1, consisting of a switching power converter feeding three types of loads represented by impedances. Z_N represents a nominal load, i.e., the load that is considered during the design stage of the converter and which remains connected in the implementation. Z_k, $k = 1, \ldots, L$, represents a *switched impedance*, i.e., a finite amount of loads that can be connected or disconnected arbitrarily and which are not necessarily known during the design stage, e.g., domestic/commercial (dis-)connectable loads, (dis-)connectable electric vehicles, etc. Finally, Z_{CPL} represents the negative impedance of a switching power converter behaving as a constant power load (CPL), which is a potential destibilizer of the network (see [8]). The CPL is modeled according to [17] as a negative impedance in parallel with a constant current source.

Fig. 11.1 Energy distribution network

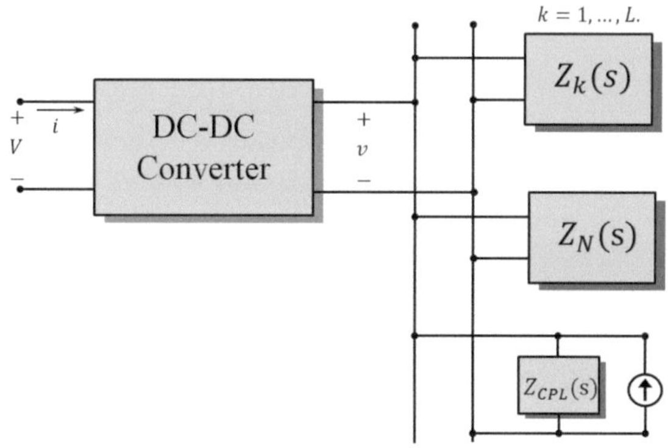

Note that the complexity of the network is neither initially bounded nor fixed, i.e., the McMillan degree associated to each impedance depends on their constitutive reactive elements which in the case of Z_k, $k = 1, \ldots, L$, may change depending on the loads that are connected during certain intervals of time. In the following sections, we discuss a natural modeling approach that deals effectively with this type of network.

11.3.1 Modeling of Loads as Impedances

When we study systems consisting of interconnections of port-driven electrical networks, e.g., transmission lines with points of common coupling, filters, loads, etc., we are compelled to adopt the calculus of m-port impedances for simplification of computations, see, e.g., [6, 14, 15, 21, 29]. In the case of energy distribution networks, this is also a common approach for the study of stability, see, e.g., [8, 20, 25, 30].

Models based on impedance matrices describe the "input–output dynamics" of the network in terms of the variables $V := \text{col}(v_1, \ldots, v_m)$ and $I := \text{col}(i_1, \ldots, i_m)$, corresponding, respectively, to the voltages across and currents through each port. Let $P\left(\frac{d}{dt}\right) V = Q\left(\frac{d}{dt}\right) I$, with $P, Q \in \mathbb{R}^{m \times m}[s]$, be an input–output representation (see [16]) of the network obtained by applying current and voltage laws. Adopting $\mathfrak{C}^\infty$ as the solution space, the external behavior of the network is defined as

$$\mathfrak{B} := \left\{ \text{col}(V, I) \in \mathfrak{C}^\infty(\mathbb{R}, \mathbb{R}^{2m}) \;\middle|\; P\left(\frac{d}{dt}\right) V = Q\left(\frac{d}{dt}\right) I \right\}. \tag{11.1}$$

The impedance $Z \in \mathbb{R}^{m \times m}(s)$ associated to the external behavior is defined as $Z(s) := P(s)^{-1}Q(s)$. If the behavior $\mathfrak{B}$ is controllable (see Chap. 5 of [16]), i.e., $R(s) := \begin{bmatrix} Q(s) & -P(s) \end{bmatrix}$ is such that rank $R(s)$ is equal to rank $R(\lambda)$ for all $\lambda \in \mathbb{C}$, then it admits an image representation

$$\begin{bmatrix} I \\ V \end{bmatrix} = \begin{bmatrix} U\left(\frac{d}{dt}\right) \\ Y\left(\frac{d}{dt}\right) \end{bmatrix} z \tag{11.2}$$

where $z \in \mathfrak{C}^\infty(\mathbb{R}, \mathbb{R}^{\mathtt{z}})$ corresponds to a latent variable and $U, Y \in \mathbb{R}^{m \times m}[s]$ are such that $Z(s) = Y(s)U(s)^{-1}$. Moreover, if $M(\lambda)$ is of full column rank for all $\lambda \in \mathbb{C}$, we conclude that the latent variable z is observable from $w := \text{col}(V, I)$ and its number of components corresponds to the number of inputs, i.e., $\mathtt{z} = m$. A controllable behavior always admits an observable image representation (see [28], Sect. VI-A).

Assuming controllability, the dynamic model of a network described as (11.2) can be obtained in a simple way by *series-* and *parallel computations*, since any complex m-port impedance matrix Z consists of the interconnection of impedances

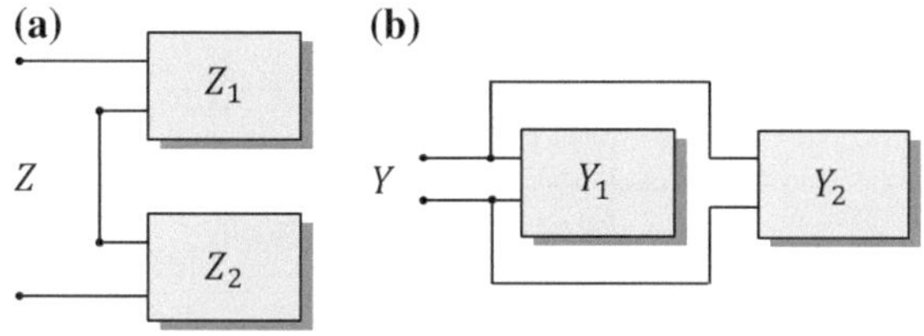

Fig. 11.2 Series/parallel interconnection of impedances/admittances

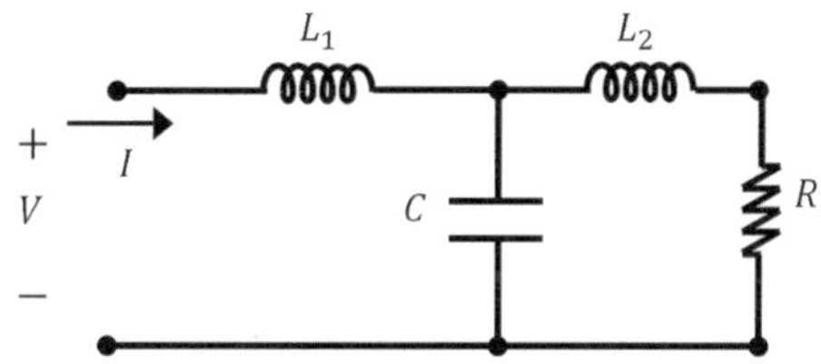

Fig. 11.3 Port-driven electrical circuit

of lower complexity. The simplest components are 1-port impedances corresponding to inductors, resistors, and capacitors, i.e., $Z_L(s) = Ls$, $Z_R(s) = R$, $Z_c(s) = \frac{1}{Cs}$. The inverse of an impedance, if exists, is equal to an admittance denoted by Y, i.e., $Y = Z^{-1}$.

Consider for instance the n-port networks in Fig. 11.2, whose terminals represent an m number of terminal pairs. The resultant m-port impedance/admittance due to series (Fig. 11.2a) and parallel (Fig. 11.2b) interconnections is computed as $Z = Z_1 + Z_2$ and $Y = Y_1 + Y_2$, respectively.

Example 11.1 Consider the 1-port electrical circuit in Fig. 11.3. The 1-port impedance of the circuit can be computed by series and parallel operations as

$$Z(s) = L_1 s + \frac{(L_2 s + R)\left(\frac{1}{C_1 s}\right)}{(L_2 s + R) + \left(\frac{1}{C_1 s}\right)} = \frac{L_1 L_2 C_1 s^3 + R L_1 C_1 s^2 + (L_1 + L_2)s + R}{L_2 C_1 s^2 + R C_1 s + 1}, \tag{11.3}$$

which corresponds to the input–output description

$$L_1 L_2 C_1 \frac{\mathrm{d}^3}{\mathrm{d}t^3} I + R L_1 C_1 \frac{\mathrm{d}^2}{\mathrm{d}t^2} I + (L_1 + L_2) \frac{\mathrm{d}}{\mathrm{d}t} I + R I = L_2 C_1 \frac{\mathrm{d}^2}{\mathrm{d}t^2} V + R C_1 \frac{\mathrm{d}}{\mathrm{d}t} V + V .$$

Let for simplicity $R = 1\ \Omega$, $L_1 = L_2 = 1\ H$ and $C = 1\ F$, then

$$\underbrace{\begin{bmatrix} V \\ I \end{bmatrix}}_{=:w} = \underbrace{\begin{bmatrix} \frac{\mathrm{d}^3}{\mathrm{d}t^3} + \frac{\mathrm{d}^2}{\mathrm{d}t^2} + 2\frac{\mathrm{d}}{\mathrm{d}t} + 1 \\ \frac{\mathrm{d}^2}{\mathrm{d}t^2} + \frac{\mathrm{d}}{\mathrm{d}t} + 1 \end{bmatrix}}_{=:M\left(\frac{\mathrm{d}}{\mathrm{d}t}\right)} z ,$$

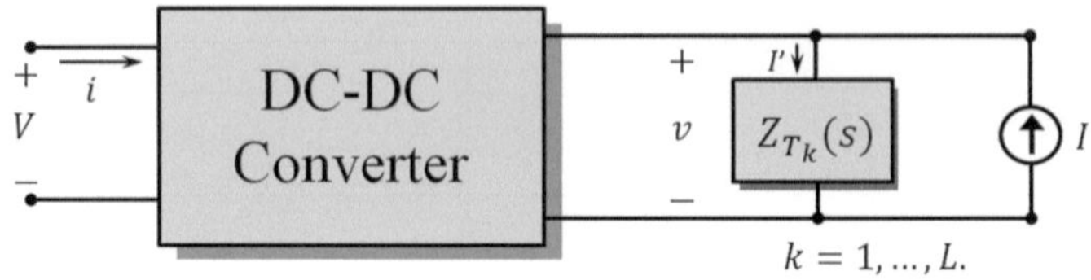

Fig. 11.4 Simplification of the energy distribution network in Fig. 11.1

where z is a latent variable corresponding to the current through the inductor L_2. Since $M(\lambda)$ is of full column rank for all $\lambda \in \mathbb{C}$, we conclude that the latent variable z is observable from w. □

The calculus of impedances facilitates our analysis, for instance the energy distribution network in Fig. 11.1 can be simplified by computing Z_{T_k}, $k = 1, \ldots, L$, as

$$\frac{1}{Z_{T_k}(s)} = \frac{1}{Z_k(s)} + \frac{1}{Z_N(s)} + \frac{1}{Z_{CPL}(s)} \; ; \quad k = 1, \ldots, L \; .$$

The simplified network is depicted in Fig. 11.4.

Remark 11.2 It is important to emphasize that Z_{CPL}, and consequently Z_{T_k}, $k = 1, \ldots, L$, do not necessarily correspond to impedances of passive networks as in traditional circuit theory, since Z_{CPL} corresponds to the local approximation of a constant power load which is by definition nonpassive (i.e., it is not positive-real in the sense of [14]), tipically modeled as a negative resistor [17].

We have illustrated the modeling of loads as impedances, that gives rise in a natural way to higher order descriptions. In the following section, we discuss a modeling approach that permits the study of switching dynamics induced by the DC–DC converter and the switched impedance Z_{T_k}, $k = 1, \ldots, L$, directly in higher order terms.

11.3.2 Switched Linear Differential Systems Framework

We now introduce the SLDS framework. We illustrate the main concepts of this approach by modeling a switching power converter.

Definition 11.3 ([10]) A *switched linear differential system (SLDS)* Σ is a quadruple $\Sigma = \{\mathcal{P}, \mathcal{F}, \mathcal{S}, \mathcal{G}\}$ where

- $\mathcal{P} = \{1, \ldots, N\} \subset \mathbb{N}$, is the set of *indices*;
- $\mathcal{F} = \{\mathfrak{B}_1, \ldots, \mathfrak{B}_N\}$, with $\mathfrak{B}_i$ a linear differential behavior and $i \in \mathcal{P}$, is the *bank of behaviors*;

- $\mathcal{S} = \{s : \mathbb{R} \to \mathcal{P}\}$, with s piecewise constant and right-continuous, is the set of admissible *switching signals*; and
- $\mathcal{G} = \left\{(G^-_{k\to j}(s), G^+_{k\to j}(s)) \in \mathbb{R}^{\bullet\times \mathtt{w}}[s] \times \mathbb{R}^{\bullet\times \mathtt{w}}[s] \mid 1 \leq k, j \leq N\ ,\ k \neq j\right\}$, is the set of *gluing conditions*.

The set of *switching instants* associated with $s \in \mathcal{S}$ is defined by $\mathbb{T}_s := \{t \in \mathbb{R} \mid s(t^-) \neq s(t^+)\} = \{t_1, t_2, \dots\}$, where $t_i < t_{i+1}$.

The set of all admissible trajectories satisfying the laws of the mode behaviors and the gluing conditions is the *switched behavior*, and is the central object of study in our framework.

Definition 11.4 ([10]) Let $\Sigma = \{\mathcal{P}, \mathcal{F}, \mathcal{S}, \mathcal{G}\}$ be a SLDS, and let $s \in \mathcal{S}$. The s *-switched linear differential behavior* $\mathfrak{B}^s$ is the set of trajectories $w : \mathbb{R} \to \mathbb{R}^{\mathtt{w}}$ that satisfy the following two conditions:

1. for all $t_i, t_{i+1} \in \mathbb{T}_s$, $w|_{[t_i,t_{i+1})} \in \mathfrak{B}_{s(t_i)}|_{[t_i,t_{i+1})}$;
2. w satisfies the gluing conditions $\mathcal{G}$ at the switching instants for each $t_i \in \mathbb{T}_s$, i.e.,

$$G^+_{s(t_{i-1})\to s(t_i)}\left(\frac{\mathrm{d}}{\mathrm{d}t}\right) w(t_i^+) = G^-_{s(t_{i-1})\to s(t_i)}\left(\frac{\mathrm{d}}{\mathrm{d}t}\right) w(t_i^-)\ . \tag{11.4}$$

The *switched linear differential behavior (SLDB)* $\mathfrak{B}^\Sigma$ of Σ is defined by $\mathfrak{B}^\Sigma := \bigcup_{s\in\mathcal{S}} \mathfrak{B}^s$.

The trajectories in $\mathfrak{B}^\Sigma$ are *piecewise infinitely differentiable functions* from $\mathbb{R}$ to $\mathbb{R}^{\mathtt{w}}$ denoted by $\mathfrak{C}^\infty_p(\mathbb{R}, \mathbb{R}^{\mathtt{w}})$, i.e., smooth when a mode is active and possibly discontinuous at switching instants.

Example 11.5 Consider the *high-voltage switching power converter* presented in [2] and depicted in Fig. 11.5a. For practical purposes such as voltage/current/power regulation, we are particularly interested in the dynamics at the input/output terminals. Consequently, we define the external variable (the set of variables of interest) as $w := \mathrm{col}(E, i_L, v_2, i_o)$.

By means of a switching signal, we can arbitrarily induce two possible electrical configurations that occur when the transistor is in either *closed* (see Fig. 11.5b) or *open* (see Fig. 11.5c) operation. Considering a standard modeling of two-port impedances for each case, we can derive the following physical laws describing the dynamics of the power converter.

$$\text{Mode 1}: \begin{cases} L\frac{\mathrm{d}}{\mathrm{d}t}i_L + R_L i_L - E = 0 \\ (C_1 + C_2)\frac{\mathrm{d}}{\mathrm{d}t}v_2 + \frac{1}{R}v_2 - i_o = 0 \end{cases}.$$

$$\text{Mode 2}: \begin{cases} LC_1\frac{\mathrm{d}^2}{\mathrm{d}t^2}i_L + R_L C_1\frac{\mathrm{d}}{\mathrm{d}t}i_L - C_1\frac{\mathrm{d}}{\mathrm{d}t}E + i_L = 0 \\ C_2\frac{\mathrm{d}}{\mathrm{d}t}v_2 + \frac{1}{R}v_2 - i_o = 0 \end{cases}.$$

The mode behaviors are defined as $\mathfrak{B}_j := \ker\ R_j\left(\frac{\mathrm{d}}{\mathrm{d}t}\right)$, $j = 1, 2$, where

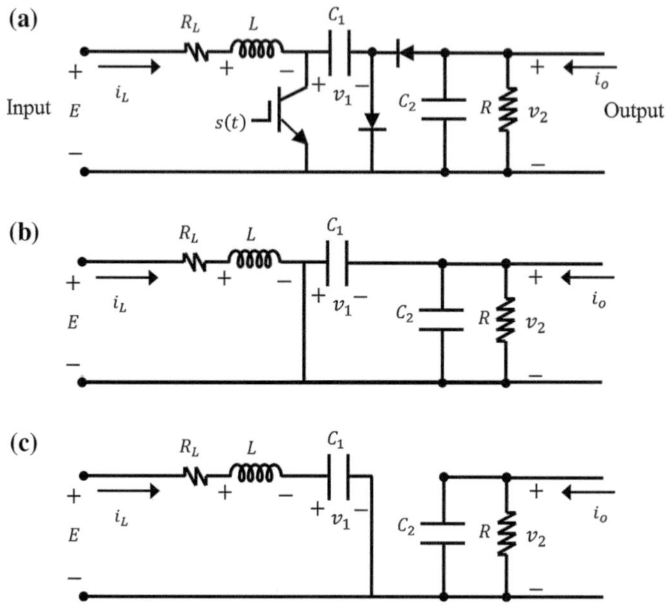

Fig. 11.5 **a** High-voltage switching power converter, **b** electrical configuration when the transistor is closed, **c** electrical configuration when the transistor is open

$$R_1\left(\frac{\mathrm{d}}{\mathrm{d}t}\right) := \begin{bmatrix} -1 & L\frac{\mathrm{d}}{\mathrm{d}t}+R_L & 0 & 0 \\ 0 & 0 & (C_1+C_2)\frac{\mathrm{d}}{\mathrm{d}t}+\frac{1}{R} & -1 \end{bmatrix};$$

$$R_2\left(\frac{\mathrm{d}}{\mathrm{d}t}\right) := \begin{bmatrix} -C_1\frac{\mathrm{d}}{\mathrm{d}t} & LC_1\frac{\mathrm{d}^2}{\mathrm{d}t^2}+R_LC_1\frac{\mathrm{d}}{\mathrm{d}t}+1 & 0 & 0 \\ 0 & 0 & C_2\frac{\mathrm{d}}{\mathrm{d}t}+\frac{1}{R} & -1 \end{bmatrix}.$$

As we show later, the physical constraints imposed by physics at switching instants can be modeled using gluing conditions. □

According to Definition 11.3 gluing conditions are algebraic constraints on the trajectories of the dynamical modes at switching instants and in real-life situations their selection is motivated by physical laws. For instance, at switching instants *conservation principles* forbid instantaneous changes in *conserved quantities* (see [13]) such as charge, flux, momentum, molar mass, volume, etc. Another well-known example of this type of constraints is the case of *state reset maps* in multicontroller systems that re-initialize a bank of switched controllers interconnected to a plant.

Example 11.6 (Cont'd from Example 11.5) At switching instants, the physical laws of the circuit impose constraints to the trajectories of the external variable at switching instants. By inspecting the circuits in Fig. 11.5 and using the *principle of conservation of charge* (see [13], Sect. 3.3.3), we find the following conditions at switching instants.

When switching from $\mathfrak{B}_1$ to $\mathfrak{B}_2$ at t_i:

$$\begin{aligned}
&i_L(t_i^+) = i_L(t_i^-),\\
&\underbrace{E(t_i^+) - R_L i_L(t_i^+) - L\frac{\mathrm{d}}{\mathrm{d}t} i_L(t_i^+)}_{v_1(t_i^+)} = v_2(t_i^-),\\
&v_2(t_i^+) = v_2(t_i^-).
\end{aligned} \tag{11.5}$$

When switching from $\mathfrak{B}_2$ to $\mathfrak{B}_1$ at t_i:

$$\begin{aligned}
i_L(t_i^+) &= i_L(t_i^-),\\
(C_1 + C_2)v_2(t_i^+) &= \underbrace{C_1 E(t_i^-) - C_1 R_L i_L(t_i^-) - LC_1 \frac{\mathrm{d}}{\mathrm{d}t} i_L(t_i^-)}_{C_1 v_1(t_i^-)} + C_2 v_2(t_i^-).
\end{aligned} \tag{11.6}$$

Consequently, the gluing conditions can be defined as

$$G_{1\to 2}^{+}\left(\frac{\mathrm{d}}{\mathrm{d}t}\right) := \begin{bmatrix} 0 & 0 & 1 & 0 \\ 1 & 0 & (-R_L - L\frac{\mathrm{d}}{\mathrm{d}t}) & 0 \\ 0 & 0 & 0 & 1 \end{bmatrix} ; \quad G_{1\to 2}^{-}\left(\frac{\mathrm{d}}{\mathrm{d}t}\right) := \begin{bmatrix} 0 & 0 & 1 & 0 \\ 0 & 0 & 0 & 1 \\ 0 & 0 & 0 & 1 \end{bmatrix} ;$$

$$G_{2\to 1}^{+}\left(\frac{\mathrm{d}}{\mathrm{d}t}\right) := \begin{bmatrix} 0 & 0 & 1 & 0 \\ 0 & 0 & 0 & (C_1 + C_2) \end{bmatrix} ; \quad G_{2\to 1}^{-}\left(\frac{\mathrm{d}}{\mathrm{d}t}\right) := \begin{bmatrix} 0 & 0 & 1 & 0 \\ C_1 & 0 & (-C_1 R_L - LC_1 \frac{\mathrm{d}}{\mathrm{d}t}) & C_2 \end{bmatrix} .$$

Equations (11.5) and (11.6) can be compactly written as

$$\begin{aligned}
G_{1\to 2}^{+}\left(\frac{\mathrm{d}}{\mathrm{d}t}\right) w(t_i^+) &= G_{1\to 2}^{-}\left(\frac{\mathrm{d}}{\mathrm{d}t}\right) w(t_i^-) ;\\
G_{2\to 1}^{+}\left(\frac{\mathrm{d}}{\mathrm{d}t}\right) w(t_i^+) &= G_{2\to 1}^{-}\left(\frac{\mathrm{d}}{\mathrm{d}t}\right) w(t_i^-) .
\end{aligned}$$

□

A realistic set of gluing conditions are *well-defined* and *well-posed*. In order to introduce these concepts, we use the notion of *state maps*.

Definition 11.7 Let Σ be a SLDS and let $X_j \in \mathbb{R}^{\mathrm{n}(\mathfrak{B}_j)\times \mathrm{w}}[s]$, induce minimal state maps for $\mathfrak{B}_j$, $j = 1, \ldots, N$. The gluing conditions are *well-defined* if there exist constant matrices $F_{j\to k}^{-}$ and $F_{j\to k}^{+}$, with $j, k = 1, \ldots, N$, $j \neq k$, such that $G_{j\to k}^{-}(s) = F_{j\to k}^{-} X_j(s)$ and $G_{j\to k}^{+}(s) = F_{j\to k}^{+} X_k(s)$, with $j, k = 1, \ldots, N$, $j \neq k$.

If $\mathcal{G} := \{(F_{j\to k}^{-} X_j(s),\ F_{j\to k}^{+} X_k(s))\}_{j,k=1,\ldots,N,\ j\neq k}$ are well-defined, we call them *well-posed* if for all $k, j = 1, \ldots, N$ with $k \neq j$, there exists a *re-initialisation* map

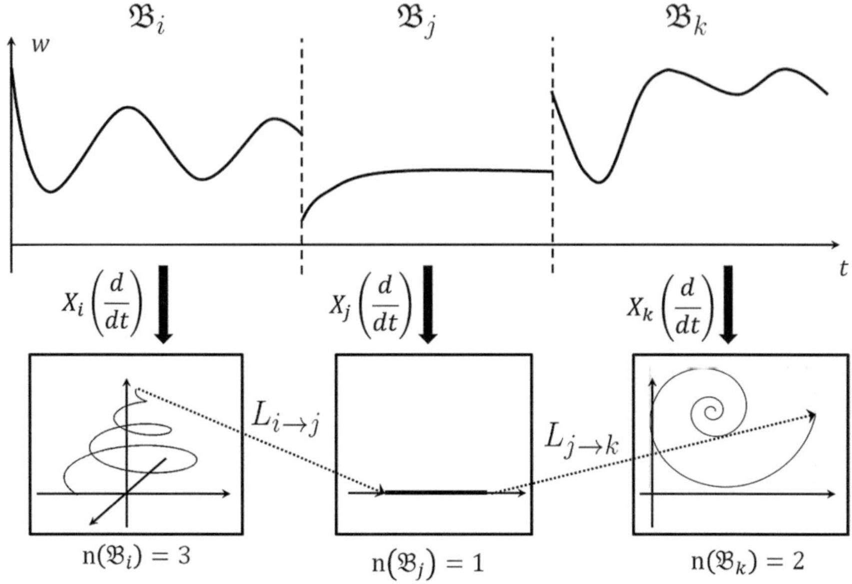

Fig. 11.6 Example: Switching between dynamical modes with different state space and well-posed gluing conditions

$L_{j\to k} : \mathbb{R}^{n(\mathfrak{B}_j)} \to \mathbb{R}^{n(\mathfrak{B}_k)}$ such that given a switching signal $s \in \mathcal{S}$ such that $s(t_{i-1}) = j$ and $s(t_i) = k$; for all $t_i \in \mathbb{T}_s$ and all admissible $w \in \mathfrak{B}^{\Sigma}$ with associated latent variable trajectories, it holds that $X_j\left(\frac{d}{dt}\right) w(t_i^+) = L_{k\to j} X_k\left(\frac{d}{dt}\right) w(t_i^-)$.

Well-defined and well-posed gluing conditions imply that if a transition occurs between $\mathfrak{B}_j$ and $\mathfrak{B}_k$ at t_i, and if an admissible trajectory ends at a "final state" $v_j := X_j\left(\frac{d}{dt}\right) w(t_i^-)$, then there exists at most one "initial state" for $\mathfrak{B}_k$, defined by $X_k\left(\frac{d}{dt}\right) w(t_i^+) =: v_k$, compatible with the gluing conditions. Moreover, the matrix $L_{j\to k}$ determines the reinitialization of the state space of $\mathfrak{B}_k$ as a linear function of that of $\mathfrak{B}_j$, as illustrated in Fig. 11.6.

Example 11.8 (Cont'd Example 11.5) Consider the following state maps for $\mathfrak{B}_1$ and $\mathfrak{B}_2$ respectively.

$$X_1\left(\frac{d}{dt}\right) := \begin{bmatrix} 0 & 0 & 1 & 0 \\ 0 & 0 & 0 & 1 \end{bmatrix}; \quad X_2\left(\frac{d}{dt}\right) := \begin{bmatrix} 0 & 0 & 1 & 0 \\ 1 & 0 & (-R_L - L\frac{d}{dt}) & 0 \\ 0 & 0 & 0 & 1 \end{bmatrix};$$

inducing the states $X_1\left(\frac{d}{dt}\right) w = \text{col}(i_L, v_2)$ and $X_2\left(\frac{d}{dt}\right) w = \text{col}(i_L, v_1, v_2)$. The gluing conditions can be written as

$$G^{+}_{1\to 2}\left(\frac{\mathrm{d}}{\mathrm{d}t}\right) := I_3 X_2\left(\frac{\mathrm{d}}{\mathrm{d}t}\right) \; ; \; G^{-}_{1\to 2}\left(\frac{\mathrm{d}}{\mathrm{d}t}\right) := \begin{bmatrix} 1 & 0 \\ 0 & 1 \\ 0 & 1 \end{bmatrix} X_1\left(\frac{\mathrm{d}}{\mathrm{d}t}\right) \; ;$$

$$G^{+}_{2\to 1}\left(\frac{\mathrm{d}}{\mathrm{d}t}\right) := \begin{bmatrix} 1 & 0 \\ 0 & (C_1 + C_2) \end{bmatrix} X_1\left(\frac{\mathrm{d}}{\mathrm{d}t}\right) \; ; \; G^{-}_{2\to 1}\left(\frac{\mathrm{d}}{\mathrm{d}t}\right) := \begin{bmatrix} 1 & 0 & 0 \\ 0 & C_1 & C_2 \end{bmatrix} X_2\left(\frac{\mathrm{d}}{\mathrm{d}t}\right) .$$

It is thus a matter of straightforward verification to conclude that the gluing conditions are well-defined and well-posed according to Definition 11.7. □

The properties of well-definedness and well-posedness are in general satisfied for common implementations of energy networks, consider for example the following proposition.

Proposition 11.9 *Assume that switching among the dynamical modes of a switched electrical network does not involve short-circuiting of voltage sources or open-circuiting of current sources. Then the gluing conditions are well-defined.*

Proof If switching between modes does not involve short- or open-circuiting sources, no constraints on the input variables of the system are imposed at the switching instants. Consequently, the gluing conditions only impose constraints on the output variables of the modes, which are linear functions of the state variables. The claim follows. □

Well-posed gluing conditions (see Definition 11.7) guarantee that after a switching instant, only one initial state for the new dynamical regime is specified from the final state of the previous one. Such property holds since the switching cannot cause any increase in the total amount of charge or flux stored in the system. On this issue, see [13] where the analysis of a wide variety of physical systems exhibiting discontinuities is presented, and [4, 5, 22]. In the rest of this paper, we assume that *the gluing conditions are well-posed.*

11.3.3 Latent Variables

As discussed in the previous section, controllable mode behaviors can be described using observable image representations $w = M_j\left(\frac{\mathrm{d}}{\mathrm{d}t}\right) z_j$, $j = 1, \ldots N$. It follows that every trajectory of the latent variable z_j corresponds to a unique trajectory of the external variable w when the jth mode is active. In the rest of this chapter we adopt the use of image representations, where $w := (u, y)$ has m inputs and m outputs, denoted by u and y, respectively, and corresponding to port-voltages and currents, as discussed in Sect. 11.3.1.

Example 11.10 (Cont'd from Example 11.5) Recall that $w := \mathrm{col}(E, i_L, v_2, i_o)$. It can be verified that the mode behaviors $\mathfrak{B}_j$, $i = 1, 2$, are *controllable* and thus can be described by $w = M_j\left(\frac{\mathrm{d}}{\mathrm{d}t}\right) z_j$, $j = 1, 2$, where

$$M_1\left(\frac{\mathrm{d}}{\mathrm{d}t}\right) := \begin{bmatrix} L\frac{\mathrm{d}}{\mathrm{d}t} + R_L & 0 \\ 0 & (C_1 + C_2)\frac{\mathrm{d}}{\mathrm{d}t} + \frac{1}{R} \\ 1 & 0 \\ 0 & 1 \end{bmatrix} ;$$

$$M_2\left(\frac{\mathrm{d}}{\mathrm{d}t}\right) := \begin{bmatrix} LC_1\frac{\mathrm{d}^2}{\mathrm{d}t^2} + R_L C_1\frac{\mathrm{d}}{\mathrm{d}t} + 1 & 0 \\ 0 & C_2\frac{\mathrm{d}}{\mathrm{d}t} + \frac{1}{R} \\ C_1\frac{\mathrm{d}}{\mathrm{d}t} & 0 \\ 0 & 1 \end{bmatrix} ;$$

and $z_1 := \mathrm{col}(i_L, v_2)$, $z_2 := \mathrm{col}(v_1, v_2)$. Moreover, since $M_j(\lambda)$, $j = 1, 2$, are full column rank for all $\lambda \in \mathbb{C}$ we conclude that the latent variables z_j, $j = 1, 2$ are observable from w. □

According to Definitions 11.3 and 11.4, the gluing conditions are algebraic constraints acting on the external variables at switching instants; however, they can be rewritten in terms of latent variables z_j, $j = 1, \ldots N$, in the following manner. Define

$$\overline{G}^{+}_{s(t_i-1)\rightarrow s(t_i)}\left(\frac{\mathrm{d}}{\mathrm{d}t}\right) := \left(G^{+}_{s(t_i-1)\rightarrow s(t_i)} M_{s(t_i)}\right)\left(\frac{\mathrm{d}}{\mathrm{d}t}\right) ,$$

$$\overline{G}^{-}_{s(t_i-1)\rightarrow s(t_i)}\left(\frac{\mathrm{d}}{\mathrm{d}t}\right) := \left(G^{-}_{s(t_i-1)\rightarrow s(t_i)} M_{s(t_i-1)}\right)\left(\frac{\mathrm{d}}{\mathrm{d}t}\right) ,$$

with $s \in \mathcal{S}$. Consequently, if w and z_j are related by an observable image representation $w = M_j\left(\frac{\mathrm{d}}{\mathrm{d}t}\right) z_j$, the gluing conditions in (11.4) can be equivalently written as

$$\overline{G}^{+}_{s(t_i-1)\rightarrow s(t_i)}\left(\frac{\mathrm{d}}{\mathrm{d}t}\right) z_{s(t_i)}(t_i^{+}) = \overline{G}^{-}_{s(t_i-1)\rightarrow s(t_i)}\left(\frac{\mathrm{d}}{\mathrm{d}t}\right) z_{s(t_i-1)}(t_i^{-}) .$$

Example 11.11 (Cont'd from Example 11.10) Given the gluing conditions in Example 11.8, we can reformulate them in terms of latent variables using $M_1\left(\frac{\mathrm{d}}{\mathrm{d}t}\right)$ and $M_2\left(\frac{\mathrm{d}}{\mathrm{d}t}\right)$ as follows.

$$\overline{G}^{-}_{1\rightarrow 2}\left(\frac{\mathrm{d}}{\mathrm{d}t}\right) := (G^{-}_{1\rightarrow 2}M_1)\left(\frac{\mathrm{d}}{\mathrm{d}t}\right) = \begin{bmatrix} 1 & 0 \\ 0 & 1 \\ 0 & 1 \end{bmatrix} ,$$

$$\overline{G}^{+}_{1\rightarrow 2}\left(\frac{\mathrm{d}}{\mathrm{d}t}\right) := (G^{+}_{1\rightarrow 2}M_2)\left(\frac{\mathrm{d}}{\mathrm{d}t}\right) = \begin{bmatrix} C_1\frac{\mathrm{d}}{\mathrm{d}t} & 0 \\ 1 & 0 \\ 0 & 1 \end{bmatrix}^{\top} ,$$

$$\overline{G}^{-}_{2\rightarrow 1}\left(\frac{\mathrm{d}}{\mathrm{d}t}\right) := (G^{-}_{2\rightarrow 1}M_2)\left(\frac{\mathrm{d}}{\mathrm{d}t}\right) = \begin{bmatrix} C_1\frac{\mathrm{d}}{\mathrm{d}t} & 0 \\ C1 & C2 \end{bmatrix} ,$$

$$\overline{G_{2\to 1}^{+}}\left(\frac{\mathrm{d}}{\mathrm{d}t}\right) := (G_{2\to 1}^{+}M_1)\left(\frac{\mathrm{d}}{\mathrm{d}t}\right) = \begin{bmatrix} 1 & 0 \\ 0 & C_1 + C_2 \end{bmatrix}.$$

□

11.4 Modularity

One of the main features of this framework is its *modularity*; every time a dynamic mode is added to the underlying bank, there is no need to modify the mathematical description of the existing modes. In the case of the energy distribution network in Fig. 11.4, the dynamic modes of the converter and the loads can be individually modeled and linked in a single model by the elimination of auxiliary variable. Consider the following proposition.

Proposition 11.12 *Consider the energy distribution network in Fig. 11.4. Assume that the dynamical modes of the switching power converter can be described in image form* $w = M_j\left(\frac{\mathrm{d}}{\mathrm{d}t}\right) z_j$, *where* $M_j \in \mathbb{R}^{4\times 2}[s]$; $z_j = \mathrm{col}(z_{1,j}, z_{2,j}) \in \mathfrak{C}_p^{\infty}(\mathbb{R}, \mathbb{R}^2)$; $j = 1, 2$; *and* $w := \begin{bmatrix} V & I & i & v \end{bmatrix}^{\top}$. *Let* $z_k \in \mathfrak{C}_p^{\infty}(\mathbb{R}, \mathbb{R})$, $k = 1, \dots, L$, *then there exist* $\widehat{M}_{j,k} \in \mathbb{R}^{4\times 2}[s]$ *such that the mode behaviors can be described by image representations*

$$\begin{bmatrix} V \\ I \\ i \\ v \end{bmatrix} = \widehat{M}_{j,k}\left(\frac{\mathrm{d}}{\mathrm{d}t}\right)\begin{bmatrix} z_{1,j} \\ z'_k \end{bmatrix}, \tag{11.7}$$

with $j = 1, 2$, *and* $k = 1, \dots, L$.

Proof The impedance Z_{T_k}, $k = 1, \dots, L$, is described by a one-port, and consequently can also be represented in observable image representation by $M' \in \mathbb{R}^{2\times 1}[s]$ with external variables $w' := \begin{bmatrix} I' & v \end{bmatrix}^{\top}$ and a one-dimensional latent variable denoted by z'_k. It follows from the elimination theorem (see Sect. 6 of [16]) that after the elimination of the latent variable $z_{2,j}$, $j = 1, 2$, the interconnection of this one-port with the switching power converter has a number $2L$ of dynamic modes that can be described as two-ports, corresponding to the image representations (11.7). □

Example 11.13 Consider the energy distribution network in Fig. 11.4, where the DC-DC converter is that of Fig. 11.5. Let $p_k, q_k \in \mathbb{R}[s]$, $k = 1, \dots, L$, define $Z_k(s) := \frac{n_k(s)}{d_k(s)}$, $k = 1, \dots, L$. The mode dynamics with $w := \mathrm{col}(E, I, i_L, v)$ are described by $w = M_{j,k}\left(\frac{\mathrm{d}}{\mathrm{d}t}\right) z_k$, where $z_1 := \mathrm{col}(i_1, z'_k)$, $z_2 := \mathrm{col}(v_1, z'_k)$, $k = 1, \dots, L$, and $j = 1, 2$.

$$M_{1,k}\left(\frac{\mathrm{d}}{\mathrm{d}t}\right) := \begin{bmatrix} R_L + L\frac{\mathrm{d}}{\mathrm{d}t} & 0 \\ 0 & d_k\left(\frac{\mathrm{d}}{\mathrm{d}t}\right) + (C_1 + C_2)\frac{\mathrm{d}}{\mathrm{d}t}n_k\left(\frac{\mathrm{d}}{\mathrm{d}t}\right) \\ 1 & 0 \\ 0 & n_k\left(\frac{\mathrm{d}}{\mathrm{d}t}\right) \end{bmatrix};$$

$$M_{2,k}\left(\frac{\mathrm{d}}{\mathrm{d}t}\right) := \begin{bmatrix} LC_1\frac{\mathrm{d}^2}{\mathrm{d}t^2} + R_L C_1\frac{\mathrm{d}}{\mathrm{d}t} + 1 & 0 \\ 0 & d_k\left(\frac{\mathrm{d}}{\mathrm{d}t}\right) + C_2\frac{\mathrm{d}}{\mathrm{d}t}n_k\left(\frac{\mathrm{d}}{\mathrm{d}t}\right) \\ C_1\frac{\mathrm{d}}{\mathrm{d}t} & 0 \\ 0 & n_k\left(\frac{\mathrm{d}}{\mathrm{d}t}\right) \end{bmatrix};$$

with $k = 1, \ldots, L$. The gluing conditions can be obtained by defining the impedances Z_k, $k = 1, \ldots, L$ and following the procedure exemplified in Examples 11.5 and 11.11. □

As illustrated in Example 11.13, each mode can be modeled independently, i.e., we compute the laws of each two-port network that depends on the mode of operation of the converter and the model of the switched impedance $Z_k, 1, \ldots, L$. It can be easily verified that the McMillan degree of each mode behavior is not fixed and depends on the degree of the denominator of $Z_k, 1, \ldots, L$. However each mode exhibits only the required level of complexity to describe each dynamic mode. This is in sharp contrast with the traditional approach where the dynamic modes are represented by $\frac{\mathrm{d}}{\mathrm{d}t}x = A_i x$, with $A_i \in \mathbb{R}^{n\times n}$, i.e., considering a global state space and where n is the highest possible McMillan degree. The latter approach results in more complex dynamic models (with more variables and more equations), which has an impact also on the complexity of stability analysis, simulation, control, etc. Moreover, there is no compelling reason to resort to such non-parsimonious approach if we can study the dynamic properties of the network directly in higher order terms, as shown in the following section.

11.5 Passivity

The concept of passivity will be crucial for the development of stability conditions and stabilization methods discussed in this chapter. To define passive SLDS, we first introduce the following notation. Since we require the integration of functionals acting on $w \in \mathfrak{B}^{\Sigma}$, we assume that they involve piecewise infinitely differentiable trajectories of compact support whose set is denoted by $\mathfrak{D}_p(\mathbb{R}, \mathbb{R}^{\mathtt{w}})$. Thus the trajectories which we will be considering in the following belong to $\mathfrak{B}^{\Sigma} \cap \mathfrak{D}_p(\mathbb{R}, \mathbb{R}^{\mathtt{w}})$.

Let $s \in \mathcal{S}$ be a fixed switching signal, with associated set of switching instants $\mathbb{T}_s := \{t_1, t_2, \ldots, t_n, \ldots\}$. We denote by $|\mathbb{T}_s|$ the total number of switching instants, possibly infinite, in $\mathbb{T}_s$. Let

$$\Phi := \frac{1}{2}\begin{bmatrix} 0_{m\times m} & I_m \\ I_m & 0_{m\times m} \end{bmatrix}, \tag{11.8}$$

and let $w \in \mathfrak{B}^{\Sigma} \cap \mathfrak{D}_p(\mathbb{R}, \mathbb{R}^{\mathtt{w}})$. If $|\mathbb{T}_s| = \infty$, define

$$\int Q_{\Phi}(w) := \int_{-\infty}^{t_1^-} Q_{\Phi}(w)\,\mathrm{d}t + \int_{t_1^+}^{t_2^-} Q_{\Phi}(w)\,\mathrm{d}t + \cdots + \int_{t_n^+}^{t_{n+1}^-} Q_{\Phi}(w)\,\mathrm{d}t + \cdots .$$

If $0 < |\mathbb{T}_s| < \infty$, then define

$$\int Q_{\Phi}(w) := \int_{-\infty}^{t_1^-} Q_{\Phi}(w)\,\mathrm{d}t + \sum_{k=2}^{|\mathbb{T}_s|} \int_{t_{k-1}^+}^{t_k^-} Q_{\Phi}(w)\,\mathrm{d}t + \int_{t_{|\mathbb{T}_s|}^+}^{\infty} Q_{\Phi}(w)\,\mathrm{d}t.$$

If $|\mathbb{T}_s| = 0$, i.e., no switching takes place, then

$$\int Q_{\Phi}(w) := \int_{-\infty}^{+\infty} Q_{\Phi}(w)\,\mathrm{d}t.$$

The definition of passive SLDS is as follows.

Definition 11.14 Let Σ be a SLDS and define Φ as in (11.8). Σ is *passive* if $\int Q_{\Phi}(w) \geq 0$ for all $w \in \mathfrak{B}^{\Sigma} \cap \mathfrak{D}_p(\mathbb{R}, \mathbb{R}^{\mathtt{w}})$.

In the previous definition, the quadratic differential form Q_{Φ} can be interpreted as the power that is oriented into the system, and consequently its integral over the real line measures the energy that is being supplied to, or flows out from the SLDS. If the net flow of energy is nonnegative, then we call the SLDS passive. Passivity implies input–output stability (see e.g., [26]), in the sense that unbounded output trajectories cannot occur as a consequence of bounded input trajectories (see also Sect. V of [10] for further elaboration).

In the SLDS framework, the concept of storage function arises in a natural way, describing the energy stored in each individual dynamical mode.

Definition 11.15 Let Σ be a SLDS and let $s \in \mathcal{S}$. An N-tuple $\left(Q_{\Psi_1}, \ldots, Q_{\Psi_N}\right)$ is a *multiple storage function* for Σ with respect to Q_{Φ} if

(1) $\frac{\mathrm{d}}{\mathrm{d}t} Q_{\Psi_i} \overset{\mathfrak{B}_i}{\leq} Q_{\Phi}, i = 1, \ldots, N$.
(2) $\forall\, w \in \mathfrak{B}^{\Sigma}$ and $\forall\, t_k \in \mathbb{T}_s$, it holds $Q_{\Psi_{s(t_{k-1})}}(w)(t_k^-) - Q_{\Psi_{s(t_k)}}(w)(t_k^+) \geq 0$.

We now prove that the existence of a multiple storage function implies that the SLDS is passive.

Theorem 11.16 *Let Σ be a SLDS and let $\Phi := \frac{1}{2}\begin{bmatrix} 0_{m\times m} & I_m \\ I_m & 0_{m\times m}\end{bmatrix}$. Assume that there exists a multiple storage function as in Definition 11.15. Then Σ is passive.*

Proof Let $t_0 := -\infty$ and let $s_w \in \mathcal{S}$ denote the switching signal that corresponds to a given trajectory $w \in \mathfrak{B}^{\Sigma}$. We consider the three possible cases, i.e., (1) $|\mathbb{T}_s| = \infty$, (2) $0 < |\mathbb{T}_s| < \infty$ and (3) $|\mathbb{T}_s| = 0$. It follows from Theorem 4.3 of [24], that since there exists Q_{Ψ_i} such that $\frac{d}{dt} Q_{\Psi_i} \overset{\mathfrak{B}_i}{\leq} Q_\Phi$, $i = 1, \ldots, N$, then $\mathfrak{B}_i$, $i = 1, \ldots, N$, is passive (i.e., dissipative with respect to Q_Φ).

Let $a < b$, then for all $w \in \mathfrak{B}^\Sigma$ with $s_w(t) = i$ for $t \in [a, b]$, it holds that $\int_a^b Q_\Phi(w)\ dt \geq Q_{\Psi_i}(w)(b) - Q_{\Psi_i}(w)(a)$, corresponding to the integration over $t \in [a, b]$ of $Q_{\Psi_i} \leq Q_\Phi$, for $w \in \mathfrak{B}_i \in \mathfrak{D}_p(\mathbb{R}, \mathbb{R}^{\mathrm{w}})$.

Since $\lim_{t \to \pm\infty} w(t) = 0$ for all $w \in \mathfrak{B}^\Sigma \cap \mathfrak{D}_p(\mathbb{R}, \mathbb{R}^{\mathrm{w}})$ we obtain the following expressions for cases (1) and (2), where $s = s_w$:

$$
\begin{aligned}
(1) \int Q_\Phi(w) &\geq (Q_{\Psi_{s(t_0)}}(w)(t_1^-) - Q_{\Psi_{s(t_1)}}(w)(t_1^+)) + \cdots \\
&\quad + (Q_{\Psi_{s(t_{n-1})}}(w)(t_n^-) - Q_{\Psi_{s(t_n)}}(w)(t_n^+)) + \cdots . \\
(2) \int Q_\Phi(w) &\geq (Q_{\Psi_{s(t_0)}}(w)(t_1^-) - Q_{\Psi_{s(t_1)}}(w)(t_1^+)) \\
&\quad + \Sigma_{k=2}^{|\mathbb{T}_s|-1} (Q_{\Psi_{s(t_{k-1})}}(w)(t_k^-) - Q_{\Psi_{s(t_k)}}(w)(t_k^+)) \\
&\quad + (Q_{\Psi_{s(|\mathbb{T}_s|-1)}}(w)(t_{|\mathbb{T}_s|}^-) - Q_{\Psi_{s(|\mathbb{T}_s|)}}(w)(t_{|\mathbb{T}_s|}^+)).
\end{aligned}
$$

Since $Q_{\Psi_{s(t_{k-1})}}(w)(t_k^-) - Q_{\Psi_{s(t_k)}}(w)(t_k^+) \geq 0$, $\forall\ t_k \in \mathbb{T}_s$, we conclude that in both cases $\int Q_\Phi(w) \geq 0$.

Finally the claim for (3) when no switching takes place, i.e., $s_w(t) = i$ for all t, follows readily from the existence of a storage function Q_{Ψ_i} and Theorem 4.3 of [24]. □

The conditions for the existence of a multiple storage function can be expressed in terms of linear matrix inequalities according to the following result (see Theorem 4 of [10]) providing an LMI-based test for passivity of SLDS. In the following, the *coefficient matrix* of $F(s) = \sum_{i=0}^N F_i s^i \in \mathbb{R}^{q_1 \times q_2}[s]$ is defined by

$$
\widetilde{F} := \begin{bmatrix} F_0 & F_1 & \ldots & F_N \end{bmatrix} . \tag{11.9}
$$

Note that $F(s) = \widetilde{F} \begin{bmatrix} I_{q_2} & sI_{q_2} & \ldots & I_{q_2} s^N \end{bmatrix}^\top$.

Theorem 11.17 *Let Σ be a SLDS with $\mathcal{G}$ well-defined and well-posed. Let $X_k \in \mathbb{R}^{\mathrm{n}(\mathfrak{B}_k) \times \mathrm{z}}[s]$ be a minimal state map for $\mathfrak{B}_k$, acting on the latent variable z_k, $k = 1, \ldots, N$, and let $L_{i \to j} \in \mathbb{R}^{\mathrm{n}(\mathfrak{B}_j) \times \mathrm{n}(\mathfrak{B}_i)}$ for all $i, j \in \mathcal{P}$, $i \neq j$, be the re-initialisations maps of Σ. Denote the coefficient matrix of $M_k(s)$ by $\widetilde{M}_k := \begin{bmatrix} M_{k,0} & \ldots & M_{k,L_k} \end{bmatrix}$; then there exist $X_{k,j} \in \mathbb{R}^{\mathrm{n}(\mathfrak{B}_k) \times m}$, $k = 1, \ldots, N$, $j = 1, \ldots, L_k - 1$ such that $X_k(s)$ can be written as $\widetilde{X}_k := \begin{bmatrix} X_{k,0} & \ldots & X_{k,L_k-1} \end{bmatrix}$.*

If there exist $K_k = K_k^\top \in \mathbb{R}^{\mathrm{n}(\mathfrak{B}_k) \times \mathrm{n}(\mathfrak{B}_k)}$, $k = 1 \ldots, N$, such that

$$
\widetilde{M}_k^\top \Phi \widetilde{M}_k - \begin{bmatrix} 0_{m \times \mathrm{n}(\mathfrak{B}_k)} \\ \widetilde{X}_k^\top \end{bmatrix} K_k \begin{bmatrix} \widetilde{X}_k & 0_{\mathrm{n}(\mathfrak{B}_k) \times m} \end{bmatrix} - \begin{bmatrix} \widetilde{X}_k^\top \\ 0_{m \times \mathrm{n}(\mathfrak{B}_k)} \end{bmatrix} K_k \begin{bmatrix} 0_{\mathrm{n}(\mathfrak{B}_k) \times m} & \widetilde{X}_k \end{bmatrix} \geq 0 , \tag{11.10}
$$

and moreover, if for $k, j = 1, \ldots, N$, $k \neq j$, it holds that

$$K_k - L_{k\to j}^{\top} K_j L_{k\to j} \geq 0\,, \tag{11.11}$$

then Σ is passive.

Based on this result, in the following section we develop a stabilization technique for energy distribution networks.

11.6 Energy-Based Stabilization

To deal with instability of energy distribution networks, we use *passive damping* (see, e.g., [1]), where a passive load (filter) is interconnected to the system in order to guarantee stability.

We consider the case where the energy distribution network is unstable due to the presence of constant power loads (see [17]). We proceed to design a filter that guarantees stability when interconnected to the converter, see Fig. 11.7.

For ease of exposition, we consider only one impedance $Z_T(s)$ and the filter as an additional load in the array depicted in Fig. 11.7. The impedance function of the filter is given by

$$Z_f(s) = \frac{p(s)}{q(s)}\,; \tag{11.12}$$

with an associated image form representation

$$\begin{bmatrix} i_f \\ v \end{bmatrix} = \begin{bmatrix} p(\frac{\mathrm{d}}{\mathrm{d}t}) \\ q(\frac{\mathrm{d}}{\mathrm{d}t}) \end{bmatrix} z'\,, \tag{11.13}$$

and whose parameters need be computed. The interconnection of impedances (11.12) and $Z_T(s)$ in Fig. 11.7 yields

$$Z_{int}(s) := \frac{Z_T(s) Z_f(s)}{Z_T(s) + Z_f(s)} = \frac{n(s)}{d(s)}\,. \tag{11.14}$$

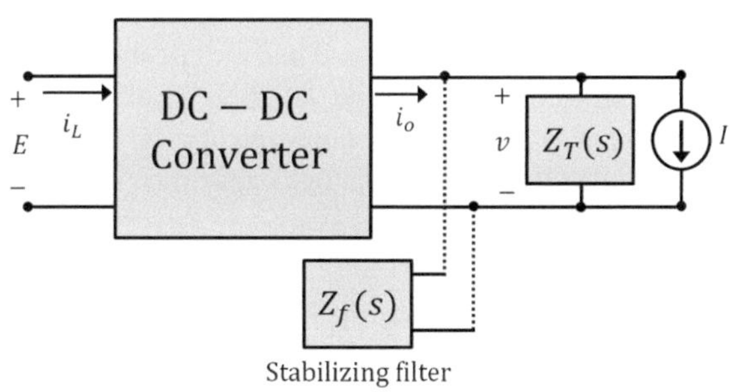

Fig. 11.7 Energy distribution network with a stabilizing filter

The first step in our procedure is to obtain image representations $w = M_k\left(\frac{\mathrm{d}}{\mathrm{d}t}\right) z_k$, $i = 1, \ldots, N$, describing each mode as in Proposition 11.12, and exemplified in Example 11.13. Similarly, we model the corresponding gluing conditions and compute reinitialization maps as in Definition 11.7.

The second step in our procedure is the setting up of a system of matrix inequalities corresponding to the conditions of Theorem 11.17. To make explicit the linear dependence on the parameters of Z_{int}, in the following we write $M_k(s)$ and their corresponding state maps $X_k(s)$, respectively, as $M_{k,\widetilde{n},\widetilde{d}}(s)$, and $X_{k,\widetilde{n},\widetilde{d}}(s)$, where $\widetilde{n}$, $\widetilde{d}$ are the coefficient matrices of the numerator and denominator of Z_{int}, that also involve the coefficients of the passive filter:

$$\begin{aligned}
&\widetilde{M}_{k,\widetilde{n},\widetilde{d}}^{\top} \Phi \widetilde{M}_{k,\widetilde{n},\widetilde{d}} - \begin{bmatrix} 0_{m\times \mathrm{n}(\mathfrak{B}_k)} \\ \widetilde{X}_{k,\widetilde{n},\widetilde{d}}^{\top} \end{bmatrix} K_k \left[\widetilde{X}_{k,\widetilde{n},\widetilde{d}} \; 0_{\mathrm{n}(\mathfrak{B}_k)\times m}\right] \\
&\qquad - \begin{bmatrix} \widetilde{X}_{k,\widetilde{n},\widetilde{d}}^{\top} \\ 0_{m\times \mathrm{n}(\mathfrak{B}_k)} \end{bmatrix} K_k \left[0_{\mathrm{n}(\mathfrak{B}_k)\times m} \; \widetilde{X}_{k,\widetilde{n},\widetilde{d}}\right] \geq 0\,, \quad k = 1, \ldots, N\,, \\
&K_k - L_{k\to j}^{\top} K_j L_{k\to j} \geq 0\,, k, j = 1, \ldots, N\,, k \neq j\,.
\end{aligned} \tag{11.15}$$

The third step is to formalize the requirement that the filter is passive. Define

$$\Phi' := \frac{1}{2}\begin{bmatrix} 0 & 1 \\ 1 & 0 \end{bmatrix}, \quad M'(s) := \begin{bmatrix} p(s) \\ q(s) \end{bmatrix}, \quad X'(s) := \begin{bmatrix} 1 \\ s \\ \vdots \\ s^{\deg(p)-1} \end{bmatrix}, \tag{11.16}$$

and denote the coefficient matrices of M' and X' by $\widetilde{M'}_{\widetilde{p},\widetilde{q}}$ and $\widetilde{X}'$, respectively. With these positions, it follows from the positive-real lemma that $\frac{q}{p}$ is positive-real if and only if there exists $K' = K'^{\top} \in \mathbb{R}^{\deg(p)\times\deg(p)}$ such that

$$\begin{aligned}
&\widetilde{M}'^{\top}_{\widetilde{p},\widetilde{q}} \Phi' \widetilde{M}'_{\widetilde{p},\widetilde{q}} - \begin{bmatrix} 0_{1\times\deg(p)} \\ \widetilde{X}'^{\top} \end{bmatrix} K' \left[\widetilde{X}' \; 0_{\deg(p)\times 1}\right] \\
&\qquad - \begin{bmatrix} \widetilde{X}'^{\top} \\ 0_{1\times\deg(p)} \end{bmatrix} K' \left[0_{\deg(p)\times 1} \; \widetilde{X}'\right] \geq 0\,.
\end{aligned} \tag{11.17}$$

If values of the parameters $\widetilde{p}$ and $\widetilde{q}$ exist such that the matrix inequalities (11.15), (11.17) are satisfied for some K_k, $k = 1, \ldots, N$ and K', then the interconnection of Fig. 11.7 is passive, and consequently i/o stable. Moreover, the filter $\frac{q}{p}$ can be implemented using only resistors, capacitors, inductors, and transformers (see [14]).

We close this section with a numerical example.

Example 11.18 (Cont'd from Example 11.13) We consider the implementation in Fig. 11.8, with $R_L = 0.1\,\Omega$; $L = 880\,\mu\mathrm{H}$; $C_1 = C_2 = 220\,\mu\mathrm{F}$; $R = 500\,\Omega$.

According to (11.12), we define the impedance of the filter $Z_f(s) := \frac{p(s)}{q(s)}$ with $p(s) = a_0 s + a1$ and $q(s) = 1$, for which the a-parameters will be computed.

We consider the total impedance as a constant power load, i.e., $Z_T(s) = -R_{CP}$ with $-R_{CP} = -300\,\Omega$. Considering (11.14), we obtain $n(s) = 300(a_0 + a_1 s)$ and $d(s) = 300 - a_0 - a_1 s$. We thus substitute $n\left(\frac{\mathrm{d}}{\mathrm{d}t}\right)$ and $d\left(\frac{\mathrm{d}}{\mathrm{d}t}\right)$ in the dynamic models computed in Example 11.13. Define state maps for each dynamical mode acting, respectively, on the latent variables z_1 and z_2 as

$$X_1\left(\frac{\mathrm{d}}{\mathrm{d}t}\right) := \begin{bmatrix} 1 & 0 \\ 0 & n\left(\frac{\mathrm{d}}{\mathrm{d}t}\right) \\ 0 & d\left(\frac{\mathrm{d}}{\mathrm{d}t}\right) \end{bmatrix}, \quad X_2 := \begin{bmatrix} C_1\frac{\mathrm{d}}{\mathrm{d}t} & 0 \\ 1 & 0 \\ 0 & n\left(\frac{\mathrm{d}}{\mathrm{d}t}\right) \end{bmatrix},$$

then for every $t_k \in \mathbb{T}_s$, the gluing conditions can be expressed as $X_2\left(\frac{\mathrm{d}}{\mathrm{d}t}\right) z_2(t_k^+) = L_{1\to 2} X_1\left(\frac{\mathrm{d}}{\mathrm{d}t}\right) z_1(t_k^-)$ and $X_1\left(\frac{\mathrm{d}}{\mathrm{d}t}\right) z_1(t_k^+) = L_{2\to 1} X_2\left(\frac{\mathrm{d}}{\mathrm{d}t}\right) z_2(t_k^-)$, where

$$L_{1\to 2} := \begin{bmatrix} 1 & 0 & 0 \\ 0 & 1 & 0 \\ 0 & 1 & 0 \\ 0 & 0 & 1 \end{bmatrix}, \quad L_{2\to 1} := \begin{bmatrix} 1 & 0 & 0 & 0 \\ 0 & \frac{C_1}{C_1+C_2} & \frac{C_2}{C_1+C_2} & 0 \\ 0 & 0 & 0 & 1 \end{bmatrix}.$$

We now solve simultaneously the bilinear matrix inequalities (11.15) and (11.17) using standard solvers such as `Yalmip`. We thus obtain a solution $a_0 = 377$, $a_1 = 293 \times 10^{-6}$, $b_2 = 377$. Finally, the realization of the filter with impedance $Z_f(s) = 293 \times 10^{-6}s + 377$ is shown in Fig. 11.9. □

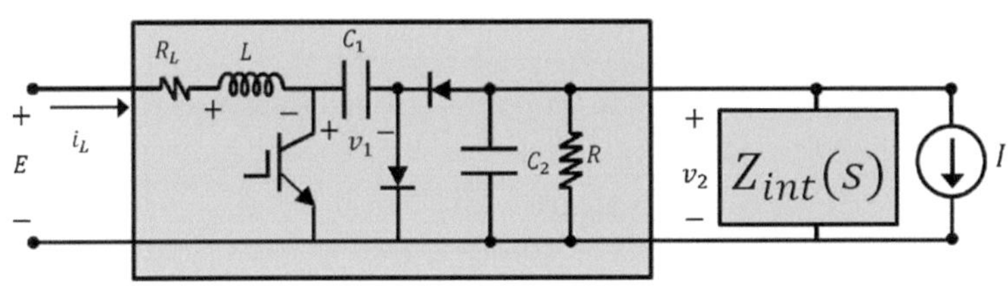

Fig. 11.8 Stable interconnection of a DC–DC converter with a passive filter and a constant power load

Fig. 11.9 Realization of the stabilizing filter

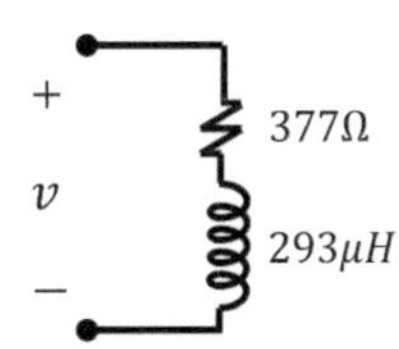

11.7 Conclusions

We introduced a modeling approach for energy distribution networks based on the switched linear differential framework in [12]. We also introduce the concept of passive SLDS and we study its relevance in the study of networks, deriving a stabilization method for switching power converters feeding potential destabilizers such as constant power loads. We have shown that elements of behavioral system theory such as linear differential behaviors and quadratic differential forms provide suitable tools to study the network using higher order differential models obtained directly from first principles.

References

1. M. Cespedes, L. Xing, J. Sun, Constant-power load system stabilization by passive damping. IEEE Trans. Power Electron. **26**(7), 1832–1836 (2011)
2. Z. Dongyan, A. Pietkiewicz, S. Cuk, A three-switch high-voltage converter. IEEE Trans. Power Electron. **14**(1), 177–183 (1999)
3. W. Du, J. Zhang, Y. Zhang, Z. Qian, Stability criterion for cascaded system with constant power load. IEEE Trans. Power Electron. **28**(4), 1843–1851 (2013)
4. R. Frasca, M.K. Çamlibel, I.C. Goknar, L. Iannelli, F. Vasca, Linear passive networks with ideal switches: consistent initial conditions and state discontinuities. IEEE Trans. Circuits Syst. I, Reg. Papers **57**(12), 3138–3151 (2010)
5. I. Goknar, Conservation of energy at initial time for passive RLCM network. IEEE Trans. Circuit Theory **19**(4), 365–367 (1972)
6. S. Krishna, *An Introduction to Modelling of Power System Components* (Springer, New York, 2014)
7. D. Liberzon, *Switching in Systems and Control* (Birkhauser, Boston, 2003)
8. H. Lin, P.J. Antsaklis, Stability and stabilizability of switched linear systems: a survey of recent results. IEEE Trans. Autom. Control **54**(2), 308–322 (2009)
9. J.C. Mayo-Maldonado, P. Rapisarda, On Positive-realness and Stability of Switched Linear Differential Systems, in *Proceedings of the 53nd IEEE CDC* (2013)
10. J.C. Mayo-Maldonado, P. Rapisarda, Dissipative switched linear differential systems. IEEE Trans. Autom. Control (2014), submitted
11. J.C. Mayo-Maldonado, P. Rapisarda, Modelling of Switching Dynamics in Electrical Systems, in *Proceedings of the Mathematical Theory of Networks and Systems Symposium (MTNS)* (2014)
12. J.C. Mayo-Maldonado, P. Rapisarda, P. Rocha, Stability of switched linear differential systems. IEEE Trans. Autom. Control **59**(8), 2038–2051 (2014)
13. P.J. Mosterman, G. Biswas, A theory of discontinuities in physical system models. J. Franklin Inst. **335**(3), 401–439 (1998)
14. R. Newcomb, *Linear Multiport Synthesis* (McGraw-Hill, New York, 1966)
15. L. Peng, B. Lehman, Performance prediction of dc-dc converters with impedances as loads. IEEE Trans. Power Electron. **19**(1), 201–209 (2004)
16. J.W. Polderman, J.C. Willems, *Introduction to Mathematical System Theory: A Behavioral Approach* (Springer, Berlin, 1997)
17. A.M. Rahimi, A. Emadi, Active damping in DC/DC power electronic converters: a novel method to overcome the problems of constant power loads. IEEE Trans. Industr. Electron. **56**(5), 1428–1439 (2009)

18. P. Rapisarda, P. Rocha, Positive Realness and Lyapunov Functions. in *Proceedings of the 20th MTNS*, Melbourne, Australia, 2012
19. P. Rocha, J.C. Willems, P. Rapisarda, D. Napp, On the Stability of Switched Behavioral Systems, in *Proceedings of the 50th IEEE CDC-ECC* (2011), pp. 1534–1538
20. J. Sun, Impedance-based stability criterion for grid-connected inverters. IEEE Trans. Power Electron. **26**(11), 3075–3078 (2011)
21. N. Tleis, *Power Systems Modelling and Fault Analysis: Theory and Practice* (Newnes, Woburn, 2007)
22. J. Tolsa, M. Salichs, Analysis of linear networks with inconsistent initial conditions. IEEE Trans. Circuits Syst. I, Fundam. Theory **40**(12), 885–894 (1993)
23. S. Trenn, in *Switched Differential Algebraic Equations*. eds. by F. Vasca, L. Iannelli. Dynamics and Control of Switched Electronic Systems. Chapter 6 (Springer, New York, 2012)
24. H.L. Trentelman, J.C. Willems, Every storage function is a state function. Syst. Control Lett. **32**(5), 249–259 (1997) (System and Control Theory in the Behavioral Framework)
25. S. Vesti, T. Suntio, J.A. Oliver, R. Prieto, J.A. Cobos, Impedance-based stability and transient-performance assessment applying maximum peak criteria. IEEE Trans. Power Electron. **28**(5), 2099–2104 (2013)
26. J.C. Willems, K. Takaba, Dissipativity and stability of interconnections. Int. J. Robust Nonlin. **17**(5–6), 563–586 (2007)
27. J.C. Willems, H.L. Trentelman, On quadratic differential forms. SIAM J. Control Optim. **36**(5), 1703–1749 (1998)
28. J.C. Willems, H.L. Trentelman, Synthesis of dissipative systems using quadratic differential forms: part I. IEEE Trans. Autom. Control **47**(1), 53–69 (2002)
29. O. Wing, *Classical Circuit Theory* (Springer, New York, 2008)
30. F. Xiaogang, L. Jinjun, F.C. Lee, Impedance specifications for stable DC distributed power systems. IEEE Trans. Power Electron. **17**(2), 157–162 (2002)
31. Y. Xinghuo, C. Cecati, T. Dillon, M.G. Simes, The new frontier of smart grids. IEEE Industr. Electron. Mag. **5**(3), 49–63 (2011)
32. Q.C. Zhong, T. Hornik, *Control of Power Inverters in Renewable Energy and Smart Grid Integration* (Wiley, New York, 2012)

Chapter 12
Nonlinear Controller Design Based on Invariant Manifold Theory

Noboru Sakamoto

Abstract The role of invariant manifold in nonlinear control theory is reviewed. First, stable, center-stable and center manifold algorithms to compute flows on these manifolds are presented. Next, application results of the computational methods are illustrated for optimal stabilization, optimal output regulation and periodic orbit design problems.

12.1 Introduction

Invariant manifold plays a central role in the analysis of the geometric structure of a dynamical system and its history dates back to Poincare [2, 3, 8]. Especially, stable, center-stable, center unstable and unstable manifolds around a fixed point locally decompose phase space and gives a clear understanding of the dynamical system.

In nonlinear control theory, the significance of stable manifold of Hamiltonian systems is recognized in [15, 16] and this line of research is extended in [10, 12]. Huang and Rugh [5, 6], Isidori and Byrnes [7] and present solvability conditions for output regulation problem via center manifold theory, while, in [11], it is shown that center-stable manifold is directly related to a solution for optimal output regulation via Hamiltonian systems.

In this paper, we review the recent developments of nonlinear control design methods based on invariant manifold theory. First, we show the computational algorithms for stable, center-stable and center manifolds. These methods consists of solving ordinary differential equations iteratively, which is easy to implement with a computer, and provide flows on the manifolds. We first present the iterative algorithms for computing the three types of invariant manifolds in Sect. 12.2 and then present their

To Arjan van der Schaft on the occasion of his 60th birthday

N. Sakamoto (✉)
Faculty of Science and Engineering, Nanzan University, Yamasato-cho,
Showa-ku, Nagoya, Japan
e-mail: noboru.sakamoto@nanzan-u.ac.jp

M.K. Camlibel et al. (eds.), *Mathematical Control Theory I*,
Lecture Notes in Control and Information Sciences 461,
DOI 10.1007/978-3-319-20988-3_12

applications in nonlinear control problems. The applications include optimal swing up and stabilization of an inverted pendulum [4], design of optimal servo system [11] and computation of periodic orbits in the restricted three body problem [9].

N. Sakamoto had an opportunity to work with Arjan van der Schaft during the years of 2005–2006 in Groningen. The work became a publication [12], from which fundamentals of this paper came out.

12.2 Invariant Manifolds and Their Computational Algorithms

Consider the following system

$$\begin{cases} \dot{x} = Ax + X(t, x, y, z) \\ \dot{y} = By + Y(t, x, y, z) \\ \dot{z} = Cz + Z(t, x, y, z) \end{cases} \tag{12.1}$$

$x \in \mathbb{R}^{n_x}$, $y \in \mathbb{R}^{n_y}$, $z \in \mathbb{R}^{n_z}$. The functions X, Y, Z are C^k functions ($k \geqslant 1$) such that $X(t, 0, 0, 0) = 0$, $DX(t, 0, 0, 0) = 0$, $Y(t, 0, 0, 0) = 0$, $DY(t, 0, 0, 0) = 0$, $Z(t, 0, 0, 0) = 0$, $DZ(t, 0, 0, 0) = 0$ for all $t \in \mathbb{R}$.

Assumption 12.1 (i) All the eigenvalues of A have negative real parts.
(ii) All the eigenvalues of B have zero real parts.
(iii) All the eigenvalues of C have positive real parts.

Under these conditions, it is known that there exist several invariant manifolds for (12.1).

Definition 12.2 C^k-manifolds W^s, W^{cs} and W^c in a neighborhood of the origin are said to be a stable, center-stable and center manifold, respectively, if W^s, W^{cs} are W^c are invariant under the flow of (12.1) as long as the solution remains in the neighborhood and W^s, W^{cs} and W^c are graphs of C^k functions $(y, z) = (v^+(t, x), w^+(t, x))$, $z = w^{*+}(t, x, y)$ and $(x, z) = (u^*(t, y), w^*(t, y))$, respectively, with $D_x v^+(t, 0) = 0$, $D_x w^+(t, 0) = 0$, $D_\nu w^{*+}(t, 0, 0) = 0$ $(\nu = x, y)$, $D_y u^*(t, 0) = 0$ and $D_y w^*(t, 0) = 0$ for all $t \in \mathbb{R}$.

The following Lemma is derived using Assumption 12.1.

Lemma 12.3 *There exist constants* $\alpha > 0$, $K_A > 1$ *and* $K_C > 1$ *such that*

$$|e^{At}x| \leqslant K_A e^{-\alpha t}|x| \; (t \geqslant 0),$$
$$|e^{Ct}z| \leqslant K_C e^{\alpha t}|z| \; (t \leqslant 0),$$

for all $x \in \mathbb{R}^{n_x}$, $z \in \mathbb{R}^{n_z}$. *Moreover, for every* $0 < \varepsilon < \alpha$ *there exists a* $K_\varepsilon > 1$ *such that*

$$|e^{Bt}y| \leqslant K_\varepsilon e^{\varepsilon|t|}|y| \text{ for all } y \in \mathbb{R}^{n_y}. \tag{12.2}$$

Assuming that appropriate cut-off functions [14] are already multiplied to (12.1) to ensure the uniqueness of the invariant manifolds, the following Lemma can be shown.

Lemma 12.4 *There exists a continuous nonnegative nondecreasing function* κ : $[0, \infty) \to [0, \infty)$ *with* $\kappa(0) = 0$ *such that*

$$\begin{aligned}
|X(t,x,y,z)| + |Y(t,x,y,z)| + |Z(t,x,y,z)| &\leqslant \kappa(|x|+|y|+|z|)(|x|+|y|+|z|),\\
|X(t,x,y,z) - X(t,x',y',z')| &\leqslant \kappa(\delta)(|x-x'|+|y-y'|+|z-z'|),\\
|Y(t,x,y,z) - Y(t,x',y',z')| &\leqslant \kappa(\delta)(|x-x'|+|y-y'|+|z-z'|),\\
|Z(t,x,y,z) - Z(t,x',y',z')| &\leqslant \kappa(\delta)(|x-x'|+|y-y'|+|z-z'|),
\end{aligned}$$

for all x, y, z, x', y', z' *and* t.

12.3 Algorithms for Invariant Manifold Computation

(i) Stable manifold algorithm:

$$\begin{cases}
x_{k+1}(t,x_0) = e^{At}x_0 + \int_0^t e^{A(t-s)}X(s,x_k(s),y_k(s),z_k(s))\,ds\\
y_{k+1}(t,x_0) = -\int_t^\infty e^{B(t-s)}Y(s,x_k(s),y_k(s),z_k(s))\,ds\\
z_{k+1}(t,x_0) = -\int_t^\infty e^{C(t-s)}Z(s,x_k(s),y_k(s),z_k(s))\,ds
\end{cases} \tag{12.3}$$

with $x_1(t) = e^{At}x_0$, $y_1(t) = 0$, $z_1(t) = 0$.

(ii) Center-stable manifold algorithm:

$$\begin{cases}
x_{k+1}(t,x_0,y_0) = e^{At}x_0 + \int_0^t e^{A(t-s)}X(s,x_k(s),y_k(s),z_k(s))\,ds\\
y_{k+1}(t,x_0,y_0) = e^{Bt}y_0 + \int_0^t e^{B(t-s)}Y(s,x_k(s),y_k(s),z_k(s))\,ds\\
z_{k+1}(t,x_0,y_0) = -\int_t^\infty e^{C(t-s)}Z(s,x_k(s),y_k(s),z_k(s))\,ds
\end{cases} \tag{12.4}$$

with $x_1(t) = e^{At}x_0$, $y_1(t) = e^{Bt}y_0$, $z_1(t) = 0$.

(iii) Center manifold algorithm:

$$\begin{cases} x_{k+1}(t, y_0) = \int_{-\infty}^{t} e^{A(t-s)} X(s, x_k(s), y_k(s), z_k(s))ds \\ y_{k+1}(t, y_0) = e^{Bt}y_0 + \int_{0}^{t} e^{B(t-s)} Y(s, x_k(s), y_k(s), z_k(s))ds \\ z_{k+1}(t, y_0) = -\int_{t}^{+\infty} e^{C(t-s)} Z(s, x_k(s), y_k(s), z_k(s))ds \end{cases} \tag{12.5}$$

with $x_1(t) = 0$, $y_1(t) = e^{Bt}y_0$, $z_1(t) = 0$.

Theorem 12.5 *(i) For sufficiently small* $|x_0|$, $x_k(t, x_0)$, $y_k(t, x_0)$, $z_k(t, x_0)$ *in* (12.3) *converge, uniformly with respect to* x_0, *to a solution of* (12.1). *The limit functions represent a flow on the stable manifold of* (12.1).

(ii) For sufficiently small $|x_0|$ *and* $|y_0|$, $x_k(t, x_0, y_0)$, $y_k(t, x_0, y_0)$, $z_k(t, x_0, y_0)$ *in* (12.4) *converge, uniformly with respect to* x_0 *and* y_0, *to a solution of* (12.1). *The limit functions represent a flow on the center-stable manifold of* (12.1).

(iii) For sufficiently small $|y_0|$, $x_k(t, y_0)$, $y_k(t, y_0)$, $z_k(t, y_0)$ *in* (12.5) *converge, uniformly with respect to* y_0, *to a solution of* (12.1). *The limit functions represent a flow on the center manifold of* (12.1).

This theorem is obtained originally in [11, 12].

12.4 Optimal Regulator and Hamilton-Jacobi Equations

Hamilton-Jacobi equations (HJEs) are one of the fundamental equations in control theory. HJEs are nonlinear partial differential equations of first order of the following form

$$\text{(HJ)} \quad H(x, p) := p^T f(x) - \frac{1}{2}p^T R(x)p + \frac{1}{2}x^T Qx = 0$$

where $p_1 = \partial V/\partial x_1, \ldots, p_n = \partial V/\partial x_n$. The role of HJEs is first recognized in optimal regulator problem and later on, it is found that HJEs are related with system dissipativity [17, 18], H^∞ control [15, 16] and balanced realization theory [13].

Definition 12.6 A solution $V(x)$ of (HJ) is said to be a stabilizing solution if $x = 0$ is an asymptotically stable equilibrium of a vector field $\frac{\partial H}{\partial p}(x, p(x))$, where $p(x) = (\partial V/\partial x)^T(x)$.

For example, let us consider the problem of optimal regulator:

$$\Sigma \quad \begin{cases} \dfrac{dx}{dt} = f(x(t)) + g(x(t))u(t) \\ y = h(x(t)) \end{cases}$$
$$J = \frac{1}{2}\int_0^\infty y(t)^T y(t) + u(t)^T \bar{R} u(t)\, dt$$

where $f : \mathbb{R}^n \to \mathbb{R}^n$, $g : \mathbb{R}^n \to \mathbb{R}^{n\times m}$ are sufficiently smooth and $\bar{R} > 0$. If one finds a stabilizing solution $V(x)$ for the Hamilton-Jacobi equation

$$p^T f(x) - \frac{1}{2} p^T g(x)\bar{R}^{-1} g(x)^T p + \frac{1}{2} h(x)^T h(x) = 0, \quad p = \frac{\partial V}{\partial x}^T,$$

then, the optimal regulator for (Σ, J) is state feedback controller

$$u = -g(x)^T \frac{\partial V}{\partial x}(x)^T.$$

Suppose $V(x)$ is a (not necessarily stabilizing) solution of (HJ). Then, the set

$$\Lambda_V = \{(x, p) | p = \partial V / \partial x(x)\}$$

is invariant under the flow of the associated Hamiltonian system derived from (HJ):

$$\dot{x} = \frac{\partial H}{\partial p}(x, p), \quad \dot{p} = -\frac{\partial H}{\partial x}(x, p) \tag{12.6}$$

Conversely, if an n-dimensional manifold Λ in the (x, p)-space is invariant under (12.6) and, at a point (x_0, p_0), the canonical projection $\pi : (x, p) \mapsto x$ on Λ is surjective, then Λ possesses Lagrangian submanifold property and there exists a solution $V(x)$ of (HJ) around x_0. If $(x_0, p_0) = (0, 0)$ and the above conditions such as the Lagrangian property hold, then the Hamiltonian flow of (12.6) on Λ is convergent to the origin and therefore Λ is a stable manifold of (12.6). The function $V(x)$ defined on a neighborhood $\mathcal{U}$ of the origin satisfies

$$\Lambda \cap \pi^{-1}(\mathcal{U}) = \{(x, p) | p = \partial V / \partial x(x),\ x \in \mathcal{U}\}.$$

A sufficient condition for the local existence of the stable manifold, or equivalently, the local stabilizing solution for (HJ) is obtained in [15]. It is a natural condition based on a linearization argument. For the algebraic Riccati equation

$$\text{(RIC)} \quad PA + A^T P - PR(0)P + Q = 0,$$

which is the linearization of (HJ), a symmetric matrix P is said to be the stabilizing solution of (RIC) if it is a solution of (RIC) and $A - R(0)P$ is stable.

Theorem 12.7 ([15]) *If the Riccati equation has the stabilizing solution P, there exists, locally around the origin, the stabilizing solution $V(x)$ to* (HJ) *with* $(\partial^2 V/\partial x^2)(0) = P$.

The theorem also says that if the stabilizing solution to (RIC) exists, an n-dimensional stable manifold, from which the stabilizing solution to (HJ) can be derived, locally exists around the origin.

Let us assume that the following algebraic Riccati equation has the stabilizing solution P

$$PA + A^T P - PR(0)P + Q = 0. \tag{12.7}$$

The associated Hamiltonian system (12.6) is written as

$$\begin{pmatrix} \dot{x} \\ \dot{p} \end{pmatrix} = \begin{pmatrix} \dfrac{\partial H}{\partial p} \\ -\dfrac{\partial H}{\partial x} \end{pmatrix} = \begin{pmatrix} A & -R(0) \\ -Q & -A^T \end{pmatrix} \begin{pmatrix} x \\ p \end{pmatrix} + \begin{pmatrix} N_1'(x, p) \\ N_2'(x, p) \end{pmatrix} \tag{12.8}$$

where $N_1'(x, p)$, $N_2'(x, p)$ are higher order terms. Using the stabilizing solution P for (12.7) and (12.8) can be put in a block-diagonalized form

$$\begin{pmatrix} \dot{x} \\ \dot{p} \end{pmatrix} = \begin{pmatrix} A - R(0)P & 0 \\ & -(A - R(0)P)^T \end{pmatrix} \begin{pmatrix} x \\ p \end{pmatrix} + \begin{pmatrix} N_1(x, p) \\ N_2(x, p) \end{pmatrix} \tag{12.9}$$

with appropriate linear coordinate transformation and therefore, it is shown that (12.8) has a stable manifold and the stable manifold algorithm can be applied to (12.9).

Example: Inverted pendulum optimal swing up and stabilization
The equations of motion of the inverted pendulum in Fig. 12.1 are

$$\begin{cases} (M + m)\ddot{x} + ml(\ddot{\theta}\cos\theta - \dot{\theta}^2 \sin\theta) + b\dot{x} = u \\ ml^2\ddot{\theta} + ml\ddot{x}\cos\theta - mgl\sin\theta = 0. \end{cases}$$

In our experiment, input voltage u has a limitation and to explicitly consider it, we put the above equations into the state-space equation with input saturation

$$\dot{x} = f(x) + g(x)\mathrm{sat}(u)$$

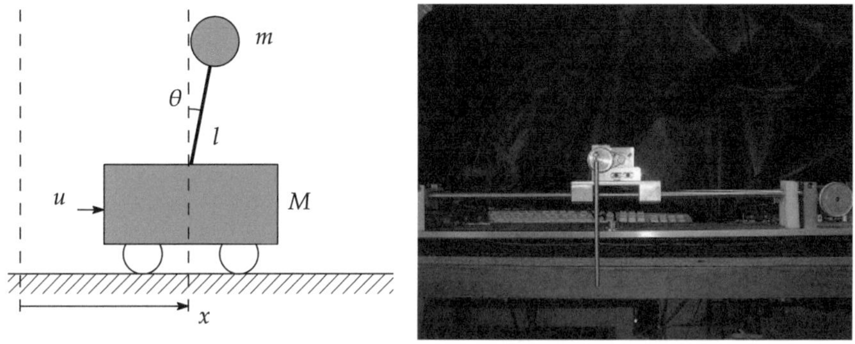

Fig. 12.1 Inverted pendulum on a cart (*left* schematic model, *right* experiment apparatus)

with

$$f(x) = \begin{bmatrix} x_2 \\ \dfrac{mlx_4^2 \sin x_3 - bx_2 - mg\cos x_3 \sin x_3}{M + m\sin^2 x_3} \\ x_4 \\ \dfrac{(M+m)g\sin x_3 + bx_2\cos x_3 - mlx_4^2 \sin x_3 \cos x_3}{l(M+m\sin^2 x_3)} \end{bmatrix}, \quad g(x) = \begin{bmatrix} 0 \\ \dfrac{1}{M+m\sin^2 x_3} \\ 0 \\ \dfrac{-\cos x_3}{l(M+m\sin^2 x_3)} \end{bmatrix}$$

The optimal control is sought for the cost function

$$J = \int_0^\infty x^T Q x + u^T R u \, dt, \quad Q = \begin{pmatrix} 0 & 0 & 0 & 0 \\ 0 & 0 & 0 & 0 \\ 0 & 0 & 4 & 0 \\ 0 & 0 & 0 & 40 \end{pmatrix}, \quad R = 1.$$

A Hamilton-Jacobi equation including a saturation function is derived and the stable manifold iteration (12.3) is applied. The feedback controller is constructed with 7th order polynomials that interpolate the data obtained by the stable manifold algorithm. Figure 12.2 shows the responses by the optimal control with saturation limit = 18[V] while Fig. 12.3 depicts the responses with saturation = 12[V]. It is interesting to note that the controller with smaller input limitation uses more swings so that the swing up is possible efficiently using smaller input voltage and these control strategy is derived from the HJE defining the optimal control problem for the inverted pendulum.

Fig. 12.2 2 swing responses with $|u| \leqslant 18$[V] (experiment)

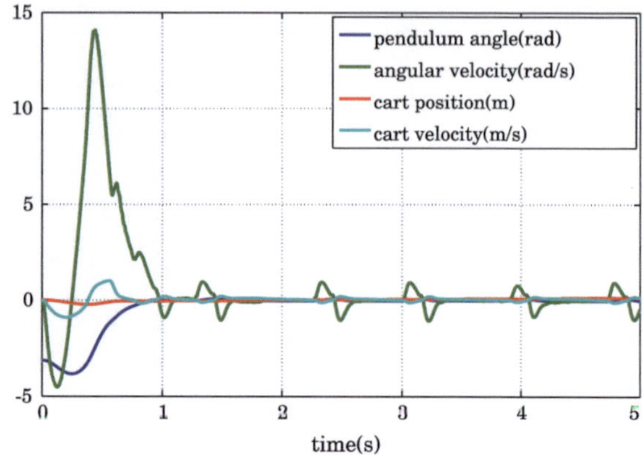

Fig. 12.3 3 swing responses with $|u| \leqslant 12$[V] (experiment)

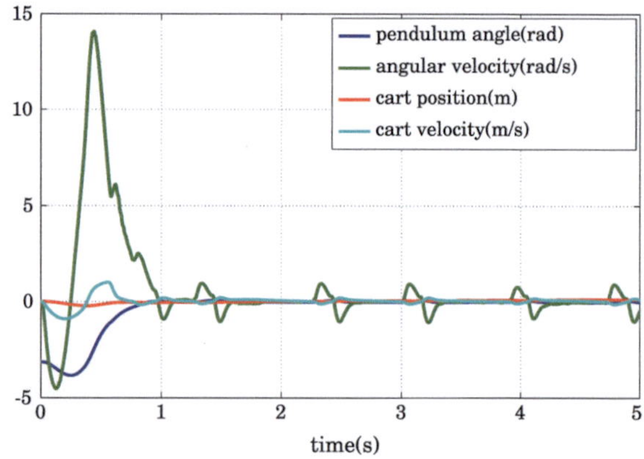

12.5 Optimal Output Regulation

Let us consider the set of equations of the form

$$\dot{x} = f(x) + g(x)u, \;\; x(t) \in \mathbb{R}^n, \;\; f(0) = 0 \tag{12.10}$$

$$e = h(x, w) \tag{12.11}$$

$$\dot{w} = s(w), \quad w(t) \in \mathbb{R}^p, \;\; s(0) = 0, \tag{12.12}$$

where $f : \mathbb{R}^n \to \mathbb{R}^n$, $g : \mathbb{R}^n \to \mathbb{R}^{n \times m}$, $s : \mathbb{R}^p \to \mathbb{R}^p$ and $h : \mathbb{R}^n \times \mathbb{R}^p \to \mathbb{R}^r$ are all sufficiently smooth. Output regulation problem is to find a controller $u = u(x, w)$ that stabilizes the system (12.10) with $w = 0$ so that the error (12.11) which consists of the system state x and the state w of exosystem (12.12) is regulated 0 as $t \to \infty$ for all initial values $x(0)$, $w(0)$ in some neighborhoods of the origin. This section presents how to introduce optimality in output regulation theory based on the results in [11].

Denote $A = \frac{\partial f}{\partial x}x(0)$, $B = g(0)$, $C = \frac{\partial h}{\partial x}(0,0)$, $S = \frac{\partial s}{\partial w}(0)$, $Q = \frac{\partial h}{\partial w}(0,0)$.

Assumption 12.8 (i) The triple $(A,\ B,\ C)$ is stabilizable and detectable.
(ii) The system is square, that is, the number of inputs and outputs are both r.
(iii) The system has relative degree 1, that is, $L_g h(0,0)$ is nonsingular.

Definition 12.9 (*Optimal output regulation problem*) Find a feedback controller $u(x,w)$ such that the closed loop system with $w=0$ is locally asymptotically stable at $x=0$ and for any initial condition $(x(0), w(0))$ in a neighborhood of the origin in $\mathbb{R}^n \times \mathbb{R}^p$, the error trajectory in (12.10) with $u(x,w)$ minimizes the cost function

$$J = \frac{1}{2}\int_0^\infty |e|^2 + |\dot{e}|^2\, dt.$$

Note that J can be written as

$$J = \frac{1}{2}\int_0^\infty |h(x,w)|^2 + \left|L_f h(x,w) + (L_g h(x,w))u + L_s h(x,w)\right|^2 dt,$$

from which one sees that Assumption (12.8-iii) is necessary for the regularity of the optimal control problem.

Assumption 12.10 (i) All the eigenvalues of S are on the imaginary axis.
(ii) The reference signal $w(t)$ is bounded.
(iii) The zero dynamics of system (12.10) with output $h(x,0)$ has a hyperbolic equilibrium at $x=0$.

The above assumptions guarantee the existence of solution $\Pi \in \mathbb{R}^{n\times p}$, $\Sigma \in \mathbb{R}^{r\times p}$ to the linear regulator equation

$$\Pi S = A\Pi + B\Sigma, \quad C\Pi + Q = 0. \tag{12.13}$$

The Hamilton-Jacobi equation for the optimal output regulation (Definition 12.9) is

$$\begin{aligned} &p_x^T \left\{ f - g(L_g h)^{-1}(L_f h + L_s h) \right\} + p_w^T s(w) \\ &\quad - \frac{1}{2} p_x^T g(L_g h)^{-1}(L_g h)^{-T} g^T p_x + \frac{1}{2}|h(x,w)|^2 = 0 \end{aligned} \tag{12.14}$$

and the Hamiltonian system associated with (12.14) is

$$\begin{cases} \dot{x} = (A - B(CB)^{-1}CA)x - B(CB)^{-1}QSw \\ \quad - B\left(B^T C^T C B\right)^{-1} B^T p_x + N_1(x,w,p_x) \\ \dot{w} = Sw + N_2(w) \\ \dot{p}_x = C^T C x - C^T Q w - (A - B(CB)^{-1}CA)^T p_x + N_3(x,w,p_x) \\ \dot{p}_w = -Q^T C x - Q^T Q w + S^T Q^T (B^T C^T)^{-1} B^T p_x \\ \quad - S^T p_w + N_4(x,w,p_x,p_w), \end{cases} \tag{12.15}$$

where N_1, N_2, N_3 and N_4 are nonlinear terms.

Lemma 12.11 *Under Assumptions 12.8 and 12.10, there exists a linear coordinate trasformation in which the linear part of* (12.15) *is written*

$$\begin{pmatrix} A_c & 0 & 0 & 0 \\ 0 & S & 0 & 0 \\ 0 & 0 & -{A_c}^T & 0 \\ 0 & 0 & 0 & -S^T \end{pmatrix}; \quad A_c = A - B(CB)^{-1}CA - B(B^TC^TCB)^{-1}B^TP,$$

where P is a stabilizing solution of a Riccati equation

$$P\bar{A} + \bar{A}^TP - PR_BP + C^TC = 0;$$
$$with\ \bar{A} = A - B(CB)^{-1}CA, \quad R_B = B(B^TC^TCB)^{-1}B^T,$$

and A_c is a stable matrix.

From Lemma 12.11, one knows that there exists a center-stable manifold in (12.15).

Theorem 12.12 *Under Assumptions 12.8 and 12.10,*

(i) there exists a center-stable manifold $p_x = p_x(x, w)$ of (12.15) around the origin $(x, w) = (0, 0)$ such that it satisfies HJE (12.14) and is represented as a derivative of a function of (x, w) defined in a neighborhood of $(x, w) = (0, 0)$,
(ii) the solution of optimal output regulation problem is given by

$$u = -(L_gh)^{-1}\left\{(L_gh)^{-T}g(x)^Tp_x(x, w) + L_fh(x, w) + L_sh(x, w)\right\},$$

where $p_x(x, w)$ represents the center-stable manifold of (12.15) around the origin.

A numerical example: Consider the example with unstable linearization

$$\begin{pmatrix} \dot{x}_1 \\ \dot{x}_2 \end{pmatrix} = \begin{pmatrix} 1 & 0.5 \\ 3 & 1.6 \end{pmatrix}\begin{pmatrix} x_1 \\ x_2 \end{pmatrix} + \begin{pmatrix} 0 \\ x_1^3 + x_2^3 \end{pmatrix} + \begin{pmatrix} 1 \\ 1 \end{pmatrix} u$$

exosystem: $\dot{w} = 0$

The goal is to design a feedback law $u = u(x, w)$ that achieves $x_1 = w$ as $t \to \infty$ in an optimal way:

$$J = \frac{1}{2}\int_0^{+\infty} (x_1 - w)^2 + (\dot{x}_1)^2 dt$$

The Hamiltonian system in (12.15) is

$$\begin{pmatrix} \dot{x}_1 \\ \dot{x}_2 \\ \dot{w} \\ \dot{p}_1 \\ \dot{p}_2 \\ \dot{p}_w \end{pmatrix} = H \begin{pmatrix} x_1 \\ x_2 \\ w \\ p_1 \\ p_2 \\ p_w \end{pmatrix} + \begin{pmatrix} 0 \\ x_1^3 + x_2^3 \\ 0 \\ -3x_1^2 p_2 \\ -3x_2^2 p_2 \\ 0 \end{pmatrix}; \quad H = \begin{pmatrix} 0 & 0 & 0 & -1 & -1 & 0 \\ 2 & 1.1 & 0 & -1 & -1 & 0 \\ 0 & 0 & 0 & 0 & 0 & 0 \\ -1 & 0 & -1 & 0 & -2 & 0 \\ 0 & 0 & 0 & 0 & -1.1 & 0 \\ -1 & 0 & -1 & 0 & 0 & 0 \end{pmatrix}.$$

The center-stable manifold iteration is applied to 10 times to the system with block-diagonalized linear part. The approximation of the center-stable manifold has been carried out to get a function $p_x(x, w)$ describing the center-stable manifold. We employed 11th order polynomials in x_1, x_2 and w for $p_x(x, w)$. The closed loop responses by the nonlinear optimal output regulation and by the linear output regulation controller are shown below. From Figs. 12.4 and 12.5, it can be seen that the nonlinear optimal regulator drives x_1 to track w while the linear optimal regulator fails to stabilize the system.

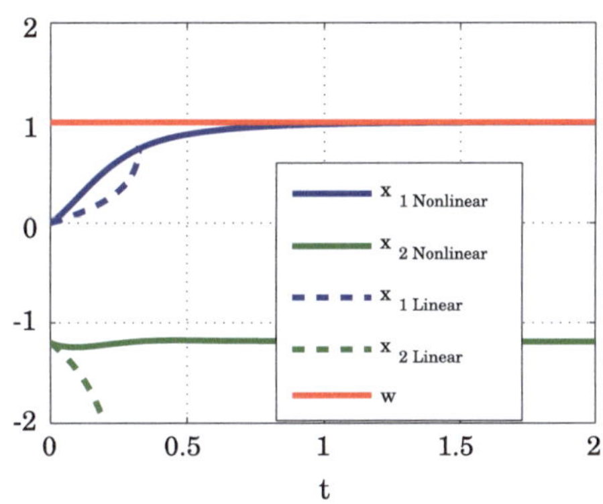

Fig. 12.4 State responses by linear and nonlinear output regulators

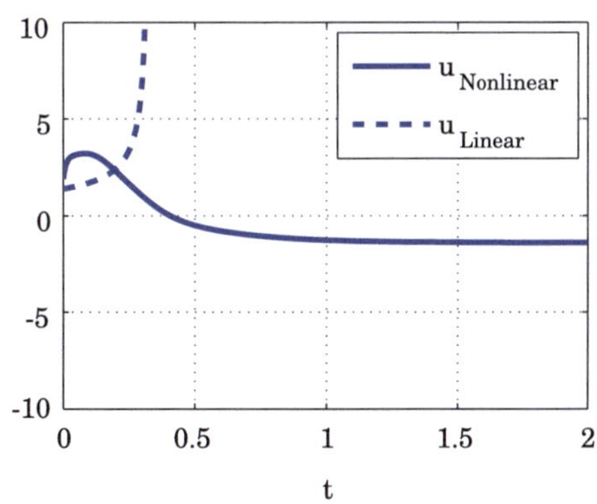

Fig. 12.5 Input responses by linear and nonlinear output regulators

12.6 Computation of Center Manifold of Periodic Orbit

Purpose of this section is to discuss a framework for computing a center manifold of a periodic orbit in a nonlinear system. This problem will be useful when analyzing geometric structure of a complicated nonlinear dynamical system.

Let us consider a nonlinear system

$$\dot{q} = f(q), \tag{12.16}$$

where $f : \mathbb{R}^n \to \mathbb{R}^n$ is a sufficiently smooth vector field. We assume that (12.16) has an unstable periodic orbit denoted as q_c satisfying $q_c(t+T) = q_c(t)$ for all $t \in \mathbb{R}$, where T is the period of q_c. Defining x as the distance of perturbed trajectory from the periodic orbit q_c, namely, $q = q_c + x$, we have

$$\dot{x} = A(t)x + N(t, x), \tag{12.17}$$

where

$$A(t) = \frac{\partial f}{\partial x}(q_c(t))$$

and $X(t, x_e)$ is a nonlinear term and given by

$$N(t, x) = f(q_c(t) + x) - f(q_c(t)) - A(t)x. \tag{12.18}$$

Let $\Phi(t)$ be the fundamental matrix of $A(t)$ satisfying $\dot{\Phi}(t) = A(t)\Phi(t)$, $\Phi(0) = I$. Noting that $A(t)$ is periodic with period T, it can be shown that there exists a constant matrix $\hat{A} \in \mathbb{R}^{n\times n}$ such that $\Phi(T) = e^{\hat{A}T}$. It is possible to assume that $\hat{A}$ is a real matrix by taking double period $2T$ if necessary. Let $L(t) = \Phi(t)e^{-\hat{A}t}$. Then, by the coordinate transformation $\hat{x} = L(t)x$, $\dot{x} = A(t)x$ can be put into $\dot{\hat{x}} = \hat{A}\hat{x}$. By this transformation, (12.17) is written as

$$\dot{\hat{x}} = \hat{A}x + \hat{N}(t, \hat{x}). \tag{12.19}$$

The matrix $\Phi(T)$ is called *monodromy matrix* and the above transformation is called Floquet-Lyapunov transformation [1]. The linear part of (12.19) is now time-invariant and the existence of a center manifold is equivalent to that $\hat{A}$ has an eigenvalue on the imaginary axis. When this condition is met, one can apply the center manifold algorithm (12.5) after block-diagonalizing (12.19).

Example: Computation of a periodic orbit in the restricted three body problem The Circular restricted three body problem (CRTBP) describes the motion of a massless object in the gravity field by two main bodies (primaries) that move in circles (see Fig. 12.6). Let $m_1, m_2, m_2 < m_1$, be the masses of the primaries and choose a rotating coordinate system with the origin at the center of mass. x–y frame is taken in such a way that x is directed from m_1 to m_2 and y is perpendicular to x with its

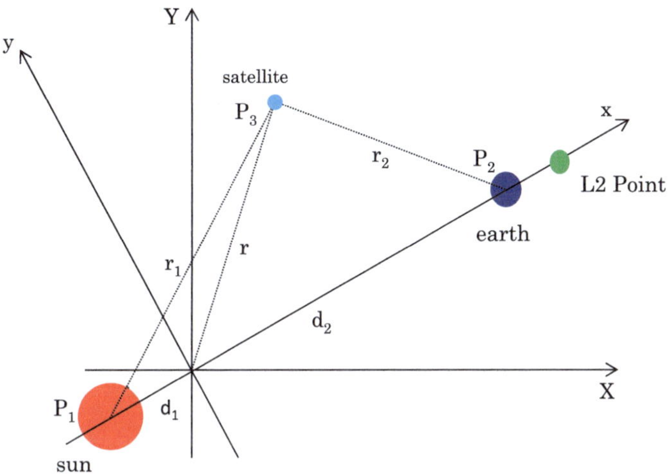

Fig. 12.6 Circular restricted three body problem

positive direction being the velocity of m_2. The z axis is taken to form the right-hand coordinate system. Normalizing the distance between m_1, m_2 and normalizing time so that the period of the circular motion around the center of mass is 2π, we get the dimensionless equations of motion as follows

$$\ddot{x} - 2\dot{y} - x = -(1-\rho)\frac{x+\rho}{{r_1}^3} - \rho\frac{x-(1-\rho)}{{r_2}^3} \tag{12.20}$$

$$\ddot{y} + 2\dot{x} - y = -(1-\rho)\frac{y}{{r_1}^3} - \rho\frac{y}{{r_2}^3} \tag{12.21}$$

$$\ddot{z} = -(1-\rho)\frac{z}{{r_1}^3} - \rho\frac{z}{{r_2}^3}, \tag{12.22}$$

where $\rho = m_2/(m_1+m_2)$, $r_1 = \sqrt{(x+\rho)^2 + y^2 + z^2}$ and $r_2 = \sqrt{(x-1+\rho)^2 + y^2 + z^2}$.

Defining a vector q as $q = [x, y, z, \dot{x}, \dot{y}, \dot{z}]^T$, (12.20)–(12.22) can be rewritten as

$$\frac{dq}{dt} = f(q), \tag{12.23}$$

where $f(q)$ is expressed as follows:

$$f(q) = \begin{bmatrix} \dot{x} \\ \dot{y} \\ \dot{z} \\ 2\dot{y} + x - (1-\rho)\frac{x+\rho}{r_1^3} - \rho\frac{x-(1-\rho)}{r_2^3} \\ -2\dot{x} + y - (1-\rho)\frac{y}{r_1^3} - \rho\frac{y}{r_2^3} \\ -(1-\rho)\frac{z}{r_1^3} - \rho\frac{z}{r_2^3} \end{bmatrix}. \tag{12.24}$$

We linearize the equation of motion of the RTBP in order to obtain closed orbits around the L2 equilibrium point. The linear system can be written as follows

$$\dot{x} = Ax. \tag{12.25}$$

Here, the matrix A is given as

$$A = \begin{bmatrix} 0 & 0 & 0 & 1 & 0 & 0 \\ 0 & 0 & 0 & 0 & 1 & 0 \\ 0 & 0 & 0 & 0 & 0 & 1 \\ 8.878 & 0 & 0 & 0 & 2 & 0 \\ 0 & -2.939 & 0 & -2 & 0 & 0 \\ 0 & 0 & -3.939 & 0 & 0 & 0 \end{bmatrix}.$$

The eigenvalues and eigenvectors of matrix A are

$$\lambda = \begin{bmatrix} -2.484 \\ 2.484 \\ 2.057i \\ 2.057i \\ 1.985i \\ 1.985i \end{bmatrix}$$

$$V = \begin{bmatrix} 0.328 & -0.328 & -0.131 & -0.131 & 0 & 0 \\ 0.179 & 0.179 & -0.417i & 0.417i & 0 & 0 \\ 0 & 0 & 0 & 0 & 0.450i & -0.450i \\ -0.814 & -0.814 & -0.269i & 0.269i & 0 & 0 \\ -0.444 & 0.444 & 0.858 & 0.858 & 0 & 0 \\ 0 & 0 & 0 & 0 & -0.893 & -0.893 \end{bmatrix}.$$

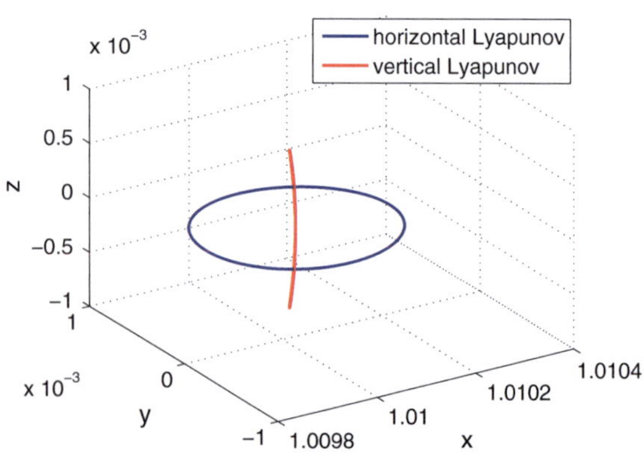

Fig. 12.7 Horizontal and vertical Lyapunov orbits

From the two center eigenvectors, v_3 and v_4, corresponding to $\lambda_3 = \lambda_4 = 2.057i$, one can obtain a closed orbit for (12.25) in the x-y plane. From the other two center eigenvectors, v_5 and v_6 corresponding to $\lambda_5 = \lambda_6 = 1.985i$, on the other hand, one finds a closed orbit for (12.25) in the z-direction. We take these orbit as q_c in (12.16) and apply the center manifold algorithm to compute flows on the center manifolds of q_c. It is seen that the pairs (v_3, v_4) and (v_5, v_6) are directions for shifting to horizontal and vertical center manifolds, respectively. The results are shown in Fig. 12.7.

References

1. S. Bittanti, P. Colaneri, *Periodic Systems Filtering and Control* (Springer, London, 2009)
2. J. Carr, *Applications of Centre Manifold Theory* (Springer, New York, 1981)
3. S.-N. Chow, J.K. Hale, *Methods of Bifurcation Theory* (Springer, New York, 1982)
4. R. Fujimoto, N. Sakamoto, The Stable Manifold Approach for Optimal Swing Up and Stabilization of an Inverted Pendulum with Input Saturation. in *Proceedings of IFAC World Congress* (2011)
5. J. Huang, W.J. Rugh, An approximation method for the nonlinear servomechanism problem. IEEE Trans. Autom. Control **37**, 1395–1398 (1992)
6. J. Huang, W.J. Rugh, Stabilization on zero-error manifolds and the nonlinear servomechanism problem. IEEE Trans. Autom. Control **37**, 1009–1013 (1992)
7. A. Isidori, C.I. Byrnes, Output regulation of nonlinear systems. IEEE Trans. Autom. Control **35**, 131–140 (1990)
8. A.L. Kelley, The stable, center-stable, center, center-unstable, unstable manifolds. J. Differ. Equ. **3**, 546–570 (1967)
9. K. Nagata, N. Sakamoto, Y. Habaguchi, Center Manifold Method for the Orbit Design of the Restricted Three Body Problem. in *54th IEEE Conference on Decision and Control* (2015). Submitted
10. N. Sakamoto, Case studies on the application of the stable manifold approach for nonlinear optimal control design. Automatica **49**, 568–576 (2013)
11. N. Sakamoto, B. Rehák, Iterative Methods to Compute Center and Center-Stable Manifolds with Application to the Optimal Output Regulation Problem. in *Proceeings of 48th IEEE Conference on Decision and Control* (2011), pp. 4640–4645
12. N. Sakamoto, A.J. van der Schaft, Analytical approximation methods for the stabilizing solution of the Hamilton-Jacobi equation. IEEE Trans. Autom. Control **53**, 2335–2350 (2008)
13. J.M.A. Scherpen, Balancing for nonlinear systems. Syst. Control Lett. **21**, 143–153 (1993)
14. J. Sijbrand, Properties of center manifolds. Trans. Am. Math. Soc. **289**, 431–469 (1985)
15. A.J. van der Schaft, On a state space approach to nonlinear H_∞ control. Syst. Control Lett. **16**, 1–18 (1991)
16. A.J. van der Schaft, L_2-gain analysis of nonlinear systems and nonlinear state feedback H_∞ control. IEEE Trans. Autom. Control **37**, 770–784 (1992)
17. J.C. Willems, Least squares stationary optimal control and the algebraic Riccati equation. IEEE Trans. Autom. Control **16**, 621–634 (1971)
18. J.C. Willems, Dissipative dynamical systems-Part I, II. Arch. Ration. Mech. Anal. **45**, 321–393 (1972)

Chapter 13
On Geometric Properties of Triangularizations for Nonlinear Control Systems

Markus Schöberl and Kurt Schlacher

Abstract We consider triangular decompositions for nonlinear control systems. For systems that are exactly linearizable by static feedback it is well known that a triangular structure exists in adapted coordinates using the Frobenius theorem to straighten out a nested sequence of involutive distributions. This triangular form is based on explicit ordinary differential equations from which it can be easily seen that exactly linearizable systems are also flat. We will analyze this triangularization also from a dual perspective using a Pfaffian system representation. This point of view allows the introduction of a triangular form corresponding to implicit ordinary differential equations. For systems that are flat but not exactly linearizable by static feedback, this modified triangular form turns out to be useful in setting up a constructive algorithm to compute so-called 1-flat outputs.

13.1 Introduction

In mathematical systems theory, the structural analysis of dynamical systems plays a prominent role. One of the main objectives is the representation of systems in a beneficial way, in order to make certain system properties visible. For nonlinear control systems differential geometric concepts such as distributions, codistributions, and the Frobenius theorem can be used to construct system decompositions to answer delicate problems such as controllability, observability, disturbance decoupling, input–output decoupling and exact linearization. In this context, outstanding results have been reported, e.g., in [5, 6, 8, 10, 13, 19, 20], to name just a few.

Dedicated to the 60th birthday of Arjan van der Schaft.

M. Schöberl (✉) · K. Schlacher
Institute of Automatic Control and Control Systems Technology,
Johannes Kepler University, Linz, Austria
e-mail: markus.schoeberl@jku.at

K. Schlacher
e-mail: kurt.schlacher@jku.at

M.K. Camlibel et al. (eds.), *Mathematical Control Theory I*,
Lecture Notes in Control and Information Sciences 461,
DOI 10.1007/978-3-319-20988-3_13

Furthermore, the famous textbooks [7, 11] provide an excellent introduction into this geometric point of view regarding nonlinear control systems.

In what follows, we aim to characterize special decompositions of nonlinear control systems, namely triangular structures, in the context of the exact linearization problem with static feedback and the flatness analysis. The paper [20] by Arjan van der Schaft proposes a triangularization for general nonlinear systems regarding the exact linearization problem. This was very inspiring for us and encouraged us to dig deeper into that structure with the desire to analyze more general triangular structures that can be used for systems that are not exactly linearizable by static feedback but possess the property to be flat. In the nonlinear control literature, much attention has been put to the so-called affine multi-input systems of the form

$$\dot{x} = a(x) + B(x)u \tag{13.1}$$

and to general nonlinear systems

$$\dot{x} = f(x, u) \tag{13.2}$$

with n states $x = (x^1, \ldots, x^n)$ and m inputs $u = (u^1, \ldots, u^m)$. The systems (13.1) and (13.2) are termed exactly linearizable by static feedback, if in new coordinates z for the state and v for the control, they can be represented as pure integrator chains (Brunovsky canonical form). Besides the question whether a system possesses the property to be exactly linearizable by static feedback also the more general question regarding the flatness property is of interest in control and systems theory. A system is termed flat if it enjoys the characteristic feature that the (time) evolution of the state and input (control) variables can be recovered from that of the flat output without integration, see [3, 4]. Hence, a system in Brunovsky form is trivially flat, as the flat output are the top entries of the integrator chains. Furthermore, systems that are flat but not exactly linearizable by static feedback, can be put to Brunovsky form by using dynamic feedback—thus one derives the integrator chains on an extended state manifold. A systematic test whether a system is exactly linearizable by static feedback is known since more than three decades and is based on the involutivity of certain distributions. Contrary, a general systematic (reasonably practicable) procedure that checks a system for flatness has not been discovered yet.

For the exact linearization problem of the affine input system (13.1) the distributions

$$\Lambda_i = \sum_{k=1}^{i} \mathrm{span}\{\mathrm{ad}_a^{k-1}(b_l)\}_{l=1}^{m} \tag{13.3}$$

where b_j denotes the jth column of the matrix $B(x)$ play a key role.

Remark 13.1 For the repeated application of the Lie-bracket of two vector fields v_1 and v_2, the following common notation will be used

$$\mathrm{ad}_{v_1}^0(v_2) = v_2\,, \quad \mathrm{ad}_{v_1}^k(v_2) = [v_1, \mathrm{ad}_{v_1}^{k-1}(v_2)]\,, \quad k \geq 1.$$

Similarly, for the system (13.2) the distributions

$$\Delta_i = \sum_{k=0}^{i} \mathrm{span}\{\mathrm{ad}_f^k(\partial_{u^l})\}_{l=1}^m = \sum_{k=0}^{i} \mathrm{span}\{\mathrm{ad}_f^k(\partial_u)\}$$

are of importance, with $\partial_{u^l} = \partial/\partial_{u^l}$. Additionally, $\rho_1 \geq \rho_2 \geq \cdots \geq \rho_r$ is met, where

$$\rho_j = \dim(\Delta_j) - \dim(\Delta_{j-1}), \quad j = 1, \ldots, r,$$

with $\rho_0 = m$ and for simplicity we assume $\rho_1 = m$ (no redundant inputs). As (13.1) is only a special case of (13.2), we now focus on general nonlinear systems. In the paper [20] by Arjan van der Schaft and in the textbook [11] authored by Henk Nijmeijer and Arjan van der Schaft, the following triangular form

$$\begin{bmatrix} \dot z_r \\ \vdots \\ \dot z_i \\ \vdots \\ \dot z_3 \\ \dot z_2 \\ \dot z_1 \end{bmatrix} = \begin{bmatrix} f_r(z_r, z_{r-1}) \\ \vdots \\ f_i(z_r, z_{r-1}, \ldots, z_{i-1}) \\ \vdots \\ f_3(z_r, z_{r-1}, \ldots, z_3, z_2) \\ f_2(z_r, z_{r-1}, \ldots, z_3, z_2, z_1) \\ f_1(z_r, z_{r-1}, \ldots, z_3, z_2, z_1, v) \end{bmatrix} \tag{13.4}$$

where the new state is partitioned into r blocks $z = (z_1, \ldots, z_r)$ with

$$\mathrm{rank}(\frac{\partial f_i}{\partial z_{i-1}}) = \dim(z_i) = \rho_i \,, \quad z_i = (z_i^1, \ldots, z_i^{\rho_i})$$

is introduced ($z_0 = v$), which is based on the involutivity of the distributions Δ_i and the existence of a real number r such that $\dim(\Delta_r) = m + n$. The triangular form (13.4) can be obtained (locally) by a state transformation $x = \phi(z)$ together with a static feedback $u = \psi(v, z)$, where the Frobenius theorem was heavily used in straightening out the involutive distributions Δ_i. A slightly modified triangular form can be obtained as

$$\begin{bmatrix} \dot z_r \\ \vdots \\ \dot z_i \\ \vdots \\ \dot z_3 \\ \dot z_2 \\ \dot z_1 \end{bmatrix} = \begin{bmatrix} f_r(z_r, \hat z_{r-1}) \\ \vdots \\ f_i(z_r, z_{r-1}, \ldots, \hat z_{i-1}) \\ \vdots \\ f_3(z_r, z_{r-1}, \ldots, z_3, \hat z_2) \\ f_2(z_r, z_{r-1}, \ldots, z_3, z_2, \hat z_1) \\ f_1(z_r, z_{r-1}, \ldots, z_3, z_2, z_1, v) \end{bmatrix} \tag{13.5}$$

which accounts for the fact that in (13.4) the inputs for the system $(f_r, \dots, f_i)$ are z_{i-1}. As $\rho_i \geq \rho_{i+1}$, we can introduce new coordinates z_i such that we have a decomposition of the variables in the form $z_i = (\hat{z}_i, y_i)$ and

$$\text{rank}(\frac{\partial f_i}{\partial \hat{z}_{i-1}}) = \dim(z_i) = \rho_i$$

is met if $\rho_i > \rho_{i+1}$. Otherwise, if $\rho_i = \rho_{i+1}$ the rank condition is also fulfilled but we have $z_i = \hat{z}_i$ with $\dim(\hat{z}_i) = \dim(\rho_{i+1})$.

Remark 13.2 In [11] based on the normal form (13.4), the system is transformed directly to a Brunovsky like representation. However, to be able to compare our results regarding a more general triangular form, we have introduced the intermediate structure as in (13.5).

From the representation (13.5), the flatness of the system can be easily checked. Indeed, the flat output is $(y_r, y_{r-1}, \dots y_1)$ with $z_r = y_r$ where (some of) the y_i, $i = r-1, \dots, 1$ can be empty. This follows (locally) by the implicit function theorem when the triangular form (13.5) is worked through from the top to the bottom. It can easily be checked that $\dim(y) = \dim(y_r, \dots, y_1) = m$ is met, due to a dimension argument regarding the ρ_i, see also [16], as $\dim(y_i) = \rho_i - \rho_{i+1}$, $i = r-1, \dots, 1$ and $\dim(y_r) = \dim(z_r) = \rho_r$.

Remark 13.3 As a consequence, if $\dim(z_r) < m$, $(z_r = y_r)$ then in some lower blocks additional y_i have to appear.

In the following, we want to present two examples that are not exactly linearizable by static feedback, but their flat outputs can also be read off by utilizing a triangular form which is more general than (13.4) or (13.5), respectively.

13.1.1 Motivating Example I

We consider the model of an induction motor, which is of the class (13.1) with $n = 5$ and $m = 2$, reading as

$$\begin{aligned} \dot{\omega} &= (\mu\psi_d i_q - \tfrac{\tau_L}{J}), & \dot{i}_d &= v_d \\ \dot{\psi}_d &= \eta(M i_d - \psi_d), & \dot{i}_q &= v_q \\ \dot{\rho} &= (n_p\omega + \eta M \tfrac{i_q}{\psi_d}) \end{aligned} \qquad (13.6)$$

where an input transformation has already been applied,[1] see also [2, 9, 16]. The stator currents are i_d and i_q, the rotor angular velocity is ω, whereas ψ_d and ρ are functions of the rotor fluxes and the new inputs are v_d and v_q. The parameters n_P, μ, M, J, η are assumed to be constant. Furthermore, τ_L is the load-torque, and for

[1]From (13.6) it is readily observed that the model shows a singularity if $\psi_d = 0$.

simplicity we assume τ_L to be a known function of ω. Let us consider a decomposition of (13.6) as

$$\begin{aligned} \dot{\omega} &= (\mu\psi_d i_q - \tfrac{\tau_L}{J}) \\ \dot{\rho} &= (n_p\omega + \eta M \tfrac{i_q}{\psi_d}) \end{aligned}, \quad \begin{aligned} \dot{\psi}_d &= \eta(M i_d - \psi_d) \\ \dot{i}_q &= v_q \end{aligned}, \quad \dot{i}_d = v_d \tag{13.7}$$

and a different decomposition which is obtained by combining the differential equations for $\dot{\omega}$ and $\dot{\rho}$ in the form

$$\dot{\omega} - \tfrac{\mu\psi_d^2}{\eta M}\dot{\rho} = -\tfrac{\mu\psi_d^2}{\eta M} n_p\omega - \tfrac{\tau_L}{J}, \quad \begin{aligned} \dot{\psi}_d &= \eta(M i_d - \psi_d) \\ \dot{\rho} &= (n_p\omega + \eta M \tfrac{i_q}{\psi_d}) \end{aligned}, \quad \begin{aligned} \dot{i}_d &= v_d \\ \dot{i}_q &= v_q \end{aligned}. \tag{13.8}$$

From the first decomposition (13.7), we deduce that when we consider the top block (which is the most left system) that we can choose ω and ρ in an arbitrary manner (as functions of time) from which i_q and ψ_d follow without any integration. Continuing with the next subsystem (the block in the center), we can derive i_d and v_q again by pure differentiation as ψ_d and i_q are already prescribed. Finally from the last subsystem v_d follows. It is remarkable that (13.7) is not in the form (13.4) as the dimensions of the subsystems are not decreasing from the bottom to the top (from right to left). To obtain a triangular structure that meets this dimension requirement, one can consider (13.8) but the block at the top is an implicit differential equation. Nevertheless, also from the representation (13.8) it can be checked that (ω, ρ) is a flat output, as from the first subsystem ψ_d can be computed without integration when ω and ρ are prescribed. From the center block i_d and i_q follow and finally from the block at the bottom (the most right one) one obtains v_d and v_q. From this example, it is clear that for the flatness analysis the important structural property is the triangular form, and not the fact that the ordinary differential equations are of explicit nature—also implicit differential equations can be put into a triangular shape. This will be heavily exploited also in the next example.

13.1.2 Motivating Example II

Let us consider the following academic example

$$\begin{aligned} \dot{x}^1 &= u^1 \\ \dot{x}^2 &= u^2 \\ \dot{x}^3 &= \sqrt{u^1 u^2} \end{aligned} \tag{13.9}$$

belonging to the class (13.2) with $n = 3$ and $m = 2$. The system is not linearizable by static feedback and contrary to the foregoing example, a pure regrouping or combination of the equations does not allow to read off the flat outputs. However, we can apply the transformation

$$
\begin{aligned}
x^1 &= \hat{z}_2^1 \hat{z}_1^1, & u^1 &= e^{\hat{z}_0^1} \hat{z}_2^1 \\
x^2 &= \hat{z}_1^1 + y_3^1, & u^2 &= e^{\hat{z}_0^1} \\
x^3 &= \sqrt{\hat{z}_2^1 \hat{z}_1^1} + y_3^2
\end{aligned}
$$

using the new coordinates $(y_3^1, y_3^2, \hat{z}_2^1, \hat{z}_1^1, \hat{z}_0^1)$.

Remark 13.4 The new variables are grouped into four blocks $z = (z_3, z_2, z_1, z_0)$ with $z_i = (\hat{z}_i, y_i)$. We have $\dim(z_3) = 2$ where $\hat{z}_3$ is empty ($z_3 = y_3$). The three remaining blocks z_i, $i = 2, 1, 0$ are of dimension one, where y_i is empty. It should be noted, that the superscript indices are superficial for $\hat{z}_2^1, \hat{z}_1^1, \hat{z}_0^1$ as the corresponding blocks are of dimension one. However, to be consistent with what follows, we also indicate these indices at this stage already.

Combining the differential equations, we obtain in the new coordinates the implicit system

$$
\begin{aligned}
\dot{y}_3^1 \sqrt{\hat{z}_2^1} - 2\dot{y}_3^2 &= 0 \\
\dot{y}_3^1 \hat{z}_2^1 - \dot{\hat{z}}_2^1 \hat{z}_1^1 &= 0 \\
\dot{\hat{z}}_1^1 + \dot{y}_3^1 &= e^{\hat{z}_0^1}
\end{aligned}
\tag{13.10}
$$

from which the flat output (y_3^1, y_3^2) can be read off. Indeed, from the first differential equation one can express $\hat{z}_2^1$ once y_3^1 and y_3^2 are specified. The second differential equation allows to derive $\hat{z}_1^1$ as at that stage y_3^1 and $\hat{z}_2^1$ are already specified, and finally from the last equation $\hat{z}_0^1$ can be obtained. From the inverse transformation

$$
\begin{aligned}
\hat{z}_0^1 &= \ln(u^2), & y_3^1 &= x^2 - x^1 \tfrac{u^2}{u^1} \\
\hat{z}_1^1 &= x^1 \tfrac{u^2}{u^1}, & y_3^2 &= x^3 - x^1 \sqrt{\tfrac{u^2}{u^1}} \\
\hat{z}_2^1 &= \tfrac{u^1}{u^2}
\end{aligned}
$$

we obtain the flat output (y_3^1, y_3^2) in the original coordinates (x, u). The flat output is called 1-flat as it depends on the state x and on the control u but not on the derivatives of the control, see also Sect. 13.4. Furthermore, it should be noted that the transformation $(x^1, x^2, x^3, u^1, u^2) = \varphi(y_3^1, y_3^2, \hat{z}_2^1, \hat{z}_1^1, \hat{z}_0^1)$ is not a state transformation together with static feedback, as all coordinates (state and control) are mixed up—hence implicit differential equations are the outcome.

13.2 Mathematical Framework

In this section, we introduce the mathematical machinery needed for our analysis. The main tools in the derivation of the normal form (13.4) were distributions, the Lie-bracket, and the theorem of Frobenius. In the following, we will also use the dual object, namely codistributions that allow to derive the same representation (13.4) using the so-called derived flag of a Pfaffian system. Furthermore, the use of Pfaffian systems also allows to analyze dynamical systems which are described by implicit differential equations—hence, in order to generalize the triangularization (13.4) from the case of explicit to implicit differential equations we will make heavy use of a system representation as a Pfaffian system based on codistributions. The interested reader is referred to [1] for an extensive discussion on Pfaffian systems and exterior algebra. In the next section, we summarize the main tools that will be needed subsequently.

13.2.1 Exterior Algebra

We consider a manifold M with local coordinates (x^i), $i = 1, \dots, n_x$ and $\dim(M) = n_x$. The tangent bundle $T(M)$ and the cotangent bundle $T^*(M)$ possess the bases $\{\partial_i\}$ and $\{\mathrm{d}x^i\}$, respectively, with $\partial_i = \partial/\partial x^i$. A Pfaffian system P on M can be identified with a codistribution on M

$$P = \mathrm{span}\{\omega^1, \dots, \omega^{n_P}\}, \ \omega^i = a^i_k(x)\mathrm{d}x^k$$

with $\dim(P) = n_P$, where the Einstein convention on sums has been applied (summation over k). The Pfaffian system P is called integrable iff $\mathrm{d}P = 0 \bmod P$ is met, which is equivalent to the existence of 1-forms α^i_j such that $\mathrm{d}\omega^i = \alpha^i_j \wedge \omega^j$ is met for $i = 1, \dots, n_P$, where d denotes the exterior derivative. Integrable Pfaffian systems allow a representation as

$$P = \mathrm{span}\{\mathrm{d}h^1, \dots, \mathrm{d}h^{n_P}\}$$

with $h \in C^\infty(M)$. This implies the existence of coordinates (z^i) such that $P = \mathrm{span}\{\mathrm{d}z^1, \dots, \mathrm{d}z^{n_P}\}$ is met, or equivalently $P^\perp = \mathrm{span}\{\partial_{n_P+1}, \dots, \partial_{n_x}\}$ for the annihilator $P^\perp$ which is a distribution on M meeting [2] $v \rfloor \omega = 0$ for all $\omega \in P$, $v \in P^\perp$.

Remark 13.5 This is a variant of the well-known Frobenius theorem. P is involutive (integrable) iff $P^\perp$ is involutive and vice versa. A distribution D is involutive iff $[v_i, v_j] \in D$ for all $v_i, v_j \in D$, where $[\cdot, \cdot]$ denotes the Lie-bracket. For a d-dimensional involutive distribution D, the Frobenius theorem guarantees the existence of local coordinates such that $D = \mathrm{span}\{\partial_1, \dots, \partial_d\}$ is met—in this context the Frobenius theorem is also recognized as straightening out theorem.

The following construction generates the largest involutive subcodistribution of a Pfaffian system.

[2] With $\rfloor$ we denote the contraction (hook) operator which in this case is the natural pairing between vector fields and 1-forms.

Definition 13.6 The derived flag of a Pfaffian system P is the following nested sequence of systems

$$P = P^{(0)} \supset P^{(1)} \supset P^{(2)} \supset \cdots$$

with $P^{(k+1)} := \{\omega \in P^{(k)} \big| \, \mathrm{d}\omega = 0 \bmod P^{(k)}\}$.

If there is a k^* such that $P^{(k^*+1)} = P^{(k^*)}$ is met then $P^{(k^*)}$ is integrable. $P^{(k^*)} = \{0\}$ means that P does not contain a nontrivial integrable subsystem.

Remark 13.7 Using distributions, the sequence from above can also be generated from

$$P^{(k+1)} = \{F_k + [F_k, F_k]\}^{\perp}, \quad F_k = (P^{(k)})^{\perp}, \tag{13.11}$$

with $[F_k, F_k] = \mathrm{span}\{[v_i, v_j]\}$, $v_i, v_j \in F_k$.

To answer the question whether a Pfaffian system P is described with the minimal number of variables one can construct so-called Cauchy characteristic vector fields.

Definition 13.8 The vector field v is called a *Cauchy characteristic vector field* of P, if

$$v \rfloor P = 0 \,, \quad v \rfloor \mathrm{d}P \subset P \tag{13.12}$$

is met. The distribution formed by all independent Cauchy characteristic vector fields of P is denoted as $C(P)$ which is an involutive distribution on M by construction, see, e.g., [1] for a proof.

If $\dim(C(P)) = n_c$ then the Pfaffian system P can be represented with $n_x - n_c$ coordinates. This minimal representation can be achieved by straightening out the involutive distribution $C(P)$ by using the Frobenius theorem.

Example 13.9 The system (13.9) can be represented as a Pfaffian system on a manifold with coordinates $(t, x^1, x^2, x^3, u^1, u^2)$ as $P = \mathrm{span}\{\omega^1, \omega^2, \omega^3\}$ with

$$\omega^1 = \mathrm{d}x^1 - u^1 \mathrm{d}t \,, \ \omega^2 = \mathrm{d}x^2 - u^2 \mathrm{d}t \,, \ \omega^3 = \mathrm{d}x^3 - \sqrt{u^1 u^2} \mathrm{d}t . \tag{13.13}$$

It should be noted that time t, the state variables (x^1, x^2, x^3) as well as the inputs (u^1, u^2) are coordinates on a six-dimensional manifold. However, also in a Pfaffian representation the character of a dynamical system can be made visible as discussed in the next section, where we additionally will make use of the condition that $\mathrm{d}t \neq 0$ has to be met. Hence, we will introduce a fibration with respect to time.

13.2.2 Dynamical Systems

To study dynamical systems, we consider the fibred manifold $Z \times \mathbb{R} \rightarrow \mathbb{R}$ with local coordinates (z^i) for Z and the time t. The time coordinate plays an distinguished role, as the solutions of dynamical systems can be parametrized using the time coordinate, e.g., as $z(t)$. Consequently, a time-invariant dynamical system can be represented as

$$P = \operatorname{span}\{\omega^i\}, \quad \omega^i = a^i_k(z)\mathrm{d}z^k - b^i(z)\mathrm{d}t\,, \quad \dim(z) \geq \dim(P) \tag{13.14}$$

with the 1-forms ω^i and we assume that the matrix $[a^i_k]$ is of maximal rank.

Due to the fibration with respect to the time coordinate, we can consider the vertical annihilator $V(P)^\perp$ consisting of vector fields that annihilate P and which are tangential to the fibers of $Z \times \mathbb{R} \to \mathbb{R}$. Hence, $V(P)^\perp = \operatorname{span}\{P, \mathrm{d}t\}^\perp$ or equivalently

$$V(P)^\perp := \{v \in T(Z) | \, v \rfloor \omega = 0, \ \forall \omega \in P\}.$$

Given a dynamical system (13.14) (corresponding to implicit differential equations in general) it is of interest whether it is possible to represent it in explicit form as (13.2) which in Pfaffian description reads as

$$P = \operatorname{span}\{\vartheta^i\}, \quad \vartheta^i = \mathrm{d}x^i - f^i(x, u)\mathrm{d}t \tag{13.15}$$

by means of a diffeomorphism $(x, u) = \psi(z)$.

Theorem 13.10 *The diffeomorphism $(x, u) = \psi(z)$, which allows to find a generator of the system (13.14) in the form as in (13.15), exists iff $V(P)^\perp$ is involutive.*

Necessity can be seen as $V(P)^\perp$ is already involutive in the representation (13.15), indeed, $V(P)^\perp = \operatorname{span}\{\partial_u\}$, and the transformation $(x, u) = \psi(z)$ follows by using the theorem of Frobenius, by straightening out $V(P)^\perp$.

Remark 13.11 Instead of computing $V(P)^\perp$ to check involutivity also the examination whether $\{P, \mathrm{d}t\}$ is integrable is possible.

It is readily observed that there is an (intrinsic) one-to-one correspondence between a Pfaffian system (13.14) and the affine functions $\omega^i_e = a^i_k(z)\dot{z}^k - b^i(z)$, see also [12]. Of particular importance will be the concept of parameterizable Pfaffian systems.

Definition 13.12 The Pfaffian system (13.14) is called parameterizable with respect to $\hat{x}$ with $\dim(\hat{x}) = n_P$ if in adapted coordinates $(x, \hat{x})$ it takes the following special form

$$\bar{\omega}^i = m^i_l(x, \hat{x})\mathrm{d}x^l - n^i(x, \hat{x})\mathrm{d}t \tag{13.16}$$

and the Jacobian matrix $[\partial_{\hat{x}}\bar{\omega}^i_e]$ has maximal rank $n_x = n_P$ (where $\bar{\omega}^i_e = 0$).

If the requirements of Definition 13.12 can be fulfilled, we can locally solve $\bar{\omega}^i_e = 0$ for $\hat{x} = g(x, \dot{x})$. Hence, the variables $\hat{x}$ are termed *nonderivative* and their evolution can be obtained (locally) by means of the implicit function theorem (rank criteria). It should be noted that $\partial_{\hat{x}} \in V(P)^\perp$ is met.

Remark 13.13 The crucial observation regarding systems that are parameterizable with respect to some nonderivative variables $\hat{x}$ is that the time evolution of $\hat{x}$ can be obtained by pure differentiation, when the time evolution of x is prescribed. For a system described by explicit differential equations, the inputs are nonderivative variables and the system is parameterizable if we have as many independent inputs as differential equations.

Example 13.14 (Example 13.9 continued). For the Pfaffian system (13.13), we have $V(P)^{\perp} = \mathrm{span}\{\partial_{u^1}, \partial_{u^2}\}$ when we consider a fibration with respect to the time coordinate t. Furthermore, $C(P) = \{\}$ as no redundant variables are involved. From this, we observe that for explicit systems $V(P)^{\perp}$ corresponds to the input distribution.

13.3 Exact Linearization: Pfaffian Approach

Naturally related to the nonlinear system (13.2) is the m-dimensional input distribution $\partial_u = \{\partial_{u^1}, \ldots, \partial_{u^m}\}$ as well as the vector field

$$f_e = \partial_t + f = \partial_t + f^i(x,u)\partial_i$$

describing the right-hand side of the ordinary differential equation (13.2) together with $\dot{t} = 1$. The check for the exact linearization property of the system (13.2) can equivalently be performed using the Pfaffian representation of the nonlinear control system which is given in (13.15). The following theorem that uses the derived flag (see Definition 13.6) to determine conditions of whether a nonlinear control system is exactly linearizable by static feedback has been proved in [14, 18].

Theorem 13.15 *The system (13.15) is exactly linearizable by static feedback if and only if $\{P^{(k)}, \mathrm{d}t\}$ is integrable for every k and $P^{(\mu)} = \{0\}$ for some μ and if each $P^{(k)}$ is of constant dimension.*

In [14], it is shown that the conditions of Theorem 13.15 correspond to the test using the involutivity of distributions for the affine input system (13.1). For a general nonlinear system, the equivalence can be shown as follows. The key observation is that the requirement that $\{P^{(k)}, \mathrm{d}t\}$ is integrable (involutive) for every k corresponds to the involution of the distributions Δ_i. Thus, we have to show that $\{P^{(j)}, \mathrm{d}t\}^{\perp} = \Delta_j$ or which is the same $V(P^{(j)})^{\perp} = \Delta_j$ holds. Indeed, $(P^{(0)})^{\perp} = \mathrm{span}\{\partial_u, f_e\}$ and

$$V(P^{(0)})^{\perp} = \{P^{(0)}, \mathrm{d}t\}^{\perp} = \{\partial_u\} = \Delta_0$$

where it is worth noting that $V(P^{(0)})^{\perp}$ does not include f_e in contrast to $(P^{(0)})^{\perp}$. Then using the fact that the derived flag can also be expressed using the annihilator of the corresponding distributions, we have $(P^{(1)})^{\perp} = \mathrm{span}\{f_e, \partial_u, [f, \partial_u]\}$ as well as

$$V(P^{(1)})^{\perp} = \{P^{(1)}, \mathrm{d}t\}^{\perp} = \mathrm{span}\{\partial_u, [f, \partial_u]\} = \Delta_1. \tag{13.17}$$

We furthermore have $(P^{(2)})^{\perp} = \mathrm{span}\{f_e, \partial_u, [f, \partial_u], [f, [f, \partial_u]]\}$ since Δ_1 is involutive. Then

$$\{P^{(2)}, \mathrm{d}t\}^{\perp} = \mathrm{span}\{\partial_u, [f, \partial_u], [f, [f, \partial_u]]\} = \Delta_2$$

and by continuing this procedure we see that

$$V(P^{(j)})^{\perp} = \{P^{(j)}, \mathrm{d}t\}^{\perp} = \Delta_j \tag{13.18}$$

holds. Since $\dim(\Delta_r) = m + n$ and the derived flag is computed on a manifold of dimension $m+n+1$ it follows that $P^{(r)} = \{0\}$ with $P^{(i)} = \{f_e, \Delta_i\}^{\perp}, \quad i = 0, \ldots, r.$

13.3.1 Triangularization

For a system that is exactly linearizable by static feedback, there exist adapted coordinates according to the involutive distributions Δ_i in order to represent the system in the triangular structure (13.4). The Pfaffian representation of the normal form reads as

$$\begin{aligned} \omega_r^{i_r} &= \mathrm{d}z_r^{i_r} - f_r^{i_r}(z_r, z_{r-1})\mathrm{d}t \\ &\vdots \\ \omega_2^{i_2} &= \mathrm{d}z_2^{i_2} - f_2^{i_2}(z_r, \ldots, z_2, z_1)\mathrm{d}t \\ \omega_1^{i_1} &= \mathrm{d}z_1^{i_1} - f_1^{i_1}(z_r, \ldots, z_2, z_1, v)\mathrm{d}t. \end{aligned} \tag{13.19}$$

Thus, with

$$\omega_i^{l_i} = \mathrm{d}z_i^{l_i} - f_i^{l_i}(z_r, \ldots, z_i, z_{i-1})\mathrm{d}t\,, \quad l_i = 1, \ldots, \rho_i \tag{13.20}$$

the Pfaffian system P is of the form $P = \mathrm{span}\{\omega_r^{l_r}, \ldots, \omega_1^{l_1}\}$ so that the basis is already adapted to the sequence of the derived flags since $P^{(i)} = \mathrm{span}\{\omega_r^{l_r}, \ldots, \omega_{i+1}^{l_{i+1}}\}$ for $i = 0, \ldots, r-1$. Hence, the derived flag of the system (13.15) generates the triangular structure, when the coordinates are adapted to the distributions Δ_i.

Remark 13.16 Let us consider for instance the distribution $\Delta_1 = \mathrm{span}\{\partial_u, [f, \partial_u]\}$. As $\Delta_0 = \mathrm{span}\{\partial_\mathrm{u}\}$ and Δ_1 are involutive, then also $[f, \partial_u]$ is involutive and a basis exists that $[f, \partial_u]$ does not depend on the control u. The coordinates z_1 are adapted to the involutive distribution $[f, \partial_u]$. From $P^{(1)} = \{f_e, \Delta_1\}^{\perp}$, we see that the first-derived flag of P consists of differential forms in P that additionally annihilate $[f, \partial_u]$. In adapted coordinates $[f, \partial_u]$ corresponds to ∂_{z_1} and thus $P^{(1)}$ does not include the forms ω_1 as $\mathrm{d}z_1$ would violate the condition.

On the contrary, starting with the system P as in (13.15) it is readily observed that $P^{(1)}$ is an implicit system. But, as $V(P^{(1)})^{\perp} = \Delta_1$ is involutive, straightening out Δ_1 gives again an explicit control system (see Theorem 13.10) for which the procedure can be continued. If in each step of the derived flag additionally the Cauchy characteristic vector fields are taken into account—one represents the system with the minimal number of variables—then the sequence of the derived flag can be represented as

$$\begin{aligned} \omega_r^{i_r} &= \mathrm{d}z_r^{i_r} - f_r^{i_r}(z_r, \hat{z}_{r-1})\mathrm{d}t \\ &\vdots \\ \omega_2^{i_2} &= \mathrm{d}z_2^{i_2} - f_2^{i_2}(z_r, \dots, z_2, \hat{z}_1)\mathrm{d}t \\ \omega_1^{i_1} &= \mathrm{d}z_1^{i_1} - f_1^{i_1}(z_r, \dots, z_2, z_1, v)\mathrm{d}t \end{aligned} \tag{13.21}$$

where the redundant variables have been eliminated, using $C(P^{(i)})$ if necessary.

13.4 Flat Systems and Triangularization

We consider a Pfaffian system $\mathrm{d}x^i - f^i(x, u)\mathrm{d}t$ and we wish to derive a normal form such that so-called 1-flat outputs can be read off easily. A 1-flat output $y = \psi(x, u)$ consists of functions y^i with $\dim(y) = m$ depending on the state and the control but not on the derivatives of the control, such that for the Pfaffian system the evolution of x and u can be expressed by y and its derivatives without any integration. We are looking for a coordinate transformation $z = \phi(x, u)$ such that in the new coordinates z, the flat outputs which are among the z coordinates can be easily determined.

Motivated by the explicit triangular form (13.21), we can introduce the following more general representation, corresponding to implicit ordinary differential equations

$$\begin{aligned} \Xi_r &: A_{r,l_r}^{r,j_r}\mathrm{d}z_r^{l_r} - b_r^{j_r}\mathrm{d}t \\ &\vdots \\ \Xi_2 &: A_{2,l_r}^{r,j_2}\mathrm{d}z_r^{l_r} + \cdots + A_{2,l_2}^{2,j_2}\mathrm{d}z_2^{l_2} - b_2^{j_2}\mathrm{d}t \\ \Xi_1 &: A_{1,l_r}^{r,j_1}\mathrm{d}z_r^{l_r} + \cdots + A_{1,l_2}^{2,j_1}\mathrm{d}z_2^{l_2} + A_{1,l_1}^{1,j_1}\mathrm{d}z_1^{l_1} - b_1^{j_1}\mathrm{d}t \end{aligned} \tag{13.22}$$

with

$$A_{i,l_k}^{k,j_i} = A_{i,l_k}^{k,j_i}(z_r, \dots, z_i, \hat{z}_{i-1}), \quad b_i^{j_i} = b_i^{j_i}(z_r, \dots, z_i, \hat{z}_{i-1})$$

where again $z_i = (\hat{z}_i, y_i)$ is met, see also [17].

Remark 13.17 (Notation), e.g., A_{2,l_r}^{r,j_2} are the components of a matrix A_2^r, where the subscript index corresponds to Ξ_2 (indicating the stage in the triangular form) and the superscript index is in accordance with the derivative variables $\mathrm{d}z_r$. The index l_r is a summation index regarding $\mathrm{d}z_r^{l_r}$ with $l_r = 1, \dots, \dim(z_r)$. The index $j_2 = 1, \dots, \dim(\Xi_2)$ is indicating the number of equations (differential forms) in Ξ_2.

Furthermore, $\dim(\Xi_{e,i}) = \dim(\hat{z}_{i-1})$ has to be met and the Jacobian matrices $[\partial_{\hat{z}_{i-1}}\Xi_{e,i}]$ have to be regular (where $\Xi_{e,i} = 0$) for all $i = 1, \ldots, r$, where $\Xi_{e,i}$ denotes the implicit differential equation corresponding to Ξ_i.

Remark 13.18 Hence the implicit differential equations (in matrix vector notation) take the form

$$\begin{array}{l} A_r^r(z_r, \hat{z}_{r-1})\dot{z}_r - b_r(z_r, \hat{z}_{r-1}) \\ \quad\vdots \qquad\qquad\qquad\qquad\qquad \ddots \\ A_2^r(z_r, \ldots, \hat{z}_1)\dot{z}_r + \cdots + A_2^2(z_r, \ldots, \hat{z}_1)\dot{z}_2 - b_2(z_r, \ldots, \hat{z}_1) \\ A_1^r(z_r, \ldots, \hat{z}_0)\dot{z}_r + \cdots + A_1^2(z_r, \ldots, \hat{z}_0)\dot{z}_2 + A_1^1(z_r, \ldots, \hat{z}_0)\dot{z}_1 - b_1(z_r, \ldots, \hat{z}_0). \end{array} \tag{13.23}$$

Setting $S_{d,k} = \mathrm{span}\{\Xi_r, \ldots, \Xi_{k+1}\}$, it can be deduced from the triangular structure that (13.22) enjoys the following properties that will be utilized in designing an algorithm to transform flat systems into this form.

1. For the system $S_{d,k}$ we have that $\mathrm{span}\{\partial_{\hat{z}_k}\} = D_k \subset V(S_{d,k})^{\perp}$, where D_k are involutive distributions, and furthermore $D_k \subset C(S_{d,k+1})$ for $k = 0, \ldots, r-1$.
2. If we have a nontrivial decomposition for z_k of the form $z_k = (\hat{z}_k, y_k)$ with $\dim(y_k) > 0$, then $\partial_{y_k} \subset C(S_{d,k})$.
3. Each subsystem Ξ_k is parameterizable with respect to the nonderivative variable $\hat{z}_{k-1}$, i.e.,
$$\hat{z}_{k-1} = h_{k-1}(z_r, \ldots, z_k, \dot{z}_r, \ldots, \dot{z}_k)$$
for $k = 1, \ldots, r$.

Furthermore, from the implicit triangular decomposition (13.22), we observe that we have a decomposition of $S_{d,0}$ into a sequence of Pfaffian systems

$$S_{d,0} \supset S_{d,1} \supset S_{d,2} \supset \cdots \tag{13.24}$$

as well as splittings of the form $S_{d,i} = S_{d,i+1} \oplus S_{d,i+1,c}$ where all the $S_{d,i+1,c}$ are parameterizable with respect to the corresponding nonderivative variables $\hat{z}$.

Remark 13.19 It should be noted that for systems that are exactly linearizable by static feedback, the sequence of the derived flag corresponds to (13.24), where additionally in each step the Pfaffian system $\{S_{d,i}, \mathrm{d}t\}$ is integrable, which guarantees a sequence of explicit control systems in adapted coordinates.

From the representation (13.22), it can be seen that the y coordinates again qualify for the flat outputs, where the system has to be analyzed from the top to the bottom and using the implicit function theorem. Hence, we derive a sufficient condition for control systems to be 1-flat, this is to say that if we find a transformation $z = \phi(x, u)$ in order to represent the system (13.15) in the form (13.22) then the system is obviously 1-flat. It should be noted, however, that this is no necessary criteria and that the algorithm we propose in the next section is of constructive nature only, thus a failure of our constructive scheme does in general not prove that the system is not 1-flat.

13.4.1 Constructive Algorithm

In the following, we wish to derive the decompositions $S_k = S_{k+1} \oplus S_{k+1,c}$ based on the system S_k that occurs during the constructive scheme when starting with $S_0 = P$ as in (13.15). To this end, the following steps need to be performed

1. Computation of $V(S_k)^{\perp}$, as these elements correspond to nonderivative variables. The choice of an involutive $D_k \subset V(S_k)^{\perp}$ corresponds to a selection of nonderivative variables called $\hat{z}_k$. (This correspondence becomes obvious in an adapted coordinate chart to be constructed by means of the Frobenius theorem.)
2. Construction of a splitting $S_k = S_{k+1} \oplus S_{k+1,c}$ such that $D_k \subset C(S_{k+1})$, since this guarantees that S_{k+1} is independent of $\hat{z}_k$.
3. Check, if $S_{k+1,c}$ is parameterizable with respect to the $\hat{z}_k$, which is possible only if $\dim(S_k) = \dim(S_{k+1}) + \dim(D_k)$ holds.

Starting with $k = 0$ we seek for a decomposition $S_k = S_{k+1} \oplus S_{k+1,c}$ and then the whole procedure will be continued with S_{k+1}. If in S_{k+1} further redundant variables appear (i.e., $C(S_{k+1})$ is nontrivial apart from the $\hat{z}_k$ variables), then in $S_{k+1,c}$ additional free variables appear (called y_{k+1}) corresponding to possible flat outputs. The adapted coordinates can be constructed since D_k is involutive. Hence, in new coordinates

$$D_k = \mathrm{span}\{\partial_{\hat{z}_k}\}, \quad \hat{z}_k = (\hat{z}_k^1, \ldots, \hat{z}_k^{n_{\hat{z}_k}})$$

with $n_{\hat{z}_k} = \dim(D_k)$. Consequently, if the system $S_{k+1,c}$ is parameterizable with respect to $\hat{z}_k$ and if the system S_{k+1} possesses a nontrivial Cauchy characteristic, then it is clear that in $S_{k+1,c}$ these redundant variables are candidates for possible flat outputs. Hence, a solution for S_{k+1} leads to a solution for $S_{k+1,c}$ by pure differentiation as $S_{k+1,c}$ is parameterizable with respect to $\hat{z}_k$. This constructive scheme has to be continued (if possible) until a parameterizable system is obtained, such that no further decomposition is necessary in order to read off the flat outputs.

Hence, we have to construct $D_k \subset V(S_k)^{\perp}$ and $S_{k+1} \subset S_k$ such that $D_k \rfloor \mathrm{d}S_{k+1} \subset S_{k+1}$ is met. Then also the necessary condition

$$D_k \rfloor \mathrm{d}S_{k+1} \subset S_k \tag{13.25}$$

has to be fulfilled. It should be noted that given $V(S_k)^{\perp}$ and S_k (a distribution and a codistribution) we search for an involutive [3] subdistribution D_k, then the condition $D_k \rfloor \mathrm{d}S_{k+1} \subset S_{k+1}$ leads to partial differential equations, as $S_{k+1} \subset S_k$. Using the

---

[3] $\dim(D_k) = 1$ always guarantees involutivity but we can also search for higher dimensional ones.

necessary condition (13.25) instead, only algebraic equations appear and based on this observation we will demonstrate our constructive scheme in the next section using examples. Furthermore, it should be mentioned that the constructive scheme, if successful, does, in general, not lead to a unique decomposition of the system, as branching points may appear—those branching points possibly require iterations of the constructive scheme.

13.5 Examples

Let us consider the two motivating examples from Sect. 13.1. We will represent the systems in a Pfaffian fashion, and we will sequentially derive the triangular form (13.19) by applying the proposed machinery. It can be deduced that the induction motor example is rather trivial compared to the academic example where in every step based on the necessary condition (13.25) an involutive D_k has to be constructed as well as a subcodistribution which is independent of the desired variables.

13.5.1 Induction Motor

The system of the induction motor (13.6) can be stated in a Pfaffian representation as

$$S_0 \quad \begin{aligned} \omega_0^1 &= \mathrm{d}\omega - (\mu\psi_d i_q - \tfrac{\tau_L}{J})\mathrm{d}t \\ \omega_0^2 &= \mathrm{d}\psi_d - \eta(M i_d - \psi_d)\mathrm{d}t \\ \omega_0^3 &= \mathrm{d}\rho - (n_p\omega + \eta M \tfrac{i_q}{\psi_d})\mathrm{d}t \\ \omega_0^4 &= \mathrm{d}i_d - v_d\mathrm{d}t \\ \omega_0^5 &= \mathrm{d}i_q - v_q\mathrm{d}t. \end{aligned}$$

We clearly have $V(S_0)^{\perp} = \mathrm{span}\{\partial_{v_d}, \partial_{v_q}\}$, and thus, obviously

$$S_1 \quad \begin{aligned} \omega_1^1 &= \mathrm{d}\omega - (\mu\psi_d i_q - \tfrac{\tau_L}{J})\mathrm{d}t \\ \omega_1^2 &= \mathrm{d}\psi_d - \eta(M i_d - \psi_d)\mathrm{d}t \\ \omega_1^3 &= \mathrm{d}\rho - (n_p\omega + \eta M \tfrac{i_q}{\psi_d})\mathrm{d}t \end{aligned}$$

with $S_{1,c} = \mathrm{span}\{\omega_0^4, \omega_0^5\}$. Then we observe that $V(S_1)^{\perp} = \mathrm{span}\{\partial_{i_d}, \partial_{i_q}\}$ and we seek for $D_1 \subset V(S_1)^{\perp}$ such that $D_1 \rfloor \mathrm{d}S_2 \subset S_2$ where $S_2 \subset S_1$. A possible choice for S_2 is

$$\omega_2^1 = \mathrm{d}\omega - \frac{\mu\psi_d^2}{\eta M}\mathrm{d}\rho + (\frac{\mu\psi_d^2}{\eta M} n_p\omega + \frac{\tau_L}{J})\mathrm{d}t$$

where $\omega_2^1 = \omega_1^1 - \frac{\mu\psi_d^2}{\eta M}\omega_1^3$. The complement $S_{2,c}$ can be represented as

$$\begin{aligned}\omega_{2,c}^2 &= \mathrm{d}\psi_d - \eta(Mi_d - \psi_d)\mathrm{d}t \\ \omega_{2,c}^3 &= \mathrm{d}\rho - (n_p\omega + \eta M\tfrac{i_q}{\psi_d})\mathrm{d}t\end{aligned}$$

which is parameterizable with respect to i_d and i_q. Then, since S_2 is parameterizable with respect to ψ_d, indeed

$$\frac{\eta M}{\mu}\frac{(\dot{\omega} + \frac{\tau_L}{J})}{(\dot{\rho} - n_p\omega)} = \psi_d^2,$$

a possible flat output is (ω, ρ). The triangular form consists of the systems S_2, $S_{2,c}$ and $S_{1,c}$ and is in accordance with the decomposition (13.8).

Remark 13.20 A different flat output which has also been achieved in [9] by an alternative approach follows as

$$\rho - \omega\frac{\eta M}{\mu\psi_d^2}, \quad \psi_d$$

and can be easily derived with the presented machinery, as the form ω_2^1 is stated in the two derivative variables ρ and ω. However, choosing

$$v_2 = \partial_\rho + \frac{\mu\psi_d^2}{\eta M}\partial_\omega$$

(where $v_2 \in V(S_2)^\perp$) and using the coordinate transformation $(\rho, \omega, \psi_d) = \varphi_2(\hat{\rho}, \bar{\rho}, \psi_d)$ straightening out v_2 as

$$\begin{aligned}\rho &= \bar{\rho} + \hat{\rho} \\ \omega &= \frac{\mu\psi_d^2}{\eta M}\hat{\rho} \\ \psi_d &= \psi_d\end{aligned}$$

we obtain

$$\omega_2^1 = \mathrm{d}(\frac{\mu\psi_d^2}{\eta M})\hat{\rho} - \frac{\mu\psi_d^2}{\eta M}\mathrm{d}\bar{\rho} + ((\frac{\mu\psi_d^2}{\eta M})^2\hat{\rho}n_p + \frac{\tau_L}{J})\mathrm{d}t$$

which is parameterizable with respect to $\hat{\rho}$, and thus the alternative flat output follows, since $\bar{\rho}$ and ψ_d can be assigned freely, if τ_L is assumed to be known.

13.5.2 Academic Example

Let us consider again the system (13.9) formulated as a Pfaffian system $S_0 = \mathrm{span}\{\omega_0^1, \omega_0^2, \omega_0^3\}$ with

$$\begin{aligned}\omega_0^1 &= \mathrm{d}x^1 - u^1\mathrm{d}t\\ \omega_0^2 &= \mathrm{d}x^2 - u^2\mathrm{d}t\\ \omega_0^3 &= \mathrm{d}x^3 - \sqrt{u^1u^2}\mathrm{d}t\end{aligned}$$

that has been considered also in [15] using a different approach. It can easily be seen that $V(S_0)^\perp = \mathrm{span}\{\partial_{u^1}, \partial_{u^2}\}$. Choosing

$$v_0 = u^1\partial_{u^1} + u^2\partial_{u^2}$$

we observe that $v_0 \in C(S_1)$ where $S_1 = \mathrm{span}\{\omega_1^1, \omega_1^2\}$

$$\begin{aligned}\omega_1^1 &= \sqrt{u^1u^2}\mathrm{d}x^1 - u^1\mathrm{d}x^3\\ \omega_1^2 &= \sqrt{u^1u^2}\mathrm{d}x^2 - u^2\mathrm{d}x^3.\end{aligned}$$

It should be noted that v_0 and S_1 have been constructed using the necessary condition (13.25), hence $v_0 \rfloor S_1 \subset S_0$, but it can be readily observed that also $v_0 \rfloor S_1 \subset S_1$ is met as desired. Straightening out v_0 is based on the flow of v_0 and we derive the transformation $(x^1, x^2, x^3, u^1, u^2) = \varphi_0(w^1, w^2, w^3, w^4, \hat{z}_0^1)$ reading as

$$\begin{aligned}x^1 &= w^1\ , & u^1 &= e^{\hat{z}_0^1}w^4\\ x^2 &= w^2\ , & u^2 &= e^{\hat{z}_0^1}\\ x^3 &= w^3. & &\end{aligned}\tag{13.26}$$

Thus, in the new coordinates we obtain a basis for the system S_1 as

$$\begin{aligned}\omega_1^1 &= \sqrt{w^4}\mathrm{d}w^1 - w^4\mathrm{d}w^3\\ \omega_1^2 &= \sqrt{w^4}\mathrm{d}w^2 - \mathrm{d}w^3.\end{aligned}$$

The complement $S_{1,c}$ is, e.g., given by the single form

$$\omega_{1,c}^3 = \mathrm{d}x^2 - u^2\mathrm{d}t = \mathrm{d}w^2 - e^{\hat{z}_0^1}\mathrm{d}t,$$

and it is clearly seen that the coordinate $\hat{z}_0^1$ is not appearing in S_1 and that $S_{1,c}$ is parameterizable with respect to $\hat{z}_0^1$. Then, with $v_1 \in V(S_1)^\perp$ where

$$v_1 = w^4\partial_1 + \partial_2 + \sqrt{w^4}\partial_3$$

we derive a further splitting, i.e,. $v_1 \in C(S_2)$ with $S_2 = \mathrm{span}\{\omega_2^1\}$ where

$$\omega_2^1 = -\mathrm{d}w^1 - w^4\mathrm{d}w^2 + 2\sqrt{w^4}\mathrm{d}w^3.$$

We have that $S_2 \subset S_1$ since

$$\omega_2^1 = -\frac{1}{\sqrt{w^4}}\omega_1^1 - \sqrt{w^4}\omega_1^2.$$

The complement $S_{2,c}$ can be chosen as

$$\omega_{2,c}^2 = \sqrt{w^4}\mathrm{d}w^2 - \mathrm{d}w^3.$$

Consequently, we use the flow with respect to v_1 to straighten out v_1 and derive

$$\begin{aligned} w^1 &= q^4\hat{z}_1^1 \\ w^2 &= \hat{z}_1^1 + q^2 \\ w^3 &= \sqrt{q^4}\hat{z}_1^1 + q^3 \\ w^4 &= q^4 \end{aligned} \tag{13.27}$$

such that S_2 reads as

$$\omega_2^1 = -q^4\mathrm{d}q^2 + 2\sqrt{q^4}\mathrm{d}q^3.$$

From S_2 the flat output can be read off. Indeed, $(y_3^1,\ y_3^2)$ with $y_3^1 = q^2$ and $y_3^2 = q^3$ is the flat output.

Remark 13.21 The complement $S_{2,c}$ in the new coordinates is given as

$$\omega_{2,c}^2 = \sqrt{q^4}\mathrm{d}q^2 - \mathrm{d}q^3 - \frac{\hat{z}_1^1}{2\sqrt{q^4}}\mathrm{d}q^4.$$

Thus, $S_{2,c}$ is clearly parameterizable with respect to $\hat{z}_1^1$.

From $(x, u) = \varphi_0(w, \hat{z}_0^1)$ and $w = \varphi_1(q, \hat{z}_1^1)$ using (13.26) and (13.27) and by setting $q^4 = \hat{z}_2^1$ we obtain the coordinate transformation which we have presented in Sect. 13.1. Additionally, we easily derive

$$\begin{gathered} -\hat{z}_2^1\mathrm{d}y_3^1 + 2\sqrt{\hat{z}_2^1}\mathrm{d}y_3^2 \\ \sqrt{\hat{z}_2^1}\mathrm{d}y_3^1 - \mathrm{d}y_3^2 - \frac{\hat{z}_1^1}{2\sqrt{\hat{z}_2^1}}\mathrm{d}\hat{z}_2^1 \\ \mathrm{d}\hat{z}_1^1 + \mathrm{d}y_3^1 - e^{\hat{z}_0^1}\mathrm{d}t \end{gathered}$$

based on the calculations from above (consisting of the systems S_2, $S_{2,c}$ and $S_{1,c}$ in new coordinates) which is of the desired triangular shape as in (13.19). It is readily observed that this triangular form is also in accordance with the representation (13.10) by slightly rearranging the expressions.

References

1. R.L. Bryant, S.S. Chern, R.B. Gardner, H.L. Goldschmidt, P.A. Griffiths, *Exterior Differential Systems* (Springer, New York, 1991)
2. J. Chiasson, A new approach to dynamic feedback linearization control of an induction motor. IEEE Trans. Autom. Control **43**, 391–397 (1998)
3. M. Fliess, J. Lévine, P. Martin, P. Rouchon, Flatness and defect of nonlinear systems: introductory theory and examples. Int. J. Control **61**, 1327–1361 (1995)
4. M. Fliess, J. Lévine, P. Martin, P. Rouchon, A lie-backlund approach to equivalence and flatness of nonlinear systems. IEEE Trans. Autom. Control **44**, 922–937 (1999)
5. R. Hermann, A. Krener, Nonlinear controllability and observability. IEEE Trans. Autom. Control **22**(5), 728–740 (1977)
6. A. Isidori, A.J. Krener, C. Gori-Giorgi, S. Monaco, Nonlinear decoupling via feedback: a differential geometric approach. IEEE Trans. Autom. Control **26**, 331–345 (1981)
7. A. Isidori, *Nonlinear Control Systems* (Springer, London, 1995)
8. B. Jakubczyk, W. Respondek, On linearization of control systems. Bull. Acad. Polonaise Sci. Ser. Sci. Math. **28**, 517–522 (1980)
9. F. Nicolau, W. Respondek, Flatness of two-input control-affine systems linearizable via onefold prolongation, in *Proceedings 9th IFAC Symposium on Nonlinear Control Systems* (*NOLCOS*), pp. 493–498 (2013)
10. H. Nijmeijer, A.J. van der Schaft, Controlled invariance for nonlinear systems. IEEE Trans. Autom. Control **27**(1), 904–914 (1982)
11. H. Nijmeijer, A.J. van der Schaft, *Nonlinear Dynamical Control Systems* (Springer, New York, 1990)
12. M. Rathinam, R.M. Murray, Differential flatness of two one-forms in arbitrary number of variables. Syst. Control Lett. **36**, 317–326 (1999)
13. W. Respondek, On decomposition of nonlinear control systems. Syst. Control Lett. **1**(5), 301–308 (1982)
14. S. Sastry, *Nonlinear Systems, Analysis, Stability, and Control* (Springer, New York, 1999)
15. K. Schlacher, M. Schöberl, A jet space approach to check pfaffian systems for flatness, *in Proceedings 52nd IEEE Conference on Decision and Control* (*CDC*), pp. 2576–2581 (2013)
16. M. Schöberl, *Contributions to the Analysis of Structural Properties of Dynamical Systems in Control and Systems Theory—A Geometric Approach* (Shaker, Aachen, 2014)
17. M. Schöberl, K. Schlacher, On an implicit triangular decomposition of nonlinear control systems that are 1-flat—a constructive approach. Automatica **50**, 1649–1655 (2014)
18. D. Tilbury, S.S. Sastry, On goursat normal forms, prolongations, and control systems. in *Proceedings 33rd Conference on Decision and Control* (*CDC*), pp. 1797–1802 (1994)
19. A.J. van der Schaft, Observability and controllability for smooth nonlinear systems. SIAM J. Control Optim. **20**(3), 338–354 (1982)
20. A.J. van der Schaftm, Linearization and input–output decoupling for general nonlinear systems. Syst. Control Lett. **5**, 27–33 (1984)

Chapter 14
Online Frequency Estimation of Periodic Signals

Riccardo Marino and Patrizio Tomei

Abstract The problem of estimating online the unknown period of a periodic signal is considered, with no a priori information on the period: this is a crucial problem in the design of learning and synchronizing controls, in fault detection, and for the attenuation of periodic disturbances. Given a measurable continuous, bounded periodic signal, with nonzero first harmonic in its Fourier series expansion, a dynamic algorithm is proposed which provides an online globally exponentially convergent estimate of the unknown period. The period estimate converges from any initial condition to a neighborhood of the true period whose size is explicitly characterized in terms of the higher order harmonics contained in the signal. The accuracy of the frequency estimation can be arbitrarily improved by increasing the order of a prefilter which is incorporated in the estimation algorithm, at the expense of reducing the rate of the exponential convergence. This online frequency estimation algorithm can be used to design hybrid disturbance attenuation controllers for periodic disturbances with unknown period.

14.1 Introduction

Arjan van der Schaft visited the University of Rome Tor Vergata during the summer 1990. We had the pleasure of collaborating with him and Witold Respondek on several aspects of almost disturbance decoupling for nonlinear systems [19] and on more theoretical issues involving transformations of nonlinear systems into prime forms [20]. Our collaboration started in 1986 during a summer visit of the first author at Twente University: at that time, high-gain feedback was investigated to solve almost input-output decoupling and almost disturbance decoupling problems [17, 18]. The

R. Marino · P. Tomei (✉)
Department of Electronic Engineering, University of Rome Tor Vergata,
Via Del Politecnico 1, 00133 Roma, Italy
e-mail: tomei@ing.uniroma2.it

R. Marino
e-mail: marino@ing.uniroma2.it

M.K. Camlibel et al. (eds.), *Mathematical Control Theory I*,
Lecture Notes in Control and Information Sciences 461,
DOI 10.1007/978-3-319-20988-3_14

first author first met Arjan at a conference on Differential Geometric Control Theory in Michigan in 1982 and still remembers with pleasure an adventurous car trip from northern Michigan to St. Louis, Missouri, where he was completing his Ph.D.

This paper is related to the design of feedback controls to attenuate the influence of disturbances. While the strategy in [18, 19] is to reduce the L_2-gain from the disturbance to the output possibly by high-gain feedback since the disturbance is totally unknown, in this paper we explore what can be done if the disturbance is known to be periodic even though its period is unknown. The key step is clearly the online estimation of the unknown period, according to the internal model principle.

Online frequency estimation of a periodic signal is a fundamental problem in several engineering and scientific disciplines. The classical Fourier analysis estimates the frequencies and the amplitudes of a periodic signal provided that the signal can be stored and processed off-line. As far as feedback control is concerned, learning [32] and synchronizing [27] control design and the attenuation of unmeasured periodic disturbances require online frequency estimation. If the period is known, learning controls can track periodic references for classes of linear and nonlinear systems [14, 32]. According to the internal model principle, which was formulated in [6] for linear systems, an error feedback control which is capable of tracking and/or rejecting unknown sinusoidal signals must necessarily be able to reproduce such signals: hence it should estimate their frequencies online. Fault detectors which are based on frequency estimators require online algorithms as well.

Online frequency estimation algorithms can be divided into two classes: the local ones, which converge for sufficiently close initial frequency estimates and the global ones, which converge for any initial frequency estimate. Their convergence may be either exponential or only asymptotic [10] and their domain of attraction may be either global or only local. In addition, the convergence may occur for a suitable tuning of the algorithm parameters or for any parameter value. These differences are apparent in the comparison of several algorithms which are now available for the online frequency estimation of a single sinusoidal signal with no a priori information on the frequency. In [3] a continuous time version of the discrete-time notch filter proposed by [28], which was inspired by commonly used Phase Locked Loop (PLL) algorithms in signal processing, is shown to be locally asymptotically convergent. The adaptation strategy for the frequency estimator presented in [3] was normalized in [9] in order to obtain a globally asymptotically convergent algorithm, provided that the adaptation gain is chosen to be sufficiently small depending on a known bound on the amplitude of the sinusoidal signal. The adaptive notch filter proposed in [9] was extensively analyzed in [4] and was further modified in [24, 25] to show, by means of averaging theory (see for instance [10, 30]) that the frequency estimate will asymptotically converge to a neighborhood of the fundamental frequency even when the signal is periodic but not purely sinusoidal, provided that the adaptation gain and the higher order harmonics are sufficiently small. A similar result was presented in [33] using a different algorithm based on gradient descent methods. Globally exponentially convergent frequency estimation algorithms were obtained both for a single sinusoid and for biased multiple sinusoids in 2002 by [22, 26, 31] without any restriction on the algorithm design parameters. The key techniques are adaptive observers (see [21]) in [22, 26] and adaptive identifiers (see [29]) in [31], while the state space representation of the measured signal allows for a linear parameterization of the unknown frequencies; amplitudes and phases can be recovered as well (see [7, 8] for a detailed analysis) from the estimation of the state variables. Different frequency estimation techniques for sinusoidal signals are still actively studied with the aim of exploring the robustness with respect to unaccounted disturbances (see [1, 2, 5]).

We will follow the adaptive observer approach introduced in [22] in order to address the global frequency estimation of periodic signals. Only local asymptotic frequency estimators for periodic signals have been so far obtained by using an adaptive notch filter with sufficiently small adaptation gain in [24, 25]: the stability analysis has been carried out interpreting the adaptation gain as a small parameter and applying the averaging theorems [10, 30].

In this paper, given a measurable continuous, bounded periodic signal, with nonzero first harmonic in its Fourier series expansion, a dynamic adaptive algorithm is proposed which provides an online globally exponentially convergent estimate of the unknown frequency for any tuning of its parameters, including the adaptation gain. No a priori information on the period is required. The frequency estimate converges from any initial condition to a neighborhood of the true frequency whose size is explicitly characterized in terms of the higher order harmonics contained in the periodic signal. By increasing the order of a prefilter which is incorporated in the estimation algorithm, the accuracy of the frequency estimation can be arbitrarily improved, at the expense of reducing the rate of the exponential convergence. The global stability analysis is carried out using Lyapunov functions and the property of persistency of excitation which lead to a robust exponential convergence of the estimation algorithm. When the periodic signal is a biased sinusoid, the unknown frequency is exactly estimated from any initial condition and for any value of the prefilter order, thus recovering a well-known result with improved robustness. Two examples are carried out and simulated. In the first one, the proposed method is tested on a complex signal and compared to the adaptive notch filter in [25]. In the second one, the frequency estimator is used in conjunction with a disturbance rejection compensator to attenuate a periodic disturbances with unknown frequency. Since the frequency of the disturbance compensator is updated at every predefined time interval, the overall disturbance compensator is of hybrid type. Preliminary results have been presented in [15, 16] for robust compensation of periodic disturbances.

14.2 Main Results

Consider the bounded periodic signal $y(t)$, $y \in \mathbb{R}$, of unknown period T, which is available for measurements. Assume that $y(t)$ is continuous so that it can be represented by its Fourier series expansion

$$
\begin{aligned}
y(t) &= \frac{a_0}{2} + \sum_{k=1}^{\infty} \left[a_k \cos\left(\frac{2\pi k t}{T}\right) + b_k \sin\left(\frac{2\pi k t}{T}\right) \right] \\
&\triangleq \frac{a_0}{2} + a_1 \cos\left(\frac{2\pi t}{T}\right) + b_1 \sin\left(\frac{2\pi t}{T}\right) + r_y(t)
\end{aligned} \tag{14.1}
$$

in which $r_y(t)$ contains the higher order harmonics of $y(t)$; the first harmonic of unknown frequency $1/T$ is assumed to be different from zero, i.e., $(a_1^2 + b_1^2) > 0$. Let us consider the signal $y_{fl}(t)$ obtained by filtering $y(t)$ through the stable linear filter of order $l \geq 0$

$$
\begin{aligned}
\dot{y}_{f1} &= -\lambda_f y_{f1} + \lambda_f y, \; y_{f1} \in \mathbb{R} \\
\dot{y}_{fk} &= -\lambda_f y_{fk} + \lambda_f y_{f,k-1}, \; y_{fk} \in \mathbb{R}, \; 2 \leq k \leq l
\end{aligned} \tag{14.2}
$$

in which λ_f is an arbitrary positive real. Let y_{fp} be the steady-state periodic component of $y_{fl}(t)$ in (14.2) which is obtained as the solution of

$$
\begin{aligned}
\dot{y}_{fp1} &= -\lambda_f y_{fp1} + \lambda_f y, \; y_{fp1} \in \mathbb{R} \\
\dot{y}_{fpk} &= -\lambda_f y_{fpk} + \lambda_f y_{fp,k-1}, \; y_{fpk} \in \mathbb{R}, \; 2 \leq k \leq l \\
y_{fp} &= y_{fpl}
\end{aligned} \tag{14.3}
$$

with suitable initial conditions. The signal y_{fp} may be rewritten as

$$
y_{fp}(t) = \eta_1(t) + r(t) \tag{14.4}
$$

where $\eta_1(t)$ is the sum of the bias (if any) and of the first harmonic component of frequency ω (which is different from zero by assumption) while $r(t)$ contains all remaining harmonics at higher frequencies $k\omega$, $k \geq 2$. The differences

$$
\begin{aligned}
\tilde{y}_{fi} &= y_{fi} - y_{fpi}, \; 1 \leq i \leq l \\
\tilde{y}_f &= \tilde{y}_{fl}
\end{aligned} \tag{14.5}
$$

converge exponentially to zero. The signal $\eta_1(t)$ may be equivalently generated by the exogenous system (exosystem)

$$
\begin{aligned}
\dot{\eta}_1 &= \eta_2 \\
\dot{\eta}_2 &= -\theta \eta_1 + \eta_3 \\
\dot{\eta}_3 &= 0
\end{aligned} \tag{14.6}
$$

with suitable initial conditions, in which the parameter $\theta = (2\pi/T)^2 = \omega^2$ is defined and $\eta = [\eta_1, \eta_2, \eta_3]^T \in \mathbb{R}^3$. Let us introduce the unitary gain first-order stable filter

$$
\dot{\chi} = -\lambda \chi + \lambda y_{fl}, \; \chi \in \mathbb{R} \tag{14.7}
$$

in which λ is an arbitrary positive real and y_{fl} is generated by (14.2). From (14.4), (14.6), (14.2) and (14.7), in the new state coordinates $\eta_E = [\chi, \lambda\eta^T]^T \in \mathbb{R}^4$, we have

$$\begin{aligned} \dot{\eta}_E &= A_c\eta_E - E_1\lambda\chi + r\lambda(E_1 + \theta E_3) \\ &\quad + \lambda E_1\tilde{y}_f - \lambda y_{fp}\theta E_3 \\ \chi &= C_c\eta_E \end{aligned} \tag{14.8}$$

in which E_i denotes the ith column of an identity matrix of suitable dimension and

$$A_c = \begin{bmatrix} 0\,1\,0\,0 \\ 0\,0\,1\,0 \\ 0\,0\,0\,1 \\ 0\,0\,0\,0 \end{bmatrix}.$$

Note that the unknown parameter $\theta = \omega^2$ now appears linearly in the dynamic equations (14.8). Make the time-varying change of coordinates (which is called 'filtered transformation' in [21])

$$\begin{aligned} z &= \eta_E - \begin{bmatrix} 0 \\ \theta\xi(t) \end{bmatrix} \\ \dot{\xi} &= D\xi - \lambda E_2 y_{fl},\ \xi \in \mathbb{R}^3 \\ \mu &= C_c\xi,\ \mu \in \mathbb{R} \end{aligned} \tag{14.9}$$

in which

$$D = \begin{bmatrix} -d_2\ 1\ 0 \\ -d_3\ 0\ 1 \\ -d_4\ 0\ 0 \end{bmatrix} \tag{14.10}$$

is an arbitrary Hurwitz matrix, and

$$d_A = \begin{bmatrix} d_3 - d_2^2 + d_2\lambda \\ d_4 - d_2d_3 + d_3\lambda \\ -d_2d_4 + d_4\lambda \end{bmatrix}, \quad C_c = \begin{bmatrix} 1\,0\,0 \end{bmatrix}.$$

From (14.8) and (14.9), we obtain

$$\begin{aligned} \dot{z} &= A_c z - E_1\lambda\chi + r\lambda(E_1 + \theta E_3) \\ &\quad + d\mu\theta + \lambda(E_1 + \theta E_3)\tilde{y}_f \\ \chi &= C_c z. \end{aligned} \tag{14.11}$$

The further change of coordinates

$$w_i = z_{i+1} - d_{i+1} z_1, \quad 1 \le i \le 3 \tag{14.12}$$

transforms (14.11) into ($w = [w_1, w_2, w_3]^T$)

$$\begin{aligned} \dot{w} &= Dw + d_A \chi + d_B r + r\lambda\theta E_2 + (d_B + \lambda\theta E_2)\tilde{y}_f \\ \dot{\chi} &= w_1 + (d_2 - \lambda)\chi + r\lambda + \theta\mu + \lambda\tilde{y}_f \end{aligned} \tag{14.13}$$

in which $d_B = [d_2, d_3, d_4]^T$. Note that in (14.13) the unknown parameter θ appears in the dynamics of the known signal χ multiplied by the known signal μ. The parameter θ also appears in the w-dynamics where it is multiplied by the exponentially decaying term $\tilde{y}_f$ and by $r(t)$, which is viewed as a disturbance. Let us introduce the adaptive observer for (w, χ, θ) in (14.13)

$$\begin{aligned} \dot{\hat{w}} &= D\hat{w} + d_A \chi, \ \hat{w} \in \mathbb{R}^3 \\ \dot{\hat{\chi}} &= C_c \hat{w} + (d_2 - \lambda)\chi + \hat{\theta}\mu + k_o(\chi - \hat{\chi}), \ \hat{\chi} \in \mathbb{R} \\ \dot{\hat{\theta}} &= \gamma\mu(\chi - \hat{\chi}), \ \theta \in \mathbb{R} \end{aligned} \tag{14.14}$$

in which γ is the positive adaptation gain and k_o is the positive observer gain. The dynamics for the estimate $\hat{\theta}$ of the parameter $\theta = \omega^2$ is defined (see Fig. 14.1) in terms of the signal μ generated by the linear filters (14.2) and (14.9) and of the error $\chi - \hat{\chi}$ generated by (14.2), (14.7) and (14.14). Defining the error signals $\tilde{\chi} = \chi - \hat{\chi}$, $\tilde{w} = w - \hat{w}$, $\tilde{\theta} = \theta - \hat{\theta}$, from (14.5), (14.6), (14.9), (14.13) and (14.14), we obtain the error dynamics

$$\begin{aligned} \dot{\tilde{w}} &= D\tilde{w} + \bar{d}_B r + \bar{d}_B \tilde{y}_f \\ \dot{\tilde{\chi}} &= -k_o\tilde{\chi} + r\lambda + \tilde{w}_1 + \tilde{\theta}\mu + \lambda\tilde{y}_f \\ \dot{\tilde{\theta}} &= -\gamma\mu\tilde{\chi} \\ \dot{\xi} &= D\xi - \lambda E_2 \tilde{y}_f - \lambda E_2 y_{fp} = D\xi - \lambda E_2 y_{fl} \\ \mu &= C_c \xi \end{aligned} \tag{14.15}$$

in which $\bar{d}_B = d_B + \lambda\theta E_2$. Since (14.9) is a linear dynamic system driven by y_{fl}, the signal μ in (14.9) may be decomposed as

$$\mu = \mu_p + \tilde{\mu} \tag{14.16}$$

in which μ_p is the periodic output of the system

$$\begin{aligned} \dot{\xi}_p &= D\xi_p - \lambda E_2 y_{fp} \\ \mu_p &= C_c \xi_p \end{aligned} \tag{14.17}$$

with proper initial condition $\xi_p(0)$. Now, we are able to state and prove the following theorem which characterizes the convergence properties of the estimation error $\tilde{\theta}$.

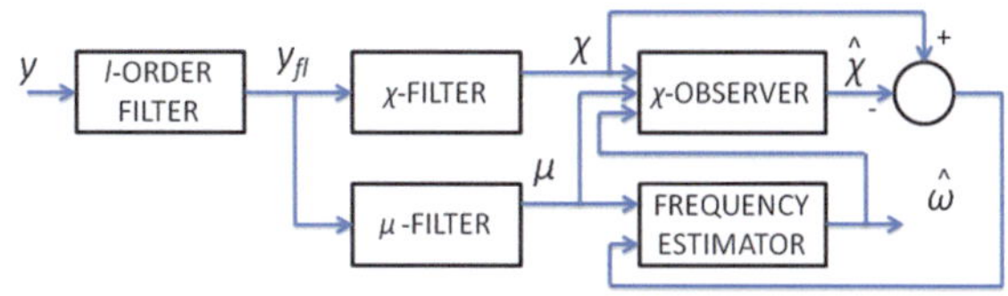

Fig. 14.1 Block diagram for the frequency estimator (14.18)

Theorem 14.1 *Let $y(t)$ be a measurable continuous, bounded periodic signal of unknown period T, with $a_1^2 + b_1^2 > 0$ in its Fourier series expansion (14.1). The following online frequency estimator of order $l + 9$ in which $\hat{\omega}$ denotes the estimate of $\omega = 2\pi/T$ (see the block diagram in Fig. 14.1):*

$$\begin{aligned}
\dot{y}_{f1} &= -\lambda_f y_{f1} + \lambda_f y, \; y_{f1} \in \mathbb{R} \\
\dot{y}_{fk} &= -\lambda_f y_{fk} + \lambda_f y_{f,k-1}, \; y_{fk} \in \mathbb{R}, \; 2 \le k \le l \\
\dot{\chi} &= -\lambda\chi + \lambda y_{fl}, \; \chi \in \mathbb{R} \\
\dot{\xi} &= D\xi - \lambda E_2 y_{fl}, \; \xi \in \mathbb{R}^3 \\
\mu &= C_c\xi, \; \mu \in \mathbb{R} \\
\dot{\hat{w}} &= D\hat{w} + d_A\chi, \; \hat{w} \in \mathbb{R}^3 \\
\dot{\hat{\chi}} &= C_c\hat{w} + (d_2 - \lambda)\chi + \hat{\theta}\mu + k_o(\chi - \hat{\chi}), \; \hat{\chi} \in \mathbb{R} \\
\dot{\hat{\theta}} &= \gamma\mu(\chi - \hat{\chi}), \; \theta \in \mathbb{R} \\
\hat{\omega} &= \begin{cases} \sqrt{\hat{\theta}} & if \hat{\theta} > 0 \\ 0 & otherwise \end{cases}
\end{aligned} \tag{14.18}$$

is such that for any initial condition $y_{f1}(0), \ldots, y_{fl}(0), \chi(0), \xi(0), \hat{w}(0), \hat{\chi}(0), \hat{\theta}(0)$, for any integer $l \ge 0$, for any $\lambda_f > 0$, $\lambda > 0$, $k_o > 3$, $\gamma > 0$ and for any Hurwitz matrix D:

(i) all signals are bounded for any $t \ge 0$;
(ii)

$$|\tilde{\theta}(t)| \le f(\|\tilde{x}(0)\|)e^{-\beta_1 t} + \beta_2 \left[\frac{1}{T}\int_0^T r_y^2(\tau)d\tau\right]^{1/2}, \; \forall t \ge 0$$

in which f is a class-k function [10] of $\tilde{x} = [\tilde{w}^T, \tilde{\chi}, \tilde{\theta}, \xi^T - \xi_p^T, \tilde{y}_{f1}, \ldots, \tilde{y}_{fl}]^T$ and β_1, β_2 are positive reals which tend to zero as l tends to infinity with

$$\beta_1 = O\left[\left(\frac{\lambda_f^2}{\lambda_f^2 + \omega^2}\right)^l\right]$$

$$\beta_2 = O\left[\left(\frac{\lambda_f^2 + \omega^2}{\lambda_f^2 + 4\omega^2}\right)^{l/2}\right].$$

Proof The signal $\mu_p(t)$ in (14.17) is unbiased since the transfer function of the linear system in (14.17) has a zero in the origin, while the signal $\tilde{\mu}(t)$ is exponentially decaying and given by ($\tilde{\xi} = \xi - \xi_p$)

$$\begin{aligned} \dot{\tilde{\xi}} &= D\tilde{\xi} - \lambda E_2 \tilde{y}_f,\ \tilde{\xi}(0) = \xi(0) - \xi_p(0) \\ \tilde{\mu} &= C_c \tilde{\xi}. \end{aligned} \tag{14.19}$$

Note that (see [12], p. 494 and Abel's Lemma [11])

$$\begin{aligned} \sup_{\tau\in[0,T]} |\mu_p(\tau)| &\leq c_1 \frac{\lambda_f^l}{(\lambda_f^2 + \omega^2)^{l/2}} y_M \\ \sup_{\tau\in[0,T]} |\dot{\mu}_p(\tau)| &\leq c_2 \frac{\lambda_f^l}{(\lambda_f^2 + \omega^2)^{l/2}} y_M \end{aligned} \tag{14.20}$$

where

$$y_M = \sup_{\tau\in[0,T]} |y(\tau)|$$

and c_1, c_2 are positive constants independent on the filter parameters l and λ_f. The signal $\mu_p(t)$ may, in turn, be decomposed as

$$\mu_p(t) = \mu_{p1}(t) + \mu_{pr}(t) \tag{14.21}$$

where μ_{p1} is the first unbiased harmonic at frequency ω and μ_{pr} contains all other higher order harmonics. We can write

$$\begin{aligned} \mu_{p1}(t) = |H_1(j\omega)| &\frac{\lambda_f^l}{(\lambda_f^2 + \omega^2)^{l/2}} \\ &\cdots (a_1^2 + b_1^2)^{1/2} \cos(\omega t + \varphi_0) \end{aligned} \tag{14.22}$$

in which $H_1(s) = C_c(sI - D)^{-1} E_2 \lambda$ and φ_0 is a suitable angle. From (14.21) and (14.22), we have

$$\begin{aligned} &\int_t^{t+T} \mu_p^2(\tau)d\tau = \int_0^T [\mu_{p1}^2(\tau) + \mu_{pr}^2(\tau)]d\tau \\ &\geq \int_0^T \mu_{p1}^2(\tau)d\tau = |H_1(j\omega)|^2 \\ &\times (a_1^2 + b_1^2) \frac{T}{2} \frac{\lambda_f^{2l}}{(\lambda_f^2 + \omega^2)^l}. \end{aligned} \tag{14.23}$$

Define

$$\bar{\mu}_p = \mu_p/\alpha,\ \alpha = \frac{\lambda_f^l}{(\lambda_f^2+\omega^2)^{l/2}}. \tag{14.24}$$

From (14.20), we have

$$\begin{aligned}\sup_{\tau\in[0,T]}|\bar{\mu}_p(\tau)| &\le c_1 y_M \triangleq \bar{\mu}_{pM}\\ \sup_{\tau\in[0,T]}|\dot{\bar{\mu}}_p(\tau)| &\le c_2 y_M \triangleq \dot{\bar{\mu}}_{pM}.\end{aligned} \tag{14.25}$$

Since by assumption $a_1^2+b_1^2>0$, from (14.23) and (14.24) we have

$$\begin{aligned}\int_t^{t+T}\bar{\mu}_p^2(\tau)d\tau &\ge |H_1(j\omega)|^2(a_1^2+b_1^2)\frac{T}{2}\\ &\triangleq k_p>0,\ \forall t\ge 0.\end{aligned} \tag{14.26}$$

Define $q(t)$ as the solution of the scalar differential equation

$$\dot{q} = -q+\bar{\mu}_p^2,\quad q(0)=e^{-T}k_p \tag{14.27}$$

so that, by construction,

$$\sup_{\tau\in[0,T]}\bar{\mu}_p^2(\tau)\ge q(t)\ge k_p e^{-2T},\quad \forall t\ge 0. \tag{14.28}$$

With reference to the first three equations in the error system (14.15), consider the Lyapunov function

$$V=\frac{1}{2}\left[\tilde{\chi}^2+\frac{\tilde{\theta}^2}{\gamma}+\tilde{w}^T P\tilde{w}+\gamma_0(\alpha q\tilde{\theta}-\bar{\mu}_p\tilde{\chi})^2\right] \tag{14.29}$$

where γ_0, γ are positive reals and $P>0$ satisfies the Lyapunov matrix equation $D^TP+PD=-2I$, in which D given by (14.10) is a Hurwitz matrix. From (14.29) and (14.15), differentiating V with respect to time along the solutions of (14.15), we obtain

$$\begin{aligned}\dot{V} &= -k_o\tilde{\chi}^2+r\lambda\tilde{\chi}+\tilde{w}_1\tilde{\chi}+\tilde{\chi}\tilde{\theta}\mu+\lambda\tilde{y}_f\tilde{\chi}\\ &\quad -\tilde{\theta}\mu\tilde{\chi}-\tilde{w}^T\tilde{w}+\tilde{w}^TP\bar{d}_Br+\tilde{w}^TP\bar{d}_B\tilde{y}_f\\ &\quad +\gamma_0(\alpha q\tilde{\theta}-\bar{\mu}_p\tilde{\chi})[-\alpha q\tilde{\theta}+\alpha\tilde{\theta}\bar{\mu}_p^2-\alpha^2 q\gamma\bar{\mu}_p\tilde{\chi}\\ &\quad -\alpha q\gamma\tilde{\mu}\tilde{\chi}-\dot{\bar{\mu}}_p\tilde{\chi}-\bar{\mu}_p(-k_o\tilde{\chi}+r\lambda+\tilde{w}_1\\ &\quad +\alpha\tilde{\theta}\bar{\mu}_p+\tilde{\theta}\tilde{\mu}+\lambda\tilde{y}_f)]\\ &= -k_o\tilde{\chi}^2-\tilde{w}^T\tilde{w}-\gamma_0\alpha^2q^2\tilde{\theta}^2\end{aligned}$$

$$
\begin{aligned}
&+\tilde{w}_1\tilde{\chi} + \gamma_0\alpha q\tilde{\theta}(-\alpha^2 q\gamma\bar{\mu}_p - \dot{\bar{\mu}}_p + \bar{\mu}_p k_o)\tilde{\chi} \\
&+\gamma_0\tilde{\chi}^2(\alpha^2 q\bar{\mu}_p^2\gamma + \bar{\mu}_p\dot{\bar{\mu}}_p - \bar{\mu}_p^2 k_o) - \gamma_0\alpha q\bar{\mu}_p\tilde{\theta}\tilde{w}_1 \\
&+\gamma_0\bar{\mu}_p^2\tilde{\chi}\tilde{w}_1 + \gamma_0(\alpha q\tilde{\theta} - \bar{\mu}_p\tilde{\chi})(-\alpha q\gamma\tilde{\chi} - \bar{\mu}_p\tilde{\theta})\tilde{\mu} \\
&+[\gamma_0(\alpha q\tilde{\theta} - \bar{\mu}_p\tilde{\chi})(-\bar{\mu}_p\lambda) + \lambda\tilde{\chi} + \tilde{w}^T P\bar{d}_B]\tilde{y}_f \\
&+r[\lambda\tilde{\chi} + \tilde{w}^T P\bar{d}_B - \gamma_0(\alpha q\tilde{\theta} - \bar{\mu}_p\tilde{\chi})\bar{\mu}_p\lambda].
\end{aligned}
\tag{14.30}
$$

By using Young's inequality ($2ab \leq a^2/k^2 + k^2b^2$), we can write

$$
\dot{V} \;\leq -\phi^T Q(t)\phi + \|\phi\|^2\rho_1(t) + \rho_2(t) + r^2\rho_3(t) \tag{14.31}
$$

in which

$$
\begin{aligned}
\phi &= \left[\, |\tilde{\chi}| \; |\tilde{\theta}| \; \|\tilde{w}\| \,\right]^T \\
\rho_1 &= \gamma_0 \left\| \begin{bmatrix} \alpha q\gamma|\bar{\mu}_p| & (\bar{\mu}_p^2 + \alpha^2 q^2\gamma)/2 \\ (\bar{\mu}_p^2 + \alpha^2 q^2\gamma)/2 & |\bar{\mu}_p|\alpha q \end{bmatrix} \right\| |\tilde{\mu}| \\
\rho_2 &= [\alpha^2 q^2\lambda^2\bar{\mu}_p^2 + (\gamma_0\bar{\mu}_p\lambda + \lambda)^2 + 4\|P\bar{d}_B\|^2]\tilde{y}_f^2 \\
\rho_3 &= \frac{1}{2}(\lambda^2 + \bar{\mu}_p^4\lambda^2 + q^2\bar{\mu}_p^2\lambda^2 + 2\|P\bar{d}_B\|^2)
\end{aligned}
\tag{14.32}
$$

and $Q(t)$ is a (3×3) symmetric matrix whose elements q_{ij} are given by

$$
\begin{aligned}
q_{11} &= k_o(1 + \gamma_0\bar{\mu}_p^2) - \gamma_0\alpha^2\gamma q\bar{\mu}_p^2 - \gamma_0|\bar{\mu}_p\dot{\bar{\mu}}_p| - \frac{1}{2} - \frac{\gamma_0^2}{2} - \frac{1}{4} \\
q_{22} &= \gamma_0\alpha^2 q^2 - \frac{\gamma_0^2\alpha^2}{2} - \frac{\gamma_0^2}{4} \\
q_{33} &= 1 - \frac{1}{4} - \frac{1}{16} \\
q_{12} &= -\frac{1}{2}\gamma_0\alpha q(\alpha^2 q\gamma|\bar{\mu}_p| + |\dot{\bar{\mu}}_p| + k_o|\bar{\mu}_p|) \\
q_{13} &= -\frac{1}{2}(1 + \gamma_0\bar{\mu}_p^2) \\
q_{23} &= -\frac{1}{2}\gamma_0\alpha q|\bar{\mu}_p|.
\end{aligned}
$$

By using again Young's inequality, we can write

$$
\inf{}_{\tau\in[0,T]}\lambda_{\min}[Q(\tau)] \geq \min_{1\leq k\leq 3} \bar{q}_{kk} \triangleq Q_m \tag{14.33}
$$

with

$$\bar{q}_{11} = k_o - \frac{5}{4} - \gamma_0\left(\alpha^2\gamma\bar{\mu}_{pM}^4 + \bar{\mu}_{pM}\dot{\bar{\mu}}_{pM} + \frac{\gamma_0}{2}\right)$$
$$\bar{q}_{22} = \gamma_0\alpha^2\{e^{-2T}k_p^2 - \gamma_0[\frac{1}{2} + \bar{\mu}_{pM}^4(\alpha^2\bar{\mu}_{pM}^3\gamma + \dot{\bar{\mu}}_{pM}$$
$$+k_o\bar{\mu}_{pM})^2 + \bar{\mu}_{pM}^6]\} - \frac{\gamma_0^2}{4}$$
$$\bar{q}_{33} = \frac{3}{16} - \gamma_0^2\bar{\mu}_{pM}^4.$$

Since by definition (14.24) $0 < \alpha < 1$, by choosing

$$k_o \geq 3$$
$$\gamma_0 \leq \min\left\{1, \frac{\sqrt{3}}{4\bar{\mu}_{pM}^2}, c_1, c_2\right\}$$

with

$$c_1 = \frac{e^{-2T}k_p^2}{0.5 + \bar{\mu}_{pM}^4(\alpha^2\bar{\mu}_{pM}^3\gamma + \dot{\bar{\mu}}_{pM} + k_{ob}\bar{\mu}_{pM})^2 + \bar{\mu}_{pM}^6 + 0.25\alpha^{-2}}$$
$$c_2 = \frac{7}{4[\alpha^2\gamma\bar{\mu}_{pM}^4 + \bar{\mu}_{pM}\dot{\bar{\mu}}_{pM} + 0.5]}$$

it follows that $Q(t)$ is positive definite with

$$Q_m = O(\alpha^2). \tag{14.34}$$

Now, note that we can write for $V(t)$

$$V \leq \frac{1}{2}\phi^T\begin{bmatrix} 1 + 2\gamma_0\bar{\mu}_p^2 & 0 & 0 \\ 0 & \frac{1}{\gamma} + \gamma_0\alpha^2 q^2 & 0 \\ 0 & 0 & \|P\| \end{bmatrix}\phi$$
$$\leq \frac{1}{2}\|\phi\|^2 c_{VM} \tag{14.35}$$

with (recall that $0 < \alpha < 1$)

$$c_{VM} = \max\left\{1 + 2\gamma_0\bar{\mu}_{pM}^2, \frac{1}{\gamma} + \gamma_0\bar{\mu}_{pM}^2, \|P\|\right\}. \tag{14.36}$$

Since, from (14.29)

$$V \geq \frac{1}{2} c_{Vm} \|\phi\|^2 \tag{14.37}$$

in which $c_{Vm} = \min\left\{1, \frac{1}{\gamma}, \lambda_{\min}(P)\right\}$, from (14.31) and (14.35) we can write

$$\begin{aligned} \dot{V} &\leq -2\frac{Q_m}{c_{VM}} V + 2c_{Vm} V \rho_1(t) + \rho_2(t) + r^2 \rho_3(t) \\ &\stackrel{\triangle}{=} -cV + \bar{\rho}_1(t) V + \rho_2(t) + r^2 \bar{\rho}_3 \end{aligned} \tag{14.38}$$

where

$$\begin{aligned} c &= 2\frac{Q_m}{c_{VM}} \\ \bar{\rho}_1(t) &= 2c_{Vm}\rho_1(t) \\ \bar{\rho}_3 &= \frac{1}{2}(\lambda^2 + \bar{\mu}_{pM}^6 \lambda^2 + \bar{\mu}_{pM}^4 \lambda^2 + 2\|P\bar{d}_B\|^2). \end{aligned} \tag{14.39}$$

Recalling (14.2), (14.3), (14.5), (14.19), (14.32) and (14.39), we can write for any $t \geq 0$

$$\begin{aligned} \rho_2(t) &\leq \rho_{20}(\|[\tilde{y}_{f1}(0), \ldots, \tilde{y}_{fl}(0)]\|) e^{-2\lambda_f t} \\ \bar{\rho}_1(t) &\leq \bar{\rho}_{10}(\|[\tilde{y}_{f1}(0), \ldots, \tilde{y}_{fl}(0), \tilde{\xi}^T(0)]\|) e^{-\lambda_m t} \end{aligned} \tag{14.40}$$

in which ρ_{20}, $\bar{\rho}_{10}$ are class-k functions and

$$\lambda_m = \min_{1 \leq i \leq 3} \{-\mathrm{Re}[\lambda_i(D)]\}$$

with λ_i being the ith eigenvalue of matrix D. By applying the comparison principle and the variation of constants formula (see [13, 23]), from (14.38) we can write (recall that $\bar{\rho}_1$ is exponentially decaying)

$$\begin{aligned} V(t) &\leq e^{\|\bar{\rho}_1\|_1} \left[V(0)e^{-ct} + \int_0^t e^{-c(t-\tau)} \rho_2(\tau) d\tau \right. \\ &\qquad \left. + \bar{\rho}_3 \int_0^t e^{-c(t-\tau)} r^2(\tau) d\tau \right] \\ &\leq e^{\|\bar{\rho}_1\|_1} \left[V(0)e^{-ct} + \int_0^t e^{-c(t-\tau)} \rho_2(\tau) d\tau \right. \\ &\qquad \left. + \bar{\rho}_3 \sum_{k=0}^{N} e^{-ckT} \int_0^T r^2(\tau) d\tau \right] \end{aligned} \tag{14.41}$$

with $\|\bar{\rho}_1\|_1 = \int_0^\infty |\bar{\rho}_1(\tau)|d\tau$ and N such that $0 \leq t - NT < T$ and $\|\bar{\rho}_1\|_1 = \int_0^\infty \bar{\rho}_1(\tau)d\tau$. Since $\rho_2(t)$ is exponentially decaying, from (14.41) we can conclude that all signals are bounded. Recalling (14.40) and (14.29), from (14.41) we have

$$\begin{aligned}\tilde{\theta}^2(t) \leq 2\gamma e^{\|\bar{\rho}_1\|_1}\left[V(0)e^{-ct} + \frac{\rho_{20}}{c}e^{-2\lambda_f t}\right]\\ +2\gamma e^{\|\bar{\rho}_1\|_1}\bar{\rho}_3\frac{1}{1-e^{-cT}}\int_0^T r^2(\tau)d\tau.\end{aligned} \quad (14.42)$$

Now, note that by defining

$$\beta = \begin{bmatrix} a_2 \\ b_2 \\ \vdots \\ a_k \\ b_k \\ \vdots \end{bmatrix}^T \quad \Phi(t) = \begin{bmatrix} \cos(2\omega t) \\ \sin(2\omega t) \\ \vdots \\ \cos(k\omega t) \\ \sin(k\omega t) \\ \vdots \end{bmatrix}^T$$

$$R = \text{block diag}\left[\, R_2 \cdots R_k \cdots \right]$$

$$R_k = M_k \begin{bmatrix} \cos\psi_k & -\sin\psi_k \\ \sin\psi_k & \cos\psi_k \end{bmatrix}$$

$$M_k = \frac{\lambda_f^l}{(\lambda_f^2 + k^2\omega^2)^{l/2}}, \quad \psi_k = l\arctan\frac{-k\omega}{\lambda_f}$$

we can write for $r_y(t)$ in (14.1) and $r(t)$ in (14.4),

$$\begin{aligned} r_y(t) &= \Phi^T(t)\beta \\ r(t) &= \Phi^T(t)R\beta. \end{aligned} \quad (14.43)$$

Since

$$\|R\| = (\lambda_{MAX}(R^T R))^{1/2} = \frac{\lambda_f^l}{(\lambda_f^2 + 4\omega^2)^{l/2}}$$

and, by Parseval Theorem,

$$\begin{aligned}\frac{1}{T}\int_0^T r^2(\tau)d\tau &= \frac{1}{2}\beta^T R^T R\beta \leq \frac{1}{2}\frac{\lambda_f^{2l}}{(\lambda_f^2 + 4\omega^2)^l}\beta^T\beta \\ &= \frac{\lambda_f^{2l}}{(\lambda_f^2 + 4\omega^2)^l}\frac{1}{T}\int_0^T r_y^2(\tau)d\tau \end{aligned} \quad (14.44)$$

from (14.44) and (14.42), we obtain statement (ii) with

$$
\begin{aligned}
f(\|x(0)\|) &= \left\{2\gamma e^{\|\bar{\rho}_1\|_1}\left[V(0) + \frac{\rho_{20}}{c}\right]\right\}^{1/2} \\
\beta_1 &= \min\left\{\frac{c}{2}, \lambda_f\right\} \\
\beta_2 &= \left[2\gamma\bar{\rho}_3 e^{\|\bar{\rho}_1\|_1}\frac{1}{1-e^{-cT}}\right]^{1/2}\frac{\lambda_f^l}{(\lambda_f^2+4\omega^2)^{l/2}}.
\end{aligned}
\tag{14.45}
$$

Since by (14.39) and (14.34), c is $O(\alpha^2)$, for sufficiently small α (and, consequently, for sufficiently high order l) we can write

$$
\frac{1}{1-e^{-cT}} \simeq \frac{1}{cT} \tag{14.46}
$$

which implies that

$$
\beta_2 = O\left[\left(\frac{\lambda_f^2+\omega^2}{\lambda_f^2+4\omega^2}\right)^{l/2}\right] \tag{14.47}
$$

and

$$
\beta_1 = O\left[\left(\frac{\lambda_f^2}{\lambda_f^2+\omega^2}\right)^{l}\right].
$$

The case $l = 0$ can be simply treated by considering $y(t)$ in place of $y_{fl}(t)$ and adjusting, accordingly, the various steps of the proof. □

Corollary 14.2 *If $y(t)$ is a biased sinusoidal signal with no higher order harmonics, then the estimate $\hat{\omega}(t)$ provided by the frequency estimator (14.18) in Theorem 14.1 is such that, for any integer $l \geq 0$, $|\tilde{\theta}(t)| \leq f(\|x(0)\|)e^{-\beta_1 t}$, $\forall t \geq 0$, in which f is a class-k function.*

Proof It follows directly from statement (ii) in Theorem 14.1, since $r_y(t) = 0$ in (14.1). □

Remark 14.3 If in Theorem 14.1, the hypothesis $a_1^2 + b_1^2 > 0$ is not satisfied but the signal $y(t)$ is not constant, then the algorithm (14.18) guarantees properties similar to (i) and (ii) for the first nonzero harmonic in the signal $y(t)$.

Remark 14.4 The frequency estimator (14.18) may be compared to the adaptive notch filter proposed in [25] in the special case in which $a_0 = 0$ in (14.1):

$$\begin{aligned}\dot{x}_1 &= x_2\\ \dot{x}_2 &= -\hat{\omega}^2 x_1 - 2\zeta\hat{\omega}x_2 + 2\zeta\hat{\omega}^2 y\\ \dot{\hat{\omega}} &= -\gamma x_1(\hat{\omega}^2 y - \hat{\omega}x_2).\end{aligned} \tag{14.48}$$

They are both adaptive linear filters whose input is the periodic signal $y(t)$ and whose output is the estimate $\hat{\omega}$: while (14.18) is a $(l+9)$-order adaptive linear filter in which $\hat{\omega}^2$ is the adapted filter parameter, the algorithm (14.48) is a third-order filter in which $\hat{\omega}$ is the adapted filter parameter. They both guarantee the convergence of the estimate $\hat{\omega}$ into a neighborhood of the true value $\omega = 2\pi/T$: while (14.48) guarantees an asymptotic convergence for sufficiently small initial errors, higher order harmonics and adaptation gain γ, the algorithm (14.18) guarantees exponential convergence for any initial condition and any parameters choice. For both algorithms $\hat{\omega}$ converges to the true value ω if there are no higher order harmonics in (14.1).

Remark 14.5 From (14.4) and (14.6), an estimate of the amplitude and phase of the first biased harmonic term can also be obtained. If we write $\eta_1(t)$ as

$$\eta_1(t) = \theta_1 \sin(\omega t) + \theta_2 \cos \omega t + \theta_3$$

the parameters θ_i may be estimated using the gradient method (see [29]) as

$$\begin{aligned}\begin{bmatrix}\dot{\hat{\theta}}_1\\ \dot{\hat{\theta}}_2\\ \dot{\hat{\theta}}_3\end{bmatrix} &= \gamma_1(\eta_1 - \eta_I)\begin{bmatrix}\sin \omega t\\ \cos \omega t\\ 1\end{bmatrix}\\ \eta_I &= \hat{\theta}_1 \sin(\omega t) + \hat{\theta}_2 \cos(\omega t) + \hat{\theta}_3.\end{aligned} \tag{14.49}$$

in which $\gamma_1 > 0$. Since, however, η_1 and ω are not known, their estimates provided by the frequency estimator (14.18) are used, so that in place of (14.49) we use

$$\begin{aligned}\begin{bmatrix}\dot{\hat{\theta}}_1\\ \dot{\hat{\theta}}_2\\ \dot{\hat{\theta}}_3\end{bmatrix} &= \gamma_1(\hat{\eta}_1 - \hat{\eta}_I)\begin{bmatrix}\sin \hat{\omega} t\\ \cos \hat{\omega} t\\ 1\end{bmatrix}\\ \hat{\eta}_I &= \hat{\theta}_1 \sin(\hat{\omega} t) + \hat{\theta}_2 \cos(\hat{\omega} t) + \hat{\theta}_3.\end{aligned}$$

The recursive least square method could be also used [29].

14.3 Examples

As a first example, we consider the problem of estimating the period of the periodic signal $y(t)$ of frequency $\omega = 3$ given by (see Fig. 14.2)

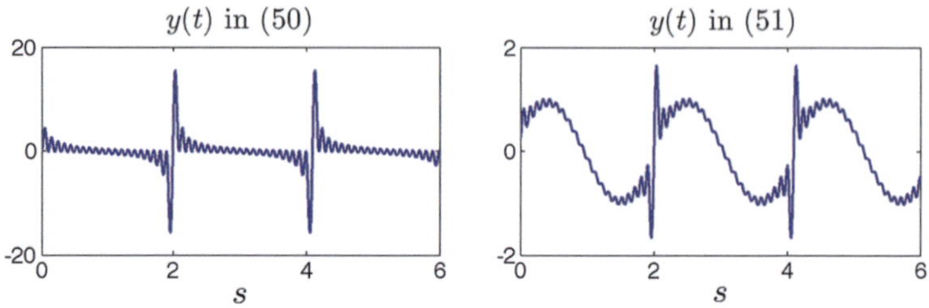

Fig. 14.2 Periodic signals: *left curve*, $y(t)$ in (14.50); *right curve*, $y(t)$ in (14.51)

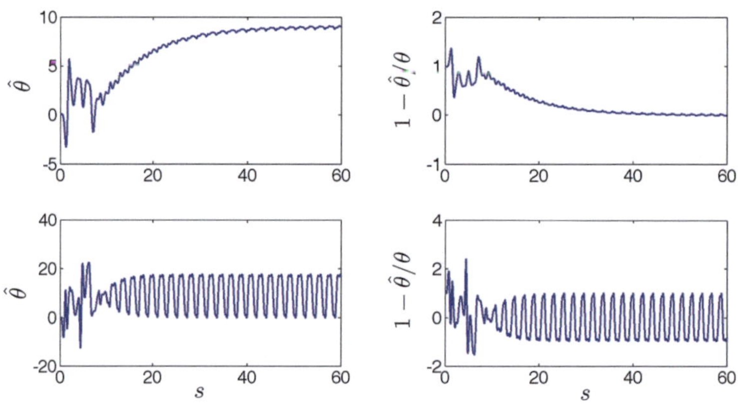

Fig. 14.3 Frequency estimator (14.18): *upper curves*, $l = 2$; *lower curves*, $l = 1$

$$y(t) = \sum_{k=1}^{21} \sin(3kt + 0.3k) \tag{14.50}$$

by means of the frequency estimator (14.18). The results are illustrated in Fig. 14.3 in which the following time histories are reported: the value of $\hat{\theta}(t)$ as obtained by the algorithm (14.18), the relative error $1 - \hat{\theta}(t)/\theta$ between the true and the estimated square of the frequency. The following parameters and initial conditions have been adopted: $\gamma = 30000$, $d_2 = 4$, $d_3 = 5$, $d_4 = 2$, $\lambda_f = 1$, $\lambda = 1$, $k_o = 1$, $\hat{\theta}(0) = 0.1$ and all other initial conditions set to zero. The upper curves refer to the case in which a second-order filter is adopted ($l = 2$) while the lower curves report the results obtained with $l = 1$. Figure 14.3 shows that in the case $l = 1$ the rate of convergence is increased while the accuracy is worse with respect to the case $l = 2$. The previous results may be compared to those obtained by the adaptive notch filter (14.48) illustrated in Remark 14.4 which are reported in Fig. 14.4. As suggested by the authors in [25], the parameters used in the algorithm (14.48) are: $\gamma = 0.1$, $\zeta = 0.35$ while the initial value for $\hat{\omega}$ was $\hat{\omega}(0) = 2.8$ (10 % less than the true value) and null initial conditions. The upper curves in Fig. 14.4 report the time histories of the square of the frequency estimate $\hat{\omega}^2(t)$ and of the relative error $[\omega^2 - \hat{\omega}^2(t)]/\omega^2$

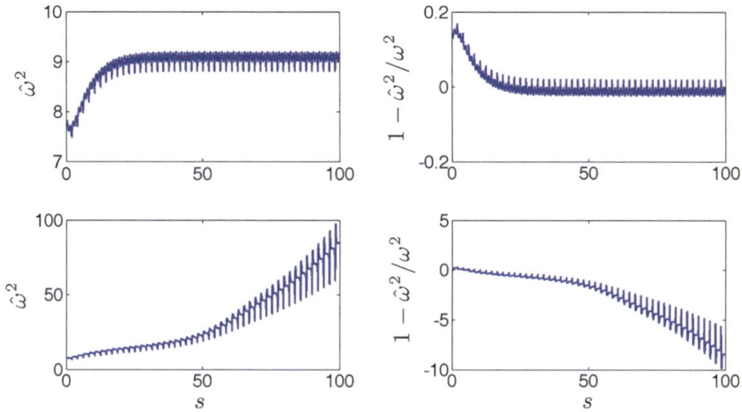

Fig. 14.4 Adaptive notch filter (14.48): *upper curves*, $y(t)$ as in (14.51), *lower curves*, $y(t)$ as in (14.50)

in the case in which the higher order harmonics in (14.50) are reduced by 90 %, i.e., the following signal is applied (see Fig. 14.2)

$$y(t) = \sin(3t + 0.3) + 0.1 \sum_{k=2}^{21} \sin(3kt + 0.3k). \tag{14.51}$$

The lower curves illustrates the performance achieved when the complete signal (14.50) was used. Figure 14.4 shows that while the adaptive notch filter has good performance when both the higher order harmonics and the initial estimate error are small, a divergent behavior occurs when the complete signal (14.50) is applied. Note that the initial frequency estimate error for the adaptive notch filter is much smaller than the corresponding initial error for the frequency estimator (14.18).

As a second example, we consider the problem of attenuating a periodic disturbance assuming that it is the output of an unknown stable system $D(s)$ whose input $y(t)$ is measurable (see Fig. 14.5)

Fig. 14.5 Block diagram for the disturbance compensator

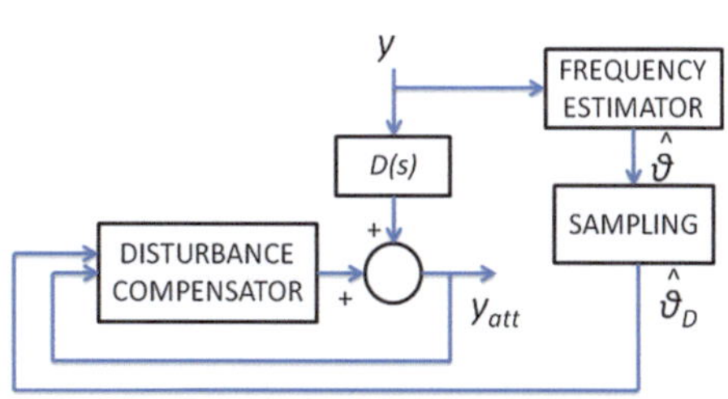

$$y(t) = \sum_{k=1}^{5} \frac{1}{k} \sin(3kt + 0.3k). \tag{14.52}$$

Note that the scheme in Fig. 14.5 applies to active noise cancelation if $D(s)$ is the transfer function between the source of noise and the listener. First of all, assuming that the frequency of the periodic signal is known and given by $\hat{\theta}_D$, the following disturbance compensator is designed

$$\begin{aligned}
\dot{x}_1 &= x_2 - k_c y_{att} \\
\dot{x}_2 &= -\hat{\theta}_D x_1 \\
\dot{x}_3 &= x_4 - k_c y_{att} \\
\dot{x}_4 &= -4\hat{\theta}_D x_3 \\
\dot{x}_5 &= x_6 - k_c y_{att} \\
\dot{x}_6 &= -9\hat{\theta}_D x_5 \\
y_{att} &= y + x_1 + x_3 + x_5
\end{aligned} \tag{14.53}$$

which is capable of cutting the first three harmonics in the periodic signal $D(s)y(s)$ when $\hat{\theta}_D = 3$. We select $D(s) = 1$ for the simulation set-up. Then, the frequency estimator (14.18) is used, with $l = 2$ and the same parameters used in the first example (with the exception of $\gamma = 3000$), to update every time interval $T = 4$ s the value of $\hat{\theta}_D$ in (14.53), so that the overall disturbance compensator is hybrid. The results of the simulation are illustrated by Fig. 14.6 in which are reported the time histories of the disturbance $y(t)$, the attenuated disturbance $y_{att}(t)$, the discrete-time estimate $\hat{\theta}_D(t)$

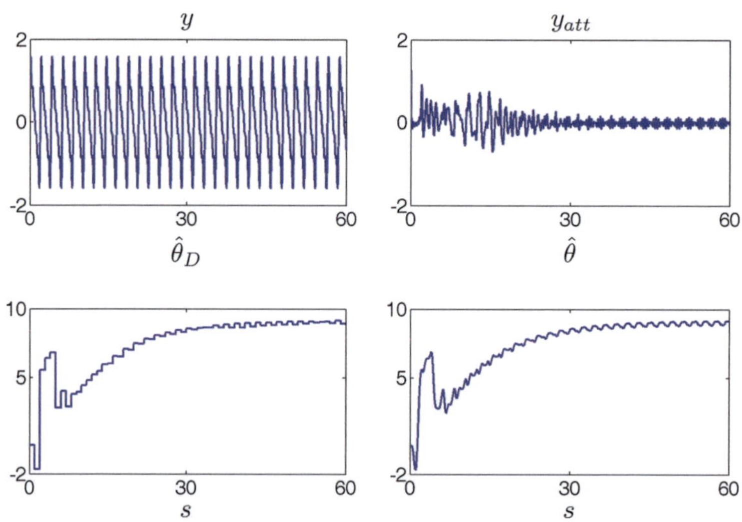

Fig. 14.6 Hybrid disturbance compensator

and the continuous-time estimate $\hat{\theta}(t)$. It can be noted that, even though there is a nonzero error in the frequency estimate, the attenuated signal $y_{att}(t)$ is much smaller than the original disturbance $y(t)$. The residual error is due to two different causes: the mismatch between the true and estimated period, and the remaining fourth and fifth harmonics which are not blocked in (14.53).

14.4 Conclusions

The adaptive $(l+9)$-order frequency estimator (14.18) has been proposed to estimate the period of a measured bounded continuous periodic signal: l denotes the order of the linear prefilter. No a priori information on the period is required. Theorem 14.1 establishes that for any initial condition, the parameter estimation error converges exponentially into a closed interval whose size depends on the higher order harmonics in (14.1). By increasing the order l of the prefilter, the accuracy of the frequency estimation can be arbitrarily improved, at the expense of reducing the rate of the exponential convergence. If there are no higher order harmonics in (14.1), that is (14.1) is a biased sinusoidal signal, then the frequency estimation error converges exponentially to zero for any value of l, including $l = 0$. This result improves the widely studied [3, 4, 9, 24, 25, 28] adaptive notch filter (14.48) whose frequency estimate convergence into a neighborhood of the true value is proved to be asymptotic and local in [25], provided that the adaptation gain is sufficiently small. Moreover, the frequency estimator may be also used to provide, at each predefined time interval, updated frequency estimates to disturbance compensators operating with constant frequency, as it is shown in the included example.

References

1. S. Aranovskiy, A. Bobtsov, A. Kremlev, N. Nikolaev, O. Slita, Identification of frequency of biased harmonic signal. Eur. J. Control **16**, 129–139 (2010)
2. A. Bobtsov, New approach to the problem of globally convergent frequency estimator. Int. J. Adaptive Control Signal Proc. **22**, 306–317 (2008)
3. M. Bodson, S.C. Douglas, Adaptive algorithms for the rejection of sinusoidal disturbances with unknown frequency. Automatica **33**, 2213–2221 (1997)
4. D.W. Clarke, On the design of adaptive notch filters. Int. J. Adaptive Control Signal Proc. **15**, 715–744 (2001)
5. G. Fedele, A. Ferrise, Non adaptive second-order generalized integrator for identification of a biased sinusoidal signal. IEEE Trans. Autom. Control **57**, 1838–1842 (2012)
6. B.A. Francis, W.M. Wonham, The internal model principle of control theory. Automatica **12**, 457–465 (1976)
7. M. Hou, Amplitude and frequency estimation of a sinusoid. IEEE Trans. Autom. Control **50**, 855–858 (2005)
8. M. Hou, Estimation of sinusoidal frequencies and amplitudes using adaptive identifier and observer. IEEE Trans. Autom. Control **52**, 493–499 (2007)

9. L. Hsu, R. Ortega, G. Damm, A globally convergent frequency estimator. IEEE Trans. Autom. Control **44**, 698–713 (1999)
10. H.K. Khalil, *Nonlinear Systems*, 3rd edn. (Prentice Hall, Upper Saddle River, 2002)
11. G.A. Korn, T.M. Korn, *Mathematical Handbook for Scientists and Engineers* (McGraw-Hill, New York, 1968)
12. M. Krstic, I. Kanellakopoulos, P.V. Kokotovic, *Nonlinear and Adaptive Control Design* (Wiley, New York, 1995)
13. V. Laksmikantham, S. Leela, *Differential and Integral Inequalities* (Academic Press, New York, 1969)
14. S. Liuzzo, R. Marino, P. Tomei, Adaptive learning control of nonlinear systems by output error feedback. IEEE Trans. Autom. Control **52**, 1232–1248 (2007)
15. R. Marino, P. Tomei, Frequency estimation of periodic signals. In *European Control Conference 2014*, pp. 7–12 (Strasbourg, France, 2014)
16. R. Marino, P. Tomei, Robust adaptive compensation of periodic disturbances with unknown frequency, in *IEEE 52nd Conference on Decision and Control*, pp. 7528–7533 (Florence, Italy, 2013)
17. R. Marino, W. Respondek, A.J. van der Schaft, A direct approach to almost disturbance and almost input-output decoupling. Int. J. Control **48**, 353–383 (1988)
18. R. Marino, W. Respondek, A.J. van der Schaft, Almost disturbance decoupling for single-input single-output nonlinear systems. IEEE Trans. Autom. Control **34**, 1013–1017 (1989)
19. R. Marino, W. Respondek, A.J. van der Schaft, P. Tomei, Nonlinear H_∞ almost disturbance decoupling. Syst. Control Lett. **23**, 159–168 (1994)
20. R. Marino, W. Respondek, A.J. van der Schaft, Equivalence of nonlinear systems to input-output prime forms. SIAM J. Control Optim. **32**, 387–407 (1994)
21. R. Marino, P. Tomei, *Nonlinear Control Design—Geometric Adaptive and Robust* (Prentice Hall, London, 1995)
22. R. Marino, P. Tomei, Global estimation of n unknown frequencies. IEEE Trans. Autom. Control **47**, 1324–1328 (2002)
23. R.K. Miller, A.N. Michel, *Ordinary Differential Equations* (Academic Press, New York, 1982)
24. M. Mojiri, A.R. Bakhshai, An adaptive notch filter for frequency estimation of a periodic signal. IEEE Trans. Autom. Control **49**, 314–318 (2004)
25. M. Mojiri, A.R. Bakhshai, Stability analysis of periodic orbit of an adaptive notch filter for frequency estimation of a periodic signal. Automatica **43**, 450–455 (2007)
26. G. Obregon-Pulido, B. Castillo-Toledo, A.G. Loukianov, Globally convergent estimators for n frequencies. IEEE Trans. Autom. Control **47**, 857–863 (2002)
27. A. Pikovsky, M. Rosenblum, J. Kurths, *Synchronization A Universal Concept in Nonlinear Sciences* (Cambridge University Press, New-York, 2001)
28. P.A. Regalia, An improved lattice-based IIR notch filter. IEEE Trans. Signal Proc. **39**, 2124–2128 (1991)
29. S.S. Sastry, M. Bodson, *Adaptive Control: Stability, Convergence, and Robustness* (Prentice Hall, Englewood Cliffs, 1989)
30. S. Wiggins, *Introduction to Applied Nonlinear Dynamical Systems and Chaos* (Springer-Verlag, New-York, 1990)
31. X. Xia, Global frequency estimation using adaptive identifiers. IEEE Trans. Autom. Control **47**, 1188–1193 (2002)
32. J.-X. Xu, S.K. Panda, T.H. Lee, *Real-time Iterative Learning Control* (Springer, London, 2009)
33. A.K. Ziarani, A. Konrad, A method of extraction of nonstationary sinusoids. Signal Proc. **84**, 1323–1346 (2004)

Chapter 15
Power-Based Methods for Infinite-Dimensional Systems

Krishna Chaitanya Kosaraju and Ramkrishna Pasumarthy

Abstract In this chapter we aim to extend the Brayton Moser (BM) framework for modeling infinite-dimensional systems. Starting with an infinite-dimensional port-Hamiltonian system we derive a BM equivalent which can be defined with respect to a non-canonical Dirac structure. Based on this model we derive stability and new passivity properties for the system. The state variables in this case are the "effort" variables and the storage function is a "power-like" function called the mixed potential. The new property is derived by "differentiating" one of the port variables. We present our results with the Maxwell's equations, and the transmission line with non-zero boundary conditions as examples.

15.1 Introduction

I (the second author) got exposed to Arjan's work for the first time during my master's in Systems and Control at VJTI, Mumbai. It was though a course on non-linear control, where we followed one of his books titled "Non-linear Dynamical Control Systems" which he had coauthored with Henk Nijmeijer. In the final year of my masters, I came across a paper by him on port-Hamiltonian system. Back then, I never imagined to have to Arjan as my Ph.D. advisor. I feel extremely fortunate that I could learn under his supervision. Arjan's immense knowledge and contributions in Systems Theory have always motivated and played a significant part in shaping my career. I am delighted to dedicate this piece of work to Arjan, on his birthday, to recognize him for his monumental research in Systems and Control theory. Happy Birthday, Arjan!

K.C. Kosaraju · R. Pasumarthy (✉)
Department of Electrical Engineering, Indian Institute of Technology Madras, Chennai, India
e-mail: ramkrishna@ee.iitm.ac.in

K.C. Kosaraju
e-mail: ee13d015@ee.iitm.ac.in

M.K. Camlibel et al. (eds.), *Mathematical Control Theory I*,
Lecture Notes in Control and Information Sciences 461,
DOI 10.1007/978-3-319-20988-3_15

Energy-based methods for modeling and control of complex physical systems has been an active area of research for the past two decades. In particular, the Hamiltonian-based formulation has proven to be an effective tool in modeling and control of complex physical systems from several physical domains, both finite and infinite-dimensional cases [7]. These systems are inherently passive with the Hamiltonian when bounded from below, serving as the storage function and the input and output pair are power conjugate. This resulted in development of so-called "Energy-Shaping" methods for control of physical systems. In some cases the natural power conjugate port variables do not necessarily help in achieving the control objectives due to the *dissipation obstacle* [13], motivating the search for alternate passive maps. One possible alternative which has been explored extensively in the finite-dimensional case is the "Brayton–Moser" (BM) framework for modeling of electrical networks [2, 5, 6], which has been successfully adapted towards analyzing passivity of RLC circuits [8] and for control of physical systems by "power shaping" [7]. For further details on various energy and power-based modeling techniques we refer to [9].

Most of the literature for control on the BM framework restricts to finite-dimensional case only. One of the first results, in the infinite dimensional case, appeared in [4], in which the authors present a stability theory in the BM framework for a transmission line connected to the non-linear load. However, the proposed Lyapunov functional does not preserve the *pseudogradient*-like structure of the system, which is essential for boundary control, and to derive passive maps is not very obvious. Later, in [10] the authors describe a electromagnetic fields analogue of the Brayton Moser formulation of Maxwell equations, again mostly for zero boundary conditions. In an earlier work [12], we have presented results on control by interconnection of a transmission line by "power shaping" in the BM framework.

In this chapter we present a BM analogue of an infinite-dimensional port-Hamiltonian systems, defined with respect to a constant Stokes Dirac structure [16]. The main results are deriving a new passivity property for mixed finite and infinite-dimensional systems by "differentiating" one of the port variables (possibly the boundary port) and a storage function directly related to the power of the system, while preserving the structure of the system. This new storage function is instrumental in analyzing the stability of the system. We present our results for a general Hamiltonian system, with Maxwell's equations and the transmission line with non-zero boundary conditions, as examples.

This chapter is organized as follows. In Sect. 15.2, we defined the Stokes Dirac structure and its Brayton Moser formulation. In Sect. 15.3, we use Brayton Moser framework to analyze stability and give admissible pairs for Maxwell's equation of electromagnetic fields and telegraphers equations of transmission line with zero energy flow trough boundary. In Sect. 15.4, we present the admissible pairs and stability for transmission line with non-zero energy flows through the boundary and derive new passivity properties. Finally in Sect. 15.5, we derive conservation laws and Casimirs in the BM framework.

Part of the results presented here have appeared in [11].

Notations and Math Preliminaries

Let Z be an n dimensional Riemannian manifold with a smooth $(n-1)$ dimensional boundary ∂Z. $\Omega^k(Z), k = 0, 1, \ldots, n$ denotes the space of all exterior $k-$ forms on Z. The dual space $\left(\Omega^k(Z)\right)^*$ of $\Omega^k(Z)$ can be identified with space of $n-k$ forms $\Omega^{n-k}(Z)$, the space of $(n-k)$ forms on Z. There exists a natural pairing between $\alpha \in \Omega^k(Z)$ and $\beta \in \left(\Omega^k(Z)\right)^*$ given by $\langle\beta|\alpha\rangle = \int_Z \beta \wedge \alpha$, were $\wedge$ is the usual wedge product of differential forms, resulting in the n form $\beta \wedge \alpha$. Similar pairing can be established between the boundary variables.

d denotes the exterior derivative and maps k forms on Z to $k+1$ forms on Z. The Hodge star operator $*$ (corresponding to Riemannian metric on Z) converts p forms to $(n-p)$ forms. Given $\alpha, \beta \in \Omega^k(Z)$ and $\gamma \in \Omega^l(Z)$, the wedge product $\alpha \wedge \gamma \in \Omega^{k+l}(Z)$. We additionally have the following properties (for details on theory of differential forms we refer to [1]).

$$\alpha \wedge \gamma = (-1)^{kl}\gamma \wedge \alpha\,, \quad **\alpha = (-1)^{k(n-k)}\alpha \tag{15.1}$$

$$\int_z \alpha \wedge *\beta = \int_z \beta \wedge *\alpha \tag{15.2}$$

$$\mathrm{d}\,(\alpha \wedge \gamma) = \mathrm{d}\alpha \wedge \gamma + (-1)^k \alpha \wedge \mathrm{d}\gamma \tag{15.3}$$

Given a functional $H(\alpha_p, \alpha_q)$, we compute its variation as

$$\begin{aligned} \delta H &= H(\alpha_p + \partial\alpha_p, \alpha_q + \partial\alpha_q) - H(\alpha_p, \alpha_q) \\ &= \int_z \left[\delta_p H \wedge \partial\alpha_p + \delta_q H \wedge \partial\alpha_q\right], \end{aligned} \tag{15.4}$$

where $\alpha_p,\ \ \partial\alpha_p \in \Omega^p(Z)$ and $\alpha_q,\ \ \partial\alpha_q \in \Omega^q(Z)$ and $\delta_p H \in \Omega^{n-p}(Z)$ and $\delta_q H \in \Omega^{n-q}(Z)$ are variational derivatives of $H(\alpha_p, \alpha_q)$ with respective to α_p and α_q. Further, the time derivative of $H(\alpha_p, \alpha_q)$ is

$$\frac{dH}{dt} = \int_z \left(\delta_p H \wedge \frac{\partial\alpha_p}{\partial t} + \delta_q H \wedge \frac{\partial\alpha_q}{\partial t}\right).$$

Let $G : \Omega^{n-p}(Z) \to \Omega^{n-p}(Z)$ and $R : \Omega^{n-q}(Z) \to \Omega^{n-q}(Z)$, we call $G \geq 0$, if and only if $\forall\alpha_p \in \Omega^p(Z)$

$$\int_Z \left(\alpha_p \wedge *G\alpha_p\right) \geq 0$$

G is said to be symmetric if $\langle\alpha_p|G\alpha_p\rangle = \langle G\alpha_p|\alpha_p\rangle$.

Lastly, for $Z \subset \mathbb{R}^n$, given $f(z,t) : Z \times \mathbb{R} \to \mathbb{R}$, we denote $\frac{\partial f}{\partial t}(z,t)$ as f_t, similarly $\frac{\partial f}{\partial z}(z,t)$ as f_z.

15.2 From Port-Hamiltonian to Brayton Moser Equations

The basic concept needed in the formulation of a port-Hamiltonian system is that of a Dirac structure, which is a geometric object formalizing general power conserving interconnections [15].

Definition 15.1 Let V be a an infinite-dimensional linear space. There exists on $V \times V^*$ a canonically defined symmetric bilinear form

$$\ll (f_1, e_1), (f_2, e_2) \gg := < e_1 \mid f_2 > + < e_2 \mid f_1 > \tag{15.5}$$

with $f_i \in V, e_i \in V^*, i = 1, 2$ and $<|>$ denoting the duality product between V and its dual subspace V^*. A constant Dirac structure on V is a linear subspace $D \subset V \times V^*$ such that

$$D = D^{\perp}, \tag{15.6}$$

where $\perp$ denotes the orthogonal complement with respect to the bilinear form $\ll, \gg$.

Let now $(f, e) \in D = D^{\perp}$. Then as an immediate consequence of (15.5)

$$0 = \ll (f, e), (f, e) \gg = 2 < e \mid f > .$$

Thus for all $(f, e) \in D$ we have $< e \mid f >= 0$, expressing power conservation with respect to the dual power variables $f \in V$ and $e \in V^*$
The Stokes Dirac Structure [16]: Define the linear space $\mathcal{F}_{p,q} = \Omega^p(Z) \times \Omega^q(Z) \times \Omega^{n-p}(\partial Z)$ called the space of flows and $\mathcal{E}_{p,q} = \Omega^{n-p}(Z) \times \Omega^{n-q}(Z) \times \Omega^{n-q}(\partial Z)$, the space of efforts, with integers p, q satisfying $p + q = n + 1$. Then, the linear subspace $D \subset \mathcal{F}_{p,q} \times \mathcal{E}_{p,q}$

$$\begin{aligned} D &= \Big\{ (f_p, f_q, f_b, e_p, e_q, e_b) \in \mathcal{F}_{p,q} \times \mathcal{E}_{p,q} | \\ &\begin{bmatrix} f_p \\ f_q \end{bmatrix} = \begin{bmatrix} *G & (-1)^r \mathrm{d} \\ \mathrm{d} & *R \end{bmatrix} \begin{bmatrix} e_p \\ e_q \end{bmatrix}, \begin{bmatrix} f_b \\ e_b \end{bmatrix} = \begin{bmatrix} 1 & 0 \\ 0 & -(-1)^{n-q} \end{bmatrix} \begin{bmatrix} e_p|_{\partial Z} \\ e_q|_{\partial Z} \end{bmatrix} \Big\} \end{aligned}$$

where $r = pq + 1$, is Stokes Dirac structure with dissipation, [16] with respect to the bilinear form

$$\begin{aligned} &\ll \left(f_p^1, f_q^1, f_b^1, e_p^1, e_q^1, e_b^1\right), \left(f_p^2, f_q^2, f_b^2, e_p^2, e_q^2, e_b^2\right) \gg \\ &= \int_Z (e_p^2 \wedge f_p^1 + e_p^1 \wedge f_p^2 + e_q^2 \wedge f_q^1 + e_q^1 \wedge f_q^2) + \int_{\partial Z} (e_b^2 \wedge f_b^1 + e_b^1 \wedge f_b^2). \end{aligned}$$

Consider a distributed parameter port Hamiltonian system on $\Omega^p(Z) \times \Omega^q(Z) \times \Omega^{n-p}(\partial Z)$, with energy variables $(\alpha_p, \alpha_q) \in \Omega^p(Z) \times \Omega^q(Z)$ representing two different physical energy domains interacting with each other. The Hamiltonian $H = \int_Z \mathrm{H}$, where H is the Hamiltonian density. Then the below system of equations represent an infinite-dimensional port-Hamiltonian system, with $f_p = -\frac{\partial \alpha_p}{\partial t},\ f_q = -\frac{\partial \alpha_q}{\partial t}$ and the efforts as the co-energy variables, i.e. $e_p = \delta_p H,\ e_q = \delta_q H$.

$$-\frac{\partial}{\partial t}\begin{bmatrix}\alpha_p\\ \alpha_q\end{bmatrix} = \begin{bmatrix}*G & (-1)^r \mathrm{d}\\ \mathrm{d} & *R\end{bmatrix}\begin{bmatrix}\delta_p H\\ \delta_q H\end{bmatrix};\ \begin{bmatrix}f_b\\ e_b\end{bmatrix} = \begin{bmatrix}1 & 0\\ 0 & -(-1)^{n-q}\end{bmatrix}\begin{bmatrix}\delta_p H\mid_{\partial Z}\\ \delta_q H\mid_{\partial Z}\end{bmatrix} \tag{15.7}$$

The time derivative of the Hamiltonian is computed as

$$\frac{dH}{dt} \leq \int_Z e_b \wedge f_b$$

This means that the increase in energy in the spatial domain is less than or equal to power supplied to the system through its boundary. This implies that the system is passive, with respect to the boundary variables, with the Hamiltonian H, which is assumed to be bounded from below serving as the storage function.

15.2.1 The Brayton Moser Mixed Potential

Brayton and Moser in the early 1960s [5, 6] showed that the dynamics of a class (topologically complete) of non-linear RLC-circuits can be written as

$$A(i_L, v_c)\begin{bmatrix}\frac{di_L}{dt}\\ \frac{dv_c}{dt}\end{bmatrix} = \begin{bmatrix}\frac{\partial P}{\partial i_L}\\ \frac{\partial P}{\partial v_C}\end{bmatrix} + \begin{bmatrix}B_{E_c}^\top E_c\\ -B_{J_c}^\top J_C\end{bmatrix} \tag{15.8}$$

where $A(i_L, v_C) = \mathrm{diag}\{L(i_L), -C(v_C)\}$ and i_L the vector of currents through inductors, v_C vector of capacitor voltages, $L(i_L)$ the inductance matrix, $C(v_C)$ the capacitance matrix, B_{E_c}, B_{J_c} the matrices containing the elements $\{-1, 0, 1\}$ decided by Kirchoff's voltage and current laws. E_C, J_C are respectively the controlled voltage and current sources. P is called the mixed potential function defined by

$$P(i_L, v_C) = F(i_L) - G(v_C) + i_L^\top \gamma\ v_c$$

where $x = (i_L, v_C)$ the system states. Here F is the content of all the current controlled resistors, G is the co-content of all voltage controlled resistors. Matrix γ contains elements $\{-1, 0, 1\}$ depending on the network topology. Computing the time derivative of P along the trajectories of (15.8) we have

$$\dot{P} = \dot{x}^\top\left(A(x) + A^\top(x)\right)\dot{x} + u^T y,$$

where, $u = (E_c,\ J_c)^\top$ and $y = \left(-B_{E_c}\frac{di_L}{dt},\ B_{J_c}\frac{dv_C}{dt}\right)^\top$.

From the above expression we can conclude that the system is passive if $(A(x) + A^\top(x)) \leq 0$, with P the storage functions and $u^\top y$ as the supply rate.

In case $(A(x) + A^\top(x)) \leq 0$ is not satisfied, then it is possible to find new $(\tilde{A}, \tilde{P})$ called an "admissible pair," (refer [3, 13]) satisfying $(\tilde{A}(x) + \tilde{A}^\top(x)) \leq 0$. The dynamics can then be equivalently be written as

$$\tilde{A}\begin{bmatrix}\frac{di_L}{dt}\\ \frac{dv_c}{dt}\end{bmatrix} = \begin{bmatrix}\frac{\partial \tilde{P}}{\partial i_L}\\ \frac{\partial \tilde{P}}{\partial v_C}\end{bmatrix} + \begin{bmatrix}B_{E_c}^\top E_c\\ -B_{J_c}^\top J_C\end{bmatrix} \tag{15.9}$$

Remark 15.2 Contrast to the case where the total energy of the system serves as the storage function and passivity is derived with respect to input–output variables which are power conjugate, for example, the voltage and currents [15]. In this case, making use of the mixed potential function as the storage function we derive passivity either with respect to controlled voltages and the derivatives of currents, or the controlled currents and the derivatives of the voltages.

The Infinite-Dimensional BM Formulation

We aim to write the infinite-dimensional port-Hamiltonian system, defined with respect to a Stokes Dirac structure (15.7) in an equivalent BM form. To begin with, we assume that the mapping from the energy variables (α_p, α_q) to the co-energy variables $(e_p, e_q) = (\delta_p H, \delta_q H)$ is invertible. This means the inverse transformation from the co-energy variables to the energy variables can be written as $(\alpha_p, \alpha_q) = (\delta_{e_p} H^*, \delta_{e_q} H^*)$. H^* is the co-energy of H obtained by $H^*(e_p, e_q) = \int_Z \big((e_p \wedge \alpha_p + e_q \wedge \alpha_q) - H(\alpha_p, \alpha_q)\big)$. Further, assume that the Hamiltonian H splits as $H(\alpha_p, \alpha_q) = H_p(\alpha_p) + H_q(\alpha_q)$, with the co-energy variables given by $e_p = \delta_p H_p$, $e_q = \delta_q H_q$. Consequently the co-Hamiltonian can also be split as $H^*(e_p, e_q) = H_p^*(e_p) + H_q^*(e_q)$. We can now rewrite the spatial dynamics of the infinite-dimensional port-Hamiltonian system, in terms of the co-energy variable as

$$\begin{bmatrix}\delta_p^2 H^* & 0\\ 0 & \delta_p^2 H^*\end{bmatrix}\begin{bmatrix}-\frac{\partial e_p}{\partial t}\\ -\frac{\partial e_q}{\partial t}\end{bmatrix} = \begin{bmatrix}*G & (-1)^r \mathrm{d}\\ \mathrm{d} & *R\end{bmatrix}\begin{bmatrix}\delta_p H\\ \delta_q H\end{bmatrix} \tag{15.10}$$

To begin with, we consider the case of a system which is lossless, that is when R and G are identically equal to zero in (15.7). Define P to be a functional of the form $\int_Z e_q \wedge \mathrm{d}e_p$. Its variation is given as

$$\delta P = P(e_p + \partial e_p, e_q + \partial e_q) - P = e_q \wedge \mathrm{d}\partial e_p + \partial e_q \wedge \mathrm{d}e_p + \cdots$$

Using the relation $e_q \wedge \mathrm{d}\partial e_p = (-1)^{pq}\partial e_p \wedge \mathrm{d}e_q + (-1)^{n-q}\mathrm{d}\left(e_q \wedge \partial e_p\right)$, and the identity (15.4), we have

$$\delta_{e_q} P = \mathrm{d}e_p(-1)^{(n-q)\times q}, \quad \delta_{e_p} P = (-1)^{pq}\mathrm{d}e_q(-1)^{(n-p)\times p}.$$

We can rewrite (15.10) in the following way

$$\begin{bmatrix} \delta_p^2 H^* & 0 \\ 0 & \delta_p^2 H^* \end{bmatrix} \begin{bmatrix} \frac{\partial e_p}{\partial t} \\ \frac{\partial e_q}{\partial t} \end{bmatrix} = \begin{bmatrix} *\delta_{e_q} P \\ *\delta_{e_p} P \end{bmatrix} \tag{15.11}$$

Note that the Hodge star operator in right hand side is necessary, because $(\delta_{e_q} P, \delta_{e_p} P) \in \Omega^q(Z) \times \Omega^p(Z)$, and $(\dot{e}_q, \dot{e}_p) \in \Omega^{n-q}(Z) \times \Omega^{n-p}(Z)$.

In order to incorporate dissipation we proceed as follows: Consider instead a functional P defined as

$$P(e_p, e_q) = \int_Z \left(e_q \wedge \mathrm{d}e_p + \frac{1}{2} Re_q \wedge *e_q - \frac{1}{2} Ge_p \wedge *e_p \right) \tag{15.12}$$

The variation in P is computed as

$$\begin{aligned} P &= e_q \wedge \mathrm{d}\partial e_p + \partial e_q \wedge \mathrm{d}e_p + \frac{1}{2}(e_q \wedge R * \partial e_q + \partial e_q \wedge *e_q) \\ &\quad -\frac{1}{2}(e_p \wedge G * \partial e_p + \partial e_p \wedge *e_p) \\ &= \int_Z \partial e_q \wedge \mathrm{d}e_p + \partial e_p \wedge (-1)^{pq} \mathrm{d}e_q + \frac{1}{2}(e_q \wedge R * \partial e_q + \partial e_q \wedge *e_q) \\ &\quad -\frac{1}{2}(e_p \wedge G * \partial e_p + \partial e_p \wedge *e_p) \\ &= \int_Z \partial e_q \wedge \left(\mathrm{d}e_p + R * e_q\right) + \partial e_p \wedge \left((-1)^{pq} \mathrm{d}e_q - G * e_p\right) \end{aligned}$$

where we have used the relation $e_q \wedge \mathrm{d}\partial e_p = (-1)^{pq} \partial e_p \wedge \mathrm{d}e_q + (-1)^{n-q} \mathrm{d} \left(e_q \wedge \partial e_p\right)$, together with properties of the wedge form and the star operator defined in (15.2) and (15.3). Lastly by making use of (15.4) we can write

$$\begin{bmatrix} \delta_{e_q} P \\ \delta_{e_p} P \end{bmatrix} = \begin{bmatrix} (\mathrm{d}e_p + R * e_q)(-1)^{(n-q)\times q} \\ \left((-1)^{pq} \mathrm{d}e_q - G * e_p\right)(-1)^{(n-p)\times p} \end{bmatrix}, \tag{15.13}$$

The dynamics (15.10) can now be written as

$$\begin{bmatrix} \delta_p^2 H^* & 0 \\ 0 & \delta_p^2 H^* \end{bmatrix} \begin{bmatrix} \frac{\partial e_p}{\partial t} \\ \frac{\partial e_q}{\partial t} \end{bmatrix} = \begin{bmatrix} *\delta_{e_q} P \\ *\delta_{e_p} P \end{bmatrix} \tag{15.14}$$

The dynamics are written as partial differential equations in the co-energy variables (e_p, e_q). The above equations together with the mixed potential functional as defined in (15.12) correspond to system of equations which are usually referred to as the

Brayton Moser equations, [4]. The above system of equations can be written in a concise way as follows,

$$Au_t = *\delta_u P. \tag{15.15}$$

where $u = (e_p, e_q)^\top$ and $A = \begin{bmatrix} \delta_p^2 H^* & 0 \\ 0 & \delta_p^2 H^* \end{bmatrix}$.

Boundary dynamics: The system (15.15) can be interconnected to other systems via the boundary of the infinite-dimensional system, which can either be finite or infinite-dimensional in nature. To include the dynamics arising due to the boundary we need to append the Eq. (15.15) in order to incorporate the boundary dynamics.

$$\begin{bmatrix} A & 0 \\ 0 & A_b \end{bmatrix} \begin{bmatrix} u_t \\ u_t^b \end{bmatrix} = \begin{bmatrix} *\delta_u P \\ *\delta_{u^b} P^b + (-1)^{(n-p)\times p} e_q|_{\partial Z} \end{bmatrix} \tag{15.16}$$

with a new mixed potential function

$$\mathcal{P}(e_p, e_q) = \int_Z P(e_p, e_q) + \int_{\partial Z} P^b(e_p, e_q)$$

with P^b taking into account the mixed potential function arising through the boundary dynamics. u^b represents the states of the systems interconnected at the boundary. The variation in P_d id given by,

$$\begin{aligned} \delta\mathcal{P} = &\int_Z \left(\delta_{e_q} P \wedge \partial e_q + \delta_{e_p} P \wedge \partial e_p\right) \\ &+ \int_{\partial Z} \left(\delta_{e_q} P^b \wedge \partial e_p + \left(\delta_{e_p} P^b + (-1)^{(n-p)\times p} e_q\right) \wedge \partial e_p\right) \end{aligned}$$

Now with $U = (u, u^b)^\top$ and

$$\delta_U \mathcal{P} = \begin{bmatrix} \delta_{e_q} P \\ \delta_{e_p} P \\ \delta_{e_q} P^b|_{\partial Z} \\ \left(\delta_{e_p} P^b + (-1)^{(n-p)\times p} e_q\right)|_{\partial Z} \end{bmatrix} \tag{15.17}$$

the Brayton Moser equations incorporating boundary dynamics can be written as

$$\mathcal{A} U_t = *\delta_U \mathcal{P},$$

where $\mathcal{A} = diag(A, A_b)$.

15.2.2 The Dirac Formulation

In this section we aim to find an equivalent Dirac structure formalism of the Brayton Moser equations of infinite-dimensional system. As we shall see such a formulation would result in a non-canonical Dirac structure. For the finite-dimensional version of the Dirac formalism of BM equations we refer to [7]. Denote by $f_s = -u_t$ as the space of flows within the spatial domain and $e_s = \delta_u P$, as the space of effort variables again in the spatial domain. Further denote by $f_b = -u_b$ as the space of boundary flows and $e_b = \delta_{u^b} P$ as the space of boundary efforts. Consider the following subspace

$$\mathcal{D} = ((f_s, e_s, f_b, e_b) \in \mathcal{F}_s \times \mathcal{E}_s \times \mathcal{F}_b \times \mathcal{E}_b : -Af_s = *e_s, \quad -A_b f_b = *e_b)$$

It can easily be shown that the above defined subspace constitutes a non-canonical Dirac structure, with respect to the bilinear form

$$\begin{aligned} &\ll (f_s^1, e_s^1, f_b^1, e_b^1), (f_s^2, e_s^2, f_b^2, e_b^2) \gg \\ &= \int_Z \left(e_s^1 \wedge f_s^2 + e_s^2 \wedge f_s^1 + f_s^1 \wedge *(A + A^\top) f_s^2\right) \\ &+ \int_{\partial Z} \left(e_b^1 \wedge f_b^2 + e_b^2 \wedge f_b^1 + f_b^1 \wedge *(A_b + A_b^\top) f_b^2\right) \end{aligned}$$

The above Dirac structure satisfies the power balance equation

$$0 = \int_Z \delta_u P \wedge u_t + \int_{\partial Z} \delta_{u^b} P^b \wedge u_t^b + \int_Z u_t \wedge *(A + A^\top) u_t + \int_{\partial Z} u_t^b \wedge *(A + A^\top) u_t^b$$

Remark 15.3 In the above Dirac structure formalism, we have assumed the case where $** = 1$, where $*$ is the hodge star operator. This is at least true for the case when the spatial domain is of dimension $n = 1$ and $n = 3$, which include respectively the case of the transmission line and the Maxwell's equations, which will be the two examples we will use in the rest of the chapter.

15.3 Admissible Pairs and Stability

Once we have written down the equations in the BM framework (sometimes also referred to as the pseudogradient form) we can pose the following question; does the mixed potential function serve as a storage function (or a Lyaunov function) to infer passivity (or equivalently stability) properties of the system? Below we aim to answer these questions with the aid of two examples.

15.3.1 Example: Maxwell Equations

The spatial domain $Z \subset \mathbb{R}^3$ is a three-dimensional boundary with a smooth two-dimensional boundary ∂Z. The energy variables are the electric field induction $\mathcal{D}$ and magnetic field induction $\mathcal{B}$. $\mathcal{D} = \frac{1}{2}\mathcal{D}_{ij} z_i \wedge z_j$ and $\mathcal{B} = \frac{1}{2}\mathcal{B}_{ij} z_i \wedge z_j$ are 2−forms on Z. The co-energy variables are electric field intensity $\mathcal{E}$ and Magnetic field intensity $\mathcal{H}$, their relationship with energy variables are given by,

$$* \mathcal{D} = \varepsilon \mathcal{E} \;\; , \quad * \mathcal{B} = \mu \mathcal{H}, \tag{15.18}$$

where $\varepsilon(t, z)$ denotes the electric permittivity and $\mu(t, z)$ the magnetic permeability. The co-energy variables are one-forms, linearly related to energy variables. The Hamiltonian H is written as

$$H(\mathcal{D}, \mathcal{B}) = \int_Z \frac{1}{2} (\mathcal{E} \wedge \mathcal{D} + \mathcal{H} \wedge \mathcal{B}) = \int_Z \left(\frac{1}{2\varepsilon} * \mathcal{D} \wedge \mathcal{D} + \frac{1}{2\mu} * \mathcal{B} \wedge \mathcal{B} \right) \tag{15.19}$$

Therefore $\delta_{\mathcal{D}} H = \mathcal{E}$ and $\delta_{\mathcal{B}} H = \mathcal{H}$. Taking into account dissipation term in the system, the dynamics can be written in the port-Hamiltonian form as

$$-\frac{\partial}{\partial t} \begin{bmatrix} \mathcal{D} \\ \mathcal{B} \end{bmatrix} = \begin{bmatrix} 0 & -\mathrm{d} \\ \mathrm{d} & 0 \end{bmatrix} \begin{bmatrix} \delta_{\mathcal{D}} H \\ \delta_{\mathcal{B}} H \end{bmatrix} + \begin{bmatrix} J_d \\ 0 \end{bmatrix} = \begin{bmatrix} *\sigma & -\mathrm{d} \\ \mathrm{d} & 0 \end{bmatrix} \begin{bmatrix} \delta_{\mathcal{D}} H \\ \delta_{\mathcal{B}} H \end{bmatrix}. \tag{15.20}$$

where $*J_d = \sigma \mathcal{E}$, J_d denotes the current density and $\sigma(z, t)$ is the specific conductivity of the material. In addition we define the boundary variables as $f_b{=}\delta_D H \mid_{\partial Z}$, $e_b = \delta_B H \mid_{\partial Z}$. The rate of the Hamiltonian is given as

$$\frac{d}{dt} H \leq \int_{\partial Z} \mathcal{H} \wedge \mathcal{E}$$

The Brayton Moser form of Maxwell's equations:
In order to write the Maxwell's equations in the BM form, we proceed as follows: The aim is to rewrite the equations in terms of the co-energy variables, i.e. $\mathcal{H}$ and $\mathcal{E}$.

Define the mixed potential functional corresponding to the Maxwell's equations as

$$P = \int_Z \left(\mathcal{H} \wedge \mathrm{d}\mathcal{E} - \frac{1}{2} \sigma \mathcal{E} \wedge *\mathcal{E} \right), \tag{15.21}$$

which gives us the following form of Maxwell's equations in terms of the mixed potential

$$\begin{bmatrix} -\mu I_3 & 0 \\ 0 & \varepsilon I_3 \end{bmatrix} \begin{bmatrix} \mathcal{H}_t \\ \mathcal{E}_t \end{bmatrix} = \begin{bmatrix} *\mathrm{d}\mathcal{E} \\ -\sigma\mathcal{E} + *\mathrm{d}\mathcal{H} \end{bmatrix} = \begin{bmatrix} *\delta_{\mathcal{H}} P \\ *\delta_{\mathcal{E}} P \end{bmatrix} \tag{15.22}$$

15.3.1.1 Stability Analysis

To infer stability properties of the system (15.22) let us begin with the case of zero energy flow through the boundary of the system. The mixed potential function (15.21) obtained via (15.12) is not positive definite. Hence we cannot use it as Lyapunov/ storage functional. Moreover, the rate of this function is computed as

$$\frac{\partial P}{\partial t} = \int_Z (-\mu\mathcal{H}_t \wedge *\mathcal{H}_t + \varepsilon\mathcal{E}_t \wedge *\mathcal{E}_t)$$

It can be easily seen that the right-hand side of the above equation is not sign definite, and hence P does not serve as a Lyapunov functional to infer any kind of stability (or for that matter passivity) properties of the system. We thus need to look for other possible Lyapunov functionals $\tilde{P}$, or in other words admissible pairs $\tilde{A}$, $\tilde{P}$ as in the case of finite-dimensional systems [8] which can prove stability of the system. Moreover, in order to conclude stability, the admissible pair should be such that the symmetric part of $\tilde{A}$ is negative semidefinite. This can be achieved in the following way, [4, 10]. Let

$$\tilde{P} = \lambda P + \frac{1}{2}\int_Z (\delta_{\mathcal{H}} P \wedge M_1 * \delta_{\mathcal{H}} P + \delta_{\mathcal{E}} P \wedge M_2 * \delta_{\mathcal{E}} P)$$

with λ be a arbitrary constant and symmetric M_1 and M_2 mapping from $\Omega^2(Z) \to \Omega^2(Z)$. Here the aim is to find λ, M_1 and M_2 such that

$$\frac{\partial}{\partial t}\tilde{P} = u_t^\top \tilde{A} u_t \leq -K||u_t||^2 \leq 0 \tag{15.23}$$

where $K \geq 0$ is a constant determined by the $\tilde{A}$. If we can find such a $(\tilde{P}, \tilde{A})$, which satisfies the above condition, then we can conclude stability of the system, by invoking the stability theorem in [4].

Below we present a constructive process to obtain new admissible pairs. The variation in $\tilde{P}$ defined in (15.23) is computed as

$$\begin{bmatrix} \delta_{\mathcal{H}} \tilde{P} \\ \delta_{\mathcal{E}} \tilde{P} \end{bmatrix} = \begin{bmatrix} \lambda I & M_2\mathrm{d}* \\ M_1\mathrm{d}* & (\lambda I - \sigma M_2) \end{bmatrix} \begin{bmatrix} \delta_{\mathcal{H}} P \\ \delta_{\mathcal{E}} P \end{bmatrix},$$

applying Hodge star on both sides and using (15.22) we get

$$* \begin{bmatrix} \delta_{\mathcal{H}} \tilde{P} \\ \delta_{\mathcal{E}} \tilde{P} \end{bmatrix} = \begin{bmatrix} -\mu\lambda I & \varepsilon M_2 * \mathrm{d} \\ -\mu M_1 * \mathrm{d} & \varepsilon\,(\lambda I - \sigma M_2) \end{bmatrix} \begin{bmatrix} \mathcal{H}_t \\ \mathcal{E}_t \end{bmatrix}.$$

Further, if we let

$$\tilde{A} = \begin{bmatrix} -\mu\lambda I & \varepsilon M_2 * \mathrm{d} \\ -\mu M_1 * \mathrm{d} & \varepsilon\,(\lambda I - \sigma M_2) \end{bmatrix}$$

we arrive at the following relationship

$$\tilde{A} u_t = *\delta_u \tilde{P}. \tag{15.24}$$

Next we show that $\tilde{P}$ and $\tilde{A}$ are admissible pairs if λ, M_1 and M_2 satisfy $\varepsilon M_2 = \mu M_1 \stackrel{\triangle}{=} \theta$ and $0 \leq \lambda \leq \sigma \|M_2\|_s$, where $\|\cdot\|_s$ is spectral norm. Some calculations show that the symmetric part of $\tilde{A} = diag(-\mu\lambda I, -\varepsilon\,(\sigma M_2 - \lambda I))$ is negative definite.

We note that P can be simplified to

$$\begin{aligned} P &= \int_z \mathcal{H} \wedge d\mathcal{E} - \frac{1}{2}\sigma \mathcal{E} \wedge *\mathcal{E} \\ &= \int_z -\frac{1}{2\sigma}\,[\delta_{\mathcal{E}} \mathcal{P} \wedge *\delta_{\mathcal{E}} \mathcal{P}] + \frac{1}{2\sigma} d\mathcal{H} \wedge *d\mathcal{H}, \end{aligned}$$

resulting in

$$\begin{aligned} \tilde{P} = \int_z & \delta_{\mathcal{E}} P \wedge \frac{\sigma M_2 - \lambda I}{2\sigma} * \delta_{\mathcal{E}} P + \frac{1}{2\sigma} d\mathcal{H} \wedge *d\mathcal{H} \\ & + \frac{1}{2}\,(\delta_{\mathcal{H}} P \wedge M_1 * \delta_{\mathcal{H}} P) \geq 0 \end{aligned} \tag{15.25}$$

Lastly, we choose $M_1 > 0$ and $M_2 > 0$ such that $\varepsilon M_2 = \mu M_1$. The time derivative of $\tilde{\mathcal{P}}$ is

$$\dot{\tilde{P}} = -\int_Z (\mu\lambda \mathcal{H}_t \wedge *\mathcal{H}_t + \mathcal{E}_t \wedge *(\lambda I - \sigma M_2)\mathcal{E}_t) \leq 0$$

thus implying stability.

15.3.2 Example: The Transmission Line

In this section we first derive the Brayton Moser equivalent of the dynamics of a transmission line modeled by the telegraphers equations. Similar to the case of Maxwell's equations we find the admissible pairs under zero boundary energy flow conditions and infer stability of the system.

The spatial domain in case of the transmission is $Z = [0, 1] \subset R$ with boundary $\partial Z = \{0, 1\}$. The charge $q(z, t)$ and flux densities $\phi(z, t) \in \Omega^1(Z)$ constitute the energy variables, whereas the co-energy variable are voltage $v(z, t)$ and current $i(z, t)$ $\in \Omega^0(Z)$. For simplicity, the relation between the energy and co-energy variables is assumed to be linear, and is given by

$$* q = Cv, \quad * \phi = Li \tag{15.26}$$

where C and L are, respectively, the spatial capacitance and inductance per unit length, which are assumed to be independent of z. The Hamiltonian H, which is the total energy of the system, is written as

$$H = \frac{1}{2}\int_Z (v \wedge q + i \wedge \phi) \tag{15.27}$$

Taking the dissipation term into account, the telegraphers equations written in port-Hamiltonian form as [16]

$$-\frac{\partial}{\partial t}\begin{bmatrix} q \\ \phi \end{bmatrix} = \begin{bmatrix} *G & \mathrm{d} \\ \mathrm{d} & *R \end{bmatrix}\begin{bmatrix} \delta_q H \\ \delta_\phi H \end{bmatrix} \tag{15.28}$$

where $\delta_q H = v$, $\delta_\phi H = i$ (using (15.26) and (15.27)). $R,\ G$, respectively, denote the distributed resistance and conductance per unit length of the transmission line. Further, we define the boundary variables as $f_b = \delta_q H|_{\partial Z}$ and $e_b = \delta_\phi H|_{\partial Z}$. The rate of Hamiltonian is given by

$$\frac{d}{dt}H = (v.i)|_0^1$$

The Brayton Moser form:
The dynamics of the transmission line (15.27) can be written in an equivalent Brayton Moser form as follows: Define a functional P as

$$P = \int_Z \left(-v \wedge \mathrm{d}i + \frac{1}{2}Ri \wedge *i - \frac{1}{2}Gv \wedge *v\right), \tag{15.29}$$

which will serve as the mixed potential function. Using the line voltage and current as the state variables, we can rewrite the dynamics as follows:

$$\begin{bmatrix} -L & 0 \\ 0 & C \end{bmatrix} \begin{bmatrix} i_t \\ v_t \end{bmatrix} = \begin{bmatrix} *\delta_i P \\ *\delta_v P \end{bmatrix} = \begin{bmatrix} Gv + *\mathrm{d}i \\ -Ri - *\mathrm{d}v \end{bmatrix}. \tag{15.30}$$

with $A \triangleq diag(-L, C)$ and $u = (i(z,t)\ v(z,t))^\top$.

15.3.2.1 Admissible Pairs and Stability

Similar to the case of Maxwell equations, we cannot use P and A directly to infer stability. We, therefore, need to generate new admissible pairs $\tilde{P}$ and $\tilde{A}$ satisfying (15.23) and (15.24) such that $\tilde{P} > 0$ and symmetric part of $\tilde{A} < 0$, resulting in stability. As in the case of Maxwell's equations, we propose a $\tilde{P}$ of the form

$$\tilde{P} = \lambda P + \frac{1}{2}\delta_u P \wedge M * \delta_u P \tag{15.31}$$

We choose $M = \begin{bmatrix} \frac{\alpha}{R} & m_2 \\ m_2 & \frac{\beta}{G} \end{bmatrix}$ where α, β, m_2 are positive constants satisfying $\alpha\frac{L}{R} = \beta\frac{C}{G}$ and λ is a dimensionless constant. Such a choice will be clear in the following discussions, which will eventually lead to a stability criterion. It is easy to check that $\tilde{P}$ has units of power. To simplify the calculations we define new positive constants θ, γ and ζ as follows:

$$\begin{aligned} \theta &\triangleq \alpha\frac{L}{R} = \beta\frac{C}{G}, \qquad m_2 \triangleq \frac{2\gamma}{CR + LG} \\ \zeta &\triangleq \frac{2\gamma}{\sqrt{LC}(\alpha+\beta)} \implies m_2 = \frac{\zeta\theta}{\sqrt{LC}}. \end{aligned} \tag{15.32}$$

To show that $\tilde{P} \geq 0$ we start with simplifying the right hand side of (15.31) in the following way. Define

$$\begin{aligned} \Delta_1 &\triangleq \left(\zeta\sqrt{\frac{L}{2}}(Gv + i_z) - \sqrt{\frac{C}{2}}(Ri + v_z)\right) \\ \Delta_2 &\triangleq \left(\zeta\sqrt{\frac{C}{2}}(Ri + v_z) - \sqrt{\frac{L}{2}}(Gv + i_z)\right). \end{aligned} \tag{15.33}$$

Using (15.32), (15.33), and after some calculations, we can show that

$$\begin{aligned}\frac{1}{2}\langle \delta_u P, M\delta_u P\rangle &= \Delta_1^2 + \frac{\beta}{2G}(1-\zeta^2)(Gv+i_z)^2\\ &= \Delta_2^2 + \frac{\alpha}{2R}(1-\zeta^2)(Ri+v_z)^2\end{aligned}$$

$\tilde{P}$ as defined in (15.31) can then be written as follows

$$\tilde{P} = \frac{\alpha(1-\zeta^2)-\lambda}{2R}(Ri+v_z)^2 + \Delta_2^2 + \frac{\lambda}{2R}v_z^2 + \frac{\lambda}{2}Gv^2 \tag{15.34}$$

$$= \frac{\beta(1-\zeta^2)+\lambda}{2G}(Gv+i_z)^2 + \Delta_1^2 - \frac{\lambda}{2G}i_z^2 - \frac{\lambda}{2}Ri^2 \tag{15.35}$$

which implies that $\tilde{P} \geq 0$ as long as the following conditions are satisfied

$$\begin{aligned}&0 \leq \lambda \leq \alpha(1-\zeta^2),\ 0 \leq \zeta \leq 1\\ &-\beta(1-\zeta^2) \leq \lambda \leq 0 \text{ or equilvalently}\\ &-\beta(1-\zeta^2) \leq \lambda \leq \alpha(1-\zeta^2),\ \ 0 \leq \zeta \leq 1\end{aligned} \tag{15.36}$$

Further the variational derivative of $\tilde{P}$ with respect to u is calculated as

$$\begin{aligned}\delta_u\tilde{P} &= \begin{bmatrix}(-\lambda+\alpha)(v_z+Ri) - m_2R(Gv+i_z)\\ (\lambda+\beta)(Gv+i_z) - m_2G(Ri+v_z)\end{bmatrix} - \frac{\partial}{\partial z}\begin{bmatrix}\frac{\beta}{G}(Gv+i_z) - m_2(v_z+Ri)\\ \frac{\alpha}{R}(v_z+Ri) - m_2(Gv+i_z)\end{bmatrix}\\ &= \begin{bmatrix}L(\lambda-\alpha-m_2\frac{\partial}{\partial z}) & C(Rm_2+\frac{\beta}{G}\frac{\partial}{\partial z})\\ L(Gm_2+\frac{\alpha}{R}\frac{\partial}{\partial z}) & -C(\lambda+\beta+m_2\frac{\partial}{\partial z})\end{bmatrix}\begin{bmatrix}i_t\\ v_t\end{bmatrix}.\end{aligned}$$

Therefore

$$\tilde{A} = \begin{bmatrix}L(\lambda-\alpha-m_2\frac{\partial}{\partial z}) & C(Rm_2+\frac{\beta}{G}\frac{\partial}{\partial z})\\ L(Gm_2+\frac{\alpha}{R}\frac{\partial}{\partial z}) & -C(\lambda+\beta+m_2\frac{\partial}{\partial z})\end{bmatrix} \tag{15.37}$$

satisfies the gradient form (15.24).

Noting that conjugate of $\frac{\partial}{\partial z}$ is $-\frac{\partial}{\partial z}$ and using $\alpha\frac{L}{R} = \beta\frac{C}{G}$ from (15.32), the symmetric part of $\tilde{A}$ (15.37) is simplified to be

$$\frac{\tilde{A}+\tilde{A}^*}{2} = \begin{bmatrix}L(\lambda-\alpha) & \gamma\\ \gamma & -C(\lambda+\beta)\end{bmatrix}$$

The symmetric part of $\tilde{A}$ is negative semidefinite as long as the following conditions are satisfied,

$$-\beta \leq \lambda \leq \alpha, \text{ and } (\lambda - \alpha)(\lambda + \beta) + \frac{(\alpha + \beta)^2}{4}\zeta^2 \leq 0. \tag{15.38}$$

We now present the following result:

Proposition 15.4 *If there exist non-zero α, β, λ and ζ satisfying* (15.32), (15.36), *and* (15.38) *then $\tilde{P}$ defined in* (15.31) *and $\tilde{A}$ defined in* (15.37) *with M are admissible pairs for the transmission line dynamics. Additionally if the symmetric part of $\tilde{A}$ is negative semidefinite, i.e. (15.38) holds true, them the system of equations (15.30) is stable.*

Proof From (15.32) we define $\tau \triangleq \frac{\alpha}{\beta} = \frac{RC}{LG}$. Given a transmission line R, C, L, G are fixed, therefore $\tau \geq 0$ is related to system parameters and thus can be treated as one. Let $\lambda' = \frac{\lambda}{\beta}$. Using this in (15.36) and (15.38) we get

$$-(1 - \zeta^2) \leq \lambda' \leq \tau(1 - \zeta^2) \tag{15.39}$$

$$(\lambda' - \tau)(\lambda' + 1) + \frac{(\tau + 1)^2}{4}\zeta^2 \leq 0 \tag{15.40}$$

Now we have to show that for all $\tau \geq 0$ there exists a pair of λ' and ζ that satisfies both the above equations. Given a $\zeta \in (0, 1)$ from (15.39) λ' lies between a positive value and a negative value $\forall \tau \geq 0$. If we can show that (15.40) has a positive and negative roots, then we can conclude the proof. The roots of (15.40) are

$$r_1 = \frac{1}{2}\left(\tau - 1 + (\tau + 1)\sqrt{1 - \zeta^2}\right)$$

$$r_2 = \frac{1}{2}\left(\tau - 1 - (\tau + 1)\sqrt{1 - \zeta^2}\right)$$

The aim is to find a condition on ζ such that r_1 and r_2 have a different signs, for all $\tau > 0$. For $0 < \tau < 1$ we have $r_2 < 0$. In order to make $r_1 > 0$ we need $\zeta^2 < 4\tau/(1 + \tau)^2$. Further for $\tau > 1$ we have $r_1 > 0$, in which case we require $r_2 < 0$ which leads to the same condition on ζ that is $\zeta^2 < 4\tau/(1 + \tau)^2$. Note that this is a valid condition on ζ since $\forall \tau$, $\frac{4\tau}{(1 + \tau)^2} \leq 1$. Which implies ζ is bounded,

$$0 \leq \zeta^2 \leq \frac{4LCRG}{(LG + RC)^2}. \tag{15.41}$$

Therefore $\forall \zeta \in [0, \frac{4\tau}{(1+\tau)^2}]$ there exists a λ' which satisfies (15.39) and (15.40).

Finally for any $\beta \in \mathbb{R}^+$, $\alpha = \tau\beta$, $\lambda = \lambda'\beta$ and $\zeta \in [0, \frac{4\tau}{(1+\tau)^2}]$ satisfies (15.32), (15.36) and (15.38).

15.4 Admissible Pairs and Stability for Non-zero Energy Flow Through Boundary

In this section we derive the Brayton Moser formulation of infinite-dimensional systems with non-zero energy flows through boundary. For simplicity, we limit our discussion for systems evolving on spatial domain $Z = (0, 1)$ of dimension $n = 1$ with point boundaries, $\partial Z = \{0, 1\}$. For $z \in Z$, let $u(z, t)$ be the states evolving on the spatial domain Z, further let $u_0(t)$ and $u_1(t)$ denote the states evolving at the boundary $z = 0$ and $z = 1$. Now consider the mixed potential function of the form

$$\mathcal{P}(U) = P(u) + P^0(u_0) + P^1(u_1) \tag{15.42}$$

where $u_0 = u(0, t)$, $u_1 = u(1, t)$ and $U = [u, u_0, u_1]$ with $P(u)$ of the form (15.29). P^0 and P^1 are the contributions to the mixed potential function arising form the boundary dynamics. Similar to infinite-dimensional case, we represent the overall dynamics of finite and infinite-dimensional system in Brayton Moser form. Dynamics evolving on the spatial domain (15.30) are given by (i.e. for $0 < z < 1$)

$$Au_t = \delta_u P,$$

dynamics at boundary $z = 0$ are represented by

$$A_0 u_{0t} = \left. \left(\frac{\partial P^0}{\partial u_0} - P_{u_z} \right) \right|_{z=0} + B_0 E_0$$

with B_0, E_0 representing input matrix and source at $z = 0$ respectively. Further $P_{u_z} = \frac{\partial P}{\partial u_z}$.

The dynamics at boundary $z = 1$ are represented using

$$A_1 u_{1t} = \left. \left(\frac{\partial P^1}{\partial u_0} + P_{u_z} \right) \right|_{z=1} + B_1 E_1$$

where B_1 and E_1 are input matrix and source at $z = 1$. Together they can be written compactly in Brayton Moser form as

$$\mathcal{A} U_t = \delta_U \mathcal{P} + BE \tag{15.43}$$

where $\mathcal{A} = diag\{A, A_0, A_1\}$, A, A_0 and $A_1 \in \mathbb{R}^{2\times 2}$. $B = [B_0, B_1]$ is the input matrix and $E = [E_0\ E_1]^\top$ are the inputs to the system. The variational derivative of $\mathcal{P}$ (15.42) with respect to U is

$$\delta_U \mathcal{P} = \begin{bmatrix} \delta_u P \\ \left(\dfrac{\partial P^0}{\partial u_0} - P_{u_z}\right)\Bigg|_{z=0} \\ \left(\dfrac{\partial P^1}{\partial u_1} + P_{u_z}\right)\Bigg|_{z=1} \end{bmatrix}.$$

Further the time derivative of mixed potential function (15.42) is

$$\frac{d}{dt}\mathcal{P} = \int_0^1 (\delta_u P \cdot u_t)\, dz + \left(\frac{\partial P^0}{\partial u_0} - P_{u_z}\right)\Bigg|_{z=0} \cdot u_{0t} + \left(\frac{\partial P^1}{\partial u_1} + P_{u_z}\right)\Bigg|_{z=1} \cdot u_{1t} \tag{15.44}$$

where $u_t = \dfrac{\partial u}{\partial t}$, $u_{0t} = \dfrac{\partial u_0}{\partial t}$, $u_{1t} = \dfrac{\partial u_1}{\partial t}$. Using the Brayton Moser form (15.43), $\dot{\mathcal{P}}$ can be written as

$$\begin{aligned} \frac{d}{dt}\mathcal{P} &= \int_0^1 (Au_t \cdot u_t)\, dz + A_0 u_{0t} \cdot u_{0t} + A_1 u_{1t} \cdot u_{1t} - E^\top B^\top U_t \\ &= \int_0^1 \left(u_t^\top \frac{A + A^\top}{2} u_t\right) dz + u_{0t}^\top \frac{A_0 + A_0^\top}{2} u_{0t} + u_{1t} \frac{A_1 + A_1^\top}{2} u_{1t} + E^\top y \end{aligned} \tag{15.45}$$

where $y = -B^\top U_t$. It can be seen that for a positive definite $\mathcal{P}$, and negative definite $\mathcal{A}$ the system is passive with input E and output y. In general $\mathcal{P}$ and $\mathcal{A}$ do not satisfy these conditions. This motivates us to search for new admissible pairs $\mathcal{P} \geq 0$ and $\mathcal{A} \leq 0$ which enables us derive cerain passivity/stability properties.

Definition 15.5 Admissible Pairs: We denote $\tilde{\mathcal{P}} = \tilde{P} + \tilde{P}^0 + \tilde{P}^1$ and $\tilde{\mathcal{A}} = diag\{\tilde{A}, \tilde{A}_0,$
$\tilde{A}_1\}$ Admissible pairs if they satisfy the following:

(a) $\tilde{P} \geq 0$ and $\tilde{A} \leq 0$ such that

$$\tilde{A} u_t = \delta_u \tilde{P} \tag{15.46}$$

(b) $\tilde{P}^0 \geq 0$ and $\tilde{A}_0 \leq 0$ such that

$$\tilde{A}_0 u_{0t} = \left(\frac{\partial \tilde{P}}{\partial u_0} - \tilde{P}_{u_z}\right)\Bigg|_{z=0} + B_0 E_0 \tag{15.47}$$

(c) $\tilde{P}^1 \geq 0$ and $\tilde{A}_1 \leq 0$ such that

$$\tilde{A}_1 u_{1t} = \left(\frac{\partial \tilde{P}}{\partial u_1} + \tilde{P}_{u_z} \right) \Bigg|_{z=1} + B_1 E_1 \tag{15.48}$$

(d) Together we can write them as $\tilde{\mathcal{P}} \geq 0$ and $\tilde{\mathcal{A}} \leq 0$ such that

$$\begin{aligned} \tilde{\mathcal{A}} U_t &= \delta_U \tilde{\mathcal{P}} + B E_b \\ y_b &= -B^\top U_t. \end{aligned} \tag{15.49}$$

Finally time derivative of $\tilde{\mathcal{P}}$ is

$$\dot{\tilde{\mathcal{P}}} \leq E_b^\top y_b.$$

Which implies that the system is passive with storage function $\tilde{\mathcal{P}}$ and ports E_b and y_b.
We next show how to derive these with the help of an example.

15.4.1 Example: Transmission Line with Circuit Elements at the Boundary

Consider a transmission line, whose boundary is interconnected to certain circuit elements as shown in Fig. 15.1. At $z = 0$ is a resistor R_0 in series with inductor L_0 connected to a voltage source E_0. The other end of the transmission line $z = 1$ is terminated with a resistor R_1.

This gives us the following dynamics at the boundary

$$\begin{aligned} v_0 + R_0 i_0 + L_0 \frac{di_0}{dt} &= E_0 \qquad z = 0 \\ v_1 &= R_1 i_1 \qquad z = 1 \end{aligned} \tag{15.50}$$

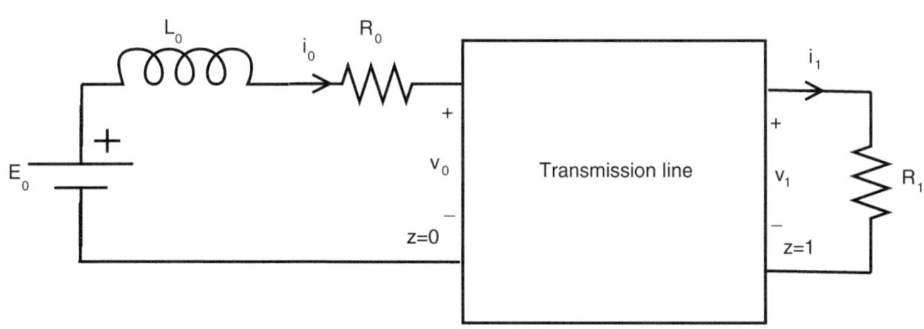

Fig. 15.1 Transmission line with boundary

where $v_0 = v(0,t)$, $i_0 = i(0,t)$ and $v_1 = v(1,t)$, $i_1 = i(1,t)$, let $U = [i, v, i_0, v_0, i_1, v_1]^\top$.

Let $u = [i, v]^\top$, $u_0 = [i_0, v_0]^\top$, $u_1 = [i_1, v_1]^\top$ and $P_{u_z} = \dfrac{\partial P}{\partial u_z}$, $u_z = \frac{\partial u}{\partial z}$.

Next we define the mixed potential function $\mathcal{P} = P + P^0 + P^1$ and $\mathcal{A}$ as follows:

$$P = \int_0^1 \left(-\frac{1}{2} R i^2 + \frac{1}{2} G v^2 + v i_z \right) \mathrm{d}z$$

$$P^0 = -\frac{1}{2} R_0 i_0^2 \qquad P^1 = -\frac{1}{2} R_1 i_1^2$$

$$\mathcal{A} = diag \left\{ \begin{bmatrix} L & -C \\ 0 & 0 \end{bmatrix}, \begin{bmatrix} L_0 & 0 \\ 0 & 0 \end{bmatrix}, \begin{bmatrix} 0 & 0 \\ 0 & 0 \end{bmatrix} \right\}$$

where P and A are defined in (15.29) and (15.30), respectively. The input matrices $B_0 = \begin{bmatrix} 0 & 0 & 1 & 0 & 0 & 0 \end{bmatrix}^\top$, $B_1 = 0$ and $E_1 = 0$. The transmission line dynamics governed by Eq. (15.28) together with the boundary dynamics given by (15.50) can be written in a compact form as

$$\begin{bmatrix} L & 0 & 0 & 0 & 0 & 0 \\ 0 & -C & 0 & 0 & 0 & 0 \\ 0 & 0 & L_0 & 0 & 0 & 0 \\ 0 & 0 & 0 & 0 & 0 & 0 \\ 0 & 0 & 0 & 0 & 0 & 0 \\ 0 & 0 & 0 & 0 & 0 & 0 \end{bmatrix} \begin{bmatrix} i_t \\ v_t \\ i_{0t} \\ v_{0t} \\ i_{1t} \\ v_{1t} \end{bmatrix} = \begin{bmatrix} -Ri - \frac{\partial v}{\partial z} \\ -Gv - \frac{\partial i}{\partial z} \\ v_0 + R_0 i_0 \\ -v_0 - R_0 i_0 \\ -v_1 + R_1 i_1 \\ v_1 - R_1 i_1 \end{bmatrix} + \begin{bmatrix} 0 \\ 0 \\ 1 \\ 0 \\ 0 \\ 0 \end{bmatrix} E_0.$$

It can easily be checked that using $\mathcal{P}$ as a storage function does not result in any kind of passivity properties of the system. Therefore, we find new admissible pairs satisfying Definition 15.5. The admissible pairs for spatial domain, found in Sect. 15.3.2.1 for zero energy flow through boundary will satisfy (15.46). Therefore $\tilde{P}$ and $\tilde{A}$ remains same for transmission line with zero or with non-zero energy flow through the boundary. For the rest of the example we choose that $\lambda = -1$, and input matrix $B_0 = \begin{bmatrix} 0 & 0 & -1 & 0 & 0 & 0 \end{bmatrix}^\top$. Next we aim to find $\tilde{A}_0$ and $\tilde{P}^0$ which satisfy (15.47). At $z = 0$ we have

$$\left. \left(\frac{\partial \tilde{P}^0}{\partial u_0} - \tilde{P}_{u_z} \right) \right|_{z=0} + B_0 E_0 = \begin{bmatrix} -m_2 L i_{0t} + \theta v_{0t} + v_0 + \frac{\partial \tilde{P}^0}{\partial i_0} - E_0 \\ \theta i_{0t} - m_2 C v_{0t} + \frac{\partial \tilde{P}^0}{\partial v_0} \end{bmatrix}$$

Let us consider $\tilde{P}^0$ of the form $\dfrac{1}{2} R_0 i_0^2$,

$$\left(\frac{\partial \tilde{P}^0}{\partial u_0} - \tilde{P}_{u_z}\right)\bigg|_{z=0} + B_0 E_0 = \begin{bmatrix} -m_2 L i_{0t} + \theta v_{0t} + v_0 + R_0 i_0 - E_0 \\ \theta i_{0t} - m_2 C v_{0t} \end{bmatrix}$$
$$= \begin{bmatrix} -m_2 L i_{0t} + \theta v_{0t} - L_0 i_{0t} \\ \theta i_{0t} - m_2 C v_{0t} \end{bmatrix}.$$

In the last step we used the boundary condition at $z = 0$, i.e. $v_0 + R_0 i_0 - E_0 = -L_0 i_{0t}$, further assuming that $\exists \zeta, \theta$ satisfying $\frac{1}{m_2 C}(1-\zeta^2)\theta^2 = L_0$,

$$\left(\frac{\partial \tilde{P}^0}{\partial u_0} - \tilde{P}_{u_z}\right)\bigg|_{z=0} + B_0 E_0 = \begin{bmatrix} -m_2 L i_{0t} + \theta v_{0t} - \left(\frac{1}{m_2 C}(1-\zeta^2)\theta^2\right) i_{0t} \\ \theta i_{0t} - m_2 C v_{0t} \end{bmatrix}$$
$$= \begin{bmatrix} \theta v_{0t} - \left(m_2 L + \frac{1}{m_2 C}(1-\zeta^2)\theta^2\right) i_{0t} \\ \theta i_{0t} - m_2 C v_{0t} \end{bmatrix}$$
$$= \begin{bmatrix} \theta v_{0t} - \frac{\theta^2}{m_2 C} i_{0t} \\ \theta i_{0t} - m_2 C v_{0t} \end{bmatrix} = \begin{bmatrix} -\frac{\theta^2}{m_2 C} & \theta \\ \theta & -m_2 C \end{bmatrix} \begin{bmatrix} i_{0t} \\ v_{0t} \end{bmatrix}$$

in the last step we used $m_2 = \frac{\zeta\theta}{\sqrt{LC}}$ (15.32). Finally, we denote

$$\tilde{A}_0 = \begin{bmatrix} -\frac{\theta^2}{m_2 C} & \theta \\ \theta & -m_2 C \end{bmatrix} \leq 0, \tag{15.51}$$
$$u_{0t}^\top \tilde{A}_0 u_{0t} \leq 0.$$

Hence $\tilde{P}^0 = \frac{1}{2} R_0 i_0^2$ and $\tilde{A}_0$ (15.52) satisfy (15.47), under the assumption that ζ and θ are chosen such that, $L_0 = \frac{1}{m_2 C}(1-\zeta^2)\theta^2$. Similarly under the assumption that $\frac{\theta}{m_2} = R_1$, we can show that for $\tilde{A}_1 = -\tilde{A}_0$ and $\tilde{P}^1 = \frac{1}{2} R_1 i_1^2$ will satisfy (15.48). But for all (i_{1t}, v_{1t}) satisfying $v_{1t} = R_1 i_{1t} = \frac{\theta}{m_2 C} i_{1t}$ we have

$$\tilde{A}_1 u_{1t} = \begin{bmatrix} \frac{\theta^2}{m_2 C} & -\theta \\ -\theta & m_2 C \end{bmatrix} \begin{bmatrix} i_{1t} \\ v_{1t} \end{bmatrix}$$
$$= \begin{bmatrix} \frac{\theta^2}{m_2 C} & -\theta \\ -\theta & m_2 C \end{bmatrix} \begin{bmatrix} i_{1t} \\ R_1 i_{1t} \end{bmatrix} = \begin{bmatrix} \frac{\theta^2}{m_2 C} & -\theta \\ -\theta & m_2 C \end{bmatrix} \begin{bmatrix} i_{1t} \\ \frac{\theta}{m_2 C} i_{1t} \end{bmatrix} = 0$$

That is we choose θ and m_2 such that u_{1t} is always in the nullspace of $\tilde{A}_1$. Which implies

$$u_{1t}^{\top}\tilde{A}_1 u_{1t} = 0 \quad \forall (i_{1t}, v_{1t}).$$

Finally for $B_b = \begin{bmatrix} 0 \; 0 \; -1 \; 0 \; 0 \; 0 \end{bmatrix}^{\top}$ and $E_b = E_0$ we get $y_b = \dfrac{di_0}{dt}$. The time derivative $\tilde{\mathcal{P}} = \tilde{P} + \tilde{P}^0 + \tilde{P}^1$ is computed as

$$\frac{d}{dt}\tilde{\mathcal{P}} \leq E_0 \frac{di_0}{dt},$$

which implies that the system is passive with respect to input E_0 and output $\dfrac{di_0}{dt}$.

Remark 15.6 Note that in Hamiltonian case the storage function is

$$H = \frac{1}{2}\int_0^1 \left(Li^2 + Gv^2\right) + \frac{1}{2}L_0 i_0^2$$

and its time derivative is calculated to be $\dfrac{d}{dt}H \leq E_0 i_0$. The system is passive with port variable E_0 and i_0.

15.5 Casimirs and Conservation Laws

We obtain conservation laws which are independent from the mixed potential function as follows [14, 16]: For simplicity, we consider the case of systems without dissipation. We further assume that the energy and the co-energy variables are related via a linear relation, given by

$$\alpha_p = *\varepsilon \; e_p \text{ and } \alpha_q = *\mu \; e_q. \tag{15.52}$$

We can write (15.10) in the following way:

$$\begin{bmatrix} -\mu & 0 \\ 0 & \varepsilon \end{bmatrix}\begin{bmatrix} \dot{e}_q \\ \dot{e}_p \end{bmatrix} = \begin{bmatrix} *\delta_{e_q} P \\ *\delta_{e_p} P \end{bmatrix} \tag{15.53}$$

Consider a function $C : \Omega^{n-p}(Z) \times \Omega^{n-q}(Z) \times Z \to \mathrm{R}$, which satisfies

$$\mathrm{d}(*\delta_{e_p} C) = 0, \quad \mathrm{d}(*\delta_{e_q} C) = 0$$

$$\frac{d}{dt}C(e_q, e_p) = \int_Z \left(\delta_{e_q} C \wedge \dot{e}_q + \delta_{e_p} C \wedge \dot{e}_p\right)$$

$$
\begin{aligned}
&= \int_Z \left(-\delta_{e_q} C \wedge * \frac{1}{\mu} \mathrm{d}e_p (-1)^{(n-q)\times q} + \delta_{e_p} C \wedge * \frac{1}{\varepsilon} (-1)^{pq} \mathrm{d}e_q (-1)^{(n-p)\times p} \right) \\
&= \int_Z \left(-(-1)^{(n-q)\times q} \frac{1}{\mu} \mathrm{d}e_p \wedge * \delta_{e_q} C + (-1)^p \frac{1}{\varepsilon} \mathrm{d}e_q \wedge * \delta_{e_p} C \right) \\
&= \int_Z \left(-(-1)^{(n-q)\times q} \frac{1}{\mu} [\mathrm{d}(e_p \wedge * \delta_{e_q} C) + (-1)^q e_p \wedge \mathrm{d}(* \delta_{e_q} C)] \right. \\
&\quad \left. + (-1)^p \frac{1}{\varepsilon} [\mathrm{d}(e_q \wedge * \delta_{e_p} C) + (-1)^p e_p \wedge \mathrm{d}(* \delta_{e_p} C)] \right) \\
&= \int_{\partial Z} \left(e_q \wedge * \delta_{e_p} C) \mid_{\partial Z} + (e_p \wedge * \delta_{e_q} C) \mid_{\partial Z} \right)
\end{aligned}
$$

In the particular case when $*\delta_{e_p} C \mid_{\partial Z} = *\delta_{e_q} C \mid_{\partial Z} = 0$, then $\frac{dC}{dt} = 0$, along the system trajectories. Such a function is called a Casimir function.

15.5.1 Example: Transmission Line

In case of the lossless transmission line, the total current

$$
C_I = \int_0^1 i(t, z) dz
$$

and the line voltage

$$
C_v = \int_0^1 v(t, z) dz
$$

are the systems conservation laws. This can easily be inferred by the following

$$
\begin{aligned}
\frac{d}{dt} C_I &= - \int_0^1 \frac{1}{l} \frac{\partial v}{\partial z} = \frac{v}{L} \mid_0 - \frac{v}{L} \mid_1 \\
\frac{d}{dt} C_v &= - \int_0^1 \frac{1}{C} \frac{\partial i}{\partial z} = \frac{i}{C} \mid_0 - \frac{i}{C} \mid_1
\end{aligned}
$$

15.5.2 Example: Maxwell's Equations

In case of Maxwell's equations with no dissipation terms, it can easily be checked that the magnetic field intensity $\int_Z \mathcal{H}$ and the electric field intensity $\int_Z \mathcal{B}$ constitute the conserved quantities. This can be seen via the following expressions:

$$\int_Z \frac{d}{dt}\mathcal{H}_t = -\int_{\partial Z} \frac{1}{\mu}\mathcal{E}$$
$$\int_Z \frac{d}{dt}\mathcal{E}_t = \int_{\partial Z} \frac{1}{\varepsilon}\mathcal{H}$$

Another class of conserved quantities can be identified in the following way: Using (15.11), the system of equations (15.53) can be rewritten as

$$\begin{bmatrix} -\mu & 0 \\ 0 & \varepsilon \end{bmatrix} \begin{bmatrix} *\dot{e}_q \\ *\dot{e}_p \end{bmatrix} = \begin{bmatrix} *\mathrm{d}e_p(-1)^{(n-q)\times q} \\ *(-1)^{pq}\mathrm{d}e_q(-1)^{(n-p)\times p} \end{bmatrix} \tag{15.54}$$

Note that

$$-\mathrm{d}\left(\mu * \dot{e}_q\right) = \mathrm{d}(\mathrm{d}e_p(-1)^{(n-q)\times q}) = 0$$
$$\mathrm{d}\left(\mu * \dot{e}_p\right) = \mathrm{d}((-1)^{pq}\mathrm{d}e_q(-1)^{(n-p)\times p}) = 0$$

This means that $\mathrm{d}(\mu * e_q)$, $\mathrm{d}(\varepsilon * e_p)$ are differential forms which do not vary with time.

In terms of *Maxwell's Equations* this would mean $\mathrm{d}(\mu * \mathcal{H})$ is a constant three-form representing the charge density and $\mathrm{d}(\varepsilon * \mathcal{E})$ is actually zero. In standard electromagnetic texts these would mean $\nabla \cdot \mathcal{D} = J$, and $\nabla \cdot \mathcal{B} = 0$, representing respectively the Gauss' electric and magnetic law.

15.6 Conclusions

The main results in this chapter deal with the Brayton Moser formulation of infinite-dimensional systems, starting from the Hamiltonian formulation of infinite-dimensional systems, defined with respect to a Stokes' Dirac structure. This formulation provides a means to generate new passive maps for infinite-dimensional systems, while preserving the pseudogradient-like structure of the Brayton Moser formulation. The preserving of the structure is key for boundary control by interconnection of infinite-dimensional systems.

References

1. R. Abraham, J. Marsden, T. Ratiu, *Manifolds, Tensor Analysis, and Applications*, 2nd edn. (Springer, Berlin, 1988)
2. G. Blankenstein, Geometric modeling of nonlinear RLC circuits. IEEE Trans. Circuits Syst. I: Regul. Pap. **52**(2), 396–404 (2005)
3. G. Blankenstein, Power balancing for a new class of non-linear systems and stabilization of RLC circuits. Int. J. Control **78**(3), 159–171 (2005)

4. R.K. Brayton, W.L. Miranker, A stability theory for nonlinear mixed initial boundary value problems. Arch. Rat. Mech. Anal. **17**(5) (1964)
5. R.K. Brayton, J.K. Moser, A theory of nonlinear networks i. Q. Appl. Math. **22**(1), 1–33 (1964)
6. R.K. Brayton, J.K. Moser, A theory of nonlinear networks ii. Q. Appl. Math. **22**(2), 81–104 (1964)
7. V. Duindam, A. Macchelli, S. Stramigioli, H. Bryuninckx (eds.), *Modeling and Control of Complex Physical Systems: The port-Hamiltonian Approach* (Springer, Berlin, 2009)
8. D. Jeltsema, R. Ortega, J.M.A. Scherpen, On passivity and power-balance inequalities of nonlinear RLC circuits, in *IEEE Transactions on Circuits and Systems-I: Fundamental Theory and Applications*, vol. 50, pp. 1174–1179, September (2003)
9. D. Jeltsema, J.M.A. Scherpen, Multidomain modeling of nonlinear networks and systems. IEEE Control Syst. Mag. (2009)
10. D. Jeltsema, A. van der Schaft, Pseudo-gradient and Lagrangian boundary control system formulation of electromagnetic field. J. Phys. A : Math. Theor. **40**, 1–17 (2007)
11. K.C. Kosaraju, R. Pasumarthy, D. Jeltsema, Alternate Passive Maps and Stability of Infinite-dimensional Systems via Mixed Potential Functions, in *Proceedings of the 5th IFAC Workshop on Lagrangian and Hamiltonian Methods for Non Linear Control*, Lyon, France, 4–7 July (2015)
12. R. Pasumarthy, K.C. Kosaraju, A. Chandrasekar, On Power Balancing and Stabilization for A Class of Infinite-dimensional Systems, in *Proceedings of the Mathematical Theory of Networks and Systems*, Groningen, The Netherlands, July, pp. 7–11 (2014)
13. R. Ortega, D. Jeltsema, J.M.A. Scherpen, Power shaping: a new paradigm for stabilization of nonlinear RLC circuits. IEEE Trans. Autom. Control **48**(10), 1762–1767 (2003)
14. R. Pasumarthy, A. van der Schaft, Achievable Casimirs and its implications on control of port-Hamiltonian systems. Int. J. Control **80**(9), 1421–1438 (2007)
15. A.J. van der Schaft, *L2-Gain and Passivity Techniques in Nonlinear Control* (Springer, London, 2000)
16. A.J. van der Schaft, B.M. Maschke, Hamiltonian formulation of distributed-parameter systems with boundary energy flow. J. Geom. Phys. **42**, 166–194 (2002)

Chapter 16
On Stabilization of Mixed Dimensional Parameter Port Hamiltonian Systems Via Energy Shaping

H. Rodríguez-Cortés

Abstract For systems described by Port-Hamiltonian (PH) equations, the Control by Interconnection method, based on the existence of Casimir functions, provides a simple and elegant procedure for stabilization of nonlinear systems with finite dissipation. This work explores the possibility of extending this technique to the case where the plant contains an infinite-dimensional subsystem. Conditions for the existence of Casimir functions reveal the constraints for the application of the design procedure. A simple example of an RLC circuit coupled with an infinite-dimensional transmission line illustrates the main ideas of this paper.

16.1 Introduction

16.1.1 Meeting Prof. Arjan van der Schaft

During my Ph.D. studies, I had the opportunity to work on nonlinear control theory, specifically on the energy-based control techniques. At that time, I was introduced to port-Hamiltonian systems and the interconnection and damping assignment control method. Thanks to the excellent relationship of my Ph.D. advisor, Prof. Romeo Ortega, with Prof. Arjan van der Schaft I had the opportunity to perform a research stay at the University of Twente. During this visit, I took the DISC course on Nonlinear Systems, and I had the opportunity to do research with Prof. Arjan van der Schaft. Prof. van der Schaft proposed to explore the possibilities to extend the Control by Interconnection technique to a system composed of finite- and infinite-dimensional systems. This research was a challenging for me; I had little experience with partial

On sabbatical leave from Sección de Mecatrónica, Departamento de Ingeniería Eléctrica, CINVESTAV-IPN.

H. Rodríguez-Cortés (✉)
Departamento de Ingeniería Industrial y Operaciones, Instituto Tecnológico Autónomo de México, Río Hondo 1, Col. Tizapan, México, D.f. México
e-mail: hrodriguez@cinvestav.mx

M.K. Camlibel et al. (eds.), *Mathematical Control Theory I*,
Lecture Notes in Control and Information Sciences 461,
DOI 10.1007/978-3-319-20988-3_16

differential equations. However, the expert guidance and humble patience of Prof. van der Schaft allows us to construct a fundamental block to extend the Control by Interconnection method to mixed finite- and infinite-dimensional systems. I always will be grateful to Prof. van der Schaft by receiving me to answer most of the times quite basic questions. However, I always came out of his office with new research directions to follow. In 2013, Prof. van der Schaft visit my workplace, CINVESTAV-IPN, to give us a seminar on his latest advances in modeling of infinite-dimensional port-Hamiltonian systems. The seminar was inspiring to continue working on modeling and control of infinite-dimensional port-Hamiltonian systems.

16.1.2 Technical Introduction

A set of lumped potential and kinetic energy storing elements together with a set of lumped energy dissipative elements and a set of lumped external interconnection ports naturally describe finite-dimensional systems. On the other hand, infinite-dimensional systems are devices that cannot be accurately described by the lumped element assumption. For instance, elevated roadways, flexible structures for outer space, transmission lines, and aircraft wing. Although, the same sets of items describe both finite- and infinite-dimensional systems; the difference is that each component is spatially distributed. Consequently, infinite-dimensional systems are modeled with partial differential equations.

For both finite- and infinite-dimension systems, energy flows through the sets of elements by means of power conserving interconnections derived from physical laws like Kirchhoff's laws and Newton's second law. To every interconnection of elements, there is associated two power variables, called flows and efforts, whose product is the power [3]. In this framework, the power conserving interconnection relates flows and efforts corresponding to the energy storing elements, the energy dissipative elements, and ports in such a way that the total incoming power is always zero.

Starting from some given energy level, determined by the initial conditions and system' sources, power will flow between elements until a point of minimal energy is reached. If the sources are zero, these points correspond to the open-loop equilibria. Motivated by the central role played by energy in the system's behavior, the energy-based control methods aim to shape the closed-loop energy [1], as well as the energy flow pattern through the control action [9].

Among the energy-based control methods, passivity-based control (PBC) design technique interprets the controller as another dynamical system interconnected with the plant through a power preserving interconnection, with the aim of reshaping the open-loop energy function [7, 8]. This energy shaping step of PBC precedes the damping injection stage, where the dissipation structure is modified to enforce attractivity of the desired equilibrium. These two fundamental principles of PBC are independent of the dimension of the system so that it is natural to look for its extension to the infinite-dimensional setting. Building a construction block to extend the Control by Interconnection version of PBC to the infinite-dimensional framework is the objective of this work. In particular for a system composed of a finite-dimensional controller interconnected to a finite-dimensional plant through an infinite-dimensional system with spatial dimension equal one.

Instrumental for the developments of this work is the introduction of the notion of a Dirac structure, which formalizes in a geometric language the concept of power preserving interconnection for PH systems [12]. Roughly speaking, a Dirac structure is the subspace of the space of efforts and flows, where their inner product, power, is equal to zero. Dirac structures for finite-dimensional implicit PH systems are reported in [12] while [6] introduces Dirac structures for infinite-dimensional systems. Dirac structures are important in our problem for two reasons: first, they provide a natural framework to interconnect finite- and infinite-dimensional systems. Second, they give a geometric interpretation in terms of subspaces of the Dirac structure to the Casimir functions, dynamic invariants [5], required for the design of the PBC. The main contribution of this paper is the derivation of the conditions of existence of the Casimir functions for the controller-infinite-dimensional system plant. Interestingly enough, these conditions consist of the well-known conditions of the finite-dimensional controller plant interconnection, plus a new set of conditions stemming from the presence of the distributed parameter subsystem.

The organization of this manuscript is as follows. Section 16.2 presents the Dirac structure associated with PH models, for ease of reference it starts with finite-dimensional systems, and then the infinite-dimensional case. Section 16.3 presents the principal contributions of this work. The interconnection between finite- and infinite-dimensional system is described above. The conditions to guarantee the existence of Casimir functions for the interconnected system. Finally, Sect. 16.3 includes a stabilization procedure for interconnected finite- and infinite-dimensional systems, based on the results of [11], as well as a simple example. Section 16.4 completes this manuscript presenting some concluding remarks. The work in [10] contains the main results of this chapter.

16.2 Dirac Structures and Port Controlled Hamiltonian Systems

The concept of power preserving interconnection can be formalized geometrically by means of the Dirac structure. In this section, we briefly introduce this notion for both finite- and infinite-dimensional systems. Specifically, for infinite-dimensional systems with scalar spatial dimension ranging in a segment $[0, l]$. The interested reader is referred to [6, 12] for further details and generalizations.

16.2.1 Finite-Dimensional Systems

To define a Dirac structure for finite-dimensional systems, we consider the finite-dimensional linear space $\mathcal{F}$ of flows f, and its dual, $\mathcal{F}^*$, the space of efforts e. Power is then defined as $P = \langle e \mid f \rangle$ with $\langle \cdot | \cdot \rangle$ denoting the duality product.[1] As shown in [12], on $\mathcal{F} \times \mathcal{F}^*$ there exists the canonical symmetric bilinear form

$$\langle\langle (f_1, e_1), (f_2, e_2) \rangle\rangle \overset{\text{def}}{=} \langle e_1 \mid f_2 \rangle + \langle e_2 \mid f_1 \rangle \tag{16.1}$$

Definition 16.1 On the finite-dimensional linear space $\mathcal{F}$, a constant Dirac structure is a linear subspace $\mathcal{S} \subset \mathcal{F} \times \mathcal{F}^*$ such that $\mathcal{S}^\perp = \mathcal{S}$, where $\mathcal{S}^\perp$ denotes the orthogonal complement of $\mathcal{S}$ with respect to the bilinear form (16.1).

Hence, for all $(f, e) \in \mathcal{D}$ the following holds $\langle e, f \rangle = 0$. As result, a Dirac structure models a power conserving interconnection. To model a finite-dimensional pH system, the space of flows is partitioned as $\mathcal{F} \overset{\text{def}}{=} \mathcal{F}_S \times \mathcal{F}_R \times \mathcal{F}_P$. $\mathcal{F}_S$ denotes the space of flows f_S connected to the energy storing elements, $\mathcal{F}_R$ denotes the space of flows f_R connected the dissipative elements and $\mathcal{F}_P$ denotes the space of external flows f_P connected to the environment. Analogously, the space of efforts becomes $\mathcal{F}^* \overset{\text{def}}{=} \mathcal{F}_S^* \times \mathcal{F}_R^* \times \mathcal{F}_P^*$, with $e_S \in \mathcal{F}_S^*$. The efforts connected to the energy storing elements, $e_R \in \mathcal{F}_R^*$ the efforts connected to dissipative elements and $e_P \in \mathcal{F}_P^*$ the efforts connected to the environment. Hence, the total power becomes

$$P = \langle e_S \mid f_S \rangle + \langle e_R \mid f_R \rangle + \langle e_P \mid f_P \rangle$$

This work considers port-Hamiltonian systems with Dirac structure in the input-output form, that is,

$$\mathcal{D} \overset{\text{def}}{=} \left\{ (f, e) \in \mathcal{F} \times \mathcal{F}^* \;\middle|\; \begin{array}{l} f_S = -\mathcal{J}(x) e_S - g_R(x) f_R - g(x) f_P, \\ e_P = g^\top(x) e_S, \, e_R = g_R^\top(x) e_S \end{array} \right\} \tag{16.2}$$

where $(f, e) \overset{\text{def}}{=} (f_S,\ f_R,\ f_P,\ e_S,\ e_R,\ e_P)$, $x \in \mathbb{R}^n$ is the state of the system, $\mathcal{J}(x) = -\mathcal{J}^\top(x)$ is the interconnection matrix, and $g(x),\ g_S(x),\ g_R(x)$ are input matrices of suitable dimensions. From (16.2) it is easy to show that $\mathcal{D} = \mathcal{D}^\perp$. Furthermore, given that for all $(f, e) \in \mathcal{D} = \mathcal{D}^\perp$, we have that

$$0 = \langle\langle (f, e), (f, e) \rangle\rangle = 2\langle f \mid e \rangle,$$

therefore $P = 0$ for all elements of $\mathcal{D}$.

If we assume that the flow and effort variables of the dissipative elements are related by $f_R = -R(x) e_R$, where $R(x) = R^\top(x) \geq 0$, the following relationship between the power variables holds

$$\begin{bmatrix} f_S \\ e_P \end{bmatrix} = \begin{bmatrix} -\mathcal{J}(x) + \mathcal{R}(x) & -g(x) \\ g^\top(x) & 0 \end{bmatrix} \begin{bmatrix} e_S \\ f_P \end{bmatrix} \tag{16.3}$$

[1] If $\mathcal{F}$ is a Hilbert space, then $\mathcal{F}^*$ can be naturally identified with $\mathcal{F}$ in such a way that for all $f \in \mathcal{F}$, $e \in \mathcal{F}^*$ we have $\langle e \mid f \rangle = \langle e, f \rangle$, where $\langle \cdot, \cdot \rangle$ is the standard inner product; see e.g., [4].

where $\mathcal{R}(x) \stackrel{\text{def}}{=} g_R(x)^\top R(x) g_R(x)$. This model is a representation in power coordinates of a PH system. Since the power equals the time derivative of energy, and the Hamiltonian function, $H(x) : \mathbb{R}^n \to \mathbb{R}$, defines the stored energy. It follows that

$$P(t) = \frac{\mathrm{d}}{\mathrm{d}t} H[x(t)] = \left\langle \frac{\partial H}{\partial x}[x(t)] \mid \dot{x}(t) \right\rangle$$

as a result, the following relationship between power and energy coordinates for energy storing elements hold[2]

$$f_S = -\dot{x}, \quad e_S = \frac{\partial H}{\partial x}(x)$$

In this way, we get the well-known model of a PH system in energy variables

$$\dot{x} = [\mathcal{J}(x) - \mathcal{R}(x)] \tfrac{\partial H}{\partial x}(x) + g(x) f_P$$

$$e_P = g^\top(x) \tfrac{\partial H}{\partial x}(x)$$

which clearly satisfies the energy balance

$$\frac{\mathrm{d}}{\mathrm{d}t} H[x(t)] = -\frac{\partial^\top H}{\partial x}[x(t)] \mathcal{R}[x(t)] \frac{\partial H}{\partial x}[x(t)] + e_P^\top(t) f_P(t)$$

To fit the standard input–output notation in the following, we denote $e_P = y_p$ and $f_P = u_p$.

16.2.2 Infinite-Dimensional Systems

A Dirac structure for an infinite-dimensional system requires an infinite-dimensional function space. Here, we consider a Dirac structure for an infinite-dimensional system with spatial dimension one; see [6] for a general differential geometric setting appropriate to multidimensional spatial domains.

We consider an infinite-dimensional function space with the following structure $\mathcal{E} \stackrel{\text{def}}{=} \mathcal{H}_{1M}(Z) \times \mathcal{H}_{1E}(Z) \times B$, with $\mathcal{H}_{1M}(Z), \mathcal{H}_{1E}(Z)$ denoting the spaces of e_M and e_E, respectively. B denotes the external boundary efforts e_b at Z. Additionally, $\mathcal{H}_1(Z)$ denotes the Sobolev space of $\mathcal{L}_2$ functions on Z with derivatives also in $\mathcal{L}_2$. Thus, $\bar{\mathcal{F}}$ is the dual space of $\mathcal{E}$ with respect to the following duality product

[2] Strictly, the power variables f_S, e_S live in the tangent and cotangent spaces to the finite-dimensional manifold of energy variables. f_S, e_S are in a no constant Dirac structure on a manifold, see [2] for details.

$$\langle (e_E, e_M, e_b), (f_E, f_M, f_b) \rangle = \int_0^\ell [f_E(z)e_E(z) + f_M(z)e_M(z)]\mathrm{d}z + e_b f_b \mid_0^\ell$$

with (f_E, f_M) the flows, both functions on Z belonging to the dual Sobolev space $\mathcal{H}_1(Z)^*$, and f_b the boundary flow.

In an analogy with (16.1), the bilinear form between two elements of $\bar{\mathcal{F}} \times \mathcal{E}$ takes the form

$$\langle\langle (f^1, e^1), (f^2, e^2) \rangle\rangle \stackrel{\text{def}}{=} \int_0^\ell \left(e_E^1 f_E^2 + e_E^2 f_E^1 + e_M^1 f_M^2 + e_M^2 f_M^1 \right) \mathrm{d}z + \left(e_b^1 f_b^2 + e_b^2 f_b^1 \right) \mid_0^\ell \tag{16.4}$$

where $(f^i, e^i) \stackrel{\text{def}}{=} \left(f_E^i, f_M^i, f_b^i, e_E^i, e_M^i, e_b^i \right) \in \bar{\mathcal{F}} \times \mathcal{E},\ i = 1, 2.$

The following result defines the Dirac structure for an infinite-dimensional systems with scalar spatial dimension.

Proposition 16.2 *The subspace*

$$\bar{\mathcal{D}} \stackrel{\text{def}}{=} \left\{ (f, e) \in \bar{\mathcal{F}} \times \mathcal{E} \mid \begin{bmatrix} f_E \\ f_M \end{bmatrix} = \begin{bmatrix} 0 & \frac{\partial}{\partial z} \\ \frac{\partial}{\partial z} & 0 \end{bmatrix} \begin{bmatrix} e_E \\ e_M \end{bmatrix}, \begin{bmatrix} f_b \\ e_b \end{bmatrix} = \begin{bmatrix} 0 & -1 \\ 1 & 0 \end{bmatrix} \begin{bmatrix} e_{Eb} \\ e_{Mb} \end{bmatrix} \right\} \tag{16.5}$$

of $\bar{\mathcal{F}} \times \mathcal{E}$ is a constant Dirac structure, with respect to the bilinear form (16.4).

Now, we verify that the infinite-dimensional Dirac structure satisfies a generalized form of power conservation. From (16.4) it follows that $\langle\langle (f, e), (f, e) \rangle\rangle = 0$, for all $(f \mid e) \in \bar{\mathcal{D}}$. Hence, one obtains the energy balance property as follows:

$$2\int_0^\ell [e_E f_E + e_M f_M]\mathrm{d}z + 2e_b f_b \mid_0^\ell = \int_0^\ell [e_E f_E + e_M f_M]\mathrm{d}z = e_b(0) f_b(0) - e_b(\ell) f_b(\ell)$$

$e_b(0) f_b(0)$ represents power coming into the domain Z at 0 while $e_b(\ell) f_b(\ell)$ represents the power going out of z at ℓ.

In this case, the infinite-dimensional PH system in power coordinates follows directly from the Dirac structure as

$$\begin{bmatrix} f_E \\ f_M \end{bmatrix} = \begin{bmatrix} 0 & \frac{\partial}{\partial z} \\ \frac{\partial}{\partial z} & 0 \end{bmatrix} \begin{bmatrix} e_E \\ e_M \end{bmatrix}, \begin{bmatrix} f_b \\ e_b \end{bmatrix} = \begin{bmatrix} 0 & -1 \\ 1 & 0 \end{bmatrix} \begin{bmatrix} e_{Eb} \\ e_{Mb} \end{bmatrix} \tag{16.6}$$

At the boundary, the following holds

$$\begin{bmatrix} f_0 \\ e_0 \end{bmatrix} = \begin{bmatrix} -e_{M0} \\ e_{E0} \end{bmatrix}; \quad \begin{bmatrix} f_\ell \\ e_\ell \end{bmatrix} = \begin{bmatrix} -e_{M\ell} \\ e_{E\ell} \end{bmatrix} \tag{16.7}$$

To express the infinite-dimensional PH system in energy coordinates, we consider the Hamiltonian density given by

$$H : \mathcal{H}_{1M}{}^{*} \times \mathcal{H}_{1E}{}^{*} \times Z \rightarrow \mathcal{L}_1$$

The total energy functional associated to the Hamiltonian density is $\mathcal{H} = \int_0^{\ell} H(\bar{q})\mathrm{d}z$, with $\bar{q} \overset{\text{def}}{=} [q_E, q_M, z]$. We assume that $\mathcal{H}$ is differentiable, with the time derivative given by [11]

$$\frac{d\mathcal{H}}{dt} = \int_0^{\ell} \begin{bmatrix} \delta_M^{\top}\mathcal{H} & \delta_E^{\top}\mathcal{H} \end{bmatrix}^{\top} \begin{bmatrix} c\dfrac{\partial q_M}{\partial t} & \dfrac{\partial q_E}{\partial t} \end{bmatrix} \mathrm{d}z$$

where $\delta_E\mathcal{H} = \frac{\delta\mathcal{H}}{\delta q_E}$, $\delta_E\mathcal{H} = \frac{\delta\mathcal{H}}{\delta q_E}$ denote the variational derivative.

Similar to the finite-dimensional framework, in the case of the infinite-dimensional framework, the relationship between the power and energy coordinates is as follows:

$$f_E = -\frac{\partial}{\partial t}q_E, \quad e_E = \delta_E\mathcal{H}$$
$$f_M = -\frac{\partial}{\partial t}q_M, \quad e_M = \delta_M\mathcal{H}$$

as a result the infinite-dimensional PH system in energy coordinates gets the following structure

$$\begin{bmatrix} \frac{\partial}{\partial t}q_E \\ \frac{\partial}{\partial t}q_M \end{bmatrix} = \begin{bmatrix} 0 & -\frac{\partial}{\partial z} \\ -\frac{\partial}{\partial z} & 0 \end{bmatrix} \begin{bmatrix} \delta_E\mathcal{H} \\ \delta_M\mathcal{H} \end{bmatrix}, \begin{bmatrix} f_b \\ e_b \end{bmatrix} = \begin{bmatrix} 0 & -1 \\ 1 & 0 \end{bmatrix} \begin{bmatrix} \delta_E\mathcal{H}\mid_b \\ \delta_M\mathcal{H}\mid_b \end{bmatrix} \tag{16.8}$$

Straightforward computations show that the infinite-dimensional PH system satisfies the energy balancing equation

$$\frac{\mathrm{d}\mathcal{H}}{\mathrm{d}t} = \delta_E^{\top}\mathcal{H}(0)\delta_M\mathcal{H}(0) - \delta_E^{\top}\mathcal{H}(\ell)\delta_M\mathcal{H}(\ell)$$

16.2.3 Example: Transmission Line

Here we present a PH model of a transmission line whose dynamics are described by the telegrapher's equations. For the sake of illustration, we first derive the transmission line following a classical approach, and then using the Dirac framework. A method to obtain an infinite-dimensional model of a dynamic system, typically starts with spatially distributed finite-dimensional systems. Then, by taking the limit as the finite-dimensional systems become infinitesimal in size, it is possible to construct an infinite-dimensional model.

Consider the transmission line of Fig. 16.1, the dynamic equations for the nth mesh are given by, see Fig. 16.2.

$$\dot{q}_n = \frac{\lambda_n}{L_n} - \frac{\lambda_{n-1}}{L_{n-1}}, \; \dot{\lambda}_n = \frac{q_{n+1}}{C_{n+1}} - \frac{q_n}{C_n} \tag{16.9}$$

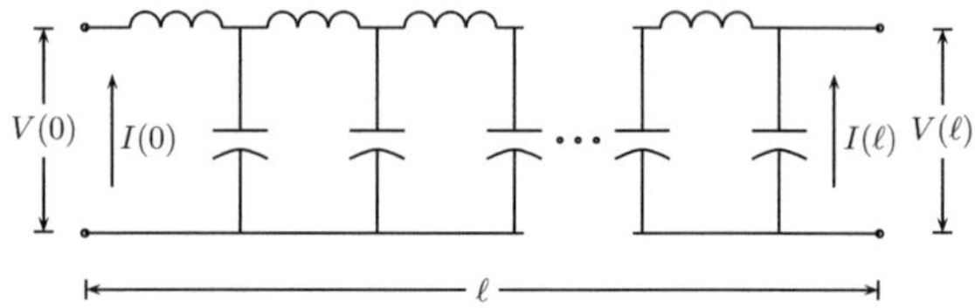

Fig. 16.1 Transmission line

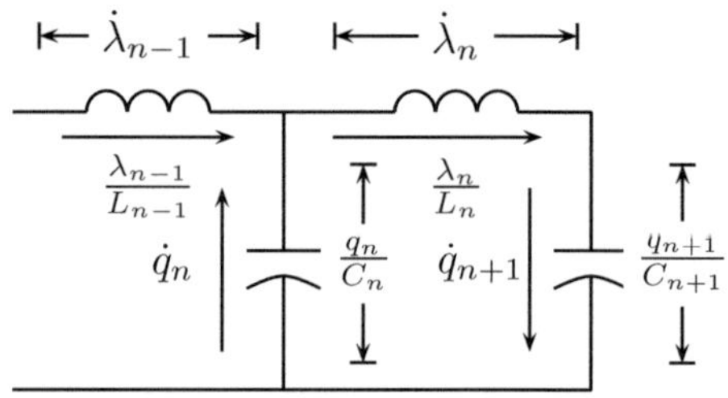

Fig. 16.2 *n*th finite-dimensional element of the transmission line

Reordering the first equation of (16.9) one gets

$$\dot{q}_n = -\frac{L_n(\lambda_{n-1} - \lambda_n) - \lambda_n(L_{n-1} - L_n)}{L_n L_{n-1}}$$

which is an approximation of the partial derivative with respect to the spatial coordinate $z \in [0, 1]$ of $\frac{\lambda(z,t)}{L_{tl}(z)}$, with $L_{tl}(z)$ the distributed inductance. Thus, we have

$$\frac{\partial}{\partial z}\frac{\lambda(z,t)}{L_{tl}(z)} \approx \frac{L_n(\lambda_{n-1} - \lambda_n) - \lambda_n(L_{n-1} - L_n)}{L_n L_{n-1}}$$

Therefore, in the limit of a differential spacing between the finite-dimensional capacitive-inductive circuits, the system of ordinary differential equations becomes a single partial differential equation of the form

$$\frac{\partial}{\partial t}q(z,t) = -\frac{\partial}{\partial z}\frac{\lambda(z,t)}{L_{tl}(z)}$$

In a similar way we obtain for the second equation of (16.9)

$$\frac{\partial}{\partial t}\lambda(z,t) = -\frac{\partial}{\partial z}\frac{q(z,t)}{C_{tl}(z)}$$

where $C_{tl}(z)$ is the distributed capacitance.

In the Dirac structure framework, the model is given as follows. The energy variables are electric charge and magnetic flux, $q_E(t) = q(z,t),\ q_M(t) = \lambda(z,t)$, respectively. The total energy functional is

$$\mathcal{H} = \frac{1}{2}\int_0^{\ell}\left[\frac{q^2(z,t)}{C_{tl}(z)} + \frac{\lambda^2(z,t)}{L_{tl}(z)}\right]\mathrm{d}z$$

with variational derivative given by $\delta\mathcal{H} = \begin{bmatrix} \frac{q(z,t)}{C_{tl}(z)} & \frac{\lambda(z,t)}{L_{tl}(z)} \end{bmatrix}^{\top}$.

Power flows through the boundaries $\{0, l\}$, the boundary variables are current and voltage. Finally, the telegrapher's equations may be expressed as an infinite-dimensional PH system of the form

$$\begin{bmatrix} \frac{\partial}{\partial t} q(z,t) \\ \frac{\partial}{\partial t}\lambda(z,t) \end{bmatrix} = \begin{bmatrix} 0 & -\frac{\partial}{\partial z} \\ -\frac{\partial}{\partial z} & 0 \end{bmatrix} \begin{bmatrix} \frac{q(z,t)}{C_{tl}(z)} \\ \frac{\lambda(z,t)}{L_{tl}(z)} \end{bmatrix}, \begin{bmatrix} f_b \\ e_b \end{bmatrix} = \begin{bmatrix} 0 & -1 \\ 1 & 0 \end{bmatrix} \begin{bmatrix} \frac{q(z,t)}{C_{tl}(z)} \big|_0^{\ell} \\ \frac{\lambda(z,t)}{L_{tl}(z)} \big|_0^{\ell} \end{bmatrix} \tag{16.10}$$

at the boundary points $\{0, \ \ell\}$ one has

$$\begin{bmatrix} e_{M0} \\ e_{E0} \end{bmatrix} = \begin{bmatrix} -\frac{\lambda(0,t)}{L_{tl}(0)} \\ \frac{q(0,t)}{C_{tl}(0)} \end{bmatrix}; \quad \begin{bmatrix} e_{M\ell} \\ e_{E\ell} \end{bmatrix} = \begin{bmatrix} -\frac{\lambda(\ell,t)}{L_{tl}(\ell)} \\ \frac{q(\ell,t)}{C_{tl}(\ell)} \end{bmatrix} \tag{16.11}$$

Telegraphers equations satisfy the following energy balance equation

$$\frac{\mathrm{d}\mathcal{H}}{\mathrm{d}t} = \frac{q(0,t)}{C_{tl}(0)}\frac{\lambda(0,t)}{L_{tl}(0)} - \frac{q(\ell,t)}{C_{tl}(\ell)}\frac{\lambda(\ell,t)}{L_{tl}(\ell)}$$

Since capacitance and inductance have upper and lower bounds on $[0 \ \ell]$, that is,

$$L_m \leq \frac{1}{L_{tl}(z)} \leq L_M, \, C_m \leq \frac{1}{C_{tl}(z)} \leq C_M, \quad L_i, C_i > 0, \quad i = M, m \tag{16.12}$$

the energy balance equation is not singular.

16.3 Control by Interconnection: Mixed Finite- and Infinite-Dimensional Case

The central building block of the Control by Interconnection technique is the existence of the Casimir functions. To explore the possibility of extending Control by Interconnection to the mixed finite- and infinite-dimensional framework, we consider a PH plant connected to a PH controller through an infinite-dimensional system, as illustrated in Fig. 16.3. In order to make clear the interconnection, we will work in power coordinates. The respective models in power coordinates are (16.3) for the plant,

$$\begin{bmatrix} f_c \\ y_c \end{bmatrix} = \begin{bmatrix} 0 & -1 \\ 1 & 0 \end{bmatrix} \begin{bmatrix} e_c \\ u_c \end{bmatrix} \tag{16.13}$$

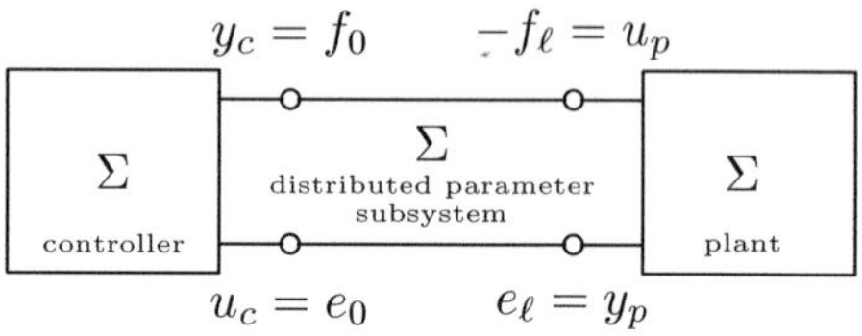

Fig. 16.3 Interconnection constraints

for the controller and (16.6) for the infinite-dimensional system. Hence, the interconnection constraints are

$$\begin{array}{ll} y_c = f_0 & u_c = e_0 \\ y_p = e_\ell & u_p = -f_\ell \end{array} \tag{16.14}$$

In order to get the closed-loop dynamics, we replace the interconnection constraints (16.14) into (16.3), (16.13) and (16.6), and we obtain

$$\begin{aligned} f_s &= -\left[\mathcal{J}(x) - \mathcal{R}(x)\right] e_s + g(x) f_\ell = -\left[\mathcal{J}(x) - \mathcal{R}(x)\right] e_p - g(x) e_{M\ell} \\ f_c &= -e_0 = -e_{E0} \\ f_E &= \frac{\partial}{\partial z} e_M \\ f_M &= \frac{\partial}{\partial z} e_E \\ y_p &= g(x)^T e_s = e_{E\ell} \\ y_c &= e_c = -e_{M0} \end{aligned}$$

where we have considered (16.7). The following equations describe the closed-loop dynamics expressed in energy coordinates,

$$\begin{aligned} \begin{bmatrix} \dot{x} \\ \dot{x}_c \\ \frac{\partial}{\partial t} q_E(z,t) \\ \frac{\partial}{\partial t} q_M(z,t) \end{bmatrix} &= \begin{bmatrix} \mathcal{J}(x) - \mathcal{R}(x) & 0 & 0 & 0 \\ 0 & 0 & 0 & 0 \\ 0 & 0 & 0 & -\frac{\partial}{\partial z} \\ 0 & 0 & -\frac{\partial}{\partial z} & 0 \end{bmatrix} \begin{bmatrix} \frac{\partial}{\partial x} H(x) \\ \frac{\partial}{\partial x_c} H_c(x_c) \\ \delta_M \mathcal{H}(\bar{q}) \\ \delta_E \mathcal{H}(\bar{q}) \end{bmatrix} \\ &+ \begin{bmatrix} g(x) & 0 \\ 0 & 1 \\ 0 & 0 \\ 0 & 0 \end{bmatrix} \begin{bmatrix} \delta_M \mathcal{H}(\bar{q}) \mid_\ell \\ \delta_E \mathcal{H}(\bar{q}) \mid_0 \end{bmatrix} \\ \begin{bmatrix} \delta_E \mathcal{H}(\bar{q}) \mid_\ell \\ \delta_M \mathcal{H}(\bar{q}) \mid_0 \end{bmatrix} &= \begin{bmatrix} g(x)^\top \frac{\partial}{\partial x} H(x) \\ -\frac{\partial}{\partial x_c} H_c(x_c) \end{bmatrix} \end{aligned} \tag{16.15}$$

In the extended space $\chi = [x, x_c, q_E(z,t), q_M(z,t)]^\top$, the closed-loop total energy function is as follows:

$$H_{cl}(\chi) = H(x) + H_c(x_c) + \mathcal{H}(\bar{q})$$

with the energy rate equals to

$$\dot{H}_{cl}(\chi) = -\frac{\partial^\top H}{\partial x}(x)\mathcal{R}(x)\frac{\partial H}{\partial x}(x)$$

16.3.1 Casimir Functions

Casimir functions are dynamic invariants [5]. Hence, a function $\mathcal{C}(\chi)$ will be a Casimir function provided the following holds

$$\begin{bmatrix} 0 \\ 0 \\ 0 \\ 0 \end{bmatrix} = \begin{bmatrix} -\mathcal{J}(x) - \mathcal{R}(x) & 0 & 0 & 0 \\ 0 & 0 & 0 & 0 \\ 0 & 0 & 0 & \frac{\partial}{\partial z} \\ 0 & 0 & \frac{\partial}{\partial z} & 0 \end{bmatrix} \begin{bmatrix} \frac{\partial}{\partial x}\mathcal{C}(\chi) \\ \frac{\partial}{\partial x_c}\mathcal{C}(\chi) \\ \delta_M \mathcal{C}(\chi) \\ \delta_E \mathcal{C}(\chi) \end{bmatrix} + \begin{bmatrix} -g(x) & 0 \\ 0 & -1 \\ 0 & 0 \\ 0 & 0 \end{bmatrix} \begin{bmatrix} \delta_M \mathcal{C}(\chi)\mid_\ell \\ \delta_E \mathcal{C}(\chi)\mid_0 \end{bmatrix}$$
$$\begin{bmatrix} \delta_E \mathcal{C}(\chi)\mid_\ell \\ \delta_M \mathcal{C}(\chi)\mid_0 \end{bmatrix} = \begin{bmatrix} g^\top(x)\frac{\partial}{\partial x}\mathcal{C}(\chi) \\ -\frac{\partial}{\partial x_c}\mathcal{C}(\chi) \end{bmatrix} \tag{16.16}$$

From the third and fourth relations of (16.16), it follows that every Casimir function of (16.15) should be linear with respect to the spatial variables, that is,

$$\begin{aligned} \delta_M \mathcal{C}(\chi) &= \text{constant as a function of } z \\ \delta_E \mathcal{C}(\chi) &= \text{constant as a function of } z \end{aligned} \tag{16.17}$$

then, one has

$$\begin{aligned} \delta_M \mathcal{C}(\chi) = \delta_M \mathcal{C}(\chi)\mid_0 = \delta_M \mathcal{C}(\chi)\mid_\ell &= -\frac{\partial}{\partial x_c}\mathcal{C}(\chi) \\ \delta_E \mathcal{C}(\chi) = \delta_E \mathcal{C}(\chi)\mid_0 = \delta_E \mathcal{C}(\chi)\mid_\ell &= g(x)^T \frac{\partial}{\partial x}\mathcal{C}(\chi) \end{aligned} \tag{16.18}$$

Now, by replacing (16.18) into (16.16) and taking into account (16.17), the conditions for a Casimir function (16.16) reduce to

$$\begin{bmatrix} \mathcal{J}(x) + \mathcal{R}(x) & -g(x) \\ g^\top(x) & 0 \end{bmatrix} \begin{bmatrix} \frac{\partial}{\partial x}\mathcal{C}(\chi) \\ \frac{\partial}{\partial x_c}\mathcal{C}(\chi) \end{bmatrix} = 0$$
$$\begin{bmatrix} \delta_E \mathcal{C}(\chi) \\ \delta_M \mathcal{C}(\chi) \end{bmatrix} = \begin{bmatrix} g^\top(x)\frac{\partial}{\partial x}\mathcal{C}(\chi) \\ -\frac{\partial}{\partial x_c}\mathcal{C}(\chi) \end{bmatrix}$$

The right Casimir functions for the Control by Interconnection method are the ones that relate the energy coordinates of the interconnected system. In particular, we consider Casimir functions of the form

$$\mathcal{C}(\chi) = -x_c + F(x) + \mathcal{F}(\bar{q}(z,t)) \tag{16.19}$$

Hence, we have

Proposition 16.3 *The functions $-x_c + F(x) + \mathcal{F}(\bar{q}(z,t)$ are Casimir functions of the interconnected PH system (16.15) if and only if the function $F(x)$ satisfies*

$$\mathcal{R}(x)\frac{\partial}{\partial x}F(x) = 0, \ \frac{\partial}{\partial x}^{\top}F(x)\mathcal{J}(x) = g^{\top}(x). \tag{16.20}$$

and the functional $\mathcal{F}(\bar{q}(z,t))$ satisfies (16.17) and

$$\delta_E\mathcal{F}(\bar{q}(z,t)) = 0, \ \delta_M\mathcal{F}(\bar{q}(z,t)) = 1 \tag{16.21}$$

if $y_p = e_\ell$ or

$$\delta_E\mathcal{F}(\bar{q}(z,t)) = 1, \ \delta_M\mathcal{F}(\bar{q}(z,t)) = 0 \tag{16.22}$$

if $y_p = f_\ell$.

Before closing this section, we analyze the interconnection shown in Fig. 16.4. In this system, a finite-dimensional subsystem with two external ports, modeled as

$$\begin{aligned} f_p &= -\left[\mathcal{J}(x) - \mathcal{R}(x)\right]e_s - g_1(x)u_{p1} - g_2(x)u_{p2} \\ y_{p1} &= g_1^{\top}(x)e_s, \quad y_{p2} = g_2^{\top}(x)e_s \end{aligned}$$

has an external port connected to the controller and the other one connected to an infinite-dimensional subsystem. In this case, the infinite-dimensional subsystem represents a non-modeled dynamics, and the interconnection constraints are as follows:

$$\begin{bmatrix} f_0 \\ e_0 \end{bmatrix} = \begin{bmatrix} 0 & 1 \\ -1 & 0 \end{bmatrix}\begin{bmatrix} y_{p2} \\ u_{p2} \end{bmatrix}, \ \begin{bmatrix} u_{p1} \\ u_c \end{bmatrix} = \begin{bmatrix} 0 & -1 \\ 1 & 0 \end{bmatrix}\begin{bmatrix} y_{p1} \\ y_c \end{bmatrix}$$

For this system, it is not possible to find Casimir functions of the form (16.19). Thus, we cannot apply the Control by Interconnection methodology.

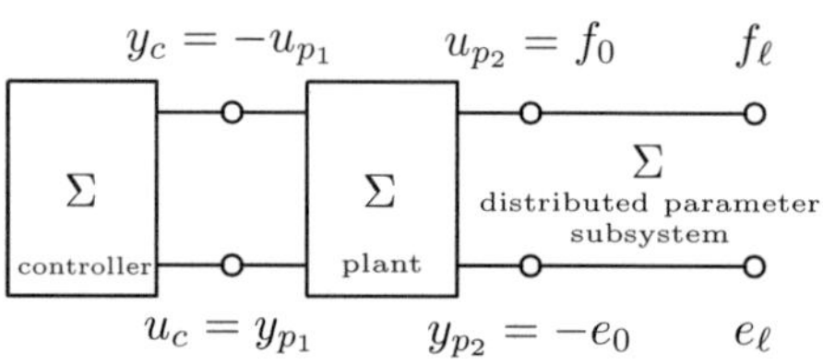

Fig. 16.4 Interconnection constraints

16.3.2 Control Design

The idea behind the stability argument for infinite-dimensional systems is the same as for finite-dimensional systems in that we wish to show that the equilibrium solution corresponds to a strict extremum of the total energy. In the case of infinite-dimensional systems, care must be taken to specify the norm associated with the stability argument. Since, stability with respect to one norm does not necessarily imply stability with respect to other.[3]

In the case of mixed lumped and distributed parameter systems, we will define stability in the sense of Lyapunov as follows:

Definition 16.4 The equilibrium point χ_{r*} of a distributed parameters system is said to be stable in the sense of Lyapunov with respect to the norm $\| \cdot \|$, if for every $\varepsilon > 0$ there exist $\delta > 0$ such that $\| \chi_r(0) - \chi_{r*} \| < \delta \Rightarrow \| \chi_r - \chi_{r*} \| < \varepsilon$ for all $t > 0$, where $\chi_r(0)$ is the initial condition of χ_r.

The underlying mathematical procedure for a proof of linear stability in the sense of Lyapunov based on the Hamiltonian structure of a mixed lumped and distributed parameters system can be summarized in the following steps [11]:

1. Define the total energy of the interconnected system restricted to the state space χ_r as the candidate Lyapunov function.

$$H_d(\chi_r) = H(x) + \mathcal{H}(\bar{q}(z,t)) + H_c(F(x) + \mathcal{F}(\bar{q}(z,t))),$$

with $H_c(\cdot)$ to be defined.
2. Show that the equilibrium point χ_{r*} satisfies the first-order necessary conditions to be an extremum of the candidate Lyapunov function, that is,

$$\nabla H_d(\chi_{r*}) \overset{\text{def}}{=} \begin{bmatrix} \frac{\partial}{\partial x}\left[H(x_*) + H_c(F(x_*) + \mathcal{F}(\bar{q}_*))\right] \\ \delta_M \left[\mathcal{H}(\bar{q}_*) + H_c(F(x_*) + \mathcal{F}(\bar{q}_*))\right] \\ \delta_E \left[\mathcal{H}(\bar{q}_*) + H_c(F(x_*) + \mathcal{F}(\bar{q}_*))\right] \end{bmatrix} \equiv 0 \tag{16.23}$$

3. Introduce the nonlinear functional

$$\mathcal{N}(\Delta\chi_r) = H_d(\chi_{r*} + \Delta\chi_r) - H_d(\chi_{r*}) \tag{16.24}$$

proportional to the second variation of $H_d(\chi_r)$ in the sense that its Taylor expansion about $\Delta\chi_r$ is given by

$$\mathcal{N}(\Delta\chi_r) \approx \frac{1}{2}\nabla^2 H_d(\chi_{r*})$$

[3] In an infinite-dimensional space not every convergent sequence on the unit, ball converges to a point on the unit ball. Unit balls on infinite-dimensional spaces need not to be compact [11].

and determine the convexity conditions, with respect to a suitable norm, that the functional (16.24) must satisfy to assure that is definite, that is,

$$c_1 \parallel \Delta\chi_r \parallel^2 \leq H_d(\chi_{r*} + \Delta\chi_r) - H_d(\chi_{r*}) \leq c_2 \parallel \Delta\chi_r \parallel^2 \tag{16.25}$$

with $c_1, c_2 > 0$. In our context, an appropriate choice of the norm is

$$\parallel \Delta\chi_r \parallel = \left(\mid \Delta x \mid^2 + \int_0^\ell \Delta q_M^2(z,t)\mathrm{d}z + \int_0^\ell \Delta q_E^2(z,t)\mathrm{d}z \right)^{\frac{1}{2}}$$

with $\mid \cdot \mid$ the usual Euclidean norm.

16.3.3 Example: RLC with a Transmission Line

In this section, we illustrate the Control by Interconnection method of mixed finite- and infinite-dimensional systems. We consider a finite-dimensional controller connected to a finite-dimensional plant, an RCL circuit through an infinite-dimensional system, the transmission line, as shown in Fig. 16.5.

The energy coordinates for the interconnected system are

$$\chi = [x_1, x_2, x_c, q(z,t), \lambda(z,t)]^\top$$

with x_1 the capacitor's electric charge, x_2 the inductor's magnetic flux, x_c the controller state, $q(z,t)$, $\lambda(z,t)$ the distributed electric charge and $\lambda(z,t)$ the distributed magnetic charge. Additionally, the power preserving interconnection constraints are given by

$$\begin{aligned} y_c = -V(0,t) = -\frac{q(0,t)}{C_{tl}}, \; u_c = I(0,t) = -\frac{\lambda(0,t)}{L_{tl}}, \\ y_p = I(\ell,t) = -\frac{\lambda(\ell,t)}{C_{tl}}, \quad u_p = V(\ell,t) = -\frac{q(\ell,t)}{L_{tl}} \end{aligned} \tag{16.26}$$

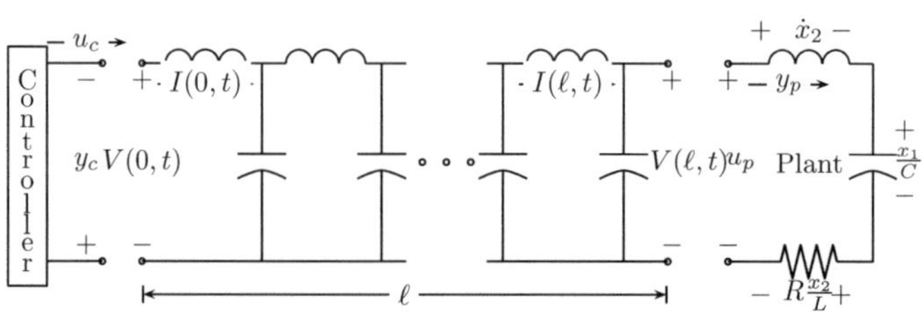

Fig. 16.5 Finite-dimensional controller, infinite-dimensional transmission line, finite-dimensional plant

with C_{tl} and L_{tl} the constants capacitance and inductance of the transmission line, respectively. Note that the interconnection constraints are well-defined in the sense that voltages couple with voltages and the same for currents.

The following equations describe the closed-loop dynamics in energy variables

$$\begin{bmatrix} \dot{x}_1 \\ \dot{x}_2 \\ \dot{x}_c \\ \frac{\partial}{\partial t} q(z,t) \\ \frac{\partial}{\partial t} \lambda(z,t) \end{bmatrix} = \begin{bmatrix} 0 & 1 & 0 & 0 & 0 \\ -1 & -R & 0 & 0 & 0 \\ 0 & 0 & 0 & 0 & 0 \\ 0 & 0 & 0 & 0 & -\frac{\partial}{\partial z} \\ 0 & 0 & 0 & -\frac{\partial}{\partial z} & 0 \end{bmatrix} \begin{bmatrix} \frac{x_1}{C} \\ \frac{x_2}{L} \\ \frac{\partial H_c}{\partial x_c}(x_c) \\ \frac{q(z,t)}{C_{tl}} \\ \frac{\lambda(z,t)}{L_{tl}} \end{bmatrix} + \begin{bmatrix} 0 & 0 \\ 1 & 0 \\ 0 & -1 \\ 0 & 0 \\ 0 & 0 \end{bmatrix} \begin{bmatrix} \frac{q(\ell,t)}{C_{tl}} \\ \frac{\lambda(0,t)}{L_{tl}} \end{bmatrix}$$

$$\begin{bmatrix} \frac{\lambda(\ell,t)}{L_{tl}} \\ \frac{q(0,t)}{C_{tl}} \end{bmatrix} = \begin{bmatrix} -\frac{x_2}{L} \\ -\frac{\partial H_C}{\partial x_c}(x_c) \end{bmatrix} \tag{16.27}$$

The total closed-loop energy function is defined by

$$H_{cl}(\chi) = \frac{1}{2}\frac{x_1^2}{C} + \frac{1}{2}\frac{x_2^2}{L} + H_c(x_c) + \frac{1}{2}\int_0^\ell \left(\frac{q^2(z,t)}{C_{tl}(z)} + \frac{\lambda^2(z,t)}{L_{tl}(z)} \right) \mathrm{d}z$$

with energy rate given by

$$\dot{H}_{cl}(\chi) = -R\left(\frac{x_2}{L}\right)^2$$

In order to apply the Control by Interconnection methodology, we will find the Casimir functions of the form (16.19) for the closed-loop dynamics (16.27). To perform this task, we follow the result in Proposition 16.3. It is easy to verify that

$$\mathcal{C}(\chi) = -x_c + x_1 + \int_0^\ell q(z,t)\,\mathrm{d}z$$

is a Casimir function for (16.27). This Casimir function defines the following invariant set $\Omega = \{\chi \mid x_c = x_1 + \int_0^\ell q(z,t)\,\mathrm{d}z\}$. Thus, the total energy function restricted to Ω takes the following form

$$H_d(\chi_r) = \frac{1}{2}\frac{x_1^2}{C} + \frac{1}{2}\frac{x_2^2}{L} + H_c\left(x_1 + \int_0^\ell q(z,t)\,\mathrm{d}z \right) + \frac{1}{2}\int_0^\ell \left(\frac{q^2(z,t)}{C_{tl}(z)} + \frac{\lambda^2(z,t)}{L_{tl}(z)} \right) \mathrm{d}z$$

with $\chi_r = [x_1, x_2, q(z,t), \lambda(z,t)]^\top$. In the following we will show that by selecting

$$H_c(x_c) = \frac{1}{2}\frac{1}{C_c}\tilde{x}_c^2 + k\tilde{x}_c,\ \ \tilde{x}_c = x_c - x_{c*}$$

with $C_c > 0$, $k \in \mathbb{R}$ and x_{c*} the equilibrium point for the state of the controller, we can shape the closed-loop energy in such a way that it has a minimum at the equilibrium point $\chi_{r*} = \left[\frac{x_{1*}}{C}, 0, q_*(z), \lambda_*(z)\right]^\top$. From the first-order conditions (16.23) one has

$$\nabla H_d(\chi_{r*}) = \begin{bmatrix} \frac{x_{1*}}{C} + k \\ 0 \\ \frac{q_*(z)}{C_{tl}} + k \\ \frac{\lambda_*(z)}{L_{tl}} \end{bmatrix} \equiv 0 \Rightarrow \begin{array}{rcl} k &=& -\frac{x_{1*}}{C} \\ q_*(z) &=& C_{tl}\frac{x_{1*}}{C} \\ \lambda_*(z) &=& 0 \end{array} \tag{16.28}$$

In order to verify the second-order conditions, we compute the functional (16.24) to this end we get

$$\begin{aligned} \mathcal{N}(\Delta\chi_r) = &\frac{1}{2}\frac{\Delta x_1^2}{C} + \frac{1}{2}\frac{\Delta x_2^2}{L} + \int_0^\ell \left(\frac{\Delta q^2(z,t)}{C_{tl}(z)} + \frac{\Delta\lambda^2(z,t)}{L_{tl}(z)}\right) \mathrm{d}z \\ &+\frac{1}{2}\frac{1}{C_c}\left(\Delta x_1 + \int_0^\ell \Delta q(z,t)\,\mathrm{d}z\right)^2 \end{aligned} \tag{16.29}$$

Now, we verify condition (16.25) with respect to the following norm

$$\| \chi_r \| = \left(\Delta x_1^2 + \Delta x_2^2 + \int_0^\ell \Delta q^2(z,t)\,\mathrm{d}z + \int_0^\ell \Delta\lambda^2(z,t)\,\mathrm{d}z\right)^{\frac{1}{2}}$$

It is easy to see that, we can find $c_{2i}, c_{\lambda i}, i = 1, 2$ which satisfy

$$\begin{gathered} c_{21}\Delta x_2^2 \leq \frac{1}{2}\frac{\Delta x_2^2}{C} \leq c_{22}\Delta x_2^2 \\ c_{\lambda 1}\int_0^\ell \Delta\lambda^2(z,t)\,\mathrm{d}z \leq \int_0^\ell \frac{\Delta\lambda^2(z,t)}{L_{tl}}\,\mathrm{d}z \leq c_{\lambda 2}\int_0^\ell \Delta\lambda^2(z,t)\,\mathrm{d}z \end{gathered}$$

therefore condition (16.25) reduces to find $c_{ai}, i = 1, 2$ such that

$$c_{a1} \| \Delta\chi_r \|^2 \leq \frac{\Delta x_1^2}{2C} + \int_0^\ell \frac{\Delta q^2(z,t)}{C_{tl}}\,\mathrm{d}z + \frac{\left(\Delta x_1 + \int_0^\ell \Delta q(z,t)\,\mathrm{d}z\right)^2}{2C_c} \leq c_{a2} \| \Delta\chi_r \|^2$$

Let us first consider the upper bound, for which we have

$$\begin{aligned} \mathcal{N}_r(\Delta\chi_r) \leq &\frac{1}{2}\frac{\Delta x_1^2}{C} + C_M\int_0^\ell \Delta q^2(z,t)\,\mathrm{d}z \\ +\frac{1}{2}\frac{1}{C_c}&\left[\Delta x_1^2 + 2\mid \Delta x_1 \mid\mid \int_0^\ell \Delta q(z,t)\,\mathrm{d}z \mid + \left(\int_0^\ell \Delta q(z,t)\,\mathrm{d}z\right)^2\right] \\ \leq &\left(\frac{1}{2C} + \frac{1}{C_c}\right)\Delta x_1^2 + \left(C_M + \frac{1}{C_c}\ell\right)\int_0^\ell \Delta q^2(z,t)\,\mathrm{d}z \end{aligned}$$

where we considered the relationship in (16.12),

$$| \Delta x_1 || \int_0^\ell \Delta q(z,t)\,\mathrm{d}z | \leq \frac{1}{2} | \Delta x_1 |^2 + \frac{1}{2} | \int_0^\ell \Delta q(z,t)\,\mathrm{d}z |^2$$

and

$$\left(\int_0^\ell \Delta q(z,t)\,\mathrm{d}z \right)^2 \leq \ell \int_0^\ell \Delta q^2(z,t)\,\mathrm{d}z$$

Therefore by choosing

$$c_{a2} > \max \left\{ \frac{1}{2C} + \frac{1}{C_c}, C_M + \frac{1}{C_c}\ell \right\}$$

we get an upper bound. Now, for the lower bound, we have that

$$\mathcal{N}_r(\Delta\chi) \geq \frac{1}{2}\frac{\Delta x_1^2}{C} + C_m \int_0^\ell \Delta q^2(z,t)\,\mathrm{d}z$$

therefore selecting

$$c_{a1} \leq \min \left\{ \frac{1}{2C}, C_m \right\}$$

we get a lower bound. Finally, we have

$$\begin{aligned} c_1 &\stackrel{\text{def}}{=} \min \left\{ \tfrac{1}{2C}, C_m, c_{21}, c_{\lambda 1} \right\} \\ c_2 &\stackrel{\text{def}}{=} \max \left\{ \tfrac{1}{2C} + \tfrac{1}{C_c}, C_M + \tfrac{1}{C_c}\ell, c_{22}, c_{\lambda 2} \right\} \end{aligned}$$

Hence, we have proved.

Proposition 16.5 *Consider the closed-loop dynamics described by (16.27) with the PH controller given by*

$$\begin{aligned} \dot{x}_c &= -\tfrac{\lambda(0,t)}{L_{tl}} \\ y_c &= \frac{1}{C_c}x_c - \left(\frac{1}{C_c} + \frac{1}{C} \right) x_{1*} - \frac{x_{1*}}{CC_c} \int_0^\ell C_{tl}\,\mathrm{d}z \end{aligned} \tag{16.30}$$

The closed-loop dynamics has an stable equilibrium in the sense of Definition 16.4 at

$$\chi_* = \left[\frac{x_{1*}}{C},\ 0,\ x_{1*} + \frac{x_{1*}}{C} \int_0^\ell C_{tl}(z)\,\mathrm{d}z,\ \frac{x_{1*}}{C} C_{tl}(z),\ 0 \right]^\top$$

16.4 Conclusions

We have presented a few more details of the Control by Interconnection of mixed finite- and infinite-dimensional systems reported in [10]. It is glad for the authors that the results presented in [10] effectively were a fundamental step for many other works on control of infinite-dimensional PH systems.

In the last years, my research focused on control of aerial vehicles. Aerial vehicles are subject to aerodynamic forces which are modeled using dimensional analysis. Hence, the aerodynamic forces have the form $F = 1/2\rho S V^2 C_F$ with F the aerodynamic force, ρ the air density, S a characteristic area of the body and C_F the force coefficient. This model does not allow to obtain a natural[4] PH model for this kind of vehicles since the model of forces has not physics behind it. However, these aerodynamic forces come from the interaction between the air flow and the vehicle geometry which obeys Navier–Stokes equations. Hence, we have an infinite-dimensional system, aerodynamic forces, connected to a finite-dimensional system, the vehicle dynamics. The work of Prof. van der Shaft on model and control of PH systems undoubtedly inspires this future work.

References

1. S. Arimoto, M. Takegaki, A new feedback method for dynamic control of manipulators. J. Dyn. Syst. Meas. Control **102**, 119–125 (1981)
2. M. Dalsmo, A. van der Schaft, On representations and integrability of mathematical structures in energy-conserving physical systems. SIAM J. Control Optim. **37**(1), 54–91 (1998)
3. D.C. Karnnop, D.L. Margolis, R. Rosenberg. *System dynamics* (Wiley, New York, 2000)
4. D.G. Luenberger, *Optimization by Vector Space Methods* (Wiley, New York, 1968)
5. J.E. Marsden, T.S. Ratiu, *Introduction to Mechanics and Symmetry: A Basic Exposition of Classical Mechanical Systems*, vol. 17 (Springer Science & Business Media, 1999)
6. B.M.J. Maschke, A.J. van der Schaft, Port Controlled Hamiltonian Representation of Distributed Parameter Systems (2000)
7. R. Ortega, A. Loria, P.J. Nicklasson, H. Sira-Ramirez, *Passivity-based Control of Euler-lagrange Systems: Mechanical, Electrical and Electromechanical Applications* (Springer Science & Business Media, 1998)
8. R. Ortega, A. van der Schaft, B. Maschke, G. Escobar, Energy-shaping of Port-controlled Hamiltonian Systems by Interconnection, in *Proceedings of the 38th IEEE Conference on Decision and Control*, vol. 2, IEEE, 1999, pp. 1646–1651
9. R. Ortega, A. van der Schaft, B. Maschke, G. Escobar, Interconnection and damping assignment passivity-based control of port-controlled hamiltonian systems. Automatica **38**(4), 585–596 (2002)
10. H. Rodríguez, A.J. van der Schaft, R. Ortega, On Stabilization of Nonlinear Distributed Parameter Port-controlled Hamiltonian Systems via Energy Shaping, in *Proceedings of the 40th IEEE Conference on Decision and Control*, vol. 1, IEEE, 2001, pp. 131–136
11. G.E. Swaters, *Introduction to Hamiltonian Fluid Dynamics and Stability Theory*, vol. 102 (CRC Press, 1999)
12. A. van der Schaft, A.J. Schaft. *L2-gain and Passivity in Nonlinear Control* (Springer, New York Inc, 1999)

[4]Natural in the sense that the energy function is the Hamiltonian function.

Chapter 17
Network Topology and Synchronization of Systems with Linear Time-Delayed Coupling

Erik Steur and Henk Nijmeijer

Abstract We consider networks of square input–output systems that interact via linear, time-delayed coupling functions. For given system dynamics, we give conditions for the construction of a (local, global) synchronization diagram. We show that a condition for (local, global) synchronization is that the coupling strength and time-delay are contained in the intersection of scaled copies of the (local, global) synchronization diagram, where the scaling factors are the nonzero eigenvalues of the symmetric Laplacian matrix.

17.1 Introduction

There are many examples of networks of interacting dynamical systems that exhibit collective behavior: Fireflies emit their light pulses at the same instants in time; crickets chirp in unison for extended periods of time; and the electrons move coherently in (arrays of) superconductive Josephson junctions, cf. [22, 30]. The most unambiguous form of collective behavior is that of *synchronization*, which refers to the state in which all systems in the network behave identically. Whether or not a network of systems will synchronize depends on, besides the specific systems' dynamics and coupling functions, the network topology. In this chapter, we consider networks of systems that interact via linear time-delay coupling functions of the form

$$u_i(t) = \sigma \sum_j a_{ij}[y_j(t-\tau) - y_i(t-\tau)] \tag{17.1}$$

E. Steur · H. Nijmeijer (✉)
Institute for Complex Molecular Systems and Department of Mechanical Engineering, Eindhoven University of Technology, P.O. Box 513, 5600 MB Eindhoven, The Netherlands
e-mail: e.steur@tue.nl

H. Nijmeijer
Department of Mechanical Engineering, Eindhoven University of Technology, P.O. Box 513, 5600 MB Eindhoven, The Netherlands
e-mail: h.nijmeijer@tue.nl

M.K. Camlibel et al. (eds.), *Mathematical Control Theory I*,
Lecture Notes in Control and Information Sciences 461,
DOI 10.1007/978-3-319-20988-3_17

and we relate conditions for synchronization of the systems to the topology of the network. In (17.1) $u_i(t)$ is the input of system i, $y_i(t-\tau)$ and $y_j(t-\tau)$ are the time-delayed outputs of systems i and j, respectively, positive constant σ is the coupling strength, and positive constants a_{ij} are defined by the network. The time-delay τ accounts for sensor and actuator dynamics, in particular, sensor and actuator delays. Such coupling functions appear in, e.g., car-following models [26], where the time-delay, which correlates with the reaction time of the driver, typically takes values between 0.6 and 2 s.

In the delay-free case, i.e., $\tau = 0$, the influence of network topology on synchronization has been studied in [2, 3, 21, 33]. In [33] a conjecture was posed that states that systems in network $\mathcal{G}_1$ synchronize for coupling strength σ_1 if and only if systems in network $\mathcal{G}_2$ synchronize for coupling strength σ_2 and the following relation holds:

$$\sigma_1\lambda_2(\mathcal{G}_1) = \sigma_2\lambda_2(\mathcal{G}_2),$$

where constant $\lambda_2(\mathcal{G})$ is the algebraic connectivity of network $\mathcal{G}$ (i.e., the Fiedler eigenvalue of the Laplacian matrix of $\mathcal{G}$) [9]. Although this conjecture was shown to be wrong [20], there is a rich class of systems for which the conjecture seems to hold true, i.e., for those systems that do not show a desynchronizing bifurcation as the coupling strength is increased. A somewhat similar method was proposed in [21], in which the concept of a *Master Stability Function* (MSF) was introduced. In this approach, the coupling parameters (i.e., coupling strength and network topology) are lumped into a single (possibly complex) parameter κ, and subsequently the stability of a linear time-varying system that describes the local dynamics around a synchronous solution is assessed as function of this parameter κ. Then if there exists a nonempty set $\mathcal{K}$ such that for $\kappa \in \mathcal{K}$ the zero solution of this linear system is stable, the condition for synchronization of a network $\mathcal{G}$ is that $\sigma\lambda_j(\mathcal{G}) \in \mathcal{K}$ for all nonzero eigenvalues λ_j of the Laplacian matrix of $\mathcal{G}$. However, it is shown in [15] that the MSF approach might fail if the isolated system (i.e., a single system without coupling) does not have an attractor. Assuming the isolated system to have an attractor might even not be sufficient to conclude that the systems synchronize; It is known that with negative Lyapunov exponents, the criteria used for stability of the MSF, a linear time-varying system may be unstable [14]. In particular, it is shown in [1, 31] that the dynamics of coupled chaotic systems might produce a specific type of intermittent behavior associated with a temporal loss of synchrony; This phenomenon, called attractor bubbling, may occur despite the Lyapunov exponents of the MSF all being negative.

In this chapter we develop a MSF-like approach, which allows the construction of a *local synchronization diagram* $\mathcal{S}$; This local synchronization diagram is the set of coupling strengths σ and time-delays τ for which the zero solution of a particular linear time-varying system is uniformly asymptotically stable. Under the assumption that the isolated system has an attractor with a neighborhood with inflowing boundary, we show that the condition for local synchronization, that is, synchronization of systems whose mutual distance in initial data is small, is that the coupling strength σ and time-delay τ are in the intersection of scaled copies of $\mathcal{S}$. Here the scaling factors are the nonzero eigenvalues of the Laplacian matrix of the network $\mathcal{G}$. See Fig. 17.1

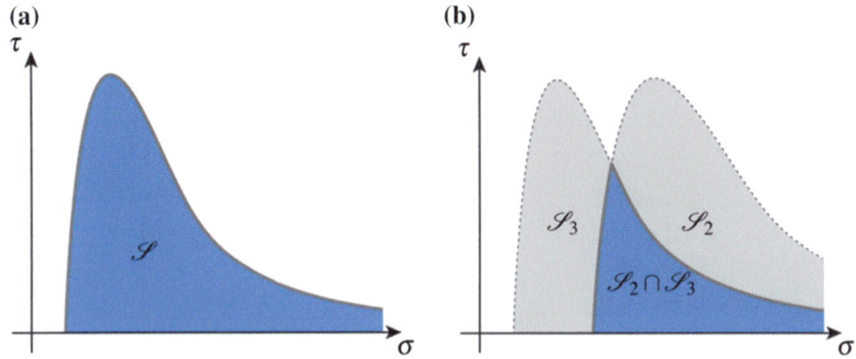

Fig. 17.1 **a** Synchronization diagram $\mathcal{S}$. **b** Two scaled copies of $\mathcal{S}$, denoted by $\mathcal{S}_2$ and $\mathcal{S}_3$, and their intersection

for a graphical example for a network of three systems, where we have assumed the network to be connected and the eigenvalues of the Laplacian matrix of that network to be real. (Under the assumption that a network is connected its Laplacian matrix has a simple zero eigenvalue.) In addition, we present a class of systems for which we are able to construct a global synchronization diagram. The intersection of scaled copies of this global synchronization diagram gives the conditions on σ and τ for which a network of systems synchronizes without requiring the mutual distances in initial data to be small.

The results we present in this chapter are, in part, reported in [27].

Notation We let $\mathbb{R} = (-\infty, \infty)$ denote the real numbers, $\mathbb{R}_+ := \{x \in \mathbb{R} \,|\, x > 0\}$ and $\overline{\mathbb{R}}_+ := \mathbb{R}_+ \cup \{0\}$. For a positive integer n, $\mathbb{R}^n$ is the n-fold Cartesian product $\mathbb{R} \times \mathbb{R} \times \cdots \times \mathbb{R}$. We let $|\cdot|$ be the Euclidean norm in $\mathbb{R}^n$: for $x \in \mathbb{R}^n$, $|x| = \sqrt{x^\top x}$ where $^\top$ denotes transposition. We denote by $\otimes$ the Kronecker (tensor) product of two matrices (cf. [13]). We let I_n be the $n \times n$ identity matrix, and $\mathbf{1}_n$ (respectively, $\mathbf{0}_n$) the n-dimensional vector with all entries equal to 1 (respectively, 0). For an $n \times n$-dimensional matrix A we let $\|A\| := \max_{|x|=1} |Ax|$ be the matrix norm induced by $|\cdot|$. Given two sets $\mathcal{X}$ and $\mathcal{Y}$, $\mathcal{C}(\mathcal{X}, \mathcal{Y})$ denotes the set of continuous functions that map $\mathcal{X}$ into $\mathcal{Y}$.

17.2 Problem Setting

Let $\mathcal{G} = (\mathcal{V}, \mathcal{E}, A)$ be an undirected weighted graph with $\mathcal{V} = \{1, 2, \ldots, N\}$ the set of vertices and $\mathcal{E} \subset \mathcal{V} \times \mathcal{V}$ the set of edges. Recall that $\mathcal{G}$ being an undirected graph means that $\mathcal{E}$ is unordered. $A = \left(a_{ij}\right)$ is the $N \times N$ weighted adjacency matrix:

$$a_{ij} = \begin{cases} w_{ij} & \text{if } (i, j) \in \mathcal{E} \\ 0 & \text{otherwise} \end{cases}$$

where w_{ij} is the weight of edge $(i, j) \in \mathcal{E}$. We suppose that $w_{ij} = w_{ji}$ such that A is symmetric. We shall assume that $\mathcal{G}$ contains no self-loops (i.e., $\mathcal{G}$ has no edges of the form (i, i)) and thus $\mathcal{G}$ is a simple graph. In addition, we shall assume that $\mathcal{G}$ is connected, that is, for every two vertices $i, j \in \mathcal{V}$ there exists a path between i and j.

Letting

$$D = \begin{pmatrix} d_1 & & & \\ & d_2 & & \\ & & \ddots & \\ & & & d_N \end{pmatrix} = \begin{pmatrix} \sum_j a_{1j} & & & \\ & \sum_j a_{2j} & & \\ & & \ddots & \\ & & & \sum_j a_{Nj} \end{pmatrix}$$

we define

$$L = D - A$$

to be the Laplacian matrix of $\mathcal{G}$. It is well-known that the Laplacian matrix of a connected graph has a simple zero eigenvalue, cf. [4]. Gerschgorin's Disc Theorem [13] implies that all other eigenvalues (which are real as L is symmetric) are positive. We always order the eigenvalues $\lambda_1, \lambda_2, \ldots, \lambda_N$ of L nondecreasingly

$$0 = \lambda_1 < \lambda_2 \leq \cdots \leq \lambda_N.$$

We assign each vertex $i \in \mathcal{V}$ the dynamics

$$\begin{cases} \dot{x}_i(t) = f(x_i(t)) + Bu_i(t) \\ y_i(t) = Cx_i(t) \end{cases} \tag{17.2}$$

with state $x_i(t) \in \mathbb{R}^n$, input $u_i(t) \in \mathbb{R}^m$ and output $y_i(t) \in \mathbb{R}^m$, $1 \leq m \leq n$, (sufficiently) smooth vectorfield $f : \mathbb{R}^n \to \mathbb{R}^n$, and matrices B and C of appropriate dimensions with CB similar to a positive definite matrix. Systems (17.2) on $\mathcal{G}$ interact via the following linear time-delay coupling law

$$u_i(t) = \sigma \sum_{j \in \mathcal{N}_i} a_{ij}[y_j(t - \tau) - y_i(t - \tau)], \tag{17.3}$$

where positive constant σ is the coupling strength, nonnegative constant τ is a time-delay, and

$$\mathcal{N}_i = \{j \in \mathcal{V} \mid (i, j) \in \mathcal{E}\}$$

is the set of neighbors of system i. Then the dynamics of the coupled systems (17.2) and (17.3) are given by the following delay-differential equation

$$\dot{x}(t) = F(x(t)) - \sigma(L \otimes BC)x(t - \tau) \tag{17.4}$$

where

$$x(t) = \begin{pmatrix} x_1(t) \\ x_2(t) \\ \vdots \\ x_N(t) \end{pmatrix}, \quad F(x(t)) = \begin{pmatrix} f(x_1(t)) \\ f(x_2(t)) \\ \vdots \\ f(x_N(t)) \end{pmatrix}.$$

The state-space of (17.4) is $\mathcal{C} = \mathcal{C}\left([-\tau, 0], \mathbb{R}^{Nn}\right)$, the space of continuous functions that map the interval $[-\tau, 0]$ into $\mathbb{R}^{Nn}$. For $\phi \in \mathcal{C}$ we let $\|\phi\| := \sup_{-\tau \leq \theta \leq 0} |\phi(\theta)|$. We remark that we also use the notation $\|\cdot\|$ for the induced matrix norm, however, no confusion should arise. Given $t \geq 0$, for $x_t \in \mathcal{C}$ we let $x_t(\theta) := x(t + \theta)$, $-\tau \leq \theta \leq 0$. For given initial data $\phi \in \mathcal{C}$ and a constant $T > 0$, a solution of (17.4) is a function $x_t = x_t(\cdot) = x_t(\cdot; \phi) \in \mathcal{C}$ such that $x_0 = \phi$ and x_t satisfies (17.4) for all $t \in [0, T)$. We assume that the solutions of our coupled systems are uniformly (ultimately) bounded (see [5] for a definition) such that $T = \infty$. Conditions for (ultimate) boundedness expressed at the level of the systems' dynamics can be found in [27, 28]. We shall write $x(t; \phi)$ instead of $x_t(0; \phi)$.

A solution x_t of the coupled systems (17.4) is a *synchronous solution* if and only if

$$x_t(\theta) = I_N \otimes s_t(\theta), \quad \forall \theta \in [-\tau, 0], \quad \forall t \geq 0,$$

where $s_t \in \mathcal{C}([-\tau, 0], \mathbb{R}^n)$. Note that, because coupling (17.3) is noninvasive, the asymptotic synchronous solution s_t satisfies the ordinary differential equation

$$\dot{s}(t) = f(s(t)).$$

The coupled systems (17.4) are said to *synchronize* if its solutions converge asymptotically to a synchronous solution:

$$\lim_{t \to \infty} \|x_t - I_N \otimes s_t\| = 0.$$

17.3 Conditions for Local Synchronization

We address first the problem of *local* synchronization, i.e., synchronization of systems with initial data that satisfy

$$\|\phi_i - \phi_j\| < \delta, \quad \phi_i, \phi_j \in \mathcal{C}\left([-\tau, 0], \mathbb{R}^n\right)$$

with δ some small positive constant. We consider the case that the isolated system

$$\dot{s}(t) = f(s(t))$$

has an attractor $\mathcal{A}$ with basin of attraction $\mathcal{B}$. We suppose that there is a neighborhood $\mathcal{U}$ of $\mathcal{A}$ contained in $\mathcal{B}$, and we let $\overline{\mathcal{U}}$ and $\partial\mathcal{U}$ be the closure of $\mathcal{U}$, respectively, the boundary of $\mathcal{U}$. We remark that in general such a neighborhood $\mathcal{U}$ does not need to exist, i.e., when $\mathcal{A}$ is a weak attractor [16]. Furthermore, we assume that $\mathcal{U}$ is *inflowing invariant* with respect to the vectorfield f [8, 32]; That is, there is a positive constant μ such that

$$\langle N(s), f(s)\rangle \leq -\mu, \quad \forall s \in \partial\mathcal{U},$$

where $N(s)$ is the outward normal of $\partial\mathcal{U}$ at point s and $\langle\cdot,\cdot\rangle$ is the innerproduct in $\mathbb{R}^n$. We denote

$$\mathcal{C}_{\mathcal{U}} = \big\{\phi \in \mathcal{C} \mid \phi(\theta) = \mathrm{col}(\phi_1(\theta), \phi_2(\theta), \ldots, \phi_N(\theta)),\\ \phi_i(\theta) \in \mathcal{U},\ i = 1, 2, \ldots, N,\ -\tau \leq \theta \leq 0\big\}.$$

Theorem 17.1 *Suppose that the isolated system* (17.2) *has an attractor $\mathcal{A}$ with an inflowing invariant neighborhood $\mathcal{U}$ contained in $\mathcal{B}$. Let there exists a nonempty set $\mathcal{S} \subset \mathbb{R}_+ \times \overline{\mathbb{R}}_+$ such that for any $(\sigma, \tau) \in \mathcal{S}$ the zero solution of the linear system*

$$\dot{\eta}(t) = J(t)\eta(t) - \sigma BC\eta(t - \tau) \tag{17.5}$$

with

$$J(t) := \frac{\partial f}{\partial x_i}(\xi(t))$$

is uniformly asymptotically stable for all $\xi \in \mathcal{C}(\mathbb{R}, \mathcal{U})$. Let

$$\mathcal{S}_j := \Big\{(\sigma, \tau) \in \mathbb{R}_+ \times \overline{\mathbb{R}}_+ \mid (\sigma\lambda_j, \tau) \in \mathcal{S}\Big\}$$

be a scaled copy of $\mathcal{S}$ with nonzero eigenvalue λ_j of L as scaling factor. If

$$(\sigma, \tau) \in \cap_{j=2}^{N}\mathcal{S}_j,$$

then there is a constant $\delta = \delta(\sigma, \tau) > 0$ such that solutions of the coupled systems (17.2) *and* (17.3), *with initial data $\phi \in \mathcal{C}_{\mathcal{U}}$ for which $\|\phi_i - \phi_j\| < \delta$ for all $i, j = 1, 2, \ldots, N$, are contained in $\mathcal{C}_{\mathcal{U}}$. Moreover, the coupled systems* (17.2) *and* (17.3) *locally synchronize.*

Proof Since L is symmetric there exists a nonsingular $(N-1)\times(N-1)$-dimensional matrix U such that

$$U \begin{pmatrix} \lambda_2 & & \\ & \ddots & \\ & & \lambda_N \end{pmatrix} U^{-1} = L_2,$$

$$\begin{pmatrix} 1 & \mathbf{0}_{N-1}^\top \\ \mathbf{1}_{N-1} & -I_{N-1} \end{pmatrix} L \begin{pmatrix} 1 & \mathbf{0}_{N-1}^\top \\ \mathbf{1}_{N-1} & -I_{N-1} \end{pmatrix} = \begin{pmatrix} 0 & L_1^\top \\ \mathbf{0}_{N-1} & L_2 \end{pmatrix}$$

with L_1 a $(N-1)$-dimensional vector. See [24] for details. We remark that L_1 has at least one nonzero entry; If not the network would not be connected. Let the zero solution of the system

$$\begin{pmatrix} \dot{\eta}_2(t) \\ \vdots \\ \dot{\eta}_N(t) \end{pmatrix} = (I_{N-1} \otimes J(t)) \begin{pmatrix} \eta_2(t) \\ \vdots \\ \eta_N(t) \end{pmatrix} - \sigma \left(\begin{pmatrix} \lambda_2 & & \\ & \ddots & \\ & & \lambda_N \end{pmatrix} \otimes BC \right) \begin{pmatrix} \eta_2(t-\tau) \\ \vdots \\ \eta_N(t-\tau) \end{pmatrix}$$

be uniformly asymptotically stable for $(\sigma, \tau) \in \cap_{j=2}^N \mathcal{S}_j$ such that, for

$$\begin{pmatrix} \zeta_2(t) \\ \vdots \\ \zeta_N(t) \end{pmatrix} = (U \otimes I_n) \begin{pmatrix} \eta_2(t) \\ \vdots \\ \eta_N(t) \end{pmatrix}$$

the zero solution of the system

$$\begin{pmatrix} \dot{\zeta}_2(t) \\ \vdots \\ \dot{\zeta}_N(t) \end{pmatrix} = (I_{N-1} \otimes J(t)) \begin{pmatrix} \zeta_2(t) \\ \vdots \\ \zeta_N(t) \end{pmatrix} - \sigma \left(L_2 \otimes BC\right) \begin{pmatrix} \zeta_2(t-\tau) \\ \vdots \\ \zeta_N(t-\tau) \end{pmatrix} \tag{17.6}$$

is uniformly asymptotically stable. We remark that the zero solution of a linear system being uniformly asymptotically stable implies the zero solution of that system to be exponentially stable, cf. Theorem 4.5 of [11]. Thus there exist positive constants α, β such that for any solution $\zeta(\cdot; \psi)$ of (17.6) through $\psi \in \mathcal{C}\left([-\tau, 0], \mathbb{R}^{(N-1)n}\right)$ the following estimate holds:

$$|\zeta(t; \psi)| \leq \beta e^{-\alpha t} \|\psi\|, \quad \forall t \geq 0.$$

Denote

$$\begin{pmatrix} \tilde{x}_1(t) \\ \tilde{x}_2(t) \\ \vdots \\ \tilde{x}_N(t) \end{pmatrix} = \begin{pmatrix} x_1(t) \\ x_1(t) - x_2(t) \\ \vdots \\ x_1(t) - x_N(t) \end{pmatrix},$$

such that

$$\dot{\tilde{x}}_1(t) = f(\tilde{x}_1(t)) - \sigma \left(L_1^\top \otimes BC\right) \begin{pmatrix} \tilde{x}_2(t-\tau) \\ \vdots \\ \tilde{x}_N(t-\tau) \end{pmatrix} \tag{17.7}$$

and

$$\begin{pmatrix} \dot{\tilde{x}}_2(t) \\ \vdots \\ \dot{\tilde{x}}_N(t) \end{pmatrix} = \begin{pmatrix} \tilde{f}(t, \tilde{x}_2(t)) \\ \vdots \\ \tilde{f}(t, \tilde{x}_N(t)) \end{pmatrix} - \sigma \left(L_2 \otimes BC\right) \begin{pmatrix} \tilde{x}_2(t-\tau) \\ \vdots \\ \tilde{x}_N(t-\tau) \end{pmatrix} \tag{17.8}$$

with $\tilde{f}(t, \tilde{x}_i(t)) := f(\tilde{x}_1(t)) - f(\tilde{x}_1(t) - \tilde{x}_i(t))$. It now follows that if $\tilde{x}_1(t) \in \mathcal{U}$ for all $t \geq 0$, then the zero solution of (17.8) is locally exponentially stable, cf. Theorem 4.6 of [11]. In particular, for $\phi \in \mathcal{C}$ with $\|\phi_i - \phi_j\| < \delta_1$, where δ_1 is small enough to ensure that the linear part of (17.8) dominates the nonlinearities, and $K = \left(1 + \frac{1}{2\alpha}\right) \beta^2 e^{2\alpha\tau}$, there is a positive constant γ such that

$$\left\| \begin{pmatrix} \tilde{x}_2(t;\phi) \\ \vdots \\ \tilde{x}_N(t;\phi) \end{pmatrix} \right\| \leq K e^{-\gamma t} \|\phi\| \leq K\delta_1, \quad \forall t \geq 0.$$

To prove the theorem we are left to show that $\tilde{x}_1(t) \in \mathcal{U}$ for all $t \geq 0$. Pick

$$\delta_2 < \frac{\mu}{\sigma K |L_1| \|BC\|}$$

and

$$\delta = \min\left(\delta_1, \delta_2\right).$$

Suppose that there is a positive constant t_1 such that $\tilde{x}_1(t_1) \in \partial\mathcal{U}$ and $\tilde{x}_1(t) \notin \overline{\mathcal{U}}$ for some $t > t_1$. Because f is inflowing invariant with constant μ, the $\tilde{x}_1$-dynamics (17.7) can only cross the boundary $\partial\mathcal{U}$ at $t = t_1$ if

$$\left| \sigma \left(L_1^\top \otimes BC\right) \begin{pmatrix} \tilde{x}_2(t-\tau) \\ \vdots \\ \tilde{x}_N(t-\tau) \end{pmatrix} \right| \geq \mu.$$

But

$$\left| \sigma \left(L_1^\top \otimes BC\right) \begin{pmatrix} \tilde{x}_2(t-\tau) \\ \vdots \\ \tilde{x}_N(t-\tau) \end{pmatrix} \right| \leq \sigma |L_1| \|BC\| K \left| \begin{pmatrix} \tilde{x}_2(t-\tau) \\ \vdots \\ \tilde{x}_N(t-\tau) \end{pmatrix} \right|$$

$$\leq \sigma |L_1| \|BC\| K < \delta\mu.$$

hence $t_1 = \infty$. □

Equation (17.5) is a MSF for the time-delay coupled systems (17.2) and (17.3). However, contrary to the MSF approach for the delay-free case presented in [21], we do assume that the isolated system has an attractor $\mathcal{A}$ with inflowing invariant neighborhood $\mathcal{U}$. In addition, we evaluate (17.5) along all possible solutions in $\mathcal{U}$ instead of a single solution on $\mathcal{A}$. However, to verify uniform asymptotic stability of the zero solution (17.5) for all possible solutions in $\mathcal{U}$, one usually has to construct a Lyapunov functional on $\mathcal{U}$. See [27] for an example. We remark that a synchronization diagram computed using the Lyapunov functional approach tends to be conservative in the sense that it is contained, but not equal to the true synchronization diagram. In case the isolated system has a fixed point or periodic orbit as attractor, we can obtain a better estimate of the true synchronization diagram $\mathcal{S}$.

Corollary 17.2 *Assume that the attractor $\mathcal{A}$ defined in Theorem 17.1 is an asymptotically stable fixed point or an orbitally stable period orbit. Let $\xi(\cdot)$ be a solution of $\dot{\xi}(t) = f(\xi(t))$ with $\xi(-\tau) \in \mathcal{A}$, i.e., $\xi(\cdot)$ is a solution of the isolated system on $\mathcal{A}$. Suppose that there exists a nonempty set $\mathcal{S} \subset \mathbb{R}_+ \times \overline{\mathbb{R}}_+$ such that for any $(\sigma, \tau) \in \mathcal{S}$ the zero solution of the linear system*

$$\dot{\eta}(t) = J(t)\eta(t) - \sigma BC\eta(t - \tau)$$

with

$$J(t) := \frac{\partial f}{\partial x_i}(\xi(t))$$

is uniformly asymptotically stable. If

$$(\sigma, \tau) \in \cap_{j=2}^{N} \mathcal{S}_j,$$

then the conclusions of Theorem 17.1 hold.

Proof Consider the linearization of (17.7) and (17.8) around the synchronous solution on $\mathcal{A}$:

$$\dot{\zeta}(t) = (I_N \otimes J(t))\zeta(t) - \left(\begin{pmatrix} 0 & L_1^\top \\ \mathbf{0}_{N-1} & L_2 \end{pmatrix} \otimes BC\right)\zeta(t - \tau).$$

As shown in the proof of Theorem 17.1, one can find new coordinates such that the matrix L_2 is the matrix above becomes diagonal. Denote this diagonal matrix by Λ_2. Thus in these new coordinates the system has a block-triangular structure. If $\mathcal{A}$ is an equilibrium, then $J(t) = J$ is a stable matrix, and it is easy to see that the conditions of the corollary imply that the characteristic equation

$$\Delta(\rho; \sigma, \tau) = \det\left(\rho I_{Nn} - (I_N \otimes J) - \sigma\left(\begin{pmatrix} 0 & L_1^\top \\ \mathbf{0}_{N-1} & \Lambda_2 \end{pmatrix} \otimes BC\right)\exp(-\rho\tau)\right)$$

has no roots in the closed right half of the complex plane. If $\mathcal{A}$ is a periodic orbit, then $J(t) = J(t+T)$ for some nonzero constant T, i.e., $J(t)$ is T-periodic. We now use Floquet theory (cf. [12]) to conclude the proof. First, we observe that the monodromy matrix of the block-triangular system has a block-triangular structure. Then our conditions imply that all Floquet multiplier except one are contained in the open unit disk in the complex plane. Moreover, as the Floquet multipliers are independent of t (cf. [12], Sect. 8.1, Lemma 1.3) it suffices to linearize around a single periodic synchronous solution. □

17.4 Example: Local Synchronization of FitzHugh-Nagumo Neurons

We consider the network shown in Fig. 17.2 with dynamics

$$f(x_i(t)) = \begin{pmatrix} \frac{2}{25}\left(x_{i,2}(t) - \frac{4}{5}x_{i,1}(t)\right) \\ x_{i,2}(t) - \frac{1}{3}x_{i,2}^3(t) - x_{i,1}(t) \end{pmatrix}, \quad B = \begin{pmatrix} 0 \\ 1 \end{pmatrix}, \quad C = \begin{pmatrix} 0 & 1 \end{pmatrix}.$$

The system above is the FitzHugh-Nagumo (FHN) neuron [10, 17], which is a model of the excitable membrane dynamics of a neuron.

Let us first show that the isolated FHN neuron has a periodic attractor. Consider the function $V : \mathbb{R}^n \to \overline{\mathbb{R}}_+$

$$V(x_i(t)) = \tfrac{25}{4}x_{i,1}^2(t) + \tfrac{1}{2}x_{i,2}^2(t).$$

Then

$$\dot{V}(x_i(t)) = -\tfrac{4}{5}x_{i,1}^2(t) - \left(\tfrac{1}{3}x_{i,2}^2(t) - 1\right)x_{i,2}^2(t),$$

Fig. 17.2 Example network. Each edge has weight 1

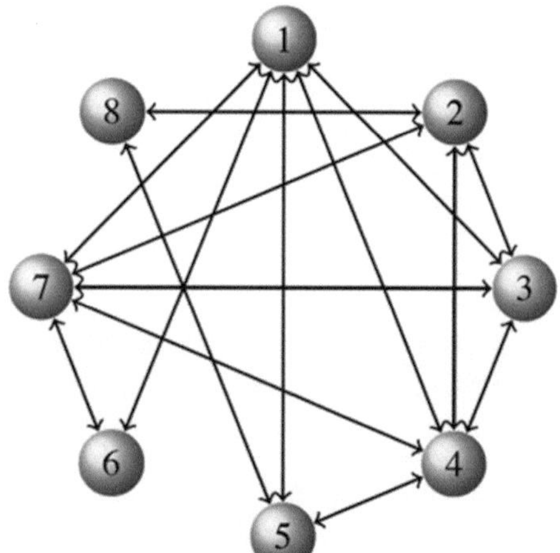

and it follows that the set

$$\Omega = \left\{ x_i(t) \in \mathbb{R}^2 \mid V(x_i(t)) \leq \tfrac{75}{4} \right\}$$

is positively invariant with respect to the dynamics of the isolated FHN neuron. One easily verifies that this system has a single equilibrium in Ω, the origin, which is unstable. Hence by the Poincaré-Bendixson theorem (cf. [29]) the isolated FHN neuron has a periodic orbit. In fact, applying Liénard's theorem (cf. [29]) to the system obtained after the well-defined change of coordinates

$$x_i(t) \mapsto \begin{pmatrix} v_i(t) \\ w_i(t) \end{pmatrix} = \begin{pmatrix} x_{i,2}(t) \\ x_{2,i}(t) - \frac{1}{3}x_{2,i}^3(t) - x_{i,1}(t) \end{pmatrix},$$

i.e.,

$$\begin{pmatrix} \dot{v}_i(t) \\ \dot{w}_i(t) \end{pmatrix} = \begin{pmatrix} w_i(t) \\ -\left(v_i^2(t) - \frac{27}{25}\right) w_i(t) - \frac{2}{25}\left(\frac{4}{15}v_i^3(t) + \frac{1}{5}v_i(t)\right) \end{pmatrix},$$

we conclude that Ω contains a unique and orbitally stable period attractor with period time T.

By Corollary 17.2, we may then determine the synchronization diagram $\mathcal{S}$ by computing the Floquet multipliers of the linear T-periodic system

$$\begin{pmatrix} \dot{\eta}_1(t) \\ \dot{\eta}_2(t) \end{pmatrix} = \begin{pmatrix} -\frac{8}{125} & \frac{2}{25} \\ -1 & 1 - \xi_2^2(t) \end{pmatrix} \begin{pmatrix} \eta_1(t) \\ \eta_2(t) \end{pmatrix} - \sigma \begin{pmatrix} 0 & 0 \\ 0 & 1 \end{pmatrix} \begin{pmatrix} \eta_1(t-\tau) \\ \eta_2(t-\tau) \end{pmatrix},$$

where $\xi_2(t) = \xi_2(t+T)$ satisfies

$$\begin{pmatrix} \dot{\xi}_1(t) \\ \dot{\xi}_2(t) \end{pmatrix} = \begin{pmatrix} \frac{2}{25}\left(\xi_2(t) - \frac{4}{5}\xi_1(t)\right) \\ \xi_2(t) - \frac{1}{3}\xi_2(t) - \xi_1(t) \end{pmatrix}$$

with initial conditions on the unique periodic attractor. The synchronization diagram, which we computed with the numerical software package DDE-Biftool [7, 25], is shown in Fig. 17.3a. The Laplacian matrix of the network shown in Fig. 17.2 is

$$L = \begin{pmatrix} 5 & 0 & -1 & -1 & -1 & -1 & -1 & 0 \\ 0 & 4 & -1 & -1 & 0 & 0 & -1 & -1 \\ -1 & -1 & 4 & -1 & 0 & 0 & -1 & 0 \\ -1 & -1 & -1 & 5 & -1 & 0 & -1 & 0 \\ -1 & 0 & 0 & -1 & 3 & 0 & 0 & -1 \\ -1 & 0 & 0 & 0 & 0 & 2 & -1 & 0 \\ -1 & -1 & -1 & -1 & 0 & -1 & 5 & 0 \\ 0 & -1 & 0 & 0 & -1 & 0 & 0 & 2 \end{pmatrix}$$

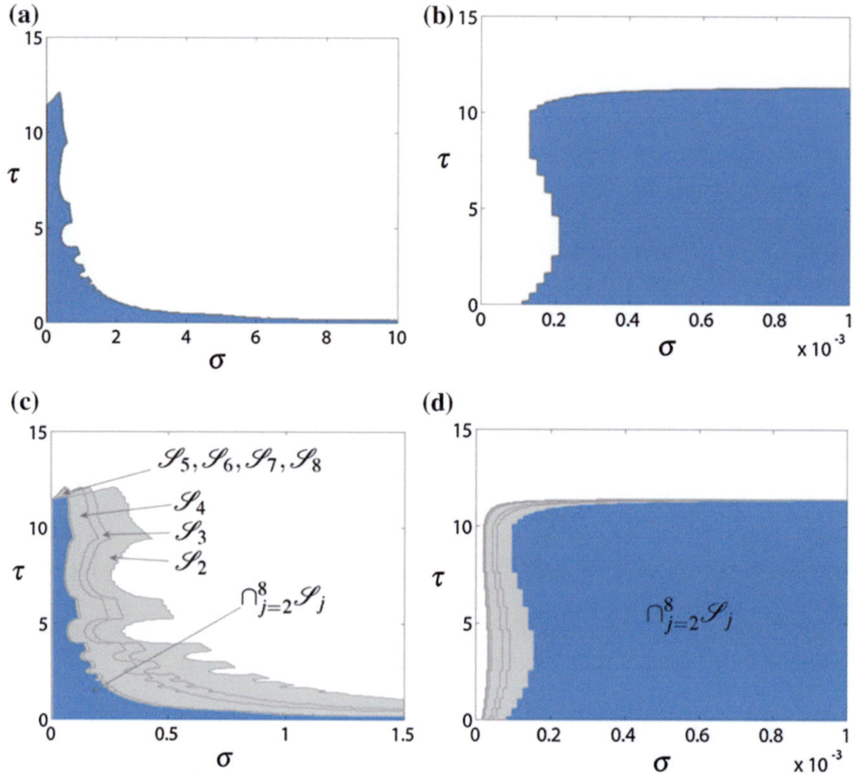

Fig. 17.3 **a** Synchronization diagram $\mathcal{S}$ for the FHN neuron. **b** Zoom of the *left* of (**a**). **c** Seven scaled copies of $\mathcal{S}$ and their intersection. **d** Zoom of the *left* of (**c**)

and has eigenvalues (approximated using Matlab®)

$$\begin{array}{cccc} \lambda_1 = 0, & \lambda_2 = 1.3643, & \lambda_3 = 2.3083, & \lambda_4 = 2.9266, \\ \lambda_5 = 4.9626, & \lambda_6 = 5.7110, & \lambda_7 = 6.2899, & \lambda_8 = 6.4374. \end{array}$$

The seven scaled copies of $\mathcal{S}$ and their intersection are shown in Fig. 17.3c. By Corollary 17.2, for any values of the coupling strength and time-delay belonging to this intersection, the network of FHN neurons locally synchronizes.

17.5 Conditions for Global Synchronization

In this section, we introduce a class of systems for which there exists a global synchronization diagram. This global synchronization diagram allows for the construction of a set of values of the coupling strength and time-delay for which a network of systems globally synchronizes. First, since we have assumed the matrix CB to be similar to a positive definite matrix, it is possible to find new coordinates

$$x_i(t) \mapsto \begin{pmatrix} z_i(t) \\ y_i(t) \end{pmatrix}$$

with $z_i(t) \in \mathbb{R}^{n-m}$. See [6, 23] for details about this transformation. In these new coordinates the systems' dynamics read as

$$\dot{z}_i(t) = q(z_i(t), y_i(t)) \tag{17.9a}$$

$$\dot{y}_i(t) = a(z_i(t), y_i(t)) + CBu_i(t) \tag{17.9b}$$

where $q : \mathbb{R}^{n-m} \times \mathbb{R}^m \to \mathbb{R}^{n-m}$ and $a : \mathbb{R}^{n-m} \times \mathbb{R}^m \to \mathbb{R}^m$ are (sufficiently) smooth vectorfields.

We shall assume that

A1. There exists a nonempty set $\mathcal{S}_B \in \mathbb{R}_+ \times \overline{\mathbb{R}}_+$ such that for $(\sigma, \tau) \in \mathcal{S}_B$ the solutions of the coupled systems are uniformly bounded with bound B that is independent of N.

In addition we assume that

A2. There exists a positive definite matrix $P = P^\top$ and a positive constant κ such that

$$\left[\frac{\partial q}{\partial z_i}(z_i, y_i)\right]^T P + P\left[\frac{\partial q}{\partial z_i}(z_i, y_i)\right] \le -\kappa I_{n-m}$$

for all $z_i \in \mathbb{R}^{n-m}$ and $y_i \in \mathbb{R}^m$.

The latter assumption implies that the system

$$\dot{z}_i(t) = q(z_i(t), y_i(t))$$

is an exponentially convergent system with respect to input $y_i(t)$ [18, 19]. Interesting is that such an exponentially convergent system has an exponentially stable steady-state solution that is solely determined by the vectorfield q and input signal $y_i(\cdot)$. It then follows that for any two input signals $y_i(\cdot)$, $y_j(\cdot)$ that satisfy

$$\lim_{t\to\infty} |y_i(t) - y_j(t)| = 0,$$

the solutions of the systems

$$\dot{z}_i(t) = q(z_i(t), y_i(t))$$

and

$$\dot{z}_j(t) = q(z_j(t), y_j(t))$$

satisfy

$$\lim_{t\to\infty} |z_i(t) - z_j(t)| = 0,$$

independent of the initial conditions of those systems.

We first give a result about global synchronization of two coupled systems.

Lemma 17.3 *Consider two coupled systems (17.9a) and* (17.3) *and let* $a_{12} = a_{21} = 1$. *Suppose that assumptions A1 and A2 hold. Then there exist two positive constants* $\bar{\sigma}$ *and* $\bar{\gamma}$ *such that if*

$$(\sigma, \tau) \in \mathcal{S}^* \cap \mathcal{S}_B,$$

where

$$\mathcal{S}^* := \left\{(\sigma, \tau) \in \mathbb{R}_+ \times \overline{\mathbb{R}}_+ \,|\, \sigma \geq \bar{\sigma} \text{ and } \sigma\tau \leq \bar{\gamma}\right\},$$

then the two coupled systems globally synchronize.

The set $\mathcal{S}^*$ is shown in Fig. 17.4. The proof of the lemma follows from the proof of the next theorem.

Theorem 17.4 *Consider a network of coupled systems (17.9a) and* (17.3) *and suppose that assumptions A1 and A2 hold. If*

$$(\sigma, \tau) \in \mathcal{S}_2^* \cap \mathcal{S}_N^* \cap \mathcal{S}_B,$$

where

$$\mathcal{S}_j^* := \left\{(\sigma, \tau) \in \mathbb{R}_+ \times \overline{\mathbb{R}}_+ \,\middle|\, \left(\frac{\lambda_j}{2}\sigma, \tau\right) \in \mathcal{S}^*\right\}, \quad j = 2, N,$$

with $\mathcal{S}^*$ *as in Lemma 17.3, then the network of coupled systems globally synchronizes.*

Proof Let

$$\tilde{y}_j(t) = y_1(t) - y_j(t), \quad \tilde{z}_j(t) = z_i(t) - z_j(t), \quad j = 2, \ldots, N,$$

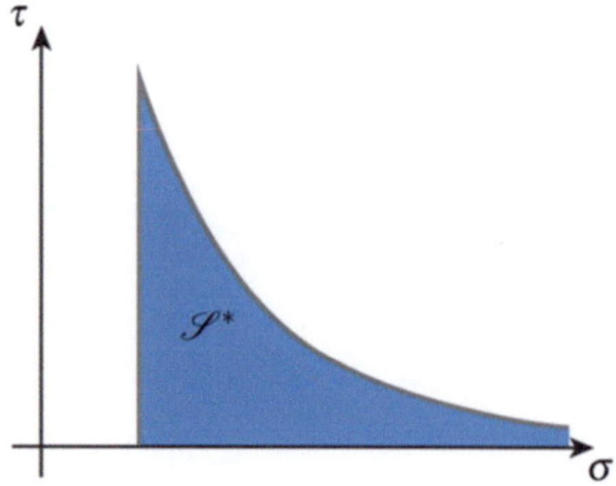

Fig. 17.4 The global synchronization diagram $\mathcal{S}^*$ for two coupled systems with its shape predicted by Lemma 17.3

$\tilde{z}(t) = \mathrm{col}(\tilde{z}_2(t), \ldots, \tilde{z}_N(t))$ and $\tilde{y}(t) = \mathrm{col}(\tilde{y}_2(t), \ldots, \tilde{y}_N(t))$, to obtain

$$\dot{\tilde{z}}(t) = \tilde{q}(z_1(t), y_1(t), \tilde{z}(t), \tilde{y}(t)) \tag{17.10a}$$

$$\dot{\tilde{y}}(t) = \tilde{a}(z_1(t), y_1(t), \tilde{z}(t), \tilde{y}(t)) - \sigma(L_2 \otimes CB)\tilde{y}(t-\tau) \tag{17.10b}$$

with

$$\tilde{q}(z_1(t), y_1(t), \tilde{z}(t), \tilde{y}(t)) := \begin{pmatrix} q(z_1(t), y_1(t)) - q(z_1 - \tilde{z}_2(t), y_1(t)) \\ \vdots \\ q(z_1(t), y_1(t)) - q(z_1 - \tilde{z}_N(t), y_1(t)) \end{pmatrix},$$

$$\tilde{a}(z_1(t), y_1(t), \tilde{z}(t), \tilde{y}(t)) := \begin{pmatrix} a(z_1(t), y_1(t)) - a(z_1 - \tilde{z}_2(t), y_1(t)) \\ \vdots \\ a(z_1(t), y_1(t)) - a(z_1 - \tilde{z}_N(t), y_1(t)) \end{pmatrix},$$

and the $(N-1) \times (N-1)$-dimensional matrix L_2 defined in the proof of Theorem 17.1. Recall that there is a matrix U such that

$$U^{-1} L_2 U = \begin{pmatrix} \lambda_2 & & \\ & \ddots & \\ & & \lambda_N \end{pmatrix}.$$

We assume without loss of generality that $\|U^{-1}\| = 1$. Using the equality

$$\tilde{y}(t-\tau) = \tilde{y}(t) - \int_{-\tau}^{0} \dot{\tilde{y}}(t+s)\mathrm{d}s$$

we obtain

$$\begin{aligned} \dot{\tilde{y}}(t) &= \tilde{a}(z_1(t), y_1(t), \tilde{z}(t), \tilde{y}(t)) - \sigma(L_2 \otimes CB)\tilde{y}(t) \\ &\quad + \sigma(L_2 \otimes CB) \int_{-\tau}^{0} [\tilde{a}(z_1(t+s), y_1(t+s), \tilde{z}(t+s), \tilde{y}(t+s)) \\ &\qquad - \sigma(L_2 \otimes CB)\tilde{y}(t+s-\tau)]\mathrm{d}s. \end{aligned} \tag{17.11}$$

We now show that the conditions of the theorem imply that the function

$$V(\tilde{z}(t), \tilde{y}(t)) = \tilde{z}^\top(t)(I_{N-1} \otimes P)\tilde{z}(t) + \tfrac{1}{2}\tilde{y}^\top(t)(U^{-\top}U^{-1} \otimes I_m)\tilde{y}(t)$$

is a Lyapunov–Razumikhin function [12], that proves uniform asymptotic stability of the origin of (17.10a) and (17.11), hence synchronization of the coupled systems. Assumption A2 implies that there exists a positive constant c_1 such that

$$[q(z_1(t), y_1(t)) - q(z_1 - \tilde{z}_j(t), y_1(t))]^\top P \\ + P[q(z_1(t), y_1(t)) - q(z_1 - \tilde{z}_j(t), y_1(t))] \leq -c_1|\tilde{z}_j(t)|^2.$$

See [24] for details. Moreover, since the solutions of the coupled systems are assumed to be bounded and the functions a and q are sufficiently smooth, there exist positive constants c_2, c_3 and c_4 such that

$$|2P[q(z_1(t) - \tilde{z}_j(t), y_1(t)) - q(z_1(t) - \tilde{z}_j(t), y_1(t) - \tilde{y}_j(t))]| \leq c_2|\tilde{y}_j(t)|,$$

and

$$\begin{aligned} &|a(z_1(t), y_1(t)) - a(z_1(t) - \tilde{z}_j(t), y_1(t) - \tilde{y}_j(t))| \\ &\qquad \leq |a(z_1(t), y_1(t)) - a(z_1(t) - \tilde{z}_j(t), y_1(t))| \\ &\qquad + |a(z_1(t) - \tilde{z}_j(t), y_1(t)) - a(z_1(t) - \tilde{z}_j(t), y_1(t) - \tilde{y}_j(t))| \\ &\qquad \leq c_3|\tilde{z}_j(t)| + c_4|\tilde{y}_j(t)|. \end{aligned}$$

Choose constant $\nu > 1$ such that if

$$\nu|\tilde{y}(t)| \geq |\tilde{y}(t + \theta)|$$

and

$$\nu^2 V(\tilde{z}(t), \tilde{y}(t)) \geq (\tilde{z}(t + \theta), \tilde{y}(t + \theta))$$

for $-2\tau \leq \theta \leq 0$, then

$$\begin{aligned} \dot{V} &\leq -W(\tilde{z}(t), \tilde{y}(t)) \\ &= -\begin{pmatrix} \tilde{z}(t) \\ \tilde{y}(t) \end{pmatrix}^\top \begin{pmatrix} c_1 & -\frac{c_2+c_4+\gamma c_4}{2} \\ -\frac{c_2+c_4+\gamma c_4}{2} & \beta_1\sigma\lambda_2 - c_3 - \gamma(c_3 + \beta_2\sigma\lambda_N) \end{pmatrix} \begin{pmatrix} \tilde{z}(t) \\ \tilde{y}(t) \end{pmatrix}, \end{aligned}$$

where $\gamma = \nu\beta_2\sigma\tau\lambda_N$, with positive constants β_1 and β_2 being the smallest, respectively, largest eigenvalue of CB. For a network of $N = 2$ systems with $a_{12} = a_{21} = 1$ we have $\lambda_2 = \lambda_N = 2$. It follows that whenever σ is sufficiently large and γ sufficiently small, i.e., $\sigma \geq \bar{\sigma}$ and $\gamma \leq \bar{\gamma}$ for some positive constants $\bar{\sigma}$ and $\bar{\gamma}$, then the function W is negative definite. This proves Lemma 17.3. Then we conclude that for any other network the function W negative definite if $(\sigma, \tau) \in \mathcal{S}_2^* \cap \mathcal{S}_N^*$. □

17.6 Example: Global Synchronization of FitzHugh–Nagumo neurons

Let us show that the FHN neurons introduced in Sect. 17.4 satisfy the conditions of Lemma 17.3. Let

$$x_i(t) = \begin{pmatrix} x_{i,1}(t) \\ x_{i,2}(t) \end{pmatrix} = \begin{pmatrix} z_i(t) \\ y_i(t) \end{pmatrix}$$

and

$$f(x_i(t)) = \begin{pmatrix} q(z_i(t), y_i(t)) \\ a(z_i(t), y_i(t)) \end{pmatrix} = \begin{pmatrix} \frac{2}{25}\left(y_i(t) - \frac{4}{5}z_i(t)\right) \\ y_i(t) - \frac{1}{3}y_i^3(t) - z_i(t) \end{pmatrix}.$$

Then one easily verifies that assumption A2 holds with $P = 1$. We will now show that assumption A1 is satisfied as well.

Proposition 17.5 *Consider N time-delay coupled FHN neurons and suppose that*

- $\max_i \sum_{j \in \mathcal{N}_i} a_{ij} = 1$;
- $\sigma\tau\left(6\sigma + \frac{39}{4}\right) \leq \frac{9}{4}$;
- *for each $i = 1, \ldots, N$, $\phi_i \in \mathcal{C}([-\tau, 0], \mathbb{R}^n)$, the initial data for the ith FHN neuron, is Lipschitz continuous on $[-\tau, 0]$ with Lipschitz constant $K \leq 12$.*

Then the set $\Omega^N := \Omega \times \Omega \times \cdots \times \Omega$ with

$$\Omega := \left\{(z_i, y_i) \in \mathbb{R}^2 \mid |z_i| \leq \tfrac{15}{4} \text{ and } |y_i| \leq 3\right\}$$

is a positively invariant set for the coupled FHN neurons.

Proof Let us consider first an isolated FHN neuron. The nulclines of this isolated neuron and the set Ω are shown in Fig. 17.5. From this picture it is clear that the coupling (17.3) can drive the solution $x_i(t) = \text{col}(z_i(t), y_i(t))$ outside of Ω though the boundaries $y_i = \bar{y}$ or $y_i = -\bar{y}$ with $\bar{y} = 3$. Consider an arbitrary solution of the coupled systems and let $t_1 \leq 0$ be such that this solution is contained in Ω^N for $t \leq t_1$. Suppose that at t_1 the solution of the ith is at the boundary $\bar{y}$, i.e. $y_i(t_1) = \bar{y}$. Write

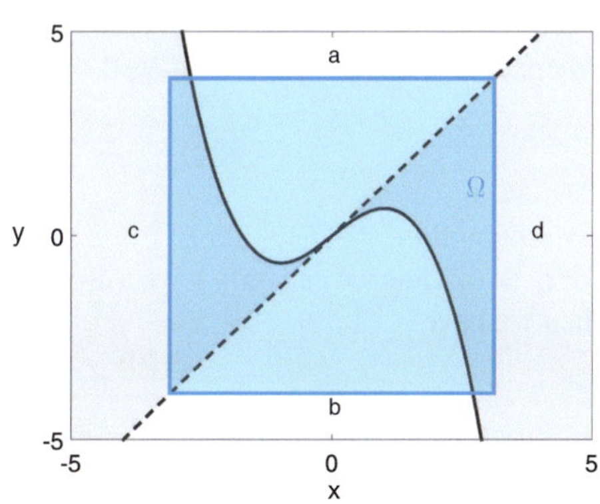

Fig. 17.5 The set Ω (in cyan) and nulclines of the isolated ($u_i = 0$) FHN neuron. Thick black line represents $\dot{y}_i = 0$, dashed black line represents $\dot{z}_i = 0$

$$u_i(t) = \sigma \sum_{j \in \mathcal{N}_i} a_{ij}[y_j(t-\tau) - y_i(t-\tau)]$$
$$= \sigma \sum_{j \in \mathcal{N}_i} a_{ij}[(y_j(t-\tau) - y_i(t)) + (y_i(t) - y_i(t-\tau))],$$

hence,

$$u_i(t_1) = \sigma \sum_{j \in \mathcal{N}_i} a_{ij}[(y_j(t_1-\tau) - \bar{y}) + (\bar{y} - y_i(t_1-\tau))] \leq \sigma(\bar{y} - y_i(t_1-\tau))$$

as $|y_j(t_1-\tau)| \leq \bar{y}$ for all j and $\sum_{j \in \mathcal{N}_i} a_{ij} \leq 1$. It then follows that $y_i(t) > \bar{y}$ for some $t > t_1$ requires

$$0 < \dot{y}_i(t_1) \leq a(z_i, \bar{y}) + \sigma(\bar{y} - y_i(t_1-\tau)) \leq -\nu + \sigma(\bar{y} - y_i(t_1-\tau)),$$

where

$$\nu = \max_{-\frac{15}{4} \leq z_i \leq \frac{15}{4}} -a(z_i, \bar{y}) = \min_{-\frac{15}{4} \leq z_i \leq \frac{15}{4}} a(z_i, -\bar{y}) = \frac{9}{4}.$$

As $|y_i(t_1-\tau)| \leq \bar{y}$ we have

$$\dot{y}_i(t_1) \leq -\nu + 2\sigma B_1,$$

where $B_1 := \bar{y} = 3$, hence to escape from Ω it is required that $\sigma > \frac{\nu}{2B_1}$. Thus let $\sigma > \frac{\nu}{2B_1}$. For $t_1 > 0$ we have

$$y_i(t_1) - y_i(t_1-\tau) = \int_{t_1-\tau}^{t_1} \left[a(z_i(s), y_i(s)) - \sigma \sum_{j \in \mathcal{N}_i} a_{ij}[y_j(s-\tau) - y_i(s-\tau)] \right] \mathrm{d}s$$
$$\leq \tau(B_2 + 2\sigma B_1),$$

where $B_2 := \max_{(z_i, y_i) \in \Omega} |a(z_i, y_i)| = \frac{39}{4}$. Hence

$$\dot{y}(t_1) \leq -\nu + \sigma\tau(B_2 + 2\sigma B_1).$$

By assumption, $\sigma\tau(B_2 + 2\sigma B_1) = \sigma\tau(\frac{39}{4} + 6\sigma) \leq \frac{9}{4} = \nu$, which gives $\dot{y}(t_1) \leq 0$ for $t_1 > 0$. Thus we can only have a crossing of $\bar{y}$ at $t_1 = 0$. If $t_1 = 0$, i.e. $\phi_1(0) = \bar{y}$, then we have

$$\phi_i(0) - \phi_i(-\tau) \leq K\tau.$$

But $K \leq B_2 + \nu = \frac{39}{4} + \frac{9}{4} = 12$ such that, as $\sigma > \frac{\nu}{2B_1}$ hence

$$K \leq B_2 + \nu < B_2 + 2B_1\sigma,$$

we have

$$\sigma\tau K \leq \sigma\tau(B_2 + 2B_1\sigma) \leq \nu,$$

which implies $\dot{y}(0) \leq 0$. The same reasoning gives that, if $y_i(t_3) = -\bar{y}$ for some $t_3 \geq 0$, then $\dot{y}(t_3) \geq 0$, hence solutions cannot escape from Ω^N. □

By Proposition 17.5, assuming the Lipschitz condition on the initial data, we conclude that assumption A1 is satisfied for all

$$(\sigma, \tau) \in \mathcal{S}_B := \left\{(\sigma, \tau) \in \mathbb{R}_+ \times \overline{\mathbb{R}}_+ \,|\, \sigma\tau\left(6\sigma + \tfrac{39}{4}\right) \leq \tfrac{9}{4}\right\}.$$

Then Lemma 17.3 implies the existence of a non-empty set $\mathcal{S}^* \cap \mathcal{S}_B$ such that for $(\sigma, \tau) \in \mathcal{S}^* \cap \mathcal{S}_B$ two time-delay coupled FHN neurons globally synchronize (in $\Omega \times \Omega$). Invoking Theorem 17.4 we derive conditions for global synchronization (in Ω^N) of any network of N time-delay coupled FHN neurons.

17.7 Discussion

We have constructed a (local, global) synchronization diagram for time-delay coupled systems and we have shown that a condition for (local, global) synchronization of a network is that the coupling strength and time-delay belong to the intersection of scaled copies of that (local, global) synchronization diagram. The scaling factors are the nonzero eigenvalues of the Laplacian matrix of the undirected, simple, and connected network. We have demonstrated our results with a network of FHN neurons.

We have assumed the network Laplacian matrix to be symmetric to ensure that the eigenvalues (and thus the scaling factors) are real valued. A natural extension of this work would be to allow for networks with asymmetric network Laplacian matrices, e.g., in case of directed networks.

An other important extension would be to consider coupling functions of the form

$$u_i(t) = \sigma \sum_{j \in \mathcal{N}_i} a_{ij}[y_j(t - \tau) - y_i(t)]. \tag{17.12}$$

There is an important difference between this type of coupling and the coupling functions considered in this chapter, i.e., coupling (17.3); coupling (17.12) is invasive whereas the coupling (17.3) is not. For invasive coupling functions, the synchronized dynamics depend on the values of the coupling strength and time-delay. Thus for coupling (17.12), one has to impose additional conditions to ensure that the synchronization manifold exists. A sufficient condition for existence of the synchronization manifold is that the network adjacency matrix $A = \left(a_{ij}\right)$ has constant row-sums, e.g.,

$$\sum_{j \in \mathcal{N}_i} a_{ij} = 1 \quad \forall i = 1, \ldots, N,$$

cf. [28]. Under the assumption above, one can easily derive that the synchronization diagram depends on σ, τ and $\sigma\lambda_j(A)$, with $\lambda_j(A)$ being any eigenvalue of the network adjacency matrix other than 1. (We remark that in case the network is connected and all rows of A sum up to 1, the matrix A has a simple eigenvalue equal to 1.) Thus for invasive coupling (17.12), the synchronization diagram and its intersections need to be drawn in a three-dimensional space.

Finally, (for both types of coupling functions) it would be valuable to extend our results to the multiple delay case.

17.8 Epilogue

This chapter is a tribute to the 60th birthday of Arjan van der Schaft. Over a period of more than 35 years, the second author has shared many ideas, papers, thoughts, running miles, cigars, and much more with Arjan. It is my expectation that this will continue for the next 35 years; I look forward to that.

References

1. P. Ashwin, J. Buescu, I. Stewart, Bubbling of attractors and synchronisation of chaotic oscillators. Phys. Lett. A **193**, 126–139 (1994)
2. I. Belykh, M. Hasler, M. Lauret, H. Nijmeijer, Synchronization and graph topology. Int. J. Bif. Chaos **15**, 3423–3433 (2005)
3. V.N. Belykh, I.V. Belykh, M. Hasler, Connection graph stability method for synchronized coupled chaotic systems. Phys. D **195**(1–2), 159–187 (2004)
4. B. Bollobás, *Modern Graph Theory, Volume 184 of Graduate Texts in Mathematics* (Springer-Verlag, New York, 1998)
5. T.A. Burton, *Stability and Periodic Solutions of Ordinary and Functional Differential Equations* (Academic Press, New York, 1985)
6. C.I. Byrnes, A. Isidori, J.C. Willems, Passivity, feedback equivalence, and the global stabilization of minimum phase nonlinear systems. IEEE trans. Auto. Control **36**(11), 1228–1240 (1991)
7. K. Engelborghs, T. Luzyanina, D. Roose, Numerical bifurcation analysis of delay differential equations using DDE-BIFTOOL. ACM Trans. Math. Softw. **28**(1), 1–21 (2002)
8. N. Fenichel, Persistence and smoothness of invariant manifolds for flows. Indiana Univ. Math. J. **21**(3), 193–226 (1972)
9. M. Fieldler, Algebraic connectivity of graphs. Czech. Math. J. **23**(98), 298–305 (1973)
10. R. FitzHugh, Impulses and physiological states in theoretic models of nerve membrane. Biophys. J. **1**, 445–466 (1961)
11. A. Halaney, *Differential Equations: Stability, Oscillations, Time Lags* (Academic Press, New York, 1966)
12. J.K. Hale, S.M. Verduyn Lunel, *Introduction to Functional Differential Equations, Volume 99 of Applied Mathematical Sciences* (Springer-Verlag, New York, 1993)

13. R.A. Horn, C.R. Johnson, *Matrix Analysis*, 6th edn. (Cambridge University Press, Cambridge, 1999)
14. G.A. Leonov, *Strange Attractors and Classical Stability Theory* (Petersburg University Press, Petersburg, 2009)
15. W. Lu, T. Chen, New approach to synchronization analysis of linearly coupled ordinary differential equations. Phys. D **213**, 214–230 (2006)
16. J. Milnor, On the concept of the attractor. Commun. Math. Phys. **99**, 177–195 (1985)
17. J.S. Nagumo, S. Arimoto, S. Yoshizawa, An active pulse transmission line simulating nerve axon. Proc. IRE **50**, 2061–2070 (1962)
18. A. Pavlov, A. Pogromsky, N.v.d. Wouw, H. Nijmeijer, Convergent dynamics, a tribute to Boris Pavlovich Demidovich. Syst. Control Lett. **52**, 257–261 (2004)
19. A. Pavlov, N.v.d. Wouw, N. Nijmeijer, Global nonlinear output regulation: convergence-based controller design. Automatica **43**(3), 456–463 (2007)
20. L.M. Pecora, Synchronization conditions and desynchronizing patterns in coupled limit-cycle and chaotic systems. Phys. Rev. E **58**(1), 347–360 (1998)
21. L.M. Pecora, T.L. Carroll, Master stability functions for synchronized coupled systems. Phys. Rev. Lett. **80**(10), 2109–2112 (1998)
22. A. Pikovsky, M. Rosenblum, J. Kurths, *Synchronization*, 2nd edn. (Cambridge University Press, Cambridge, 2003)
23. A. Pogromsky, T. Glad, H. Nijmeijer, On diffusion driven oscillations in coupled dynamical systems. Int. J. Bif. Chaos **9**(4), 629–644 (1999)
24. A. Pogromsky, H. Nijmeijer, Cooperative oscillatory behavior of mutually coupled dynamical systems. IEEE Trans. Circ. Syst. I **48**(2), 152–162 (2001)
25. J. Sieber, K. Engelborghs, T. Luzyanina, G. Samaey, D. Roose. DDE-BIFTOOL v. 3.1 manual—bifurcation analysis of delay differential equations. http://arxiv.org/abs/1406.7144
26. R. Sipahi, S. Niculescu, C.T. Abdallah, W. Michiels, K. Gu, Stability and stabilization of systems with time delay. IEEE Control Syst. **31**(1), 38–65 (2011)
27. E. Steur, W. Michiels, H.J.C. Huijberts, H. Nijmeijer, Networks of diffusively time-delay coupled systems: conditions for synchronization and its relation to the network topology. Phys. D **277**, 22–39 (2014)
28. E. Steur, H. Nijmeijer, Synchronization in networks of diffusively time-delay coupled (semi-)passive systems. IEEE Trans. Circ. Syst. I **58**(6), 1358–1371 (2011)
29. S.H. Strogatz, *Nonlinear Dynamics and Chaos* (Perseus Books Publishing, LLC, 1994)
30. S.H. Strogatz. *Sync: The Emerging Science of Spontaneous Order*, 1st edn. (Hyperion, New York, 2003)
31. R.L. Viana, C. Grebogi, S.E. de S. Pinto, S.R. Lopes, A.M. Batista, J. Kurths, Bubbling bifurcation: loss of synchronization and shadowing breakdown in complex systems. Phys. D **206**, 94–108 (2005)
32. S. Wiggins, *Normally Hyperbolic Invariant Manifolds in Dynamical Systems, Volume 105 of Applied Mathematical Sciences* (Springer-Verlag, New York, 1994)
33. C.W. Wu, L.O. Chua, On a conjecture regarding the synchronization in an array of linearly coupled dynamical systems. IEEE Trans. Circ. Syst. I **43**(2), 161–165 (1996)

Chapter 18
Examples on Stability for Infinite-Dimensional Systems

Hans Zwart

Abstract By means of examples, we study stability of infinite-dimensional linear and nonlinear systems. First we show that having a (strict) Lyapunov function does not imply asymptotic stability, even not for linear systems. Second, we show that to conclude (local) exponential stability from the linearization, care must be taken how the linearization is obtained.

18.1 Introduction

I met Arjan for the first time when he was presenting his colloquium for his Ph.D defence. He had already left for Twente, and 4 years later I would follow him. Although we were colleagues for many years, our research did not touch. Arjan worked on nonlinear system described by ordinary differential equations, and I was working on linear systems, described by partial differential equations. This changed when Arjan started to study port-Hamiltoninan systems described by partial differential equations. After some prior discussions, also together with Goran Golo, Arjan, and I joined forces in the Ph.D. project of Javier Villegas. From that time on port-Hamiltonian systems is really one of my research directions. Also inspired by Arjans work is my more recent interest in nonlinear systems. The present paper is a result of this.

For finite-dimensional systems the following two facts are well known and used regularly when studying stability. If there exists a Lyapunov function V such that $\dot{V} < 0$, then the equilibrium point is asymptotically stable. Second, if the linearization of a nonlinear differential equation around a equilibrium point is exponentially stable,

H. Zwart (✉)
Department of Applied Mathematics, University of Twente, P.O. Box 217,
7500 AE Enschede, The Netherlands
e-mail: h.j.zwart@utwente.nl

H. Zwart
Dynamics and Control, TU/e, P.O. Box 513,
5600 MB Eindhoven, The Netherlands
e-mail: H.J.Zwart@tue.nl

M.K. Camlibel et al. (eds.), *Mathematical Control Theory I*,
Lecture Notes in Control and Information Sciences 461,
DOI 10.1007/978-3-319-20988-3_18

then the equilibrium point is locally exponentially stable for the original equation. We address these questions for infinite-dimensional systems. That is, we study the following abstract differential equation

$$\dot{x}(t) = Ax(t) + f(x(t)), \qquad x(0) = x_0, \tag{18.1}$$

where A is the infinitesimal generator of a C_0-semigroup on the Hilbert space X, and $f : X \mapsto X$ is a locally Lipschitz continuous function with $f(0) = 0$. Under these conditions, the abstract differential equation possesses for every initial condition x_0 a unique (local) solution, see e.g. [2, Chap. 6], and so we can study the stability of the equilibrium point $x_{eq} = 0$.

In the following section, we show that having a Lyapunov function V satisfying $\dot{V}(x) < 0$ for every $x \neq 0$ does not have to imply that the equilibrium solution is stable. We can even construct a linear counter example.

In Sect. 18.3, we study the question whether the exponential stability of the C_0-semigroup generated by A implies the same for the nonlinear equation (18.1). We recall a positive result, but show by means of a simple example that the conditions in this theorem cannot be weakened.

We end this introduction by introducing some notation. We denote the domain of the operator A by $D(A)$, and the class of bounded, linear operators from X to X by $\mathcal{L}(X)$. Furthermore, the semigroup generated by A is denoted by $(T(t))_{t\geq 0}$. We say that the semigroup $(T(t))_{t\geq 0}$ exponentially stable, when there exists a M and $\omega_0 > 0$ such that $\|T(t)\| \leq Me^{-\omega_0 t}$. It is asymptotically (or strongly) stable when $\lim_{t\to\infty} T(t)x_0 = 0$ for all $x_0 \in X$.

18.2 Strict Lyapunov Function Does Not Imply Asymptotic Stability

Let X be the Hilbert space $L^2(0,\infty)$ equipped with the inner product

$$\langle f, g \rangle := \int_0^\infty f(\zeta)\overline{g(\zeta)}(e^{-\zeta} + 1)d\zeta,$$

and let the operators $T(t) : X \to X$, $t \geq 0$, be defined by

$$(T(t)f)(\zeta) := f(\zeta - t) \text{ for } \zeta > t \text{ and zero otherwise.}$$

Hence $T(t)$ is shifting the function f to the right. It is not hard to show that $(T(t))_{t\geq 0}$ is a C_0-semigroup on X, see e.g. [1].

Since

$$\begin{aligned}\|T(t)f\|^2 &= \int_t^\infty |f(\zeta - t)|^2(e^{-\zeta} + 1)d\zeta \\ &= \int_0^\infty |f(\zeta)|^2(e^{-\zeta+t} + 1)d\zeta \\ &\geq \int_0^\infty |f(\zeta)|^2(e^{-\zeta} + 1)d\zeta = \|f\|^2\end{aligned}$$

we see that $T(t)$ is not asymptotically stable. In fact, for every nonzero f, $T(t)f$ does not converge to zero.

The infinitesimal generator A associated to this semigroup is given by

$$Af = -\frac{df}{d\zeta}$$

with domain

$$D(A) = \{f \in X \mid f \text{ is absolutely continuous, } \frac{df}{d\zeta} \in X, \text{ and } f(0) = 0\}.$$

Consider next the standard Lyapunov function $V(x) = \|x\|^2$. Then for $x \in D(A)$,

$$\begin{aligned}\dot{V}(x) &= \langle Ax, x\rangle + \langle x, Ax\rangle \\ &= \int_0^\infty \left[-x(\zeta)'\overline{(\zeta)} - x(\zeta)\overline{x(\zeta)'}\right](e^{-\zeta} + 1)d\zeta \\ &= \left[-|x(\zeta)|^2(e^{-\zeta} + 1)\right]_0^\infty - \int_0^\infty |x(\zeta)|^2 e^{-\zeta} d\zeta \\ &= -\int_0^\infty |x(\zeta)|^2 e^{-\zeta} d\zeta,\end{aligned}$$

where we have used the boundary condition. Since the last expression is nonzero for every $x \neq 0$, we have that

$$\dot{V}(x) < 0, \qquad x \neq 0. \tag{18.2}$$

Hence we have a strict Lyapunov function, whereas the system is not (asymptotically) stable. The reason that this is possible lies in the fact that the trajectories are not precompact. That is, for any $x_0 \neq 0$, the closure of $\{x(t) \mid t \geq 0\}$ is not a compact subset of X. This lack of compactness excludes the use of LaSalle's principle, which is needed to conclude from (18.2) asymptotic stability.

18.3 Linearization and Exponential Stability

One standard technique in finite-dimensional systems to check exponential stability is to check the exponential stability of the linearization. For infinite-dimensional systems, a similar result hold. However, before stating it, we first define two concepts of derivative.

Definition 18.1 For $f : X \mapsto X$ we say that $Df(x)$ is its Fréchet derivative at x if $Df(x)$ is a bounded operator from X to X and

$$\lim_{\|h\|\to 0} \frac{\|f(x+h) - f(x) - (Df)(x)h\|}{\|h\|} = 0.$$

Furthermore, we say that $Df(x)$ is its Gateaux derivative at x if $Df(x)$ is a bounded operator from X to X and if for every $h \in X$ there holds

$$\lim_{\varepsilon\to 0, \varepsilon\in\mathbb{R}} \left\| \frac{f(x+\varepsilon h) - f(x) - \varepsilon(Df)(x)h}{\varepsilon} \right\| = 0.$$

Hence the Gateaux derivative calculates the derivative of f by looking at every direction, whereas the Fréchet derivative is uniform. It is easy to see that if f possesses a Fréchet derivative, then it also has a Gateaux derivative, and they are equal.

Using the Fréchet derivative, the linearization result for (18.1) can be formulated. For the proof, we refer to [3].

Theorem 18.2 *Let f have zero Fréchet derivative at zero. If A generates an exponentially stable semigroup on X, then (18.1) is (locally) exponentially stable around zero.*

In the above theorem, we assumed that the Fréchet derivative at the origin was zero. By means of an example, we show that this condition cannot be replaced by the condition that the Gateaux derivative at the origin must be zero.

As state space we take $X = \ell^2(\mathbb{N})$, and we consider the differential equation

$$\dot{x}(t) = -x(t) + f(x(t)), \qquad x(0) = x_0 \tag{18.3}$$

with f given by

$$(f(x))_n = 3\sqrt[n]{(|x_n|)}x_n. \tag{18.4}$$

Hence our system is a diagonal (nonlinear) system with on the diagonal

$$\dot{x}_n(t) = (-1 + 3\sqrt[n]{|x_n(t)|})x_n(t). \tag{18.5}$$

We summarize results of these scalar differential equations in a lemma. The proofs are left to the reader.

Lemma 18.3 *The differential equation (18.5) has the following properties.*

- *The equilibrium's are* $\pm 3^{-n}$ *and zero.*
- *The right-hand side of (18.4) is locally Lipschitz continuous, and for* $|x_n| \leq r$ *the Lipschitz constant can be majorized by* $3(1+\frac{1}{n})\sqrt[n]{r}$.
- *For* $x_n(0) \in (-3^{-n}, 3^{-n})$ *the state converges to zero, and for* $|x_n(0)| > 3^{-n}$ *the state diverges.*
- *For* $|x_n(0)| > 3^{-n}$ *there is a finite escape time.*
- *The linearization of (18.5) around zero is* $\dot{x}_n(t) = -x_n(t)$ *and thus exponentially stable.*

These results are used to characterize the behavior of the nonlinear system (18.3).

Theorem 18.4 *For the nonlinear system (18.3) and (18.4) the following holds.*

1. f *is (locally) Lipschitz continuous from* X *to* X.
2. f *is Gateaux differentiable but not Fréchet at the origin. The Gateaux derivative at the origin is zero.*
3. *The origin is an unstable equilibrium point.*

Proof 1. Let x, z be two elements of X with norm bounded by r. Without loss of generality, we may assume that $r > 1$. Since the norms are bounded by r, the same holds for the absolute value of every element, i.e., $|x_n|, |z_n| \leq r$. Hence we find that

$$\begin{aligned}
\|f(x) - f(z)\|^2 &= \sum_{n=1}^{\infty} \left(3\sqrt[n]{(|x_n|)}x_n - 3\sqrt[n]{(|z_n|)}z_n\right)^2 \\
&\leq \sum_{n=1}^{\infty} \left(3(1+\frac{1}{n}\sqrt[n]{r})\right)^2 (x_n - z_n)^2 \\
&\leq (6r)^2 \|x - z\|^2,
\end{aligned}$$

where we have used Lemma 18.3 and the fact that $r > 1$. Thus f is Lipschitz continuous, and so is the right-hand side of (18.3).

2. We show that the Gateaux derivative of f is zero. This implies that the (Gateaux) linearization of (18.3) is $\dot{x}(t) = -x(t)$.

For $x \in X$ and $\varepsilon \in \mathbb{R}\backslash\{0\}$ we have

$$\begin{aligned}
\|\frac{f(0+\varepsilon x) - f(0)}{\varepsilon} - 0\|^2 &= \sum_{n=1}^{\infty} 9\sqrt[n]{\varepsilon^2 x_n^2}x_n^2 \qquad (18.6)\\
&= 9\sum_{n=1}^{\infty} \sqrt[n]{\varepsilon^2}\sqrt[n]{x_n^2}x_n^2
\end{aligned}$$

Next take a $\delta \in (0, 1)$ and choose N such that $\sum_{n=N}^{\infty} x_n(t)^2 \leq \delta$. In particular, this implies that $\sqrt[n]{x_n^2} \leq 1$ for $n \geq N$. Now choose ε such that $|\varepsilon| < 1$ and $\sum_{n=1}^{N-1} \sqrt[n]{\varepsilon^2} \sqrt[n]{x_n^2} x_n^2 \leq \delta$. Combining these two gives that for this ε there holds that

$$\|\frac{f(0+\varepsilon x) - f(0)}{\varepsilon} - 0\|^2 \leq 9(\delta + \delta).$$

Since δ is arbitrarily, this show that

$$\lim_{\varepsilon \to 0} \|\frac{f(0+\varepsilon x) - f(0)}{\varepsilon} - 0\|^2 = 0$$

and so 0 is the Gateaux derivative of (18.5).

If f would be Fréchet differentiable, then its derivative would equal the Gateaux derivative, and thus zero. However, by choosing in Eq. (18.6) $\varepsilon = 1$ and $x = (x_n)_{n \in \mathbb{N}}$ with $x_n = 0$ for $n \neq N$ and $x_N = 2^{-N}$, we see that $\lim \sup_{\|x\| \to 0} \|f(x)\|/\|x\| > 0$.

3. We choose $x(0) = (x_{0n})_{n \in \mathbb{N}}$ with $x_{0n} = 0$ for $n \neq N$ and $x_{0N} = 2^{-N}$. By Lemma 18.3 we see that the Nth equation of (18.3) is unstable, and thus the state $x(t)$ diverge. Since for $N \to \infty$, there holds $\|x(0)\| \to 0$, we see that there exists an initial state arbitrarily close to zero which is unstable. Thus the nonlinear system is not stable in the origin. □

The example is this section is not uniformly Lipschitz continuous, and almost every solution of (18.3) will have finite escape time. The following simple adaptation of (18.5) gives a uniformly Lipschitz continuous differential equation on X,

$$\dot{x}_n(t) = \frac{(-1 + 3\sqrt[n]{|x_n(t)|})x_n(t)}{1 + x_n(t)^2}.$$

References

1. R.F. Curtain, H.J. Zwart, *An Introduction to Infinite-Dimensional Linear Systems Theory* (Springer, New York, 1995)
2. A. Pazy, *Semigroups of Linear Operators and Applications to Partial Differential Equations* (Springer, New York, 1983)
3. N. Kato, A principle of linearized stability for nonlinear evolution equations. Trans. AMS **347** (1995)

Chapter 19
Model Reduction by Generalized Differential Balancing

Yu Kawano and Jacquelien M.A. Scherpen

Abstract In this chapter, we give a generalization of differential balancing method for model reduction of nonlinear systems in the direction to computation. We generalize concepts of differential controllability and observability functions, then use them for model reduction. We show some stability properties are preserved under the model reduction and estimate the error bound by the model reduction.

19.1 Introduction

For the second author, the work in this paper finds its roots in early work, [15], which I did as a Ph.D. student under the supervision of Arjan at the University of Twenty. It is my pleasure to write in the book of my teacher and mentor at the occasion of his 60th birthday. During my Ph.D. research Arjan was an inspiring researcher, teacher, and supervisor, allowing me to pursue a research direction different from the original plan. Even though I was impressed by his knowledge and ideas, I felt he was always available for questions and open discussions, with or without the many (international) visitors who came to spent time in the group in Twenty. Being one of the leaders in the field of nonlinear control, Arjan contributed significantly to the bustling scientific atmosphere in the group, greatly influencing my perspective on scientific life. After years at different universities, we are now colleagues in Groningen. Ever since I started in Groningen, we have been collaborating again, we share ideas and have jointly supervised a few Ph.D. students. I very much appreciate these encounters,

Y. Kawano
Graduate School of Informatics, Kyoto University, Sakyo-ku, Kyoto 606-8501, Japan
e-mail: ykawano@i.kyoto-u.ac.jp

J.M.A. Scherpen (✉)
Engineering and Technology Institute Groningen, Jan C. Willems Center for Systems and Control, ENTEG-DTPA, University of Groningen, Nijenborgh, 4, 9747AG Groningen, The Netherlands
e-mail: j.m.a.scherpen@rug.nl

M.K. Camlibel et al. (eds.), *Mathematical Control Theory I*,
Lecture Notes in Control and Information Sciences 461,
DOI 10.1007/978-3-319-20988-3_19

and I am honored to organize the workshop, edit the book, and contribute a chapter for Arjan's Festschrift with a topic that finds its roots in my Ph.D. work. *Congratulations Arjan!*

Model order reduction problems have been widely studied because the reduced order models are useful for analysis, design, control, and simulation. In both linear and nonlinear control theory, a balanced realization is a useful state-space representation when studying model reduction problems [2, 7, 8, 15, 19], measuring importance of state variables based on how much energy is minimally needed to reach that state variable, and how much energy is obtained starting in that state variable. Besides balancing, also moment matching [2] is a useful tool for model reduction for control, in general computationally stronger than balanced order reduction, but not having a priori error bound, and less intuition. For nonlinear systems, this method has only been recently developed, see [3, 9]. Balancing for nonlinear systems has a longer history [15], but there are still many recent developments, i.e., there are various other types of nonlinear balancing such as a flow balancing [17, 18], incremental balancing [4], and dynamic balancing [14]. These methods are developed to take into account different properties of importance, such as incremental stability, for example [4]. In general, it depends on the system analysis and the control goal which method is best. In this paper we focus on balancing.

Recently, the authors presented a new balancing method based on contraction theory [10]. Contraction theory has been studied in recent decades, and deals with trajectories of nonlinear systems with respect to one another. One of the interesting ideas of contraction theory is considering the infinitesimal metric instead of a feasible distance function. In this setting, for instance, stability [1, 6, 11], optimal and H^∞ control [12, 13], and dissipativity [5, 16] have been studied. However, if the system order becomes large, the analysis and control becomes difficult, which motivates the study of balancing in the contraction framework, called differential balancing theory. Differential balancing theory is based on two energy functions, the so-called differential controllability and observability functions. In [10], it is shown that these two energy functions have close relationships with solutions to types of Lyapunov equations in contraction theory. That is, well-known results on controllability and observability Gramians in linear systems and control theory have partly been generalized. Moreover, a new model reduction method has been established based on the differential balancing, and this model reduction method is demonstrated for a system for which we cannot apply the incremental balancing method of [4].

As with most of the nonlinear balancing methods, computation of the differential energy functions is still not straightforward. Therefore, in this chapter, we generalize differential balancing into a direction that facilitates computations for obtaining a reduced order model based on generalized differential balancing. This generalized method relies on so-called generalized differential energy functions, which give bounds on the original differential energy functions, following similar principles as in [4, 14]. The existence of these generalized differential functions guarantees boundedness of trajectories of the variational system of the nonlinear system, which property is preserved under model reduction based on generalized differential balancing. In addition, generalized differential balancing has several advantages over other computationally feasible methods as in [4, 14]. First, generalized differential balancing does not require that the vector field of the system is an odd function in contrast to the generalized incremental balancing [4]. Second, an error bound for model reduction is estimated differently from the dynamic balancing in [14]. Moreover, generalized differential balancing can be directly applied to time-varying systems.

The remainder of this paper is organized as follows. In Sect. 19.2, we review results on differential balancing such as the differential energy functions and the differential balanced realization. In Sect. 19.3, we develop generalized differential balancing and present a model reduction method based on generalized differential balancing, which is illustrated by a system composed of 100 mass-spring-damper systems with nonlinear springs. Finally in Sect. 19.4 we conclude the paper.

Notations Let $\mathbb{R}$ be the field of real numbers. Denote $\mathbb{R}_{\geq 0} := [0, \infty) \subset \mathbb{R}$. It is said that $u : [a, b] \to \mathbb{R}^m$ is in $L_2^m[a, b]$ if $||u(t)||_{L_2^m[a,b]} := \sqrt{\int_a^b ||u(t)||^2 dt} < \infty$, where $||u(t)|| := \sqrt{u^{\mathrm{T}}(t)u(t)}$. A curve γ on $\mathbb{R}^n$ is a class C^2 mapping $\gamma : \mathbb{R} \supset [0, 1] \to \mathbb{R}^n$. For matrix $A(x, t) = (a_{ij})$, denote $\delta_f(A) := (\partial a_{ij}/\partial t + (\partial a_{ij}/\partial x) f)$. If A is invertible, we use the notation $A^{-\mathrm{T}}$ to denote $(A^{-1})^{\mathrm{T}}$. For the vector valued function $F : \mathbb{R}^n \times \mathbb{R} \to \mathbb{R}^m$, denote $\partial F(x, t)/\partial x := [\partial F(x, t)/\partial x_1, \ldots, \partial F(x, t)/\partial x_n]$, and $\partial^{\mathrm{T}} F(x, t)/\partial x := (\partial F(x, t)/\partial x)^{\mathrm{T}}$.

19.2 The Differential Balanced Realization

In this section, we review results on differential balancing [10] for nonlinear systems.

Consider the nonlinear time-varying system and its associated system of differential dynamics

$$\Sigma_{BC} : \begin{cases} \dot{x}(t) := dx(t)/dt = f(x(t), t) + B(t)u(t), \\ y(t) = C(t)x(t), \end{cases}$$

$$d\Sigma_{BC} : \begin{cases} \delta\dot{x}(t) := \dfrac{d}{dt}\delta x(t) = \dfrac{\partial(f(x(t), t) + B(t)u(t))}{\partial x}\delta x(t) + B(t)\delta u(t), \\ \delta y(t) = C(t)\delta x(t), \end{cases}$$

where $x(t) \in \mathbb{R}^n$, $u(t) \in \mathbb{R}^m$ and $y(t) \in \mathbb{R}^p$ are, respectively, the state, input and output of Σ_{BC}; $\delta x(t) \in \mathbb{R}^n$, $\delta u(t) \in \mathbb{R}^m$ and $\delta y(t) \in \mathbb{R}^p$ are, respectively, the state, input, and output of $d\Sigma_{BC}$; $f : \mathbb{R}^n \times \mathbb{R} \to \mathbb{R}^n$, $B : \mathbb{R} \to \mathbb{R}^{n\times m}$ and $C : \mathbb{R} \to \mathbb{R}^{p\times n}$ are class C^2. When $u(t) \equiv 0$ and $\delta u(t) \equiv 0$, we denote Σ_{BC} and $d\Sigma_{BC}$ by Σ_C and $d\Sigma_C$, respectively.

Remark 19.1 For each $s \in [0, 1]$, let curve $\gamma(s)$ be an initial condition for Σ_{BC} and $u(\cdot, s)$ be an input signal. If $u(\cdot, \cdot)$ is class C^2, then $x(\cdot, s)$ is a solution to the system Σ_{BC}. Define $\delta x(t) := \partial x(t, s)/\partial s$ and $\delta u(t) := \partial u(t, s)/\partial s$. Then, $\delta x(\cdot, s)$

is a solution to $d\Sigma_{BC}$ from the initial condition $\partial\gamma(s)/\partial s$. Also, the output signal is given by $\delta y(t, s)$.

For differential balancing, the following two energy functions play important roles [10].

Definition 19.2 The differential controllability function of the system Σ_{BC} is defined as

$$L_C(x_0, \delta x_0, t_0) := \inf_{\delta u \in L_2^m(-\infty, t_0]} \frac{1}{2}\int_{-\infty}^{t_0} ||\delta u(t)||^2 dt,$$

for all feasible trajectories $(x(t), u(t))$ of Σ_{BC}, where $x(t_0) = x_0 \in \mathbb{R}^n$, $\delta x(t_0) = \delta x_0 \in \mathbb{R}^n$ and $\delta x(-\infty) = 0$.

Definition 19.3 The differential observability function of the system Σ_C is defined as

$$L_O(x_0, \delta x_0, t_0) := \frac{1}{2}\int_{t_0}^{\infty} ||\delta y(t)||^2 dt,$$

for all feasible trajectories $x(t)$ of Σ_C, where $x(t_0) = x_0 \in \mathbb{R}^n$, $\delta x(t_0) = \delta x_0 \in \mathbb{R}^n$, $\delta x(\infty) = 0$.

It is not guaranteed that these two differential energy functions always exist. Note that these energy functions are the controllability and observability functions for $d\Sigma_{BC}$ and $d\Sigma_C$, respectively. In the linear case, these two functions are nothing but the controllability and observability functions, respectively. Similar to the linear case, differential controllability and observability functions are characterized by Lyapunov type of equations (note that hereafter we leave out arguments when clear from the context for ease of notation) [10].

Theorem 19.4 *Suppose that there exists a nonsingular, real symmetric, and class C^1 solution $-\infty < P(x,t) < \infty$ ($\forall x \in \mathbb{R}^n, \forall t \in \mathbb{R}$) to*

$$-\delta_f(P(x,t)) + \frac{\partial f(x,t)}{\partial x}P(x,t) + P(x,t)\frac{\partial^{\mathrm{T}} f(x,t)}{\partial x} = -B(t)B^{\mathrm{T}}(t), \quad (19.1)$$

$$-\delta_B(P(x,t)) = 0. \quad (19.2)$$

Also, suppose that for all feasible trajectories $(\hat{x}(t), \hat{u}(t))$ of $\dot{\hat{x}}(t) = -f(\hat{x}(t)) - g(\hat{x}(t))\hat{u}(t)$, the trajectory $\delta\hat{x}(t)$ of the following system is bounded for all $t \geq t_0$ and $\lim_{t\to\infty} \delta\hat{x}(t) = 0$.

$$\frac{d}{dt}\delta\hat{x}(t) = -\frac{\partial f(\hat{x}(t),t)}{\partial x}\delta\hat{x}(t) - B(t)B^{\mathrm{T}}(t)P^{-1}(\hat{x}(t),t)\delta\hat{x}(t). \quad (19.3)$$

Then, $L_C(x_0, \delta x_0, t_0) = \frac{1}{2}\delta x_0^{\mathrm{T}} P^{-1}(x_0, t_0)\delta x_0$. □

Theorem 19.5 *Suppose that for all feasible trajectories $x(t)$ of Σ_C, the trajectory $\delta x(t)$ of $d\Sigma_C$ is bounded for all $t \geq 0$ and $\lim_{t\to\infty} \delta x(t) = 0$. If there exists a real symmetric and class C^1 solution $-\infty < Q(x,t) < \infty$ ($\forall x \in \mathbb{R}^n$, $\forall t \in \mathbb{R}$) to*

$$\delta_f(Q(x,t)) + \frac{\partial^{\mathrm{T}} f(x,t)}{\partial x} Q(x,t) + Q(x,t)\frac{\partial f(x,t)}{\partial x} = -C^{\mathrm{T}}(t)C(t), \quad (19.4)$$

then $L_{\mathcal{O}}(x_0, \delta x_0, t_0) = \frac{1}{2}\delta x_0^{\mathrm{T}} Q(x_0, t_0)\delta x_0$. □

In terms of the differential controllability and observability functions, we define a differentially balanced realization for the system Σ_{BC} [10].

Definition 19.6 A realization of the associated system $d\Sigma_{BC}$ is said to be a differentially balanced realization on an open subset $D \subset \mathbb{R}^n \times \mathbb{R}$ if there exists a diagonal matrix

$$\Lambda(x,t) = \mathrm{diag}\{\sigma_1(x,t), \sigma_2(x,t), \ldots, \sigma_n(x,t)\}, \quad (19.5)$$

where $\sigma_1(x,t) \geq \sigma_2(x,t) \geq \cdots \geq \sigma_n(x,t) > 0$ holds on D, and $P(x,t) = \Lambda(x,t)$ and $Q(x,t) = \Lambda(x,t)$, respectively, satisfy (19.1), (19.2) and (19.4).

Theorem 19.7 *Let $P(x,t)$ and $Q(x,t)$ be, respectively, real symmetric and class C^1 solutions to (19.1), (19.2) and (19.4), where $0 < P(x,t) < \infty$ and $0 < Q(x,t) < \infty$ for all $(x,t) \in \mathbb{R}^n \times \mathbb{R}$. The system $d\Sigma_{BC}$ can be transformed into a differentially balanced realization on an open subset $D \subset \mathbb{R}^n \times \mathbb{R}$ by a differential coordinate transformation $\delta z = T(x,t)\delta x$. Moreover, $\sigma_i^2(x,t)$ $(i = 1, \ldots, n)$ in (19.5) are the eigenvalues of the product $P(x,t)Q(x,t)$.* □

19.3 Generalized Differential Balancing

19.3.1 Generalized Differential Energy Functions

In the previous section, balancing theory based on the contraction framework is presented, which is a natural extension of linear balancing theory. From an application perspective, it is worth constructing a computationally more feasible method. Here, we present generalized differential balancing, inspired by generalized incremental balancing as in [4].

We generalize concepts of differential energy functions as follows:

Definition 19.8 If there exists a uniformly positive definite matrix $\bar{P}(t) = \bar{P}^{\mathrm{T}}(t)$ such that

$$-\frac{d\bar{P}(t)}{dt} + \frac{\partial f(x,t)}{\partial x}\bar{P}(t) + \bar{P}(t)\frac{\partial^{\mathrm{T}} f(x,t)}{\partial x} \leq -B(t)B^{\mathrm{T}}(t) \quad (19.6)$$

for all $x \in \mathbb{R}^n, t \in \mathbb{R}$ then the function $L_{\mathcal{C}}(\delta x_0, t_0) := \frac{1}{2}\delta x_0^{\mathrm{T}} \bar{P}^{-1}(t_0)\delta x_0$, is said to be a generalized differential controllability function.

Definition 19.9 If there exists a uniformly positive definite matrix $\bar{Q}(t) = \bar{Q}^{\mathrm{T}}(t)$ such that

$$\frac{d\bar{Q}(t)}{dt} + \bar{Q}(t)\frac{\partial f(x,t)}{\partial x} + \frac{\partial^{\mathrm{T}} f(x,t)}{\partial x}\bar{Q}(t) \leq -C^{\mathrm{T}}(t)C(t) \tag{19.7}$$

for all $x \in \mathbb{R}^n, t \in \mathbb{R}$ then the function $L_{\mathcal{O}}(\delta x_0, t_0) := \frac{1}{2}\delta x_0^{\mathrm{T}} \bar{Q}(t_0)\delta x_0$, is said to be a generalized differential observability function.

Remark 19.10 If we compare (19.1) and (19.4) with (19.6) and (19.7), respectively, we notice that equalities are relaxed into inequalities.

Note that these energy functions are the generalized controllability and observability functions for $d\Sigma_{BC}$, respectively. Also, in the linear case, these two functions are nothing but the generalized controllability and observability functions, respectively. Similar to the linear case, generalized controllability and observability functions are not unique, but they provide a lower bound for the differential controllability function and an upper bound for the differential observability function.

Theorem 19.11 *Suppose that the differential controllability function $L_{\mathcal{C}}(x_0, \delta x_0, t_0)$ and a generalized differential controllability function $\bar{L}_{\mathcal{C}}(\delta x_0, t_0)$ exist. Then,*

$$\bar{L}_{\mathcal{C}}(\delta x_0, t_0) \leq L_{\mathcal{C}}(x_0, \delta x_0, t_0)$$

for all $x_0 \in \mathbb{R}^n, \delta x_0 \in \mathbb{R}^n, t_0 \in \mathbb{R}$. □

Theorem 19.12 *Suppose that the differential observability function $L_{\mathcal{O}}(x_0, \delta x_0, t_0)$ and a generalized differential observability function $\bar{L}_{\mathcal{O}}(\delta x_0, t_0)$ exist. Then,*

$$\bar{L}_{\mathcal{O}}(\delta x_0, t_0) \geq L_{\mathcal{O}}(x_0, \delta x_0, t_0)$$

for all $x_0 \in \mathbb{R}^n, \delta x_0 \in \mathbb{R}^n, t_0 \in \mathbb{R}$. □

19.3.2 Boundedness of Trajectories

Existence of the differential controllability and observability functions is not directly related to controllability and observability, which is the case for linear systems. However, existence of these differential energy functions implies boundedness of trajectories of $d\Sigma_{BC}$.

Theorem 19.13 *If there exists a generalized differential controllability function, then $\delta x(t)$ of the system $d\Sigma_{BC}$ is bounded for any $x_0, \delta x_0 \in \mathbb{R}^n$, $u, \delta u \in L_2^m[0, \infty)$.*

Proof By differentiating the generalized differential controllability function $\bar{L}_C$ $(\delta x, t)$ with respect to t, we have

$$\begin{aligned}
&\frac{d\bar{L}_C(\delta x(t), t)}{dt}\\
&= \frac{1}{2}\frac{d}{dt}\left(\delta x^{\mathrm{T}}(t)\bar{P}^{-1}(t)\delta x^{\mathrm{T}}(t)\right)\\
&= \frac{1}{2}\delta x^{\mathrm{T}}(t)\frac{d\bar{P}^{-1}(t)}{dt}\delta x(t) + \frac{1}{2}\left(\delta^{\mathrm{T}} x(t)\frac{\partial^{\mathrm{T}} f(x(t), t)}{\partial x} + \delta^{\mathrm{T}} u(t)B^{\mathrm{T}}(t)\right)\bar{P}^{-1}(t)\delta x(t)\\
&\quad + \frac{1}{2}\delta^{\mathrm{T}} x(t)\bar{P}^{-1}(t)\left(\frac{\partial f(x(t), t)}{\partial x}\delta x(t) + B(t)\delta u(t)\right)
\end{aligned}$$

From (19.6) and $d\bar{P}^{-1}(t)/dt = -\bar{P}^{-1}(t)(d\bar{P}(t)/dt)\bar{P}^{-1}(t)$, we obtain

$$\begin{aligned}
&\frac{d\bar{L}_C(\delta x(t), t)}{dt}\\
&\le -\frac{1}{2}\delta x^{\mathrm{T}}(t)\bar{P}^{-1}(t)B(t)B^{\mathrm{T}}(t)\bar{P}^{-1}(t)\delta x(t) + \frac{1}{2}\delta^{\mathrm{T}} u(t)B^{\mathrm{T}}(t)\bar{P}^{-1}(t)\delta x(t)\\
&\quad + \frac{1}{2}\delta^{\mathrm{T}} x(t)\bar{P}^{-1}(t)B(t)\delta u(t)\\
&= \frac{1}{2}||\delta u(t)||^2 - \frac{1}{2}||\delta u(t) - B(t)\bar{P}^{-1}(t)\delta x(t)||^2 \le \frac{1}{2}||\delta u(t)||^2.
\end{aligned}$$

By integrating this inequality, we have

$$\bar{L}_C(\delta x(t), t) \le \bar{L}_C(\delta x_0, t_0) + \frac{1}{2}\int_{t_0}^{t} ||\delta u(\tau)||^2 d\tau. \tag{19.8}$$

Since the right-hand side is bounded, the left-hand side is also bounded. Moreover, $\bar{P}(t)$ is uniformly positive definite, which implies that $\delta x(t)$ is bounded. □

Theorem 19.14 *If there exists a generalized differential observability function, then there exists a positive real number* α *such that* $||\delta x(t)||^2 \le \alpha||\delta x_0||^2$ *for system* $d\Sigma_C$*. Moreover,* $\lim_{t\to\infty} ||\delta y(t)||^2 = 0$ *holds.*

Proof By differentiating differential observability function $\bar{L}_O(\delta x(t), t)$ with respect to t, from its definition, we have

$$\begin{aligned}
\frac{d\bar{L}_O(\delta x(t), t)}{dt} &= \frac{1}{2}\frac{d}{dt}\left(\delta x^{\mathrm{T}}(t)\bar{Q}(t)\delta x^{\mathrm{T}}(t)\right)\\
&= \frac{1}{2}\delta x^{\mathrm{T}}(t)\frac{d\bar{Q}(t)}{dt}\delta x(t) + \frac{1}{2}\delta^{\mathrm{T}} x(t)\frac{\partial^{\mathrm{T}} f(x(t), t)}{\partial x}\bar{Q}(t)\delta x(t)\\
&\quad + \frac{1}{2}\delta^{\mathrm{T}} x(t)\bar{Q}(t)\frac{\partial f(x(t), t)}{\partial x}\delta x(t)\\
&\le -\frac{1}{2}||\delta y(t)||^2 \le 0.
\end{aligned}$$

By integrating this inequality,

$$\bar{L}_{\mathcal{O}}(\delta x(t), t) \leq \bar{L}_{\mathcal{O}}(\delta x_0, t_0) - \frac{1}{2}\int_{t_0}^{t} ||\delta y(\tau)||^2 d\tau \leq \bar{L}_{\mathcal{O}}(\delta x_0, t_0). \quad (19.9)$$

The uniform positive definiteness of $\bar{Q}(t)$ implies that there exist $\alpha_2 \geq \alpha_1 > 0$ such that

$$\alpha_1 ||\delta x(t)||^2 \leq \bar{L}_{\mathcal{O}}(\delta x(t), t) \leq \bar{L}_{\mathcal{O}}(\delta x_0, t_0) \leq \alpha_2 ||\delta x_0||^2,$$

and consequently $||\delta x(t)||^2 \leq \frac{\alpha_2}{\alpha_1} ||\delta x_0||^2$.

On the other hand, (19.9) implies

$$\frac{1}{2}\int_{t_0}^{\infty} ||\delta y(\tau)||^2 d\tau \leq \bar{L}_{\mathcal{O}}(\delta x_0, t_0) - \lim_{t\to\infty} \bar{L}_{\mathcal{O}}(\delta x(t), t) \leq \bar{L}_{\mathcal{O}}(\delta x_0, t_0). \quad (19.10)$$

Since $\bar{L}_{\mathcal{O}}(\delta x_0, t_0)$ is bounded, from Barbalat's lemma $\lim_{t\to\infty} ||\delta y(t)||^2 = 0$. □

Remark 19.15 For a generalized controllability or observability function, if there exists a positive real number α such that

$$-\frac{d\bar{P}(t)}{dt} + \frac{\partial f(x,t)}{\partial x}\bar{P}(t) + \bar{P}(t)\frac{\partial^{\mathrm{T}} f(x,t)}{\partial x} \leq -\alpha I_n$$

or

$$\frac{d\bar{Q}(t)}{dt} + \bar{Q}(t)\frac{\partial f(x,t)}{\partial x} + \frac{\partial^{\mathrm{T}} f(x,t)}{\partial x}\bar{Q}(t) \leq -\alpha I_n, \quad (19.11)$$

then $\mathbb{R}^n$ is a contraction region [11] with respect to the uniformly positive definite metric $\bar{P}(t)$ or $\bar{Q}(t)$, respectively. That is, any trajectory of the system Σ_{BC} is bounded.

19.3.3 The Generalized Differentially Balanced Realization

We are now ready to define a generalized differentially balanced realization in terms of the generalized differential controllability and observability functions.

Definition 19.16 A realization of $d\Sigma_{BC}$ is said to be a generalized differentially balanced realization on an open subset $D \subset \mathbb{R}$ if there exists a diagonal matrix

$$\bar{\Lambda}(t) = \mathrm{diag}\{\bar{\sigma}_1(t), \bar{\sigma}_2(t), \ldots, \bar{\sigma}_n(t)\}, \quad (19.12)$$

where $\bar{\sigma}_1(t) \geq \bar{\sigma}_2(t) \geq \cdots \geq \bar{\sigma}_n(t) > 0$ on D holds, and $\bar{P}(t) = \bar{\Lambda}(t)$ and $\bar{Q}(t) = \bar{\Lambda}(t)$.

Theorem 19.17 *Let $\bar{L}_C(\delta x_0, t_0)$ and $\bar{L}_O(\delta x_0, t_0)$ be generalized differential controllability and observability functions, respectively. For every system Σ_{BC}, there exists a coordinate transformation $z = T(t)x$ which transforms $d\Sigma_{BC}$ into a generalized differentially balanced realization on a domain $D \subset \mathbb{R}$. Also $\bar{\sigma}_i^2(t)$ $(i = 1, \ldots, n)$ in* (19.12) *are the eigenvalues of $\bar{P}(t)\bar{Q}(t)$.*

Proof In a similar manner as for the linear case, it can be shown that there exists a class C^1 and invertible matrix $T(t) : \mathbb{R} \to \mathbb{R}^{n\times n}$ which achieves $T(t)\bar{P}(t)T^{\mathrm{T}}(t) = \bar{\Lambda}(t)$ and $T^{-\mathrm{T}}(t)\bar{Q}(t)T^{-1}(t) = \bar{\Lambda}(t)$, where $\bar{\Lambda}(t) = \mathrm{diag}\{\bar{\sigma}_1(t), \ldots, \bar{\sigma}_n(t)\}$, and $\bar{\sigma}_i(t) > 0$ $(i = 1, \ldots, n)$. Moreover, $T(t)$ can be chosen such that $\bar{\sigma}_1(t) \geq \cdots \geq \bar{\sigma}_n(t)$ in a sufficiently small open subset $D \subset \mathbb{R}$. Finally, $\bar{P}(t)\bar{Q}(t) = T^{-1}(t)\bar{\Lambda}^2(t)T(t)$ implies that $\bar{\sigma}_i^2(t)$ $(i = 1, \ldots, n)$ are eigenvalues of $\bar{P}(t)\bar{Q}(t)$. □

19.3.4 Model Reduction and Error Bound

Now we can provide a model reduction procedure based on the generalized differentially balanced realization. Moreover, we establish and estimate of the error bound for the model reduction procedure.

In (19.12), suppose that $\bar{\sigma}_k(t) > \bar{\sigma}_{k+1}(t)$ for $k < n$, which implies that z_k is more important than z_{k+1} in the sense of generalized differential energy. Hence, z_1 until z_k are more important than z_{k+1} until z_n. A possibility to reduce the number of states is by truncation, i.e., to put $z_{k+1} = 0, \ldots, z_n = 0$. We partition the system in the z-coordinates correspondingly as follows:

$$\bar{f}(z,t) = \begin{bmatrix} \bar{f}_a(z_a, z_b, t) \\ \bar{f}_b(z_a, z_b, t) \end{bmatrix} := T(t) f(T^{-1}(t)z(t), t), \ \bar{B}(t) = \begin{bmatrix} \bar{B}_a(t) \\ \bar{B}_b(t) \end{bmatrix} := T(t)B(t),$$
$$\bar{C}(t) = \begin{bmatrix} \bar{C}_a(t) \ \bar{C}_b(t) \end{bmatrix} := C(t)T^{-1}(t),$$

where $z_a := [z_1, \ldots, z_k]^{\mathrm{T}}$ and $z_b := [z_{k+1}, \ldots, z_n]^{\mathrm{T}}$.

The reduced order system is obtained by simply substituting $z_a = \bar{z}_a$ and $z_b = 0$.

$$\Sigma^r \begin{cases} \dot{\bar{z}}_a(t) = \bar{f}_a(\bar{z}_a(t), 0, t) + \bar{B}_a(t)u(t) \\ \bar{y}_a(t) = \bar{C}_a(t)\bar{z}_a(t) \end{cases}.$$

Theorem 19.18 *The state-space realization of reduced order system Σ_{BC}^r is a generalized differential balanced realization with singular value functions $\bar{\sigma}_1(t) \geq \cdots \geq \bar{\sigma}_k(t)$.*

Proof Equations (19.6) and (19.7) in the z-coordinates are

$$
\begin{aligned}
&-\frac{d}{dt}\bar{\Lambda}(t)+\bar{\Lambda}(t)\begin{bmatrix}\frac{\partial \bar{f}_a}{\partial z_a} & \frac{\partial \bar{f}_a}{\partial z_b}\\ \frac{\partial \bar{f}_b}{\partial z_a} & \frac{\partial \bar{f}_b}{\partial z_b}\end{bmatrix}^{\mathrm{T}}(z_a,z_b,t)+\begin{bmatrix}\frac{\partial \bar{f}_a}{\partial z_a} & \frac{\partial \bar{f}_a}{\partial z_b}\\ \frac{\partial \bar{f}_b}{\partial z_a} & \frac{\partial \bar{f}_b}{\partial z_b}\end{bmatrix}(z_a,z_b,t)\bar{\Lambda}(t)\\
&\le -\begin{bmatrix}\bar{B}_a\bar{B}_a^{\mathrm{T}} & \bar{B}_a\bar{B}_b^{\mathrm{T}}\\ \bar{B}_a\bar{B}_b^{\mathrm{T}} & \bar{B}_b\bar{B}_b^{\mathrm{T}}\end{bmatrix}(t),\\
&\frac{d}{dt}\bar{\Lambda}(t)+\begin{bmatrix}\frac{\partial \bar{f}_a}{\partial z_a} & \frac{\partial \bar{f}_a}{\partial z_b}\\ \frac{\partial \bar{f}_b}{\partial z_a} & \frac{\partial \bar{f}_b}{\partial z_b}\end{bmatrix}^{\mathrm{T}}(z_a,z_b,t)\bar{\Lambda}(t)+\bar{\Lambda}(t)\begin{bmatrix}\frac{\partial \bar{f}_a}{\partial z_a} & \frac{\partial \bar{f}_a}{\partial z_b}\\ \frac{\partial \bar{f}_b}{\partial z_a} & \frac{\partial \bar{f}_b}{\partial z_b}\end{bmatrix}(z_a,z_b,t)\\
&\le -\begin{bmatrix}\bar{C}_a^{\mathrm{T}}\bar{C}_a & \bar{C}_b^{\mathrm{T}}\bar{C}_a\\ \bar{C}_b^{\mathrm{T}}\bar{C}_a & \bar{C}_b^{\mathrm{T}}\bar{C}_b\end{bmatrix}(t).
\end{aligned}
$$

Let $\bar{\Lambda}_k(t) := \mathrm{diag}\{\bar{\sigma}_1(t),\dots,\bar{\sigma}_k(t)\}$. For $z_a = \bar{z}_a$ and $z_b = 0$, the upper left $k \times k$ matrix equations become

$$
\begin{aligned}
&-\frac{d}{dt}\bar{\Lambda}_k(t)+\bar{\Lambda}_k(t)\frac{\partial^{\mathrm{T}}\bar{f}_a(\bar{z}_a,0,t)}{\partial z_a}+\frac{\partial \bar{f}_a(\bar{z}_a,0,t)}{\partial z_a}\bar{\Lambda}_k(t)\le -\bar{B}_a(t)\bar{B}_a^{\mathrm{T}}(t),\\
&\frac{d}{dt}\bar{\Lambda}_k(t)+\frac{\partial^{\mathrm{T}}\bar{f}_a(\bar{z}_a,0,t)}{\partial z_a}\bar{\Lambda}_k(t)+\bar{\Lambda}_k(t)\frac{\partial \bar{f}_a(\bar{z}_a,0,t)}{\partial z_a}\le -\bar{C}_a^{\mathrm{T}}(t)\bar{C}_a(t).
\end{aligned}
$$

Thus, $(1/2)d\bar{z}_a^{\mathrm{T}}(t_0)\bar{\Lambda}_k^{-1}(t_0)d\bar{z}_a(t_0)$ and $(1/2)d\bar{z}_a^{\mathrm{T}}(t_0)\bar{\Lambda}_k(t_0)d\bar{z}_a(t_0)$ are a generalized differential controllability and observability functions for Σ_{BC}^r, respectively. □

Remark 19.19 For the reduced order system Σ_{BC}^r, Theorems 19.13 and 19.14 hold.

Next, we estimate an error bound of the trajectories of the original and reduced system. Consider the dynamics of the error $\xi := z - \bar{z}$,

$$
\begin{cases}
\dot{\xi}_a(t) = \bar{f}_a(\xi_a(t)+\bar{z}_a(t),\xi_b(t),t) - \bar{f}_a(\bar{z}_a(t),0,t),\\
\dot{\xi}_b(t) = \bar{f}_b(\xi_a(t)+\bar{z}_a(t),\xi_b(t),t) + \bar{B}_b(t)u(t),\\
y_\xi(t) = \bar{C}(t)\xi(t),
\end{cases}
\tag{19.13}
$$

where $\xi_b \equiv z_b$. Since $\bar{z}_a(t) \in \mathbb{R}^k$ can be seen as an external function of time, the associated system of differential dynamics is

$$
\begin{cases}
\delta\dot{\xi}_a(t) = \dfrac{\partial \bar{f}_a(\xi_a(t)+\bar{z}_a(t),\xi_b(t),t)}{\partial \xi_a(t)}\delta\xi_a(t) + \dfrac{\partial \bar{f}_a(\xi_a(t)+\bar{z}_a(t),\xi_b(t),t)}{\partial \xi_b(t)}\delta\xi_b(t),\\
\delta\dot{\xi}_b(t) = \dfrac{\partial \bar{f}_b(\xi_a(t)+\bar{z}_a(t),\xi_b(t),t)}{\partial \xi_a(t)}\delta\xi_a(t) + \dfrac{\partial \bar{f}_b(\xi_a(t)+\bar{z}_a(t),\xi_b(t),t)}{\partial \xi_b(t)}\delta\xi_b(t)\\
\qquad +\bar{B}_b(t)\delta u(t),\\
\delta y_\xi(t) = \bar{C}(t)\delta\xi(t),
\end{cases}
$$

where $\delta\xi_b \equiv \delta z_b$. We can upper bound the effect of δu on δy_ξ as follows:

Theorem 19.20 *Consider the error dynamics* (19.13). *Suppose that* $\bar{\sigma}_1(t) \geq \cdots \geq \bar{\sigma}_k(t) > \bar{\sigma}_{k+1}(t) \geq \cdots \geq \bar{\sigma}_n(t) > 0$ *for all* $t \geq t_0 \in \mathbb{R}^n$; $\delta z(t_0) = \delta \bar{z}(t_0) = 0$. *Then, for all* $t \in [t_0, \infty)$,

$$||\delta y_\xi(\tau)||_{L_2^p[t_0,t]} \leq 2 \sum_{i=k+1}^{n} ||\bar{\sigma}_i(\tau)\delta u(\tau)||_{L_2^m[t_0,t]}. \tag{19.14}$$

Proof Suppose that $k = n - 1$. Consider the dynamics of $\eta := z + \bar{z}$:

$$\begin{cases} \dot{\eta}_a(t) = \bar{f}_a(\eta_a(t) - \bar{z}_a(t), \eta_b(t), t) + \bar{f}_a(\bar{z}_a(t), 0, t) + 2\bar{B}_a(t)u(t), \\ \dot{\eta}_b(t) = \bar{f}_b(\eta_a(t) - \bar{z}_a(t), \eta_b(t), t) + \bar{B}_b(t)u(t), \\ y_\eta(t) = \bar{C}(t)\eta(t), \end{cases}$$

where $z_b \equiv \eta_b$, and its associated system of differential dynamics is

$$\begin{cases} \delta\dot{\eta}_a(t) = \dfrac{\partial \bar{f}_a(\eta_a(t) - \bar{z}_a(t), \eta_b(t), t)}{\partial \eta_a(t)}\delta\eta_a(t) + \dfrac{\partial \bar{f}_a(\eta_a(t) - \bar{z}_a(t), \eta_b(t), t)}{\partial \eta_b(t)}\delta\eta_b(t) \\ \qquad +2\bar{B}_a(t)\delta u(t), \\ \delta\dot{\eta}_b(t) = \dfrac{\partial \bar{f}_b(\eta_a(t) - \bar{z}_a(t), \eta_b(t), t)}{\partial \eta_a(t)}\delta\eta_a(t) + \dfrac{\partial \bar{f}_b(\eta_a(t) - \bar{z}_a(t), \eta_b(t), t)}{\partial \eta_b(t)}\delta\eta_b(t) \\ \qquad +\bar{B}_b(t)\delta u(t), \\ \delta y_\eta(t) = C(t)\delta\eta(t). \end{cases}$$

By using $\bar{\Lambda}(t)$ in (19.12), denote two differential energy functions.

$$\begin{aligned} 2\bar{L}_C(\eta(t), \delta\eta(t), t) &:= \delta\eta^{\mathrm{T}}(t)\bar{\Lambda}^{-1}(t)\delta\eta(t), \\ 2\bar{L}_O(\xi(t), \delta\xi(t), t) &:= \delta\xi^{\mathrm{T}}(t)\bar{\Lambda}(t)\delta\xi(t). \end{aligned}$$

Since $\bar{\Lambda}$ satisfies (19.6) and (19.7), we obtain

$$\begin{aligned} 2\dot{\bar{L}}_C(\eta(t), \delta\eta(t), t) \leq& -\delta\eta^{\mathrm{T}}\bar{\Lambda}^{-1}\bar{B}\bar{B}^{\mathrm{T}}\bar{\Lambda}^{-1}\delta\eta + 2\delta u^{\mathrm{T}}\bar{B}_a^{\mathrm{T}}\bar{\Lambda}_{n-1}^{-1}\delta\eta_a \\ &+2\delta\eta_a^{\mathrm{T}}\bar{\Lambda}_{n-1}^{-1}\bar{B}_a\delta u + \delta u^{\mathrm{T}}\bar{\sigma}_n^{-1}\bar{B}_b^{\mathrm{T}}\delta\eta_b \\ &+ \delta\eta_b^{\mathrm{T}}\bar{B}_b\bar{\sigma}_n^{-1}\delta u, \\ 2\dot{\bar{L}}_O(\xi(t), \delta\xi(t), t) \leq& -\delta\xi^{\mathrm{T}}\bar{C}\bar{C}^{\mathrm{T}}\delta\xi + \delta u^{\mathrm{T}}\bar{\sigma}_n\bar{B}_b^{\mathrm{T}}\delta\xi_b + \delta\xi_b^{\mathrm{T}}\bar{B}_b\bar{\sigma}_n\delta u. \end{aligned}$$

Because of $\delta\xi_b \equiv \delta\eta_b \equiv \delta x_b$, we have

$$\begin{aligned} &2\dot{\bar{L}}_O(\xi(t), \delta\xi(t), t) + 2\bar{\sigma}_n^2(t)\dot{\bar{L}}_C(\eta(t), \delta\eta(t), t) \\ &\leq -\delta\xi^{\mathrm{T}}\bar{C}\bar{C}^{\mathrm{T}}\delta\xi - \bar{\sigma}_n^2\delta\eta^{\mathrm{T}}\bar{\Lambda}^{-1}\bar{B}\bar{B}^{\mathrm{T}}\bar{\Lambda}^{-1}\delta\eta \\ &\qquad + 2\bar{\sigma}_n^2\delta u^{\mathrm{T}}\bar{B}_a^{\mathrm{T}}\bar{\Lambda}_{n-1}^{-1}\delta\eta_a + 2\bar{\sigma}_n^2\delta\eta_a^{\mathrm{T}}\bar{\Lambda}_{n-1}^{-1}\bar{B}_a\delta u \\ &\qquad + 2\bar{\sigma}_n^2\delta u^{\mathrm{T}}\bar{\sigma}_n^{-1}\bar{B}_b^{\mathrm{T}}\delta\eta_b + 2\bar{\sigma}_n^2\delta\eta_b^{\mathrm{T}}\bar{B}_b\bar{\sigma}_n^{-1}\delta u \\ &\leq -||\delta y_\xi||^2 + 4\bar{\sigma}_n^2||\delta u||^2 - \bar{\sigma}_n^2||2\delta u - \bar{B}^{\mathrm{T}}\bar{\Lambda}^{-1}\delta\eta||^2. \end{aligned}$$

Integrating over time we obtain

$$\begin{aligned}&2\bar{L}_{\mathcal{O}}(\xi(t),\delta\xi(t),t)+2\bar{\sigma}_n^2(t)\bar{L}_{\mathcal{C}}(\eta(t),\delta\eta(t),t)-2\bar{L}_{\mathcal{O}}(\xi(t_0),\delta\xi(t_0),t_0)\\&-2\bar{\sigma}_n^2(t_0)\bar{L}_{\mathcal{C}}(\eta(t_0),\delta\eta(t_0),t_0)\\&\le\int_{t_0}^{t}\left(-||\delta y_\xi||^2+4\bar{\sigma}_n^2||\delta u||^2-\bar{\sigma}_n^2||2\delta u-\bar{B}^{\mathrm{T}}\bar{\Lambda}^{-1}\delta\eta||^2\right)dt.\end{aligned}$$

From $\delta z(t_0)=\delta\bar{z}(t_0)=0$, we obtain $\delta\eta(t_0)=\delta\xi(t_0)=0$ and thus

$$\begin{aligned}\bar{L}_{\mathcal{O}}(\xi(t_0),\delta\xi(t_0),t_0)&=0,\\\bar{L}_{\mathcal{C}}(\eta(t_0),\delta\eta(t_0),t_0)&=0.\end{aligned}$$

Because of $\bar{L}_{\mathcal{O}}(\xi(t),\delta\xi(t),t)>0$, $\bar{L}_{\mathcal{C}}(\eta(t),\delta\eta(t),t)>0$ and $\bar{\sigma}_n||2\delta u-\bar{B}^{\mathrm{T}}\bar{\Lambda}^{-1}\delta\eta||\ge 0$, we have

$$||\delta y_\xi(\tau)||_{L_p^2[t_0,t]}\le 2||\bar{\sigma}_n(\tau)\delta u(\tau)||_{L_m^2[t_0,t]}.$$

By repeating this procedure for $i=n,\dots,k$, we obtain (19.14). □

19.3.5 Example

We apply model reduction based on generalized differential balancing on a system composed by 100 mass-spring-damper systems with nonlinear springs, see Fig.19.1, where k_l and k_n are, respectively, spring constants of linear and nonlinear springs, and $m=k_l=d=1$ and $k_n=2$. The characteristic of the nonlinear springs is provided in the state-space description. The original state-space representation has 200 states, f, B and C are given by

$$\begin{aligned}&f_{2i-1}=x_{2i}\ (i=1,\dots,100),\\&f_2=-x_{2i-1}+x_{2i+1}-2(x_{2i-1}-x_{2i+1})^3-x_{2i}+x_{2i+2},\\&f_{2i}=-x_{2i-1}+x_{2i-3}-2(x_{2i-1}-x_{2i-3})^3-x_{2i-1}+x_{2i+1}-2(x_{2i-1}-x_{2i+1})^3\\&\qquad-x_{2i}+x_{2i-2}-x_{2i}+x_{2i+2}\ (i=2,\dots,99),\\&f_{200}=-x_{199}+x_{197}-2(x_{199}-x_{197})^3-x_{200}+x_{198},\\&B=\begin{bmatrix}0\cdots0\ 1\end{bmatrix}^{\mathrm{T}},\ C=\begin{bmatrix}0\cdots0\ 1\ 0\end{bmatrix},\end{aligned}$$

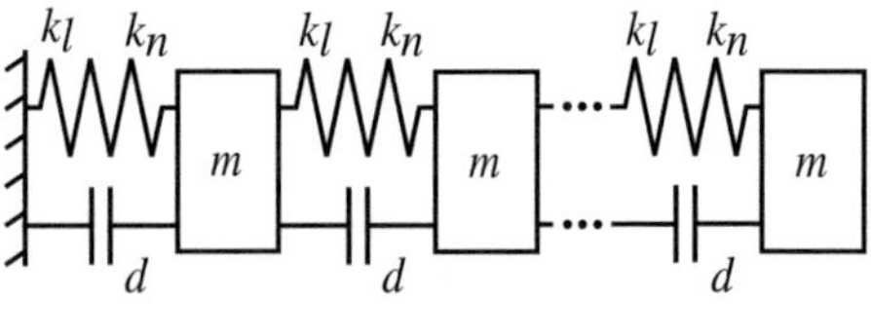

Fig. 19.1 Mass-spring-damper systems with nonlinear springs

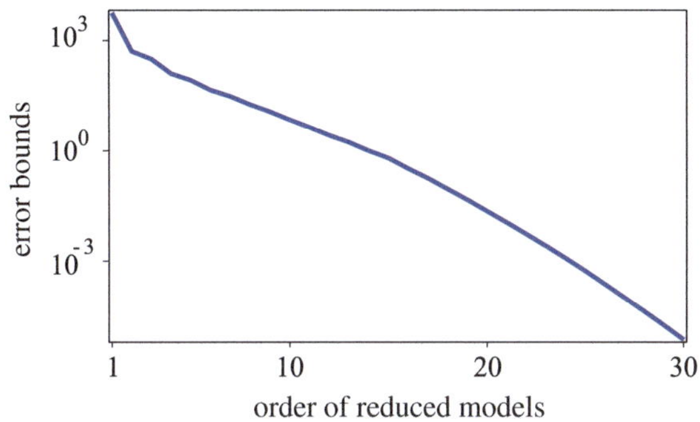

Fig. 19.2 Error bound versus order of reduced model

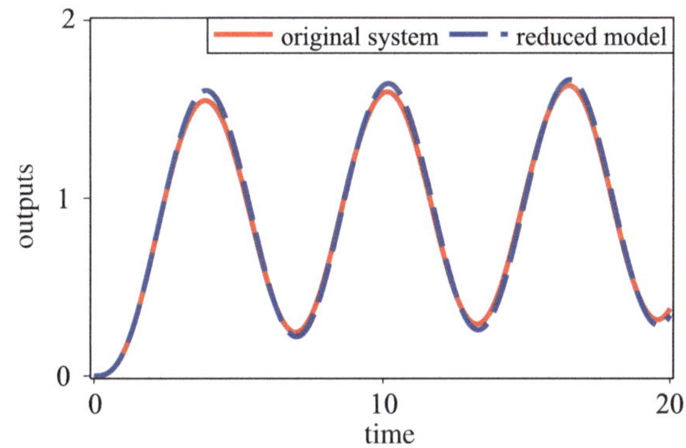

Fig. 19.3 Output trajectories of 200-dimensional original system and 20-dimensional reduced order model

where x_{2i-1} and x_{2i} ($i = 1, \ldots, 100$) are, respectively, position and velocity of the ith mass-spring-damper subsystems. By solving both (19.6) and (19.7), we obtain positive definite matrices, and consequently the system can be transformed into a generalized differential balanced realization. Thus, we can provide an error bound for model reduction using Theorem 19.20, which is shown in Fig. 19.2. For example, it can be seen that the error bound is less than 2.24×10^{-2} for the 20-dimensional reduced order model. Figure 19.3 shows output trajectories of the original system and reduced order model starting from zero initial states and input $u(t) = \sin t$.

19.4 Conclusion

In this chapter, we have presented results on generalized differential balancing for nonlinear systems, which provides an approximation method for balanced truncation with differential balancing constructed in the contraction framework. Generalized differential balancing is based on two energy functions called generalized differential controllability and observability functions. The existences of these generalized differential energy functions guarantee boundedness of trajectories of variational systems of the nonlinear systems, which is preserved under model reduction. We also provide error bounds for model reduction based on generalized differential balancing. The simulation results for a 20-dimensional reduced order model from a system composed of 100 mass-spring-damper systems show a good approximation of the original 200 order model to a sinusoidal input signal.

Acknowledgments This work of Y. Kawano was partly supported by JST CREST and JSPS KAKENHI Grant Number 15K18087. The work was partly performed while Yu Kawano was visiting researcher at the University of Groningen.

References

1. D. Angeli, A Lyapunov approach to incremental stability properties. IEEE Trans. Autom. Control **47**(3), 410–421 (2002)
2. A.C. Antoulas, *Approximation of Large-Scale Dynamical Systems* (SIAM, Philadelphia, 2005)
3. A. Astolfi, Model reduction by moment matching for linear and nonlinear systems. IEEE Trans. Autom. Control **55**(10), 2321–2336 (2010)
4. B. Besselink, N. Van De Wouw, J.M.A. Scherpen, H. Nijmeijer, Model reduction for nonlinear systems by incremental balanced truncation. IEEE Trans. Autom. Control **59**(10), 2739–2753 (2014)
5. F. Forni, R. Sepulchre. On Differentially Dissipative Dynamical Systems, in *Proceedings of the 8th IFAC Symposium on Nonlinear Control Systems*, pp. 21–25 (2013)
6. F. Forni, R. Sepulchre, A differential Lyapunov framework for contraction anlaysis. IEEE Trans. Autom. Control **59**(3), 614–628 (2014)
7. K. Fujimoto, J.M.A. Scherpen, Nonlinear input-normal realizations based on the differential eigenstructure of hankel operators. IEEE Trans. Autom. Control **50**(1), 2–18 (2005)
8. K. Fujimoto, J.M.A. Scherpen, Model reduction for nonlinear systems based on the balanced realization. SIAM J. Control Optim. **48**(7), 4591–4623 (2010)
9. T.C. Ionescu, A. Astolfi. Moment matching for nonlinear port Hamiltonian and gradient systems, in *Preprints of the 9th IFAC Symposium on Nonlinear Control Systems* (2013)
10. Y. Kawano, J.M.A. Scherpen. On Differential Balancing: Energy Functions and Balanced Realization, in *Proceedings of European Control Conference*, Linz, Austria, July 2015. To appear
11. W. Lohmiller, J.J.E. Slotine. On contraction analysis for non-linear systems. Automatica **34**(6), 683–696 (1998)
12. I.R. Manchester, J.J.E. Slotine. Control contraction metrics and universal stabilizability, in *Preprints of the 19th IFAC World Congress* (2014)
13. I.R. Manchester, J.J.E. Slotine. Control contraction metrics: Differential L^2 gain and observer duality. arXiv preprint arXiv:1403.5364v1 (2014)
14. M. Sassano, A. Astolfi, Dynamic generalized controllability and observability functions with applications to model reduction and sensor deployment. Automatica **50**(5), 1349–1359 (2014)
15. J.M.A. Scherpen, Balancing for nonlinear systems. Syst. Control Lett. **21**(2), 143–153 (1993)
16. A.J. van der Schaft, On DIfferential Passivity, in *Proceedings of the 8th IFAC Symposium on Nonlinear Control Systems*, pp. 15–20 (2013)
17. E.I. Verriest, W.S. Gray, Nonlinear balanced realizations, in *43rd IEEE Conference on Decision and Control*, vol. 2, pp. 1164–1169 (2004)
18. E.I. Verriest, W.S. Gray, Flow balancing nonlinear systems, in *14th International Symposium on Mathematical Theory of Networks and Systems*, Perpignan, France (2000)
19. K. Zhou, J.C. Doyle, K. Glover. *Robust and Optimal Control*, vol. 40. (Prentice Hall, New Jersey, 1996)

Chapter 20
Trajectory-Based Theory for Hybrid Systems

A. Agung Julius

Abstract This chapter presents a trajectory-based perspective in solving safety/reachability analysis and synthesis problems and fault diagnosability analysis in hybrid systems. The main tool used in obtaining the results presented in this chapter is the concept of trajectory robustness, which is derived from the theory of approximate bisimulation. Trajectory robustness essentially provides a guarantee on how far the system's state trajectories can deviate (in L_∞ norm) as a result of initial state variations. It further leads to the possibility of approximating the set of the system's trajectories, which is infinite, with a finite set of trajectories. This fact, in turns, allows us to pose the above problems as finitely many finite problems that can be practically solved. In addition, these finite problems can be solved in parallel.

20.1 Note from the Author

This chapter is dedicated to Arjan van der Schaft in the occasion of his 60th birthday. The work presented here germinated from seed ideas that I developed while working as a doctoral student under Arjan's mentorship. I was a graduate student at the Department of Applied Mathematics at the University of Twente in the period of 1999–2005. During this period, I had the fortune of being introduced to the (then) nascent field of hybrid systems, and exposed to the elegance of the behavioral systems theory. Under Arjan's guidance, I wrote my doctoral dissertation on essentially the intersection of these two fields. In later years, while being a postdoctoral researcher at the University of Pennsylvania, I was introduced to the seminal work of Antoine Girard and George Pappas on the approximate bisimulation theory. The trajectory-based perspective of the behavioral systems theory and the notion of invariance and metrics in the space of trajectories from the approximate bisimulation theory are largely the impetus of the results presented in this paper.

A.A. Julius (✉)
Department of Electrical, Computer, and Systems Engineering,
Rensselaer Polytechnic Institute, 110 Eighth Street, NY, Troy 12180, USA
e-mail: agung@ecse.rpi.edu

M.K. Camlibel et al. (eds.), *Mathematical Control Theory I*,
Lecture Notes in Control and Information Sciences 461,
DOI 10.1007/978-3-319-20988-3_20

20.2 Introduction

Hybrid systems are dynamical systems with interacting discrete and continuous dynamics [1]. Intuitively, one way to describe a hybrid system is to think of it as a multimodal dynamical system, where the dynamics of the continuous states depends on the discrete state of the system, which is also called the *mode* or the *location*. Because of their modeling expressivity, hybrid systems have been used in modeling of embedded systems [2–8], air traffic systems [9–14], automotive systems [15–17], electronic circuits [18–20], genetic regulatory networks [21–23], computational morphodynamics [24], and other fields.

In this chapter, we consider two types of hybrid systems, *autonomous* hybrid systems and *control* hybrid systems. More formal definitions of these systems will follow in Sect. 20.3.1. Intuitively, the autonomous hybrid systems do not admit any input. They are the hybrid systems analog of $\dot{x} = f(x)$. Hybrid control systems, on the other hand, admit both continuous and discrete control inputs. They are the hybrid systems analog of $\dot{x} = f(x, u)$. In a sense, for autonomous hybrid systems, the evolution of the states is completely determined by its initial state.[1]

Research involving autonomous hybrid systems is typically of the analysis type, i.e., they are concerned with proving whether the systems have certain properties. One of the most important analysis problems in hybrid systems is the *reachability/safety analysis*, where the question of interest is whether the system can enter an undesirable state during its execution. Reachability/safety analysis has a lot of important practical applications, for example, in the safety analysis of air traffic systems [11, 12, 14, 25, 26], design verification for electronic circuits [18–20], design verification for synthetic biology [21, 27], and model analysis for biochemical processes [28].

Another type of analysis problems that is also studied a lot is the *observability analysis* (see e.g., the editorial [29]). Here, the question of interest is whether we can infer certain properties of the state trajectories by observing certain aspects thereof. An important problem of this type is *fault diagnosis*. The central question in fault diagnosis is whether we can infer that the state trajectory is faulty (e.g., it involves a directly unobservable fault event) from partial observation (e.g., by observing only the a part of the events in the system). Fault diagnosis for hybrid systems is an active research area, with applications in embedded control systems [30], process control [31], and others.

For hybrid control systems, there is a strong research interest involving synthesis. The *synthesis* part of the safety/reachability issue deals with the construction of control laws/algorithms for systems with input and controllable events, in order to achieve executions with desired properties (e.g., safety) despite uncertainties.

In this chapter, we review some results on reachability/safety analysis and synthesis and fault diagnosis for hybrid systems. The underlying theme of the results is

[1] For simplicity, in this chapter we do not consider nondeterminism and stochasticity in the hybrid system dynamics.

that they are all trajectory based. That is, they make use of trajectories to represent the systems, and they are based on reasoning at the trajectory level, instead of at the system representation level.

20.3 Review of the Fundamentals

20.3.1 Hybrid Automata

Following [1], we define hybrid systems as hybrid automata. A hybrid automaton is expressed as an octuple $\mathcal{H} = (L, \mathcal{X}, Init, A, \mathcal{U}, E, Inv, \Sigma)$, where:

- L is a finite set of discrete states, which are also called modes or locations.
- $\mathcal{X}$ is the continuos state space.
- $Init \subset \mathcal{X} \times L$ is the set of initial states.
- A is a finite set of transition symbols.
- $\mathcal{U}$ is the space of continuous input.
- E is the set of transitions.
- $Inv : L \to 2^{\mathcal{X}}$ defines the invariant sets of each location. For an $\ell \in L$, $Inv(\ell)$ is the set in which the continuous states must remain as long as the discrete state is ℓ.
- Σ assigns each location to its continuous dynamics. For each location $\ell \in L$, we define

  $$\Sigma(\ell) : \dot{x} = F_\ell(x),\ x \in Inv(\ell), \tag{20.1}$$

  if the hybrid system is autonomous, or

  $$\Sigma(\ell) : \dot{x} = F_\ell(x, u),\ x \in Inv(\ell),\ u \in \mathcal{U}, \tag{20.2}$$

  if the hybrid system admits control inputs. Here we assume that for each location F_ℓ satisfies some conditions that guarantee well-posedness of the continuous dynamics.

Each transition in E is a pentuple $e = (\ell, \ell', Guard, R, a) \in E$, where $\ell \in L$ is the origin of the transition, $\ell' \in L$ is the target location, $Guard \subset Inv(\ell)$ is the guard set of the transition, and $R : Inv(\ell) \to Inv(\ell')$ is the reset map. The symbol $a \in A$ is the symbol associated with the transition. The semantics of the execution of a hybrid automaton can be explained as follows: (see illustration in Fig. 20.1a). An execution trajectory of $\mathcal{H}$ is a sequence

$$(\ell_0, x_0, u_0, e_0, \Delta_0), (\ell_1, x_1, u_1, e_1, \Delta_1), \ldots, (\ell_N, x_N, u_N, \varnothing, \Delta_N), \tag{20.3}$$

where for all values of i that appear here $\Delta_i \in [0, \infty)$, $e_i \in E$, and $u_i : [0, \Delta_i] \to \mathcal{U}$ is the input signal, if the hybrid system admits input. If the system does not admit

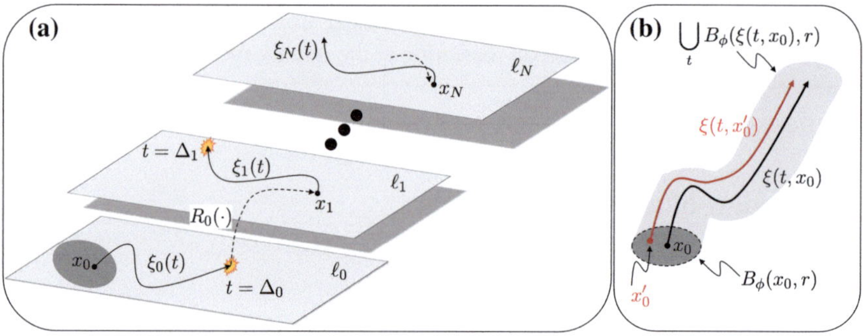

Fig. 20.1 An illustration of **(a)** the execution trajectory of a hybrid automaton, **b** the concept of trajectory robustness

any input, the execution trajectory is a sequence

$$(\ell_0, x_0, e_0, \Delta_0), (\ell_1, x_1, e_1, \Delta_1), \ldots, (\ell_N, x_N, \varnothing, \Delta_N). \tag{20.4}$$

The initial state $(x_0, \ell_0) \in Init$. Each element of the sequence is essentially an interval of execution within which the discrete state is constant. These execution intervals can be characterized recursively as follows. For the ith interval, the value of the continuous state $x(t)$ is given by $\xi_i(t)$, which satisfies $\xi_i(0) = x_i$ and the ODE given by $\Sigma(\ell_i)$. Within the time interval $[0, \Delta_i]$, $\xi_i(t) \in Inv(\ell_i)$. At time $t = \Delta_i$, the transition $e_i = (\ell_i, \ell_{i+1}, Guard_i, R_i, a)$ occurs. That means $\xi_i(\Delta_i) \in Guard_i$ and the continuous state is reset for the next interval of execution. That is, for the $(i+1)$-st interval, the continuous state is initialized at $\xi_{i+1}(0) = x_{i+1} = R_i(\xi_i(\Delta_i))$. Further, the symbol $a \in A$ is associated to the transition. If the transition is triggered externally, for example, a can be considered the discrete command that is given to the system. For the discussion in this chapter, we limit our attention to execution trajectories with finitely many intervals, and that the last interval does not terminate with a transition. Physically, the amount of time that elapses during the execution trajectory above is $\sum_{i=0}^{N} \Delta_i$. Also, we only stipulate that the transitions occur when the continuous state is in the guard set of the transition. We do not stipulate (yet) whether the transitions happen spontaneously, i.e., triggered by the system's own dynamics (example: a falling object bouncing off the floor), or they are triggered externally (example: switching gear in manual transmission).

20.3.2 Trajectory Robustness

The key ingredient in our framework is the notion of *trajectory robustness*. With the notion of trajectory robustness, we provide a guarantee on how far the system's state trajectories can deviate (in L_∞ norm) as a result of initial state variations. This concept is easily extensible to treat system parameter variation, for example, by embedding the parameters as static states in the system. Therefore, although not explicitly stated, the following discussion also applies to variations in system parameter.

We construct trajectory robustness using the theory of *approximate bisimulation*, which was developed by Girard and Pappas [32–34]. This theory was subsequently extended to stochastic hybrid systems and trajectory-based analysis of hybrid systems [35–39]. In the following, we review the application of approximate bisimulation in establishing state trajectory robustness with respect to initial condition variation for a nonlinear dynamical system

$$\Sigma : \dot{x} = F(x), \ x \in \mathcal{X}, \tag{20.5}$$

where x is the state of the system, and $\mathcal{X}$ is the state space. Suppose that we can construct a differentiable function $\phi : \mathcal{X} \times \mathcal{X} \to \mathbb{R}$ such that

$$\phi(x, x') \geq \left\| x - x' \right\|, \forall x, x' \in \mathcal{X}, \tag{20.6a}$$

$$\frac{d\phi}{dt} = \frac{\partial \phi}{\partial x} F(x) + \frac{\partial \phi}{\partial x'} F(x') \leq 0, \forall x, x' \in \mathcal{X}. \tag{20.6b}$$

Such a function $\phi(x, x')$ is called a *autobisimulation function* [32–34].

Notation We denote the solution of (20.5) with initial state x_0 as $\xi(t, x_0)$ and we define the ball

$$B_\phi(x, r) \triangleq \{y \in \mathcal{X} \mid \phi(x, y) \leq r\}, \ x \in \mathcal{X}, \ r > 0. \tag{20.7}$$

From (20.6b), we can easily conclude that the value of ϕ is nondecreasing along any two state trajectories of the system. From (20.6a), we can see that $\phi(x, x')$ is an upper bound for the Euclidean distance between the two states. Then by combining these two properties, we can conclude that for all $t \geq 0$,

$$\xi(t, x_0') \in B_\phi(\xi(t, x_0), \phi(x_0, x_0')), \tag{20.8}$$

$$\left\| \xi(t, x_0) - \xi(t, x_0') \right\| \leq \phi(x_0, x_0'), \tag{20.9}$$

for any initial state $x_0' \in \mathcal{X}$. Please refer to Fig. 20.1b for an illustration of this concept.

Remark 20.1 The concept of trajectory robustness is very related to the concept of contraction metric developed by Lohmiller and Slotine [40]. In general, there are some differences between the two concepts. For example, autobisimulation function can also be defined as a pseudometric if we are only concerned about the divergence of the state trajectories in a certain subspace. Also, as the name suggests, bisimulation functions are originally defined to bound the divergence between the state trajectories of two different systems [33]. However, as defined in this chapter, if we also require that ϕ is a metric in $\mathcal{X}$, then it can be considered as a contraction metric.

The autobisimulation function ϕ plays an essential role in establishing trajectory robustness. If (20.5) defines a stable linear affine dynamics

$$\Sigma : \dot{x} = Ax + b, \ x, b \in \mathbb{R}^n, \ A \in \mathbb{R}^{n \times n}, \tag{20.10}$$

and A is Hurwitz, then ϕ can be using a quadratic Lyapunov function as follows [33, 38].

$$\phi(x, x') = \sqrt{(x - x')^T P (x - x')}, \tag{20.11}$$

where P is a symmetric positive definite matrix satisfying the Lyapunov Linear Matrix Inequality

$$A^T P + PA \preceq 0. \tag{20.12}$$

If (20.5) defines a special class of nonlinear dynamics, the procedure above can be extended. For example, if $F(x)$ in (20.5) is polynomial, we can search for a polynomial autobisimulation function. We refer the reader for more details on this to [39], where sum-of-squares optimization technique [41] is used for this purpose. For the discussion in this chapter, it suffices to consider the linear affine case above.

20.4 Approximation with Finite Behavior

Trajectory robustness gained from approximate bisimulation theory is a very useful tool. It allows us to approximate a dynamical system or a hybrid system with a representation that has finitely many trajectories. The main idea can be explained as follows. Consider the dynamics given by (20.5), and suppose that the system is known to have an initial state in a compact set $Init \subset \mathcal{X}$. This means, any trajectory of the system can be written as $\xi(t, x_0')$, for some $x_0' \in Init$. Suppose that we have a bisimulation function ϕ that satisfies (20.6a), (20.6b). See the illustration in Fig. 20.2.

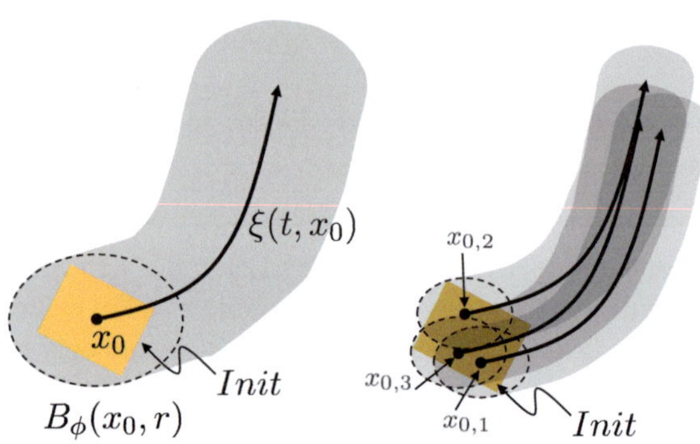

Fig. 20.2 Approximation of the sytem's trajectories with a set of one trajectory (*left*) or finitely many trajectories (*right*)

If $x_0 \in \mathcal{X}$ and $r > 0$ are such that

$$Init \subset B_\phi(x_0, r), \tag{20.13}$$

then for any trajectory of the system $\xi(t, x_0')$, we have

$$\xi(t, x_0') \in B_\phi(\xi(t, x_0), r). \tag{20.14}$$

Thus, we can think of a single trajectory $\xi(t, x_0)$ as an approximation of the entire set of the system's trajectories. Equation (20.14) essentially means that the accuracy of this approximation is given by r.

The idea above can be further generalized and stated as follows.

Theorem 20.2 *Consider a family of initial states* $x_{0,1}, x_{0,2}, \ldots, x_{0,M} \in \mathcal{X}$ *and positive numbers* $r_1, r_2, \ldots, r_M$ *such that*

$$Init \subset \bigcup_{k=1}^{M} B_\phi(x_{0,k}, r_k). \tag{20.15}$$

For any $x_0' \in Init$ *there exists* $k \in \{1, \ldots, M\}$ *such that*

$$\xi(t, x_0') \in B_\phi(\xi(t, x_{0,k}), r_k), \forall t \geq 0. \tag{20.16}$$

The main point of this theorem is that the entire set of the system's trajectories can be approximated by a finite set of trajectories. Again, the numbers $r_1, r_2, \ldots, r_M$ essentially define the accuracy of this approximation. The smaller they are, the approximation is more accurate but we can expect to need more balls to cover $Init$.

In the remainder of this section, we will discuss the extension of this idea to hybrid systems. We limit our discussion in this section to autonomous hybrid systems and leave control hybrid systems for Sect. 20.7.3. Consider a hybrid automaton $\mathcal{H}$ as defined in Sect. 20.3.1. Suppose that $L = \{\ell_0, \ell_1, \ldots, \ell_{|L|}\}$ and that for each discrete state $\ell_i \in L$ the continuous state dynamics

$$\Sigma(\ell_i) : \dot{x} = F_{\ell_i}(x)$$

admits an autobisimulation function ϕ_i. Further, we assume that the guard sets of the transitions define the boundary of the invariant sets of the locations. Also, we assume the transitions in this system occur as soon as the continuous state hits a guard of a transition. Consider an execution trajectory of such hybrid automaton, as exemplified in (20.4). For simplicity of the discussion, let us assume that there are only two intervals, i.e., the trajectory is $(\ell_0, x_0, e_0, \Delta_0), (\ell_1, x_1, \varnothing, \Delta_1)$. This is illustrated in Fig. 20.3. The transition $e_0 = (\ell_0, \ell_1, Guard_0, R_0, a_0)$. We define $\overline{Guard}$ as the union of the guards of all transitions other than e_0 that start in ℓ_0,

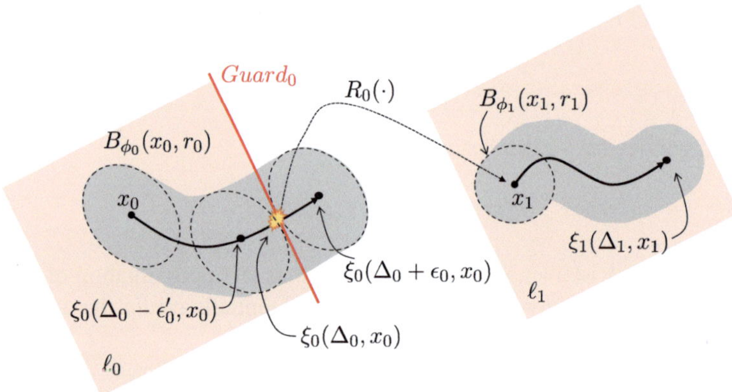

Fig. 20.3 An illustration of an execution trajectory of a hybrid automaton and the concept of trajectory robustness

$$R_0^{-1}\left(B_{\phi_1}(x_1, r_1)\right) \triangleq \left\{x \in Guard_0 \mid R_0(x) \in B_{\phi_1}(x_1, r_1)\right\},$$
$$G_0 \triangleq Guard_0 \backslash R_0^{-1}\left(B_{\phi_1}(x_1, r_1)\right).$$

We can formulate the following theorem (adapted from [38]).

Theorem 20.3 *Suppose that $r_0, r_1, \varepsilon_0, \varepsilon_0' > 0$ are such that the following are true.*

$$B_{\phi_0}(x_0, r_0) \subset Init, \tag{20.17a}$$
$$B_{\phi_0}(\xi_0(t, x_0), r_0) \not\cap (\overline{Guard} \cup G_0), \ \forall t \in [0, \Delta_0 + \varepsilon_0], \tag{20.17b}$$
$$B_{\phi_0}(\xi_0(t, x_0), r_0) \not\cap Inv(\ell_0), \tag{20.17c}$$
$$B_{\phi_0}(\xi_0(t, x_0), r_0) \not\cap Guard_0, \ \forall t \in [0, \Delta_0 - \varepsilon_0'] \tag{20.17d}$$

Then, for any $x_0' \in B_{\phi_0}(x_0, r_0)$ the following are also true:

- *The execution trajectory starting from (x_0', ℓ_0) also exits ℓ_0 through transition e_0.*
- *The transition occurs at time Δ_0', where $\Delta_0' \in [\Delta_0 - \varepsilon_0', \Delta_0 + \varepsilon_0]$.*
- *In the first interval, for all $t \in [0, \Delta_0']$, $\xi_0(t, x_0') \in B_{\phi_0}(\xi_0(t, x_0), r_0)$.*
- *After the transition e_0, the continuous state is reset into $B_{\phi_1}(x_1, r_1)$.*

This theorem can be generalized to trajectories with more intervals. Essentially, it shows that the trajectory starting at (ℓ_0, x_0) is an approximation of those starting in the location ℓ_0 with initial continuous state in the neighborhood of x_0 in the sense that (i) the divergence of the continuous state trajectory is bounded in the sense of Theorem 20.2, (ii) the sequence of transitions are preserved, and (iii) the divergence of the transition times is bounded. Following the same idea as in Theorem 20.2, if the set of initial states is compact, we can use Theorem 20.3 to approximate the set of trajectories of a hybrid automaton with a finite set of trajectories.

20.5 Safety/Reachability Analysis

Safety/reachability analysis is concerned with the question whether any of the system's state trajectories enters a predefined set of unsafe states. For a dynamical system as in (20.5) with a set of initial states $Init$, we define a set $Unsafe \subset \mathcal{X}$ and ask whether there is an initial state $x_0' \in Init$ such that $\xi(t', x_0) \in Unsafe$ for some $t' \in [0, T]$. If such initial state does not exist then the system is safe.[2] This is illustrated in Fig. 20.4. The question described above is called bounded-time safety/reachability analysis, because of the specified upper bound T. If the dynamics (20.5) admits an autobisimulation function ϕ, then the following result can be stated.

Proposition 20.4 *See the illustration in Fig. 20.4. If $r > 0$ is such that*

$$B_\phi(\xi(t, x_0), r) \not\cap Unsafe, \ \forall t \in [0, T],$$

then there exists no initial state $x_0' \in B_\phi(x_0, r)$ from which the state trajectory enters $Unsafe$ in the time interval $[0, T]$.

This proposition allows us to generalize the safety property of the trajectory initialized at x_0 to other trajectories initialized at other states in its neighborhood. Further, the safety of the entire system can be proved by analyzing the safety of finitely many trajectories, as stated below.

Theorem 20.5 *Consider a family of initial states $x_{0,1}, x_{0,2}, \ldots, x_{0,M} \in \mathcal{X}$ and positive numbers $r_1, r_2, \ldots, r_M$ such that*

$$Init \subset \bigcup_{k=1}^{M} B_\phi(x_{0,k}, r_k). \tag{20.18}$$

If for each $k \in \{1, \ldots, M\}$

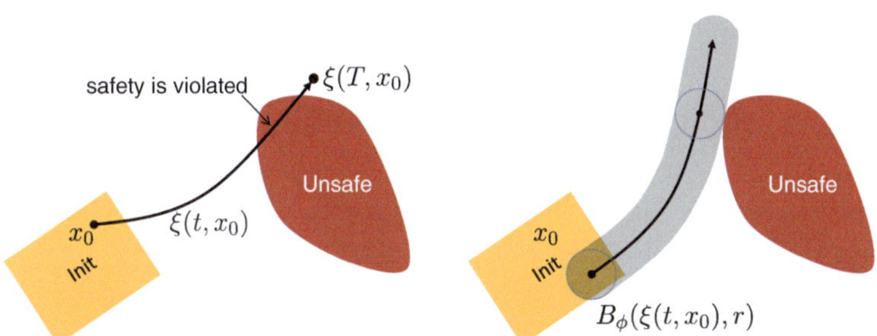

Fig. 20.4 *Left* An illustration of the safety property. *Right* How trajectory robustness can be used in safety/reachability analysis

[2]We assume that $Init$ and $Unsafe$ do not intersect. Otherwise, the problem is trivial.

$$B_\phi(\xi(t, x_{0,k}), r_k) \not\cap Unsafe, \ \forall t \in [0, T],$$

then the system is safe.

For an autonomous hybrid automaton $\mathcal{H}$, safety/reachability analyis can be setup by defining a set of unsafe states $Unsafe \subset \mathcal{X} \times L$. Again, the system is deemed safe if for any state trajectory initialized in $Init \subset \mathcal{X} \times L$, the resulting state trajectory does not enter $Unsafe$. Next, we formulate the analog of Proposition 20.4 for autonomous hybrid automaton. Consider the hybrid automaton discussed in Sect. 20.4, and its execution trajectory that is discussed in Theorem 20.3. The trajectory is $(\ell_0, x_0, e_0, \Delta_0), (\ell_1, x_1, \varnothing, \Delta_1)$.

Proposition 20.6 *Suppose that* $r_0, r_1, \varepsilon_0, \varepsilon_0' > 0$ *satisfy the conditions (20.17a), (20.17d) in Theorem 20.3. In addition, suppose that*

$$B_{\phi_0}(\xi_0(t, x_0), r_0) \not\cap Unsafe, \ \forall t \in [0, \Delta_0 + \varepsilon_0],$$
$$B_{\phi_1}(\xi_1(t, x_1), r_1) \not\cap Unsafe, \ \forall t \in [0, \Delta_1].$$

Then, for any $x_0' \in B_{\phi_0}(x_0, r_0)$ *the following are also true:*

- *The execution trajectory starting from* (x_0', ℓ_0) *is safe until transition* e_0 *that happens at time* Δ_0', *i.e.,*

$$\xi_0(t, x_0') \notin Unsafe, \ \forall t \in [0, \Delta_0'].$$

- *Afterwards, in location* ℓ_1, *the execution trajectory is safe for* Δ_1 *time units, i.e.,*

$$\xi_1(t, R_0\left(\xi_0(\Delta_0', x_0')\right)) \notin Unsafe, \ \forall t \in [0, \Delta_1].$$

This proposition can be generalized to the case where the execution trajectory has more intervals. Also, if the set of initial states $Init$ can be covered by the union of balls as in Theorem 20.5, then we can prove that the system is safe. More details on this idea is reported in [38, 42, 43].

20.6 Observability and Fault Diagnosability

Observability analysis can be intuitively explained as follows. Suppose that the set $\mathfrak{B}$ contains all trajectories of the system, and $\pi_1 : \mathfrak{B} \to \mathfrak{P}_1$ and $\pi_2 : \mathfrak{B} \to \mathfrak{P}_2$ are surjective maps with co-domains $\mathfrak{P}_1$ and $\mathfrak{P}_2$ respectively. The map π_1 can be considered as observation map that extracts information from the trajectories in $\mathfrak{B}$. If this map is bijective, then all information from the trajectories in $\mathfrak{B}$ is retained. Otherwise, multiple distinct trajectories in $\mathfrak{B}$ are mapped to the same element in $\mathfrak{P}_1$, representing the idea that some information (that distinguishes these trajectories) is lost in the observation. The map π_2 represents another aspect of the trajectories in $\mathfrak{B}$. We say that $\mathfrak{P}_2$ is *observable* from $\mathfrak{P}_1$ if the composite map $\pi_1^{-1} \circ \pi_2$ is injective, where π_1^{-1} is the set-theoretic inverse map of π_1. This means the observed information from π_1 can uniquely determine the output of π_2.

In classical linear systems theory, this (behavioral) definition of observability coincides with the notion of observability for state-space systems [44]. That is, if we define the state-space system as

$$\Sigma_{\text{lin}} : \begin{cases} \dot{x} = Ax + Bu, \ x \in \mathbb{R}^n, u \in \mathbb{R}^m, \\ y = Cx + Du, \ y \in \mathbb{R}^p, \end{cases} \tag{20.19}$$

we define $\mathfrak{B}$ to be the set of (x, u, y) trajectories that are compatible with this system description. The map π_1 takes such trajectories and retains only the x components. The map π_2 is the identity map. The characterization of observability as discussed in the previous paragraph coincides with the well-known Kalman rank condition

$$\text{rank}[C^T \ A^T C^T \ \ldots \left(A^T\right)^n C^T] = n. \tag{20.20}$$

For hybrid automata as in Sect. 20.3.1, the notion of observation can be more general. In particular, in this chapter, we consider the observation that simply retains only the discrete aspect of the trajectories. This can be made precise using the following definition.

Definition 20.7 For a hybrid automaton $\mathcal{H}$ as defined in Sect. 20.3.1, we define the function $\Gamma : E \to A$ to map any transition $e \in E$ to its transition symbol. For an execution trajectory

$$\omega \triangleq (\ell_0, x_0, e_0, \Delta_0), (\ell_1, x_1, e_1, \Delta_1), \ldots, (\ell_N, x_N, \varnothing, \Delta_N), \tag{20.21}$$

we define the map

$$\pi_{\text{discrete}}(\omega) \triangleq (\Gamma(e_0), \Delta_0), (\Gamma(e_1), \Delta_1), \ldots, (\varnothing, \Delta_N). \tag{20.22}$$

The map π_{discrete} essentially takes the execution trajectory and returns only the symbols of the transitions and the intervals between the occurrence of the symbols. Note that Γ is not necessarily injective, which implies that the transitions are not necessarily distinguishable one from another. As an extreme case, A is a singleton. That means we can only observe when a transition has occurred, but not which transition. Observability analysis for this kind of observation map has been considered, for example by Di Benedetto et al. in [45] where the question is whether the discrete state can be uniquely determined there from.

Fault diagnosability analysis is related to observability analysis. Suppose that $\mathfrak{B}$, the set of trajectories of the system, can be divided into two disjoint partitions, $\mathfrak{B}_{\text{nom}}$ and $\mathfrak{B}_{\text{fault}}$. $\mathfrak{B}_{\text{nom}}$ represents the normal behavior of the system, while $\mathfrak{B}_{\text{fault}}$ represents the faulty behavior of the system. That is, $\mathfrak{B}_{\text{fault}}$ consists of trajectories where a fault occurs. If we again define the observation map $\pi_1 : \mathfrak{B} \to \mathfrak{P}_1$ as above, then the system is *fault diagnosable* from the observation map π_1 if for any $p \in \mathfrak{P}_1$,

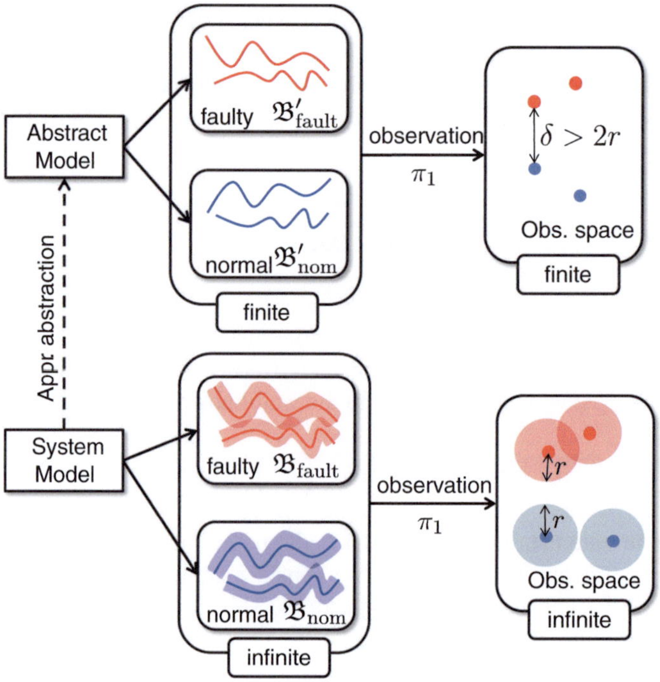

Fig. 20.5 By approximating all the system's trajectories with a set of finitely many trajectories, we can reduce fault diagnosability analysis to a finite problem

$\pi_1^{-1}(p)$ is either strictly in $\mathfrak{B}_{\text{nom}}$ or strictly in $\mathfrak{B}_{\text{fault}}$. In other words, based on the observation defined by π_1, we can always conclude whether the trajectory is normal or faulty.

Although the basic concept is easy to understand, in practice verifying fault diagnosabilty is difficult because the system typically has infinitely many trajectories. However, if we can approximate the system with another one with finitely many trajectories, as explained in Sect. 20.4, then the analysis is much simpler. This is illustrated in Fig. 20.5.

Suppose that $\mathfrak{B}$ can be approximated with $\mathfrak{B}'$ that only has finitely many trajectories. That is, there exists an injective map $\alpha : \mathfrak{B} \to \mathfrak{B}'$ such that for any trajectory $\omega \in \mathfrak{B}$, $\alpha(\omega) \in \mathfrak{B}'$ is an approximation of ω. Intuitively, ω and $\alpha(\omega)$ are close to each other. To be precise, suppose that $\mathfrak{P}_1$ is equipped with a metric $\|\cdot\|_p$ and

$$\|\pi_1(\omega) - \pi_1(\alpha(\omega))\|_p \leq r, \ \forall \omega \in \mathfrak{B}. \tag{20.23}$$

Then, if we define

$$\mathfrak{B}'_{\text{nom}} \triangleq \alpha(\mathfrak{B}_{\text{nom}}), \ \mathfrak{B}'_{\text{fault}} \triangleq \alpha(\mathfrak{B}_{\text{fault}}),$$

we have the following result.

Theorem 20.8 *If for any $\omega_1 \in \mathfrak{B}'_{\text{nom}}$ and $\omega_2 \in \mathfrak{B}'_{\text{fault}}$*

$$\|\pi_1(\omega_1) - \pi_1(\omega_2)\| > 2r \tag{20.24}$$

then the system is fault diagnosable.

Note that checking the condition (20.24) in this theorem is practically possible because both $\mathfrak{B}'_{\text{nom}}$ and $\mathfrak{B}'_{\text{fault}}$ have finitely many trajectories.

For an autonomous hybrid automaton, suppose that ω is an execution trajectory given in (20.21) and $\alpha(\omega)$ is the approximate trajectory in the sense of Theorem 20.3. From the theorem, we know that both trajectories have the same sequence of transitions, and the timing of transitions are close. That is, if the observation map is π_{discrete} and $\pi_{\text{discrete}}(\omega)$ is as given in (20.22), then

$$\pi_{\text{discrete}}(\alpha(\omega)) = \left(\Gamma(e_0), \Delta'_0\right), (\Gamma(e_1), \Delta'_1), \ldots, (\varnothing, \Delta_N), \tag{20.25}$$

$$\Delta'_k \in [\Delta_k - \varepsilon, \Delta_k + \varepsilon], \forall k \in \{0, 1, \ldots, N-1\}, \tag{20.26}$$

for some $\varepsilon > 0$. The distance between $\pi_{\text{discrete}}(\omega)$ and $\pi_{\text{discrete}}(\alpha(\omega))$ can be defined in terms of the timing difference, and hence the distance can be bounded as in (20.23). Therefore, we can use Theorem 20.8 to verify fault diagnosability for autonomous hybrid automata based on observing the timing of the transitions and the respective transition symbols. This is the underlying idea behind some recent work on fault diagnosability of some classes of hybrid systems [46] and probabilistic hybrid systems [47].

20.7 Controller Synthesis

In this section, we discuss the controller synthesis problem related to safety/ reachability properties. For a dynamical system given by

$$\Sigma : \dot{x} = F(x, u), \; x \in \mathcal{X}, \; u \in \mathcal{U}, \tag{20.27}$$

where F is well-posed, the problem can be posed as follows. Given a compact set of initial condition $Init \subset \mathcal{X}$, and a set of goal states $Goal \subset \mathcal{X}$, we want to steer the state starting from any initial state $x_0 \in Init$ such that the state trajectories enter $Goal$ at time $t = T$ and in the time interval $[0, T]$ the state remains safe (does not enter a set of states termed $Unsafe$).

The notion of trajectory robustness discussed in Sect. 20.3.2 can also be used in trajectory-based controller synthesis. The key concept in this approach is the *control autobisimulation function (CAF)* [48, 49]. A continuously differentiable function $\psi : \mathcal{X} \times \mathcal{X} \to \mathbb{R}$ is a **control autobisimulation function** of (20.5) if for any

$x, x' \in \mathcal{X}$, $\psi(x, x') \geq \|x - x'\|$, and there exists a function $k : \mathcal{X} \to \mathcal{U}$ such that

$$\frac{d\psi}{dt} = \frac{\partial \psi}{\partial x} f(x, k(x)) + \frac{\partial \psi}{\partial x'} f(x', k(x')) \leq 0. \tag{20.28}$$

Remark 20.9 The control autobisimulation function is an analog of the control Lyapunov function (CLF) [50], for trajectory robustness [33, 38]. While control Lyapunov function has been used to construct control laws that guarantee stability (e.g., [51]), we shall use the control autobisimulation function to construct control laws that guarantee trajectory robustness.

A consequence of the existence of a CAF is the existence of a feedback control law $u = k(x)$, such that the closed-loop system

$$\dot{x} = F(x, k(x)), x \in \mathcal{X}, \tag{20.29}$$

has $\psi(\cdot, \cdot)$ as a autobisimulation function (see Sect. 20.3.2). For a given dynamical system Σ in (20.27) and a control autobisimulation function ψ, the class of all feedback control laws $k(\cdot)$ that satisfy (20.28) is called the *class of admissible feedback laws*.

Notation For a given dynamical system Σ in (20.27) and a feedback control law $u = k(x)$, the closed-loop trajectory with initial condition $x(0) = x_0$ is denoted by $\xi_k(t, x_0)$. For a control autobisimulation function ψ, $x \in \mathcal{X}$, $r > 0$, we define the ball

$$B_\psi(x, r) \triangleq \{y \in \mathcal{X} \mid \psi(x, y) \leq r\}.$$

The trajectory-based controller synthesis paradigm can be stated as follows. We first construct feedback controllers from the class of feasible feedback laws. By definition, the closed-loop system will then admit a predefined autobisimulation function. This means that the trajectory robustness property discussed in Sect. 20.3.2 is guaranteed to hold. Please refer to Fig. 20.6 for an illustration.

Theorem 20.10 *Suppose that for a given initial state $x_0 \in Init$, we can design an admissible feedback law $u = k_0(x)$ that results in a closed-loop execution trajectory $\xi_{k_0}(t, x_0)$ satisfying*

$$B_\psi(\xi_{k_0}(t, x_0), r_0) \not\cap Unsafe, \ \forall t \in [0, T], \tag{20.30}$$

$$B_\psi(\xi_{k_0}(t, x_0), r_0) \subset Goal. \tag{20.31}$$

Then, for any initial state $x_0' \in B_\psi(\xi_{k_0}(t, x_0), r_0)$, the closed-loop trajectory $\xi_{k_0}(t, x_0)$ is also safe for $t \in [0, T]$ and is in the Goal set at $t = T$.

Therefore, the admissible feedback law $u = k_0(x)$ is applicable not only for the initial state x_0 but also to other initial states in its neighborhood. The controller synthesis procedure can be performed in two steps:

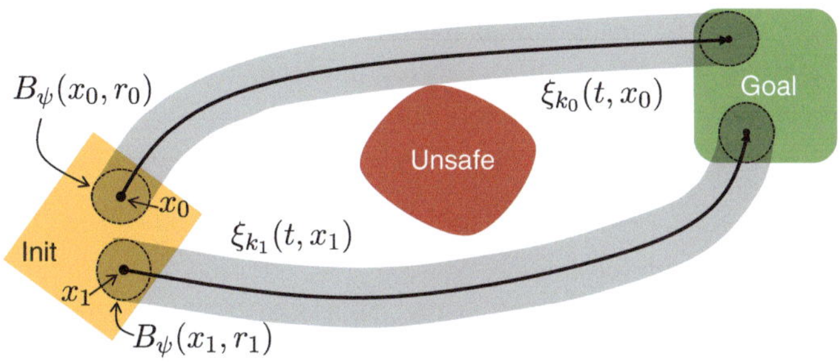

Fig. 20.6 An illustration for trajectory-based controller synthesis

Step 1 For a given initial state, synthesize an innerloop controller that endows the system with the trajectory robustness property.

Step 2 Obtain finitely many trajectories resulting from Step 1 that have the desired qualitative properties to cover $Init$. Note that the controller in Step 1 can depend on the initial state.

Effectively, the trajectory robustness approach allows us to reduce the problem of finding a control law that works for infinitely many initial states in $Init$ to a problem where this has to be done for finitely many initial states. Moreover, the control law can depend on the initial state, and the control law for each initial state can be designed independently of the others'. Steering the system from a particular initial state, is arguably an easier task than finding a control law that works for the entire $Init$ set. Thus, we break down a hard problem into a finite number of simpler and parallelizable problems.

20.7.1 Controller Synthesis for Linear Affine Dynamics

The synthesis of the CAF and the controllers for systems with linear affine dynamics is discussed below. In this case, $F(x, u)$ in (20.27) takes the form

$$F(x, u) = Ax + f + Bu, \; x \in \mathbb{R}^n, u \in \mathbb{R}^m, \tag{20.32}$$

where $A \in \mathbb{R}^{n\times n}$, $f \in \mathbb{R}^n$, and $B \in \mathbb{R}^{n\times m}$. For such systems, we again construct CAF as quadratic functions [48, 49]. That is, we assume that

$$\psi(x, x') = \sqrt{(x - x')^T P (x - x')}, \tag{20.33}$$

where $P \in \mathbb{R}^{n\times n}$ is a positive definite matrix. In this case, the inequality (20.28) becomes

$$(x - x')^T P \left(A(x - x') + B(k(x) - k(x'))\right) \leq 0. \tag{20.34}$$

We then construct a feedback law of the form $u(t) = k(x) = Kx + v(t)$, where $K \in \mathbb{R}^{m\times n}$, and $v(t) \in \mathbb{R}^m$ is a time-varying function, both to be determined later. With this controller, (20.34) becomes

$$(x - x')^T P (A + BK)(x - x') \leq 0. \tag{20.35}$$

A well-known result in control theory (see e.g., [52, 53]) states that there exist P and K such that (20.35) holds if and only if (A, B) is stabilizable. In this case, there are well-known methods to synthesize the suitable P and K. For example, we can pose (20.24) as a linear matrix inequality (LMI) [54], which can be solved efficiently using existing semidefinite programming software tools, such as SeDuMi or SDPT3. With some modification, this method can also be used to handle magnitude constraint on the input signal, $\|u\|_{L_\infty} \leq M$, for some $M > 0$ [48, 49].

Notice that given P and K that satisfy (20.35), we are still free to design $v(t)$. Whatever $v(t)$ is, the control law is admissible. The remaining task in the controller design is therefore to find $v(t)$ that steers the trajectories of the closed-loop system from $Init$ to the $Goal$ set, without entering the $Unsafe$ set. This corresponds to Step 2 in the previous section. The problem of finding such $v(t)$ for a given initial state is easier to solve than the original problem, because the control input is only required to work for that particular initial state. We can use a variety of methods for this, for example, using path planning methods from robotics [48], or by using human inputs [49].

In this chapter, we use $v(t)$ as a feedforward control input that depends on the initial state $x(0)$. It is actually possible to define $v(t)$ through a feedback control law, i.e., as a function of $x(t)$. Such feedback law can be learned from the feedforward control input, and is guaranteed to have the same safety property as the feedforward controller above. For further discussion on this topic, the reader is referred to [55].

20.7.2 Controller Synthesis for Nonlinear Dynamics

The results from the previous section can be generalized to some classes of systems with nonlinear dynamics [56]. We consider systems of the form:

$$\Sigma : \begin{cases} \frac{dx}{dt} = f(x) + g(x)u, \; x \in \mathbb{R}^n, u \in \mathbb{R}^m, \\ y = h(x), \; y \in \mathbb{R}^m, \end{cases} \tag{20.36}$$

where y is the output of the system. We assume that the safety and goal reaching properties of the system can be expressed in terms of y (instead of x).

20.7.2.1 Feedback Linearizable Systems

If (20.36) is feedback linearizable (for a comprehensive discussion, the reader is referred to standard textbooks on Nonlinear Control Systems such as [57, 58]), there exists a feedback law

$$u(t) = \kappa(x) + \lambda(x)w(t), \;\; w(\cdot) \in \mathbb{R}^m, \tag{20.37}$$

such that the closed-loop system, with new input $w(t)$ and output $y(t)$, is a linear system. The necessary and sufficient conditions for feedback linearizability and the design procedure for $\kappa(x)$ and $\lambda(x)$ are covered in the above-mentioned books. In the context of our discussion, the linearizing feedback can be implemented as an inner feedback loop. Once the system is linear, we can apply the results from the previous section for controller synthesis. This method has been applied in designing a controller for fully actuated flexible robot arms [59], whose dynamics are feedback linearizable.

20.7.2.2 Differentially Flat Systems

Differentially flat systems are widely encountered in mechanical and robotics systems. For examples and comprehensive discussion, the reader is referred to [60–62]. The system in (20.37) is differentially flat if it has flat outputs. The outputs $y = (y_1, \ldots, y_m)$ are *flat outputs* if x and u can be written as functions of y and its time derivatives,

$$x = \Xi(y, \dot{y}, \ldots, y^{(\ell)}), \;\; u = \Upsilon(y, \dot{y}, \ldots, y^{(\ell+1)}), \tag{20.38}$$

for some integer ℓ, and $(y, \dot{y}, \ldots, y^{(\ell)})$ are not constrained to satisfy a differential equation by themselves. In other words, any sufficiently smooth trajectory y is admissible as an output trajectory of the system.

A differentially flat system is related to a linear system, namely an ℓth order integrator chain, through the transformation in (20.38). In the context of our discussion, we can apply the results from the previous section for controller synthesis for the integrator chain. The controller for the nonlinear system (20.37) can then be obtained using the transformation in (20.38).

20.7.3 Controller Synthesis for Hybrid Systems

Consider the control hybrid automata defined in Sect. 20.3.1. For simplicity of the discussion, let us assume that the continuous state dynamics in each location is linear affine. That is, suppose that $L = \{\ell_0, \ell_1, \ldots, \ell_{|L|}\}$ and that for each discrete state $\ell_i \in L$ the continuous state dynamics is

$$\Sigma(\ell_i) : \dot{x} = A_{\ell_i} x + f_{\ell_i} + B_{\ell_i} u,$$

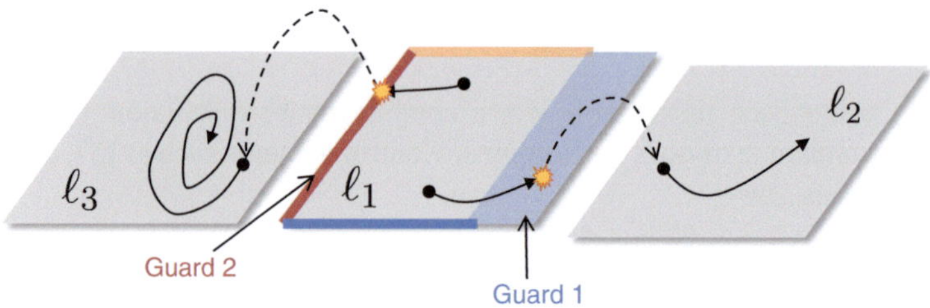

Fig. 20.7 Illustration of forcing and nonforcing guards of hybrid automata. From different initial states in location ℓ_1, the trajectories can undergo different evolution and transition to different locations. The set Guard 1 represents a nonforcing transition, while Guard 2 represents a forcing transition

where A_{ℓ_i}, B_{ℓ_i}, and f_{ℓ_i} are matrices with appropriate dimensions, and the pair (A_{ℓ_i}, B_{ℓ_i}) is stabilizable. Therefore, the continuous state dynamics in each location admits a control autobisimulation function ψ_i.

We assume there are two types of transitions, *forcing* and *nonforcing*. A forcing transition occurs immediately when the continuous state hits the guard, which is the case for autonomous hybrid automata in Sect. 20.4. See Fig. 20.7, where the guards of forcing transitions are illustrated as lines on the boundary of the invariant set of location ℓ_1. A nonforcing transition can happen at any time while the continuous state is in its guard. In Fig. 20.7, this is illustrated by the transition from location ℓ_1 to ℓ_2. In this case, the guard set is "thick," indicating that the transition can happen, but not necessarily as soon as the guard is hit. The occurrence of a nonforcing transition can be user-triggered (corresponding to a discrete control input), or externally triggered. Nonforcing transitions are useful to model events whose occurrence is not predetermined (uncertain) because it is to be triggered by the user/controller, or it is triggered externally at an a priori unknown time.

In defining the control specification, we define a set of initial state $Init \subset \mathcal{X} \times L$. We assume that there is a subset $Unsafe \subset \mathcal{X} \times L$ of unsafe states. A trajectory of the hybrid system corresponds to an unsafe execution if it enters the unsafe set. We also define the set $Goal \subset \mathcal{X} \times L$, which must be entered by the state trajectory. Again, the control problem is defined as finding the feedback control strategy that is guaranteed to bring any initial state in $Init$ to the $Goal$ set without entering the $Unsafe$ set.

Without any loss of generality, we can assume that the set $Init$ is contained in (the invariant set of) one location, called $\ell_{\text{init}} \in L$. If this is not the case, we can divide the problem into several subproblems, each with an $Init$ set contained in a specific location. Similarly, we can assume the $Goal$ is also entirely contained in one location, called $\ell_{\text{goal}} \in L$.

Controller synthesis for hybrid systems can be done using a hierarchical approach, which can be described in the following steps:

Step 1: Discrete Synthesis. We compute a discrete trajectory that starts in ℓ_{init} and ends in ℓ_{goal}. By discrete trajectory, we mean an alternating sequence of locations and transitions

$$\ell_{\text{init}} = \ell_0 \overset{e_0}{\to} \ell_1 \overset{e_1}{\to} \ell_2 \overset{e_2}{\to} \cdots \overset{e_{N-1}}{\to} \ell_N = \ell_{\text{goal}}. \tag{20.39}$$

Each transition e_i, $i \in \{0, \ldots, N-1\}$, is an element of E, originating in ℓ_i, and targeting ℓ_{i+1}. We require that each transition here is either forcing or user-triggered. Such a discrete trajectory is not necessarily unique, but at this step we only need one. The computation of such a discrete trajectory is a standard procedure in formal verification of discrete event systems [63]. For this purpose, there are many good algorithms and computational tools that can be used, such as STRIPS and PDLL [64].

Step 2: Continuous Synthesis. In this step, we synthesize the continuous controller for each of the visited locations ($\ell_{0,1,\ldots,N}$) in order to implement the computed discrete trajectory. In each location ℓ_i,, we define an initial set based on how ℓ_i is reached from ℓ_{i-1}. We then formulate the control problem of bringing the continuous state from this initial set to the interim goal set, which is the guard of transition e_i that will bring the state to location ℓ_{i+1} without entering the forbidden set. The forbidden set is defined as the union of $Unsafe$ and the guards of other forcing transitions from ℓ_i. This is thus an instance of the control problem discussed in Sect. 20.7.1. If we are able to construct a continuous controller that implements the discrete trajectory, then the hybrid control problem is solved. Otherwise, we go back to Step 1, and compute another discrete trajectory.

Remark 20.11 The control problem that we discuss in this chapter is only concerned with the safety/reachability property. In addition, it is possible to formulate an optimal control problem in which a performance objective needs to be optimized while maintaining the safety/reachability property. For further discussion on the trajectory-based approach to this problem, the reader is referred to [65].

20.8 Concluding Remarks

We review some results that allow us to use trajectory-level reasoning in solving some problems in safety/reachability analysis of hybrid systems, controller synthesis for safety/reachability, and fault diagnosability of hybrid systems. The main feature of this approach is the possibility to break down a problem involving infinitely many trajectories of the system into one that only involves finitely many of them.

While we focus solely on safety/reachability property in this chapter, the discussion is actually generalizable to verification of and controller synthesis for other properties, such as those that can be described with temporal logics. In addition, there have also been extension work that consider stochasticity in the dynamics.

Acknowledgments The author wishes to acknowledge the support from the National Science Foundation through the CAREER grant CNS-0953976 and the grant CNS-1218109 for the research leading to results presented here. The results are summarized from the author's earlier work in collaboration with George Pappas, Antoine Girard, Georgios Fainekos, Alessandro D'Innocenzo, and graduate students Sina Afshari, Andrew Winn, and Yi Deng.

References

1. A.J. van der Schaft, J.M. Schumacher, *An Introduction to Hybrid Dynamical Systems* (Springer, London, 2000)
2. P. Tabuada, G.J. Pappas, P. Lima, Compositional abstractions of hybrid control systems. Discrete Event Dyn. Syst. **14**(2), 203–238 (2005). April
3. R. Alur, R. Grosu, I. Lee, O. Sokolsky, Compositional modeling for refinement for hierarchical hybrid systems. J. Logic Algebraic Program. **68**(1–2), 105–128 (2006)
4. J. Hu, Application of Stochastic Hybrid Systems in Power Management of Streaming Data, in *Proceedings of American Control Conference*, Minneapolis, USA (2006)
5. C. Kossentini, P. Caspi, Approximation, Sampling and Voting in Hybrid Computing Systems, in *HSCC 2006*, vol. 3927, LNCS, ed. by J.P. Hespanha, A. Tiwari (Springer, Heidelberg, 2006), pp. 363–376
6. J. Kapinski, A. Donzé, F. Lerda, H. Maka, S. Wagner, B.H. Krogh, Control Software Model Checking Using Bisimulation Functions for Nonlinear Systems, in *Proceedings of IEEE Conference Decision and Control*, Cancun, Mexico (2008)
7. R. Alur, G. Weiss, Regular Specifications of Resource Requirements for Embedded Control Software, in *Proceedings of 14th IEEE Real-Time and Embedded Technology and Applications Symposium*, pp. 159–168 (2008)
8. R. Alur, A. D'Innocenzo, K. Johansson, G. Pappas, G. Weiss, Modeling and Analysis of Multi-hop Control Networks, in *Proceedings of 15th IEEE Real-Time and Embedded Technology and Applications Symposium* (2009)
9. C. Tomlin, G.J. Pappas, S. Sastry, Conflict resolution for air traffic management: a study in multi-agent hybrid systems. IEEE Trans. Autom. Control **43**(4), 509–521 (1998)
10. N. Lynch, High-level Modeling Andanalysis of An Air-traffic Management System, in *Hybrid Systems: Computation and Control*, ser. LNCS, vol. 1589. Springer, p. 3 (1999)
11. M. Prandini, J. Hu, J. Lygeros, S. Sastry, A probabilistic approach to aircraft conflict detection. IEEE Trans. Intell. Transp. Syst. **1**(4), 199–220 (2000)
12. J. Hu, M. Prandini, S. Sastry, Probabilistic Safety Analysis in Three Dimensional Aircraft Flight, in *Proceedings of 42nd IEEE Conference Decision and Control*, Maui, USA, pp. 5335–5340 (2003)
13. A. Bayen, P. Grieder, G. Meyer, C. Tomlin, Lagrangian delay predictive model for sector-based air traffic flow. AIAA J. Guidance Control. Dyn. **28**(5), 1015–1026 (2005)
14. H.A.P. Blom, J. Krystul, G.J. Bakker, A Particle System for Safety Verification of Free Flight in Air Traffic, in *Proceedings of IEEE Conference Decision and Control*, San Diego, USA (2006)
15. J. Lygeros, N. Lynch, Strings of Vehicles: Modeling and Safety Conditions, in *Hybrid Systems: Computation and Control*, ser. LNCS, vol. 1386. Springer, pp. 273–288 (1998)
16. A. Fehnker, Automotive Control Revisited: Linear Inequalities as Approximation of Reachable Sets, in *Hybrid Systems: Computation and Control*, ser. LNCS, vol. 1386. Springer, pp. 110–125, (1998)
17. A. Balluchi, F.D. Natale, A.L. Sangiovanni-Vincentelli, J.H. van Schuppen, Synthesis for Idle Speed Control of an Automotive Engine, in *HSCC 2004*, vol. 2993, LNCS, ed. by R. Alur, G.J. Pappas (Springer, Heidelberg, 2004), pp. 80–94

18. T. Dang, A. Donzé, O. Maler, Verification of Analog and Mixed-Signal Circuits Using Hybrid System Techniques, in *FMCAD 2004*, vol. 3312, LNCS, ed. by A.J. Hu, A.K. Martin (Springer, Heidelberg, 2004), pp. 21–36
19. G. Frehse, PHAVer: Algorithmic Verification of Hybrid Systems Past HyTech, in *HSCC 2005*, vol. 3414, LNCS, ed. by M. Morari, L. Thiele (Springer, Heidelberg, 2005), pp. 258–273
20. G. Frehse, B.H. Krogh, R.A. Rutenbar, O. Maler, Time domain verification of oscillator circuit properties. Electron. Notes Theoret. Comput. Sci. **153**(3), 9–22 (2006)
21. G. Batt, B. Yordanov, R. Weiss, C. Belta, Robustness analysis and tuning of synthetic gene networks. Bioinformatics **23**(18), 2415–2422 (2007)
22. S. Drulhe, G. Ferrari-Trecate, H. de Jong, The switching threshold reconstruction problem for piecewise affine models of genetic regulatory networks. IEEE Trans. Autom. Control **53**(1), 153–165 (2008)
23. E. Cinquemani, A. Milias-Argeitis, S. Summers, J. Lygeros, Local Identification of Piecewise Deterministic Models of Genetic Networks, in *HSCC 2009*, vol. 5469, LNCS, ed. by R. Majumdar, P. Tabuada (Springer, Heidelberg, 2009), pp. 105–119
24. K. Amonlirdviman, N.A. Khare, D.R.P. Tree, W.-S. Chen, J.D. Axelrod, C.J. Tomlin, Mathematical modeling of planar cell polarity to understand domineering nonautonomy. Science **307**(5708), 423–426 (2005)
25. M.L. Bujorianu, J. Lygeros, Reachability Questions in Piecewise Deterministic Markov Processes, in *HSCC 2003*, vol. 2623, LNCS, ed. by O. Maler, A. Pnueli (Springer, Heidelberg, 2003), pp. 126–140
26. A. Abate, S. Amin, M. Prandini, J. Lygeros, S.S. Sastry, Computational Approaches to Reachability Analysis of Stochastic Hybrid Systems, in *HSCC 2007*, vol. 4416, LNCS, ed. by A. Bemporad, A. Bicchi, G. Buttazzo (Springer, Heidelberg, 2007), pp. 4–17
27. G. Batt, C. Belta, R. Weiss, Temporal logic analysis of gene networks under parameter uncertainty. IEEE Trans. Autom. Control **53**(1), 215–229 (2008)
28. D. Riley, X. Koutsoukos, K. Riley, Modeling and analysis of the sugar cataract development process using stochastic hybrid systems. IET Syst. Biol. **3**(3), 137–154 (2009)
29. E. De Santis, M.D. Di Benedetto, Editorial: observability and observer-based control of hybrid systems. Int. J. Robust. Nonlinear Control **19**, 1519–1520 (2009)
30. F. Zhao, X. Koutsoukos, H. Haussecker, J. Reich, P. Cheung, Monitoring and fault diagnosis of hybrid systems. IEEE Trans. Syst. Man Cybern. Part B **35**(6), 1225–1240 (2005)
31. N. Olivier-Maget, G. Hetreux, J.M. Le Lann, M.V. Le Lann, Model-based fault diagnosis for hybrid systems: application on chemical processes. Comput. Chem. Eng. **33**(10), 1617–1630 (2009)
32. A. Girard, G.J. Pappas, Approximate Bisimulation for Constrained Linear Systems, in *Proceedings of the IEEE Conference Decision and Control*, Seville, Spain (2005)
33. A. Girard, G.J. Pappas, Approximation metrics for discrete and continuous systems. IEEE Trans. Autom. Control **52**(5), 782–798 (2007)
34. A. Girard, A.A. Julius, G.J. Pappas, Approximate simulation relations for hybrid systems. Int. J. Discrete Event Dyn. Syst. **18**, 163–179 (2008)
35. A.A. Julius, Approximate Abstraction of Stochastic Hybrid Automata, in *HSCC 2006*, vol. 3927, LNCS, ed. by J.P. Hespanha, A. Tiwari (Springer, Heidelberg, 2006), pp. 318–332
36. A.A. Julius, A. Girard, G.J. Pappas, Approximate Bisimulation for a Class of Stochastic Hybrid Systems, in *Proceedings of American Control Conference*, Minneapolis, USA (2006)
37. A.A. Julius, G.J. Pappas, Approximate abstraction of stochastic hybrid systems. IEEE Trans. Autom. Control **54**(6), 1193–1203 (2009)
38. A.A. Julius, G.E. Fainekos, M. Anand, I. Lee, G.J. Pappas, Robust Test Generation and Coverage for Hybrid Systems, in *HSCC 2007*, vol. 4416, LNCS, ed. by A. Bemporad, A. Bicchi, G. Buttazzo (Springer, Heidelberg, 2007), pp. 329–342
39. A.A. Julius, G.J. Pappas, Trajectory Based Verification Using Local Finite-Time Invariance, in *HSCC 2009*, vol. 5469, LNCS, ed. by R. Majumdar, P. Tabuada (Springer, Heidelberg, 2009), pp. 223–236

40. W. Lohmiller, J.J.E. Slotine, On contraction analysis for nonlinear systems. Automatica **34**(6), 683–696 (1998)
41. S. Prajna, A. Papachristodoulou, P. Seiler, P.A. Parillo, SOSTOOLS and its Control Application, in *Positive Polynomials In Control*. Springer (2005)
42. Y. Deng, A. Rajhans, A.A. Julius, STRONG: A Trajectory-Based Verification Toolbox for Hybrid Systems, in *QEST 2013*, vol. 8054, LNCS, ed. by K. Joshi, M. Siegle, M. Stoelinga, P.R. D'Argenio (Springer, Heidelberg, 2013), pp. 165–168
43. Y. Deng, A.A. Julius, Safe Neighborhood Computation for Hybrid System Verification, in *Proceedings of 4th Workshop on Hybrid Autonomous Systems, ser. Electronic Proceedings in Theoretical Computer Science*, vol. 174. Springer, pp. 1–12 (2014)
44. J.W. Polderman, J.C. Willems, *Introduction to Mathematical Systems Theory: a Behavioral Approach* (Springer, New York, 1998)
45. M.D. Di Benedetto, S. Di Gennaro, A. D'Innocenzo, Discrete state observability of hybrid systems. Int. J. Robust. Nonlinear Control **19**, 1564–1580 (2009)
46. Y. Deng, A.D'Innocenzo, S. Di Gennaro, M.D. Di Benedetto, A.A. Julius, Verification of hybrid automata diagnosability with measurement uncertainty, provisionally accepted to the IEEE Trans. Autom. Control (2015)
47. Y. Deng, A. D'Innocenzo, A.A. Julius, Probabilistic Diagnosability of Hybrid Systems, in *Proceedings of ACM 18th International Conference Hybrid Systems: Computation and Control*, Seattle, WA, pp. 88–97 (2015)
48. A.A. Julius, Trajectory-based Controller Design for Hybrid Systems with Affine Continuous Dynamics, in *Proceedings of IEEE Conference Automation Science and Engineering*, Toronto, Canada, pp. 1007–1012 (2010)
49. A.A. Julius, S. Afshari, Using Computer Games for Hybrid Systems Controller Synthesis, in *Proceedings of 49th IEEE Conference Decision and Control*. Atlanta, Georgia, pp. 5887–5892 (2010)
50. Z. Artstein, Stabilization with relaxed controls. Nonlinear Anal. **15**(11), 1163–1170 (1983)
51. E.D. Sontag, A 'universal' construction of Artstein's theorem on nonlinear stabilization. Syst. Control Lett. **13**(2), 117–123 (1989)
52. W.L. Brogan, *Modern Control Theory* (Prentice Hall International, New Jersey, 1991)
53. B. Friedland, *Control System Design: an Introduction to State-Space Methods*. Dover (2005)
54. S. Boyd, L. El Ghaoui, E. Feron, V. Balakrishnan, *Linear Matrix Inequalities in Systems and Control Theory* (SIAM, Philadelphia, 1994)
55. A.K. Winn, A.A. Julius, Feedback Control Law Generation for Safety Controller Synthesis, in *Proceedings of IEEE Conference Decision and Control*, Florence, Italy, pp. 3912–3917 (2013)
56. A.A. Julius, A.K. Winn, Safety Controller Synthesis Using Human Generated Trajectories: Nonlinear Dynamics with Feedback Linearization and Differential Flatness, in *Proceedings of American Control Conference*, Montreal, Canada, pp. 709–714 (2012)
57. H. Nijmeijer, A.J. van der Schaft, *Nonlinear Dynamical Control Systems* (Springer Verlag, New York, 1990)
58. H.K. Khalil, *Nonlinear Systems, 3rd edn*. (Prentice Hall, 2002)
59. S. Saha, A.A. Julius, Trajectory-based Formal Controller Synthesis for Multi-link Robots with Elastic Joints, in *Proceedings of IEEE Conference Decision and Control*, Los Angeles, CA, pp. 830–835 (2014)
60. M.J. van Nieuwstadt, R.M. Murray, Real-time trajectory generation for differentially flat systems. Int. J. Robust Nonlinear Control **8**(11), 995–1020 (1998)
61. J. Levine, *Analysis and Control of Nonlinear Systems: a Flatness Based Approach*. (Springer, 2009)
62. H. Sira-Ramirez, S. Agrawal, *Differentially Flat Systems*. (Marcel Dekker Inc., New york, 2004)
63. E.M. Clarke, O. Grumberg, D.A. Peled, *Model Checking*. (MIT Press, 1999)
64. S.J. Russell, P. Norvig, *Artificial Intelligence: A Modern Approach*. (Prentice Hall, 2003)
65. A.K. Winn, A.A. Julius, Optimization of Human Generated Trajectories for Safety Controller Synthesis, in Proceedings of American Control Conference, Washington DC, pp. 4374–4379 (2013)

Chapter 21
Controllability and Stabilizability of Discontinuous Bimodal Piecewise Linear Systems

Le Quang Thuan and Kanat Camlibel

Abstract Characterizing controllability like properties of bimodal piecewise linear systems, i.e., piecewise linear systems with two modes, is known to be a notoriously hard. In this chapter, we focus on discontinuous bimodal systems that are well-posed in the sense of existence and uniqueness of solutions. The main results of the chapter are Popov–Belevitch–Hautus-type necessary and sufficient conditions for controllability and stabilizability of such systems.

21.1 Introduction

Since conceived by Kalman, the concept of controllability has been comprehensively studied for linear systems, nonlinear systems, infinite-dimensional systems, positive systems, switching systems, hybrid systems, and behavioral systems. Easily verifiable conditions guaranteeing global controllability for large classes of systems are available only for finite-dimensional linear systems. Even for smooth nonlinear systems, however, results on controllability are local in nature and there is no optimism of obtaining general algebraic characterizations of global controllability [9]. Furthermore, it has been shown in [2] that the problem of characterizing controllability

---

Dedicated to Arjan van der Schaft on the occasion of his sixtieth birthday.

L.Q. Thuan
Department of Mathematics, Quy Nhon University, 170 An Duong Vuong, Quy Nhon, Binh Dinh, Vietnam
e-mail: lethuan2004@yahoo.com

K. Camlibel (✉)
Johann Bernoulli Institute for Mathematics and Computer Science, University of Groningen, 9700 AV Groningen, The Netherlands
e-mail: m.k.camlibel@rug.nl

M.K. Camlibel et al. (eds.), *Mathematical Control Theory I*, Lecture Notes in Control and Information Sciences 461, DOI 10.1007/978-3-319-20988-3_21 for the simplest instances of piecewise linear systems is an undecidable problem, the most undesirable category of problems from a computational complexity point of view.

In this chapter, we focus on bimodal piecewise linear systems with possibly discontinuous vector fields and investigate controllability and stabilizability properties. For bimodal piecewise linear systems, the state space is partitioned into two half-spaces separated by a hyperplane and on each region the dynamics is linear. For such systems, none of the existing results for smooth nonlinear systems can be employed due to the lack of smoothness. Neither can the existing results on the controllability of continuous piecewise affine dynamical systems [3, 5, 6, 10] be directly applied to discontinuous bimodal piecewise linear systems. In fact, the results obtained in the papers [3, 5, 6, 10] heavily rely on the continuity assumption that imposes a certain common geometric structure of the linear subsystems of a piecewise linear system. This common structure has played a key role in obtaining the algebraic necessary and sufficient conditions for classes of continuous piecewise affine dynamical systems in [3, 5, 6, 10]. For discontinuous systems, we will study first necessary conditions for the existence and uniqueness of solutions. It turns out that these conditions lead to algebraic characterizations of controllability and stabilizability for discontinuous bimodal systems by playing a similar role to that of the continuity played in [3, 5, 6, 10].

The outline of this chapter is as follows. In Sect. 21.2, after introducing the so-called bimodal piecewise linear systems, we study the existence and uniqueness of Filippov solutions. The main results will be presented in Sect. 21.3. In this section, we consider controllability and stabilizability problems of bimodal piecewise linear systems and present algebraic characterizations that are much akin to the classical Popov–Belevitch–Hautus conditions. Section 21.4 is devoted to the proofs of the main results. Finally, conclusions follow in Sect. 21.5.

21.2 Discontinuous Bimodal Piecewise Linear Systems

Consider a dynamical system given by the differential inclusion of the form

$$\dot{x}(t) \in F(x(t), u(t)) \tag{21.1}$$

where $x \in \mathbb{R}^n$ is the state, $u \in \mathbb{R}^m$ is the input, and $F : \mathbb{R}^n \times \mathbb{R}^m \rightrightarrows \mathbb{R}^n$ is a set-valued mapping defined by

$$F(x,u) = \begin{cases} \{A_1 x + B_1 u\} & \text{if } c^T x + d^T u < 0 \\ \{A_1 x + B_1 u,\ A_2 x + B_2 u\} & \text{if } c^T x + d^T u = 0 \\ \{A_2 x + B_2 u\} & \text{if } c^T x + d^T u > 0 \end{cases}$$

with $A_1, A_2 \in \mathbb{R}^{n\times n}$, $B_1, B_2 \in \mathbb{R}^{n\times m}$, $c \in \mathbb{R}^n$, and $d \in \mathbb{R}^m$. Also, consider the corresponding convexified differential inclusion

$$\dot{x}(t) \in G(x(t), u(t)) \tag{21.2}$$

where $G : \mathbb{R}^n \times \mathbb{R}^m \rightrightarrows \mathbb{R}^n$ is the set-valued mapping defined by

$$G(x,u) = \begin{cases} \{A_1x + B_1u\} & \text{if } c^Tx + d^Tu < 0 \\ \mathrm{conv}\{A_1x + B_1u, A_2x + B_2u\} & \text{if } c^Tx + d^Tu = 0 \\ \{A_2x + B_2u\} & \text{if } c^Tx + d^Tu > 0 \end{cases}$$

with $\mathrm{conv}(S)$ stands for the convex hull of the nonempty set S. Throughout this chapter, we will call the differential inclusions of the form (21.1) or (21.2) *bimodal piecewise linear systems*.

The main goal of this chapter is to investigate controllability and stabilizability problems for bimodal piecewise linear systems. In case, the implication

$$c^Tx + d^Tu = 0 \implies A_1x + B_1u = A_2x + B_2u \tag{21.3}$$

holds, the set-valued mappings F and G boil down to single-valued Lipschitz continuous functions. In this case, the system (21.1) becomes a particular type of conewise linear systems. Its controllability/stabilizability has been studied in [5, 6]. These papers exploit the structure imposed by continuity on the linear system $\Sigma(A_i, B_i, c^T, d^T)$ as well as that of the overall system, and provide necessary and sufficient conditions for controllability and stabilizability. In case (21.3) does not hold, i.e., F, or G, is not continuous, one cannot apply neither the results nor the approach of the papers [5, 6].

The present work aims at studying controllability and stabilizability problems without requiring continuity. It turns out that existence and uniqueness of solutions play a key role in the absence of continuity. In the sequel, we first define what we mean by a solution of (21.1) and then provide necessary conditions that guarantee existence and uniqueness of solutions. Later, we will exploit these necessary conditions in the context of controllability and stabilizability.

Definition 21.1 An absolutely continuous function $x : \mathbb{R} \to \mathbb{R}^n$ is said to be a *Filippov solution* of the system (21.1) for the initial state x_0 and locally integrable input u if $x(0) = x_0$ and the pair (x, u) satisfies the differential inclusion (21.2) for almost all $t \in \mathbb{R}$.

For a given locally integrable input u, the right-hand side of (21.1) can be seen as a set-valued mapping in t and x. Then, it follows from [7, Lem. 3] that the right-hand side of (21.2) is upper semicontinuous in x. This leads to the following existence result as a result of general existence theorems for upper semicontinuous differential inclusions; see for instance [7, Thm. 5].

Proposition 21.2 *The bimodal piecewise linear system* (21.1) *admits a Filippov solution for any initial state and locally integrable input.*

Definition 21.3 We say that a bimodal piecewise linear system is *well-posed* if it admits a unique Filippov solution for every initial state and locally integrable input.

Let $\mathcal{T}^*(A, B, C, D)$ and $\mathcal{V}^*(A, B, C, D)$ be the smallest input-containing conditioned and the largest output-nulling invariant subspaces of the linear system $\Sigma(A, B, C, D)$, respectively (see e.g., [1, 11]). Also let $\mathcal{K}(A, B, C, D)$ denote the "friends" of $\mathcal{V}^*(A, B, C, D)$.

The following proposition presents necessary conditions for well-posedness of the bimodal system (21.1).

Proposition 21.4 *The following conditions are necessary for the bimodal system* (21.1) *being well-posed:*

1. $\mathcal{V}_1^*(A_1, B_1, c^T, d^T) = \mathcal{V}_2^*(A_2, B_2, c^T, d^T) =: \mathcal{V}^*$.
2. $\mathcal{K}(A_1, B_1, c^T, d^T) = \mathcal{K}(A_2, B_2, c^T, d^T) =: \mathcal{K}$.
3. $(A_1 - B_1K)|_{\mathcal{V}^*} = (A_2 - B_2K)|_{\mathcal{V}^*}$ *for all* $K \in \mathcal{K}$.

Proof Let $x_0 \in \mathcal{V}_1^*(A_1, B_1, c^T, d^T)$, $K \in \mathcal{K}(A_1, B_1, c^T, d^T)$, $\tilde{u}(t) = -K e^{(A_1-B_1K)t}x_0$. Also, let x be the solution of the differential equation $\dot{x} = (A_1-B_1K)x$ with $x(0) = x_0$. Clearly, $\tilde{u}(t) = -Kx(t)$ and

$$(-1)^\nu \dot{x}(t) = A_1(-1)^\nu x(t) + B_1(-1)^\nu \tilde{u}(t) \tag{21.4a}$$
$$c^T(-1)^\nu x(t) + d^T(-1)^\nu \tilde{u}(t) = 0 \tag{21.4b}$$

for all $t \in \mathbb{R}$ and $\nu \in \{1, 2\}$. Thus, $(-1)^\nu x$ is a Filippov solution of the bimodal system (21.1) for the initial state $(-1)^\nu x_0$ and input $(-1)^\nu \tilde{u}$. On the other hand, let $\tilde{x}$ and $\tilde{y}$ be the state and the output of the system

$$\dot{\tilde{x}}(t) = A_2\tilde{x}(t) + B_2\tilde{u}(t), \qquad \tilde{x}(0) = x_0, \tag{21.5a}$$
$$\tilde{y}(t) = c^T\tilde{x}(t) + d^T\tilde{u}(t). \tag{21.5b}$$

It turns out that $\tilde{y}(t) = 0$ for all $t \in \mathbb{R}$. Indeed, if this does not hold, then by noticing that $\tilde{y}$ is real-analytic, there exist $\varepsilon > 0$ and $\nu \in \{1, 2\}$ such that

$$0 < (-1)^\nu \tilde{y}(t) = (-1)^\nu[c^T\tilde{x}(t) + d^T\tilde{u}(t)] \tag{21.6}$$

for all $t \in (0, \varepsilon)$. Then, well-posedness of the system (21.1) yields a contradiction between (21.4b) and (21.6).

Now, it follows from (21.5) and the fact "$\tilde{y} = 0$" that $x_0 \in \mathcal{V}_2^*(A_2, B_2, c^T, d^T)$ and hence $\mathcal{V}_1^*(A_1, B_1, c^T, d^T) \subseteq \mathcal{V}_2^*(A_2, B_2, c^T, d^T)$. Then, by symmetry one obtains

$$\mathcal{V}_1^*(A_1, B_1, c^T, d^T) = \mathcal{V}_2^*(A_2, B_2, c^T, d^T).$$

It also follows from (21.5) and the fact "$\tilde{y} = 0$" that $\tilde{x}$ is a Filippov solution of (21.1) for the initial state x_0 and input $\tilde{u}$. So, well-posedness implies that $x(t) = \tilde{x}(t)$ for all t. In particular, one gets $\dot{x}(0) = \dot{\tilde{x}}(0)$, i.e., $(A_1 - B_1 K)x_0 = (A_2 - B_2 K)x_0$ for all $x_0 \in \mathcal{V}^*$ and $K \in \mathcal{K}(A_1, B_1, c^T, d^T)$. It means

$$(A_1 - B_1 K)\mathcal{V}^* = (A_2 - B_2 K)\mathcal{V}^* \tag{21.7}$$

for all $K \in \mathcal{K}(A_1, B_1, c^T, d^T)$. With (21.7), it is not hard to see that $\mathcal{K}(A_1, B_1, c^T, d^T) = \mathcal{K}(A_2, B_2, c^T, d^T)$, and also the last statement holds.

We now ask ourself whether the smallest input-containing conditioned invariant subspaces of the two linear subsystems coincide as a consequence of well-posedness. In general, the answer is not affirmative as illustrated in the following example.

Example 21.5 Consider the bimodal system with scalar inputs of the form

$$\dot{x}_1(t) \in \begin{cases} \{0\} & \text{if } \; x_2(t) < 0 \\ [0, 1]u(t) & \text{if } \; x_2(t) = 0 \\ \{u(t)\} & \text{if } \; x_2(t) > 0 \end{cases} \tag{21.8a}$$

$$\dot{x}_2(t) = u(t). \tag{21.8b}$$

This system is of the form (21.1) with $c^T = \begin{bmatrix} 0 & 1 \end{bmatrix}$, $d = 0$, and

$$A_1 = A_2 = \begin{bmatrix} 0 & 0 \\ 0 & 0 \end{bmatrix}, \quad B_1 = \begin{bmatrix} 0 \\ 1 \end{bmatrix}, \quad B_2 = \begin{bmatrix} 1 \\ 1 \end{bmatrix}.$$

In terms of the set-valued relay function

$$\operatorname{sgn}(\eta) := \begin{cases} -1 & \text{if } \; \eta < 0 \\ [-1, 1] & \text{if } \; \eta = 0 \\ 1 & \text{if } \; \eta > 0, \end{cases}$$

one can rewrite the bimodal system (21.8) in the form

$$\dot{x}_1(t) \in \frac{1}{2}[\operatorname{sgn}(x_2(t))]u(t) + \frac{1}{2}u(t) \tag{21.9a}$$

$$\dot{x}_2(t) = u(t). \tag{21.9b}$$

We prove that the system (21.9) is well-posed. To do so, it suffices to show that every two Filippov solutions for the same initial state and locally integrable input coincide in a neighborhood of the initial time $t = 0$.

Let $x^0 = \text{col}(x_1^0, x_2^0)$ be an initial state and u be a locally integrable input. For these initial data, the Eq. (21.9b) clearly admits the unique Filippov solution

$$x_2(t) = x_2^0 + \int_0^t u(s)ds. \tag{21.10}$$

Substituting (21.10) into (21.9a), one obtains the following differential inclusion

$$\dot{x}_1(t) \in \frac{1}{2}[\text{sgn}\big(x_2^0 + \int_0^t u(s)ds\big)]u(t) + \frac{1}{2}u(t). \tag{21.11}$$

Due to continuity, it is easy to see that the system (21.11) locally admits the unique Filippov solution if $x_2^0 \neq 0$. It turns out that this also holds if $x_2^0 = 0$. In order to show this, we will use the following result. Its proof will be presented in Sect. 21.4.

Lemma 21.6 *For any given locally integrable function $u : \mathbb{R} \to \mathbb{R}$ and any positive number T, the set*

$$\Gamma_T := \Big\{h \in [0, T] \;\Big|\; \int_0^h u(s)ds = 0 \text{ and } u(h) \neq 0\Big\}$$

has zero measure.

By this lemma, the right-hand side of the differential inclusion (21.11) is almost everywhere equal to the single-valued locally integrable function v defined on $\mathbb{R}$ by

$$v(t) := \begin{cases} 0 & \text{if } x_2(t) < 0 \\ \dfrac{u(t)}{2} & \text{if } x_2(t) = 0 \\ u(t) & \text{if } x_2(t) > 0. \end{cases}$$

In view of this, the differential inclusion (21.11) admits the unique Filippov solution

$$x_1(t) = x_1^0 + \int_0^t v(s)ds.$$

Therefore, the bimodal system (21.8) is well-posed. However, for the bimodal system (21.8), one can easily check that the smallest input-containing conditioned invariant subspaces of two linear subsystems are different and which are

$$\mathcal{T}_1^* = \text{span}\Big\{\begin{bmatrix}0\\1\end{bmatrix}\Big\} \text{ and } \mathcal{T}_2^* = \text{span}\Big\{\begin{bmatrix}1\\1\end{bmatrix}\Big\}.$$

The above example tells us the reason why we cannot use the approach of the papers [5, 6] to solve controllability and stabilizability problems of the bimodal system (21.1) in the case that (21.3) does not fulfill. To solve these problems, we need to find a new technique.

21.3 Main Results

In this section, we present the main results of the chapter. Before doing this, we define controllability and stabilizability concepts of well-posed bimodal piecewise linear systems.

Definition 21.7 Assume that the bimodal piecewise linear system (21.1) is well-posed. We say that the system is

- *controllable* if for any two states x_0, x_f there exist a positive real number T and a locally integrable input u such that $x^u(T; x_0) = x_f$.
- *stabilizable* if for any initial state x_0 there exists a locally integrable input u such that $\lim_{t\to\infty} x^u(t; x_0) = 0$.

Here, $x^u(t; x_0)$ denotes the unique Filippov solution of the bimodal system (21.1) for the initial state x_0 and input u.

The main results of this chapter are presented in the following two theorems. The first one provides a verifiable and algebraic characterization for controllability of well-posed bimodal piecewise linear systems.

Theorem 21.8 *Suppose that $d^T + c^T(sI - A_i)^{-1}B_i$ is right invertible as a rational matrix, and also that the system* (21.1) *is well-posed. Then, the system* (21.1) *is controllable if and only if the following implications hold:*

1. $\lambda \in \mathbb{C}, z \in \mathbb{C}^n, z^*A_i = \lambda z^*, z^*B_i = 0$ *for all* $i = 1, 2 \implies z = 0$.
2. $\lambda \in \mathbb{R}, z \in \mathbb{R}^n, w_1, w_2 \in \mathbb{R}, w_1w_2 \leqslant 0,$ *and* $\begin{bmatrix} z^T & w_i \end{bmatrix}\begin{bmatrix} A_i - \lambda I & B_i \\ c^T & d^T \end{bmatrix} = 0$ *for all* $i = 1, 2 \implies z = 0$.

For stabilizabilty, we also provide an algebraic necessary and sufficient condition as follows.

Theorem 21.9 *Suppose that $d^T + c^T(sI - A_i)^{-1}B_i$ is right invertible as a rational matrix, and also that the system* (21.1) *is well-posed. Then, the system* (21.1) *is stabilizable if and only if the following implications hold:*

1. $\lambda \in \mathbb{C}, \mathrm{Re}(\lambda) \geqslant 0, z \in \mathbb{C}^n, z^*A_i = \lambda z^*, z^*B_i = 0$ *for all* $i = 1, 2 \implies z = 0$.
2. $\lambda \in \mathbb{R}, \lambda \geqslant 0, z \in \mathbb{R}^n, w_1, w_2 \in \mathbb{R}, w_1w_2 \leqslant 0, \begin{bmatrix} z^T & w_i \end{bmatrix}\begin{bmatrix} A_i - \lambda I & B_i \\ c^T & d^T \end{bmatrix} = 0$ *for all* $i = 1, 2 \implies z = 0$.

Proofs of these theorems will be given in the next section.

21.4 Proofs

21.4.1 Proof of Lemma 21.6

In order to prove this lemma, note first that the following equality

$$\Gamma_T = \bigcup_{k=1}^{\infty} (\mathcal{A}_T^k \bigcup \mathcal{B}_T^k) \tag{21.12}$$

holds where $\mathcal{A}_T^k$ and $\mathcal{B}_T^k$ are measurable sets defined by

$$\mathcal{A}_T^k := \{h \in [0, T] \mid \int_0^h u(s)ds = 0, \frac{1}{k} < u(h) < k\} \tag{21.13}$$

and

$$\mathcal{B}_T^k := \{h \in [0, T] \mid \int_0^h u(s)ds = 0, -k < u(h) < -\frac{1}{k}\} \tag{21.14}$$

for $k = 1, 2, \ldots$. In view of the equality (21.12), the zero measure property of the sets $\mathcal{A}_T^k, \mathcal{B}_T^k$ with $k \in \mathbb{N}$ would imply that Γ_T has zero measure due to the subadditivity of measure. Therefore, it suffices to show that $\mathcal{A}_T^k$ and $\mathcal{B}_T^k$ have zero measure for $k = 1, 2, \ldots$ to complete the proof. Note that $\mathcal{B}_T^k$ is very similar to $\mathcal{A}_T^k$ so that it is enough to verify only for $\mathcal{A}_T^k$.

Clearly, for any nonempty bounded open set G of $\mathbb{R}$, the numbers $h_1 := \inf(\mathcal{A}_T^k \cap G)$ and $h_2 := \sup(\mathcal{A}_T^k \cap G)$ finitely exist, and

$$\int_{h_1}^{h_2} u(s)ds = \int_0^{h_1} u(s)ds - \int_0^{h_2} u(s)ds = 0.$$

This implies that

$$\left|\int_G u(s)ds\right| = \left|\int_{G \setminus \mathcal{A}_T^k} u(s)ds\right| \leqslant \int_{G \setminus \mathcal{A}_T^k} |u(s)|ds \tag{21.15}$$

for any bounded open set G. On the other hand, since $\mathcal{A}_T^k$ is measurable and bounded, there exists a sequence of bounded open sets $\{G_\varepsilon : \varepsilon > 0\}$ such that $\mathcal{A}_T^k \subseteq G_\varepsilon$ for all ε and the measure of the set $G_\varepsilon \setminus \mathcal{A}_T^k$ tends to 0 as $\varepsilon \to 0$; see for instance [8, Lem. 3.1]. By the dominated convergence theorem of Lebesgue (see e.g., [8, Thm. 3.15]), one has

$$\lim_{\varepsilon \to 0} \int_{G_\varepsilon} u(s)ds = \int_{\mathcal{A}_T^k} u(s)ds \text{ and } \lim_{\varepsilon \to 0} \int_{G_\varepsilon \setminus \mathcal{A}_T^k} |u(s)|ds = 0.$$

Together with (21.13) and (21.15), these equalities imply that

$$\frac{1}{k}\text{meas}(\mathcal{A}_T^k) = \frac{1}{k}\int_{\mathcal{A}_T^k} ds \leqslant \int_{\mathcal{A}_T^k} u(s)ds = 0$$

where meas stands for the measure. Thus, the set $\mathcal{A}_T^k$ has zero measure for all $k \geqslant 1$.

21.4.2 Proof of Theorem 21.8

21.4.2.1 Proof of the "only if" Part

For this part, we assume that the bimodal system (21.1) is controllable. We will prove the implications 1 and 2 of Theorem 21.8 hold.

We consider the first implication and let $\lambda \in \mathbb{C}$, $z \in \mathbb{C}^n$ be such that $z^* A_i = \lambda z^*$ and $z^* B_i = 0$ for all $i = 1, 2$. Then, for any input u, the Filippov solution $x^u(t; 0)$ of the system (21.1) satisfies

$$z^* \dot{x}^u(t; 0) = \lambda z^* x^u(t; 0) \tag{21.16}$$

for all $t \in \mathbb{R}$. Then, on the one hand, by solving the differential equation (21.16) one gets

$$z^* x^u(t; 0) = e^{\lambda t} z^* x(0) = 0$$

for all $t \in \mathbb{R}$ and any input u. On the other hand, $x^u(t; 0)$ is arbitrary due to controllability. As such, one gets $z = 0$.

Now, we consider the last implication and let $\lambda \in \mathbb{R}$, $z \in \mathbb{R}^n$, and $w_1, w_2 \in \mathbb{R}$ be such that

$$w_1 w_2 \leqslant 0 \text{ and } \begin{bmatrix} z^T & w_i \end{bmatrix} \begin{bmatrix} A_i - \lambda I & B_i \\ c^T & d^T \end{bmatrix} = 0 \text{ for all } i = 1, 2. \tag{21.17}$$

Due to (21.17), for any locally integrable input u the Filippov solution $x^u(t; 0)$ of the system (21.1) satisfies either $z^T \dot{x}^u(t; 0) \geqslant \lambda z^T x^u(t; 0)$ or $z^T \dot{x}^u(t; 0) \leqslant \lambda z^T x^u(t; 0)$ for all $t \in \mathbb{R}$. From this and Gronwall-Belman inequality, one gets either $z^T x^u(t; 0) \geqslant 0$ or $z^T x^u(t; 0) \leqslant 0$ for any positive real number t and any locally integrable input u. Then, controllability implies that $z = 0$.

21.4.2.2 Proof of the 'if' Part

Let $\mathcal{T}_i^*$ and $\mathcal{V}_i^*$ be the smallest input-containing conditioned and the largest output-nulling invariant subspaces of the linear system $\Sigma(A_i, B_i, c^T, d^T)$, respectively. Also, let $\mathcal{K}_i$ be the set of all friends of the subspace $\mathcal{V}_i^*$. Since the bimodal system

(21.1) is well-posed, Proposition 21.4 implies that $\mathcal{V}_1^* = \mathcal{V}_2^* =: \mathcal{V}^*$ and $\mathcal{K}_1 = \mathcal{K}_2 =: \mathcal{K}$. By taking $K \in \mathcal{K}$ and applying the feedback control $u = -Kx + v$ to the bimodal system (21.1) where v is a new input, one obtains the following differential inclusion

$$\dot{x}(t) \in \begin{cases} \{(A_1 - B_1K)x(t) + B_1v(t)\} & \text{if } y(t) < 0 \\ \{(A_1 - B_1K)x(t) + B_1v(t), (A_2 - B_2K)x(t) + B_2v(t)\} & \text{if } y(t) = 0 \\ \{(A_2 - B_2K)x(t) + B_2v(t)\} & \text{if } y(t) > 0 \end{cases} \tag{21.18a}$$

$$y(t) = (c^T - d^TK)x(t) + d^Tv(t). \tag{21.18b}$$

Since the system (21.1) is well-posed with locally integrable inputs, under the feedback control $u = -Kx + v$ the system (21.18) must be well-posed. Moreover, controllability is invariant under this feedback control. Therefore, it suffices to prove controllability of the system (21.18) under the conditions of Theorem 21.8 to complete the proof.

The proof of controllability of the system (21.18) will finish in two steps. First, we prove that its controllability follows from the one of a certain push–pull system (see Proposition 21.10). Then, we show that controllability of the push–pull system follows from the conditions of Theorem 21.8 (see Proposition 21.11).

For the first step, note that well-posedness of the system (21.1) implies $\mathcal{V}_1^* \cap \mathcal{T}_1^* = \mathcal{V}_2^* \cap \mathcal{T}_2^*$. Also note that since $d^T + c^T(sI - A_i)^{-1}B_i$ is right invertible as a rational matrix, the state space $\mathbb{R}^n$ admits the decomposition

$$\mathbb{R}^n = \mathcal{X}_1 \oplus \mathcal{X}_2 \oplus \mathcal{X}_3^i \tag{21.19}$$

where $\mathcal{X}_2 = (\mathcal{V}_1^* \cap \mathcal{T}_1^*) = (\mathcal{V}_2^* \cap \mathcal{T}_2^*)$, $\mathcal{V}^* = \mathcal{X}_1 \oplus \mathcal{X}_2$, and $\mathcal{T}_i^* = \mathcal{X}_2 \oplus \mathcal{X}_3^i$ for $i = 1, 2$. The subspaces $\mathcal{X}_3^1$ and $\mathcal{X}_3^2$ may be different since $\mathcal{T}_1^*$ and $\mathcal{T}_2^*$ may be different as shown in Example 21.5. However, they have the same dimension, say n_3. Let n_1 and n_2 be the dimensions of $\mathcal{X}_1$ and $\mathcal{X}_2$, respectively. For each $i \in \{1, 2\}$, we choose a basis

$$\mathcal{B}_i = \{q_1^i, \dots, q_{n_1}^i, q_{n_1+1}^i, \dots, q_{n_1+n_2}^i, q_{n_1+n_2+1}^i, \dots, q_n^i\}$$

for $\mathbb{R}^n$ such that the first n_1 vectors form a basis for $\mathcal{X}_1$, the second n_2 for $\mathcal{X}_2$, and the last n_3 for $\mathcal{X}_3^i$. Clearly, one can take such bases with the property that

$$q_k^1 = q_k^2 \text{ for all } k = 1, 2, \dots, n_1 + n_2. \tag{21.20}$$

Let S_i be the matrix of basis transformation with respect to the basis $\mathcal{B}_i$. By taking (21.20) into account from now on, one clearly gets the implication

$$x \in \mathcal{V}^* \implies S_1x = S_2x. \tag{21.21}$$

With the matrices S_1 and S_2, we form a new system from (21.18) as

$$\dot{\zeta}(t) \in \begin{cases} \{S_1(A_1 - B_1K)S_1^{-1}\zeta(t) + S_1B_1v(t)\} & \text{if } y(t) < 0 \\ \{S_1(A_1 - B_1K)S_1^{-1}\zeta(t) + S_1B_1v(t), \\ S_2(A_2 - B_2K)S_2^{-1}\zeta(t) + S_2B_2v(t)\} & \text{if } y(t) = 0 \\ \{S_2(A_2 - B_2K)S_2^{-1}\zeta(t) + S_2B_2v(t)\} & \text{if } y(t) > 0 \end{cases} \tag{21.22a}$$

$$y(t) = \begin{cases} (c^T - d^TK)S_1^{-1}\zeta(t) + d^Tv(t) & \text{if } y(t) \leqslant 0 \\ (c^T - d^TK)S_2^{-1}\zeta(t) + d^Tv(t) & \text{if } y(t) \geqslant 0. \end{cases} \tag{21.22b}$$

A solution of the system (21.22) is understood to be any triple $(v, \zeta, y) \in L^1_{\text{loc}}(\mathbb{R}, \mathbb{R}^m) \times AC(\mathbb{R}, \mathbb{R}^n) \times L^1_{\text{loc}}(\mathbb{R}, \mathbb{R})$ such that it satisfies (21.22) for almost all t. The system is said to be controllable if for any two states ζ^0 and ζ^{f} there exist a positive number T and a solution (v, ζ, y) such that $\zeta(0) = \zeta^0$ and $\zeta(T) = \zeta^{\text{f}}$. It turns out that controllability of the system (21.22) suffices for the one of the system (21.18). To show this, we first reduce (21.22) into a canonical form. Let $L_i \in \mathcal{L}(\mathcal{T}_i^*)$. Since $\mathcal{V}^* \subseteq \ker(c^T - d^TK)$ and $\operatorname{im}(B_i - L_id^T) \subseteq \mathcal{T}_i^*$, one immediately gets

$$S_i(B_i - L_id^T) = \begin{bmatrix} 0 \\ \tilde{B}_i \\ \bar{B}_i \end{bmatrix} \text{ and } (c^T - d^TK)S_i^{-1} = \begin{bmatrix} 0 & 0 & c_i^T \end{bmatrix} \tag{21.23}$$

where $\tilde{B}_i$, $\bar{B}_i$ and c_i^T are $n_2 \times m$, $n_3 \times m$ and $1 \times n_3$ matrices, respectively. Note that the systems $\Sigma(A_i, B_i, c^T, d^T)$ and $\Sigma(A_i - B_iK, B_i, c^T - d^TK, d^T)$ share the same $\mathcal{V}^*$ and $\mathcal{T}_i^*$. Thus, one has

$$(A_i - B_iK - L_ic^T + L_id^TK)\mathcal{V}^* \subseteq \mathcal{V}^* \text{ and } (A_i - B_iK - L_ic^T + L_id^TK)\mathcal{T}_i^* \subseteq \mathcal{T}_i^*.$$

As such, the matrix $S_i(A_i - B_iK - L_ic^T + L_id^TK)S_i^{-1}$ must have the following form

$$S_i(A_i - B_iK - L_ic^T + L_id^TK)S_i^{-1} = \begin{bmatrix} A_{11} & 0 & 0 \\ A_{21} & A_{22} & \tilde{A}_i \\ 0 & 0 & \bar{A}_i \end{bmatrix} \tag{21.24}$$

where the row (column) blocks have n_1, n_2, and n_3 rows (columns), respectively. Let the matrix S_iL_i be also partitioned as $S_iL_i = \operatorname{col}(L_{1,i}, L_{2,i}, L_{3,i})$ where $L_{\ell,i}$ is an $n_\ell \times m$ matrix. With these partitions, the matrices $S_i(A_i - B_iK)S_i^{-1}$ and S_iB_i will have the form

$$S_i(A_i - B_iK)S_i^{-1} = \begin{bmatrix} A_{11} & 0 & L_{1,i}c_i^T \\ A_{21} & A_{22} & L_{2,i}c_i^T + \tilde{A}_i \\ 0 & 0 & L_{3,i}c_i^T + \bar{A}_i \end{bmatrix}, \quad S_iB_i = \begin{bmatrix} L_{1,i}d^T \\ L_{2,i}d^T + \tilde{B}_i \\ L_{3,i}d^T + \bar{B}_i \end{bmatrix} \tag{21.25}$$

due to (21.23) and (21.24). Furthermore, we can assume without loss of generality that

$$\begin{bmatrix} \tilde{B}_i \\ \bar{B}_i \\ d^T \end{bmatrix} = \begin{bmatrix} \tilde{B}_i^1 & \tilde{B}_i^2 \\ \bar{B}_i^1 & 0 \\ d_i^T & 0 \end{bmatrix}$$

with the matrix $\operatorname{col}(\bar{B}_i^1, d_i^T)$ is of full column rank, and the matrix $\begin{bmatrix} c_i^T & d_i^T \end{bmatrix}$ is of full row rank. For such a partition, we, respectively, partition v into two component groups as $v = \operatorname{col}(v_i^1, v_i^2)$. Let ζ be partitioned as $\zeta = \operatorname{col}(\zeta_1, \zeta_2, \zeta_3)$ with $\zeta_i \in \mathbb{R}^{n_i}$. Now, writing the system (21.22) in this representation, we obtain

$$\dot{\zeta}_1 = A_{11}\zeta_1 + \Psi^1(y) \tag{21.26a}$$

$$\dot{\zeta}_2 \in A_{21}\zeta_1 + A_{22}\zeta_2 + \Psi^2(y) + \begin{cases} \{\tilde{A}_1\zeta_3 + \tilde{B}_1^1 v_1^1 + \tilde{B}_1^2 v_1^2\} & \text{if } y < 0 \\ \{\tilde{A}_1\zeta_3 + \tilde{B}_1^1 v_1^1 + \tilde{B}_1^2 v_1^2, \\ \tilde{A}_2\zeta_3 + \tilde{B}_2^1 v_2^1 + \tilde{B}_2^2 v_2^2\} & \text{if } y = 0 \\ \{\tilde{A}_2\zeta_3 + \tilde{B}_2^1 v_2^1 + \tilde{B}_2^2 v_2^2\} & \text{if } y > 0 \end{cases} \tag{21.26b}$$

$$\dot{\zeta}_3 \in \Psi^3(y) + \begin{cases} \{\bar{A}_1\zeta_3 + \bar{B}_1^1 v_1^1\} & \text{if } y < 0 \\ \{\bar{A}_1\zeta_3 + \bar{B}_1^1 v_1^1, \bar{A}_2\zeta_3 + \bar{B}_2^1 v_2^1\} & \text{if } y = 0 \\ \{\bar{A}_2\zeta_3 + \bar{B}_2^1 v_2^1\} & \text{if } y > 0 \end{cases} \tag{21.26c}$$

$$y = \begin{cases} c_1^T \zeta_3 + d_1^T v_1^1 & \text{if } y \leqslant 0 \\ c_2^T \zeta_3 + d_2^T v_2^1 & \text{if } y \geqslant 0 \end{cases} \tag{21.26d}$$

where $\Psi^\ell (\ell = 1, 2, 3)$ is the bimodal linear function defined by

$$\Psi^\ell(y) := \begin{cases} L_{\ell,1} y & \text{if } y \leqslant 0 \\ L_{\ell,2} y & \text{if } y \geqslant 0. \end{cases}$$

From (21.26), controllability of the subsystem (21.26a), on the one hand, is clearly necessary for the one of the overall system (21.26), and on the other hand, is sufficient for controllability of the system (21.18) as stated in the following proposition.

Proposition 21.10 *If the system* (21.26a) *is controllable with y is treated as input, then the system* (21.18) *is controllable.*

Proof Let x_0 and x_f be two arbitrary states. In what follows, we will construct a solution of the system (21.18) such that $x(0) = x_0$ and $x(T) = x_f$ for some $T > 0$. To do so, we first apply input $v = 0$ to the system (21.18) for the initial state x_0. Then, since the system (21.18) is well-posed, one can find an $\varepsilon \in (0, 1)$ such that the corresponding output $y^0(t; x_0) = (c^T - d^T K)x^0(t; x_0)$ is smooth and

$$\text{either } y^0(t; x_0) \geqslant 0 \text{ for all } t \in [0, \varepsilon] \text{ or } y^0(t; x_0) < 0 \text{ for some } t \in [0, \varepsilon] \tag{21.27}$$

due to [4, Theorem 2.3]. By [5, Lemma C.4], one can extend this output to a smooth function y_0 such that

$$\text{either } y_0(t) \geqslant 0 \text{ for all } t \in [0,1] \text{ or } y_0(t) < 0 \text{ for some } t \in [0,1] \tag{21.28}$$

and

$$y_0^{(k)}(1) = 0 \text{ for all } k \geqslant 0. \tag{21.29}$$

For such y_0, it can be produced from the system (21.26) by some initial state ζ^0 and input v^0. To construct these data, we first define

$$\ell_0 = \begin{cases} 1 \text{ if } y_0(t) < 0 \text{ for some } t \in [0,1] \\ 2 \text{ if } y_0(t) \geqslant 0 \text{ for all } t \in [0,1] \end{cases}$$

and then define $\mathrm{col}(\zeta_1^0, \zeta_2^0, \zeta_3^0) = \zeta^0 := S_{\ell_0} x_0$. The function y_0, on the one hand, steers the state ζ_1^0 to a state $\tilde{\zeta}_1^0$ at time $t = 1$ when applied to the system (21.26a). On the other hand, it can be generated by the system

$$\dot{\zeta}_3(t) = \bar{A}_{\ell_0}\zeta_3(t) + \bar{B}^1_{\ell_0} v^1_{\ell_0} + \Psi^3(y(t)) \tag{21.30a}$$

$$y(t) = c^T_{\ell_0}\zeta_3(t) + d^T_{\ell_0} v^1_{\ell_0}, \tag{21.30b}$$

for the initial state ζ_3^0 and input $v^1_{\ell_0}$ satisfying $v^1_{\ell_0}(t) = 0$ for all $t \in [0, \varepsilon]$ due to [10, Lemma 4.5] and the invertibility of the system $\Sigma(\bar{A}_{\ell_0}, \bar{B}^1_{\ell_0}, c^T_{\ell_0}, d^T_{\ell_0})$. That means y_0 can be generated by the system (21.26c) and (21.26d) for initial state ζ^0 and input $v^1_{\ell_0}$. Moreover, $v^1_{\ell_0}$ steers the state ζ_3^0 to the state $\zeta_3(1) = 0$ at time $t = 1$ due to (21.29), (21.30) and again [10, Lemma 4.5]. Now, since $\mathcal{X}_2 = \mathcal{V}^*_{\ell_0} \cap \mathcal{T}^*_{\ell_0}$, this implies that $(A_{22}, \tilde{B}^2_{\ell_0})$ is controllable. Thus, for given functions ζ_1, ζ_3, y_0 and $v^1_{\ell_0}$, one can find an input $v^2_{\ell_0}$ such that when applied to the system

$$\dot{\zeta}_2(t) = A_{21}\zeta_1(t) + A_{22}\zeta_2(t) + \tilde{A}_{\ell_0}\zeta_3(t) + \tilde{B}^1_{\ell_0} v^1_{\ell_0} + \tilde{B}^2_{\ell_0} v^2_{\ell_0} + \Psi^2(y_0(t))$$

it steers ζ_2^0 to $\zeta_2(1) = 0$ at time $t = 1$. Altogether, note that the input $v^0 = \mathrm{col}(v^1_{\ell_0}, v^2_{\ell_0})$ can steer the state ζ^0 to the state $\mathrm{col}(\tilde{\zeta}_1^0, 0, 0)$ at time $t = 1$, and produces y_0 and state trajectory $\zeta^0(t)$ with $\zeta^0(1) = \mathrm{col}(\tilde{\zeta}_1^0, 0, 0)$.

By employing similar ideas in the reverse time, for any $T > 1$ one can come up with an output y_f that is smooth on $[T-1, T]$ with $y_\mathrm{f}^{(k)}(T-1) = 0$ for all $k \geqslant 0$, and an input $v^\mathrm{f} = \mathrm{col}(v^1_{\ell_1}, v^2_{\ell_1})$ such that starting from some state $\mathrm{col}(\tilde{\zeta}_1^\mathrm{f}, 0, 0)$ at time $T - 1$ this input steers it to $\zeta^\mathrm{f} := S_{\ell_1} x_\mathrm{f}$ at time T, and produces state trajectory $\zeta^\mathrm{f}(t)$ with $\zeta^\mathrm{f}(T-1) = \mathrm{col}(\tilde{\zeta}_1^\mathrm{f}, 0, 0)$.

As shown in [5, Lemma C.1], since the system (21.26a) is controllable, one can steer $\tilde{\zeta}_1^0$ to $\tilde{\zeta}_1^{\mathrm{f}}$ at a time $T_1 = h\delta$ with $\delta > 0$ and $h \in \mathbb{Z}_+$ by a smooth input y_{m} such that

$$y_{\mathrm{m}}^{(i)}(k\delta) = 0 \text{ for all } k = 0, 1, \ldots, h \text{ and for all } i \in \mathbb{N} \tag{21.31}$$

and either $y_{\mathrm{m}}(t) \geqslant 0$ for all $t \in [(k-1)\delta, k\delta]$ or $y_{\mathrm{m}}(t) < 0$ for some $t \in [(k-1)\delta, k\delta]$ ($k = 1, 2, \ldots, h$). For such y_{m}, on each interval $[(k-1)\delta, k\delta]$, one can find an index ℓ_k, an input v^k and a state trajectory $\mathrm{col}(\zeta_2^k, \zeta_3^k)$ such that (21.26b)–(21.26d) are satisfied for $y(t) = y_{\mathrm{m}}(t - (k-1)\delta)$. Moreover,

$$\zeta_2^k(0) = \zeta_2^k(\delta) = 0 \text{ and } \zeta_3^k(0) = \zeta_3^k(\delta) = 0 \tag{21.32}$$

and in the overall system (21.26) it produces state trajectory $\zeta^k(t)$.

Now, we take $T = 2 + h\delta$ and consider the function defined by

$$x(t) := \begin{cases} S_{\ell_0}^{-1}\zeta^0(t) & \text{if } 0 \leqslant t \leqslant 1 \\ S_{\ell_k}^{-1}\zeta^k(t-1-(k-1)\delta) & \text{if } 1+(k-1)\delta \leqslant t \leqslant 1+k\delta,\ k = 1, 2, \ldots, h \\ S_{\ell_{\mathrm{f}}}^{-1}\zeta^{\mathrm{f}}(t) & \text{if } 1+h\delta \leqslant t \leqslant T \end{cases}$$

Note that the condition (21.32) together with $\zeta_3^0(1) = 0$ and $\zeta_3^{\mathrm{f}}(T-1) = 0$ implies that that x is absolutely continuous. Furthermore,

$$\dot{x}(t) := \begin{cases} S_{\ell_0}^{-1}\dot{\zeta}^0(t) & \text{if } 0 < t < 1 \\ S_{\ell_k}^{-1}\dot{\zeta}^k(t-1-(k-1)\delta) & \text{if } 1+(k-1)\delta < t < 1+k\delta,\ k = 1, 2, \ldots, h \\ S_{\ell_{\mathrm{f}}}^{-1}\dot{\zeta}^{\mathrm{f}}(t) & \text{if } 1+h\delta < t < T. \end{cases}$$

In view of this, one can verify that x is a Filippov solution of the system (21.18) with $x(0) = x_0$ and $x(T) = x_{\mathrm{f}}$. Therefore, the system (21.18) is controllable.

As planed, in the next step we will prove that controllability of the push–pull system (21.26a) follows from the conditions of Theorem 21.8 as stated in the proposition below.

Proposition 21.11 *The implications 1 and 2 of Theorem 21.8 imply the system* (21.26a) *is controllable.*

Proof As shown in [5, Theorem IV.I], the system (21.26a) is controllable if and only if the following two implications hold:

a. $\lambda \in \mathbb{C}, z \in \mathbb{C}^{n_1}, z^* A_{11} = \lambda z^*, z^* \begin{bmatrix} L_{1,1} & L_{1,2} \end{bmatrix} = 0 \implies z = 0.$
b. $\lambda \in \mathbb{R}, z \in \mathbb{R}^{n_1}, z^T A_{11} = \lambda z^T, (z^T L_{1,1})(z^T L_{1,2}) \leqslant 0 \implies z = 0.$

Therefore, it suffices to show that these implications, respectively, follow from those of Theorem 21.8 to complete the proof. To do so, we first consider the first implication and note that the implication 1 of Theorem 21.8 is equivalent to the following implication

$$\lambda \in \mathbb{C}, z \in \mathbb{C}^n, z^*(A_i - B_i K) = \lambda z^*, z^* B_i = 0 \text{ for all } i = 1, 2 \implies z = 0. \tag{21.33}$$

By taking $z = S_1^T \operatorname{col}(z_1, 0, 0)$ where $z_1 \in \mathbb{C}^{n_1}$, the implication (21.33) together with (21.21) implies that the implication

$$\begin{cases} \lambda \in \mathbb{C} \\ z_1 \in \mathbb{C}^{n_1} \end{cases}, \begin{bmatrix} z_1 \\ 0 \\ 0 \end{bmatrix}^* S_i(A_i - B_i K) = \lambda \begin{bmatrix} z_1 \\ 0 \\ 0 \end{bmatrix}^* S_i, \begin{bmatrix} z_1 \\ 0 \\ 0 \end{bmatrix}^* S_i B_i = 0, i = 1, 2 \Rightarrow z_1 = 0 \tag{21.34}$$

holds. Based on (21.34) and (21.25), one can easily come up with the implication

$$\lambda \in \mathbb{C}, z_1 \in \mathbb{C}^{n_1}, z_1^* A_{11} = \lambda z_1^*, \; z_1^* L_{1,i} \begin{bmatrix} c_i^T & d_i^T \end{bmatrix} = 0 \; \forall i = 1, 2 \implies z_1 = 0$$

holds. This clearly implies that the implication (a) holds. Now, we deal with the second implication. Note that the implication 2 of Theorem 21.8 is equivalent to

$$\lambda \in \mathbb{R}, z \in \mathbb{R}^n, w_1, w_2 \in \mathbb{R}, w_1 w_2 \leqslant 0, \text{ and}$$
$$\begin{bmatrix} z^T & w_i \end{bmatrix} \begin{bmatrix} A_i - B_i K - \lambda I & B_i \\ c^T - d^T K & d^T \end{bmatrix} = 0 \text{ for all } i = 1, 2 \implies z = 0. \tag{21.35}$$

By taking $z = S_1^T \operatorname{col}(z_1, 0, 0)$ where $z_1 \in \mathbb{R}^{n_1}$ and post-multiplying the equation of (21.35) by $\operatorname{col}(S_i^{-1}, I)$, the implication (21.35) together with (21.21) implies that the system

$$\begin{bmatrix} z_1 \\ 0 \\ 0 \\ w_i \end{bmatrix}^T \begin{bmatrix} A_{11} - \lambda I_{n_1} & 0 & L_{1,i} c_i^T & L_{1,i} d_i^T & 0 \\ A_{21} & A_{22} - \lambda I_{n_2} & L_{2,i} c_i^T + \tilde{A}_i & L_{2,i} d_i^T + \tilde{B}_i^1 & \tilde{B}_i^2 \\ 0 & 0 & L_{3,i} c_i^T + \bar{A}_i - \lambda I_{n_3} & L_{3,i} d_i^T + \bar{B}_i^1 & 0 \\ 0 & 0 & c_i^T & d_i^T & 0 \end{bmatrix} = 0, \; \forall i = 1, 2$$

has no solution (λ, z_1, w_1, w_2) with $\lambda, w_1, w_2 \in \mathbb{R}, 0 \neq z_1 \in \mathbb{R}^{n_1}, w_1 w_2 \leqslant 0$. This is equivalent to the system

$$z_1^T A_{11} = \lambda z^T \tag{21.36a}$$
$$(z_1^T L_{1,i} + w_i) \begin{bmatrix} c_i^T & d_i^T \end{bmatrix} = 0 \tag{21.36b}$$

has no solution (λ, z_1, w_1, w_2) with $\lambda, w_1, w_2 \in \mathbb{R}, z_1 \in \mathbb{R}^{n_1}, z_1 \neq 0, w_1 w_2 \leqslant 0$. Moreover, by the construction, the vector $\begin{bmatrix} c_i^T & d_i^T \end{bmatrix}$ is a nonzero vector. Thus, nonexistence of solution of (21.36) is equivalent to the validity of the implication

$$\lambda \in \mathbb{R}, z_1 \in \mathbb{R}^{n_1}, z_1^T A_{11} = \lambda z_1^T, (z_1^T L_{1,1})(z_1^T L_{1,2}) \leqslant 0 \implies z_1 = 0.$$

This means the implication (b) holds.

21.4.3 Proof of Theorem 21.9

To prove the "if" part, we note that stabilizability of the systems (21.1) and (21.18) is equivalent. Thus, it is enough to prove that the system (21.18) is stabilizable. To do so, we first choose a basis for $\mathcal{X}_1$ in such a way that the matrix A_{11} is block diagonal as

$$A_{11} = \begin{bmatrix} \bar{A}_{11} & 0 \\ 0 & \tilde{A}_{11} \end{bmatrix}$$

where $\bar{A}_{11}$ is an $\bar{n}_1 \times \bar{n}_1$ matrix which has only eigenvalues with negative real parts and $\tilde{A}_{11}$ is an $\tilde{n}_1 \times \tilde{n}_1$ matrix which has only eigenvalues with nonnegative real parts. Accordingly, we decompose $\mathcal{X}_1 = \bar{\mathcal{X}}_1 \oplus \tilde{\mathcal{X}}_1$ and partition A_{21}, ζ_1, $L_{1,i}$ and $\Psi^1(y)$ as

$$A_{21} = \begin{bmatrix} \bar{A}_{21} & \tilde{A}_{21} \end{bmatrix}, \; \zeta_1 = \begin{bmatrix} \bar{\zeta}_1 \\ \tilde{\zeta}_1 \end{bmatrix}, \; L_{1,i} = \begin{bmatrix} \bar{L}_{1,i} \\ \tilde{L}_{1,i} \end{bmatrix} \text{ and } \Psi^1(y) = \begin{bmatrix} \bar{\Psi}^1(y) \\ \tilde{\Psi}^1(y) \end{bmatrix}.$$

In these coordinates, the system (21.26) boils down to the form

$$\dot{\bar{\zeta}}_1 = \bar{A}_{11}\bar{\zeta}_1 + \bar{\Psi}^1(y) \tag{21.37a}$$

$$\dot{\tilde{\zeta}}_1 = \tilde{A}_{11}\tilde{\zeta}_1 + \tilde{\Psi}^1(y) \tag{21.37b}$$

$$\dot{\zeta}_2 \in \bar{A}_{21}\bar{\zeta}_1 + \tilde{A}_{21}\tilde{\zeta}_1 + A_{22}\zeta_2 + \Psi^2(y) + \begin{cases} \{\tilde{A}_1\zeta_3 + \tilde{B}_1^1 v_1^1 + \tilde{B}_1^2 v_1^2\} & \text{if } y < 0 \\ \{\tilde{A}_1\zeta_3 + \tilde{B}_1^1 v_1^1 + \tilde{B}_1^2 v_1^2, \\ \tilde{A}_2\zeta_3 + \tilde{B}_2^1 v_2^1 + \tilde{B}_2^2 v_2^2\} & \text{if } y = 0 \\ \{\tilde{A}_2\zeta_3 + \tilde{B}_2^1 v_2^1 + \tilde{B}_2^2 v_2^2\} & \text{if } y > 0 \end{cases} \tag{21.37c}$$

$$\dot{\zeta}_3 \in \Psi^3(y) + \begin{cases} \{\bar{A}_1\zeta_3 + \bar{B}_1^1 v_1^1\} & \text{if } y < 0 \\ \{\bar{A}_1\zeta_3 + \bar{B}_1^1 v_1^1, \bar{A}_2\zeta_3 + \bar{B}_2^1 v_2^1\} & \text{if } y = 0 \\ \{\bar{A}_2\zeta_3 + \bar{B}_2^1 v_2^1\} & \text{if } y > 0 \end{cases} \tag{21.37d}$$

$$y = \begin{cases} c_1^T\zeta_3 + d_1^T v_1^1 & \text{if } y \leqslant 0 \\ c_2^T\zeta_3 + d_2^T v_2^1 & \text{if } y \geqslant 0. \end{cases} \tag{21.37e}$$

In this form, by employing very similar arguments as in the proof of Proposition 21.11, one can show that the statements 1 and 2 of Theorem 21.9 imply that the following implications hold:

1. $\lambda \in \mathbb{C}, \mathrm{Re}(\lambda) \geqslant 0, z \in \mathbb{C}^{n_1}, z^* \begin{bmatrix} \bar{A}_{11} & 0 \\ 0 & \tilde{A}_{11} \end{bmatrix} = \lambda z^*, z^* \begin{bmatrix} \bar{L}_{1,1} & \bar{L}_{1,2} \\ \tilde{L}_{1,1} & \tilde{L}_{1,2} \end{bmatrix} = 0 \implies$ $z = 0$,

2. $\lambda \in \mathbb{R}, \lambda \geqslant 0, z \in \mathbb{R}^{n_1}, z^T \begin{bmatrix} \bar{A}_{11} & 0 \\ 0 & \tilde{A}_{11} \end{bmatrix} = \lambda z^T, (z^T \begin{bmatrix} \bar{L}_{1,1} \\ \tilde{L}_{1,1} \end{bmatrix})(z^T \begin{bmatrix} \bar{L}_{1,2} \\ \tilde{L}_{1,2} \end{bmatrix}) \leqslant 0 \implies z = 0.$

Consequently, one gets the following implications

a. $\lambda \in \mathbb{C}, \mathrm{Re}(\lambda) \geqslant 0, z \in \mathbb{C}^{\tilde{n}_1}, z^* \tilde{A}_{11} = \lambda z^*, z^* \begin{bmatrix} \tilde{L}_{1,1} & \tilde{L}_{1,2} \end{bmatrix} = 0 \implies z = 0,$
b. $\lambda \in \mathbb{R}, \lambda \geqslant 0, z \in \mathbb{R}^{\tilde{n}_1}, z^T \tilde{A}_{11} = \lambda z^T, (z^T \tilde{L}_{1,1})(z^T \tilde{L}_{1,2}) \leqslant 0 \implies z = 0.$

On the other hand, since $\tilde{A}_{11}$ has only eigenvalues with nonnegative real parts, the first and the second implication above also hold for $\lambda \in \mathbb{C}$ with $\mathrm{Re}(\lambda) < 0$ and $\lambda \in \mathbb{R}$ with $\lambda < 0$, respectively. In view of this, the system (21.37b) is controllable as shown in [5, Theorem IV.I]. Now, let x_0 be an arbitrary state of the system (21.18). Since the system (21.37b) is controllable, by similar arguments as in the proof of Proposition 21.10, one can come up with an input v^0 which steers the state x_0 to a state $x_{\mathrm{m}} = S_1^{-1}\mathrm{col}(\bar{\zeta}_1^{\mathrm{m}}, 0, 0, 0)$, for some $\bar{\zeta}_1^{\mathrm{m}} \in \mathbb{R}^{\bar{n}_1}$, at a finite time by the system (21.18). After this time instant, one can apply an input of the form $v^{\mathrm{f}} = \mathrm{col}(0, v_1^{2,\mathrm{f}})$ to steer x_{m} asymptotically to the origin at infinity. To see this, we consider the system (21.37) for the initial state $\mathrm{col}(\bar{\zeta}_1^{\mathrm{m}}, 0, 0, 0)$. For this initial state, note that the input $v_1^{\mathrm{l}} = 0$ will keep the state ζ_3 being zero at all time and produce the output being identically zero by subsystem (21.37d) and (21.37e). Hence, the state $\tilde{\zeta}_1$ is also kept being zero for all time by (21.37b) while the state $\bar{\zeta}_1$ tends to zero as t tends to infinity due to the stability of the matrix $\bar{A}_{11}$. For generated functions $\bar{\zeta}_1, \tilde{\zeta}_1 = 0, \zeta_3 = 0, y = 0$ and $v_1^{\mathrm{l}} = 0$, since $(A_{22}, \bar{B}_1^2)$ is controllable, one can find an input $v_1^{2,\mathrm{f}}$ such that it keeps the state ζ_2 being zero for all time. Altogether, the input $\mathrm{col}(0, \zeta_1^{2,\mathrm{f}})$ applied to the system (21.37) for the initial state $\mathrm{col}(\bar{\zeta}_1^{\mathrm{m}}, 0, 0, 0)$ generates the trajectory $\zeta(t)$ with $\lim_{t\to\infty} \zeta(t) = 0$ and zero output. Then, note that the solution of the system (21.18) for the initial state x_{m} and the input $v^{\mathrm{f}} = \mathrm{col}(0, v_1^{2,\mathrm{f}})$ is $x^{v^{\mathrm{f}}}(t; x_{\mathrm{m}}) = S_1^{-1}\zeta(t)$. Obviously, $\lim_{t\to\infty} x^{v^{\mathrm{f}}}(t; x_{\mathrm{m}}) = S_1^{-1} \lim_{t\to\infty} \zeta(t) = 0$, i.e., the system (21.18) is stabilizable.

To prove the "only if" part, we first consider the first implication and let $\lambda \in \mathbb{C}, \mathrm{Re}(\lambda) \geqslant 0$ and $z \in \mathbb{C}^n$ be such that $z^* A_i = \lambda z^*, z^* B_i = 0$ for all $i = 1, 2$. Then, for any initial state x_0 and input u, the Filippov solution $x^u(t; x_0)$ of the bimodal system (21.1) satisfies $z^* \dot{x}^u(t; x_0) = \lambda z^* x^u(t; x_0)$ for all $t \in \mathbb{R}$. This results in $z^* x^u(t; x_0) = e^{\lambda t} z^* x_0$ for all $t \in \mathbb{R}$. Since the system (21.1) is stabilizable, $\mathrm{Re}(\lambda) \geqslant 0$ implies that $z = 0$. Now, we consider the second implication and let $\lambda \in \mathbb{R}, \lambda \geqslant 0, z \in \mathbb{R}^n$, and $w_1, w_2 \in \mathbb{R}$ such that

$$w_1 w_2 \leqslant 0, \begin{bmatrix} z^T & w_i \end{bmatrix} \begin{bmatrix} A_i - \lambda I & B_i \\ c^T & d^T \end{bmatrix} = 0 \text{ for all } i = 1, 2. \tag{21.38}$$

It can be verified that for any initial state x_0 and input u, due to (21.38) the Filippov solution $x^u(t; x_0)$ of the bimodal system (21.1) satisfies either $z^T \dot{x}^u(t; x_0) \geqslant \lambda z^T x^u(t; x_0)$ or $z^T \dot{x}^u(t; x_0) \leqslant \lambda z^T x^u(t; x_0)$ for all $t \in \mathbb{R}$. From this, by Gronwall-Belman inequality, one gets either $z^T x^u(t; x_0) \geqslant e^{\lambda t} z^T x_0$ or $z^T x^u(t; x_0) \leqslant e^{\lambda t} z^T x_0$ for all $x_0 \in \mathbb{R}^n, t \in \mathbb{R}_+$ and for any input u. Then, stabilizability implies that $z = 0$.

21.5 Conclusions

In this chapter, we have presented algebraic necessary and sufficient conditions for controllability and stabilizability of bimodal piecewise linear systems with possibly discontinuous vector fields. To come up with these conditions, we first studied well-posedness of these systems with Filippov solution concept and established necessary conditions for the system being well-posed. These conditions have a number of implications that made it possible to present full characterizations of controllability and stabilizability. Apart from the possible extensions of similar ideas to arbitrary piecewise affine systems, feedback stabilization issue is one of the directions for future work.

Acknowledgments The work of the first author is funded by Vietnam National Foundation for Science and Technology Development (NAFOSTED) under grant number 101.02-2014.32.

References

1. H. Aling, J.M. Schumacher, A nine-fold canonical decomposition for linear systems. Int. J. Control **39**, 779–805 (1984)
2. V.D. Blondel, J.N. Tsitsiklis, Complexity of stability and controllability of elementary hybrid systems. Automatica **35**, 479–489 (1999)
3. M.K. Camlibel, Popov–Belevitch–Hautus type controllability tests for linear complementarity systems. Syst. Control Lett. **56**, 381–387 (2007)
4. M.K. Camlibel, Well-posed Bimodal Piecewise Linear Systems Do Not Exhibit Zeno Behavior, in *Proceeding of 17th IFAC World Congress on Automatic Control*, Seoul, South Korea (2008)
5. M.K. Camlibel, W.P.M.H. Heemels, J.M. Schumacher, Algebraic necessary and sufficient conditions for the controllability of conewise linear systems. IEEE Trans. Automat. Control **53**, 762–774 (2008)
6. M.K. Camlibel, W.P.M.H. Heemels, J.M. Schumacher, A full characterization of stabilizability of bimodal piecewise linear systems with scalar inputs. Automatica **44**, 1261–1267 (2008)
7. A.F. Filippov, *Differential Equations with Discontinuous Right Hand Sides. Mathematics and Its Applications* (Prentice-Hall, Dordrecht, 1988)
8. K. Saxe, *Beginning Functional Analysis* (Springer, New York, 2002)
9. E.D. Sontag, Controllability is harder to decide than accessibility. SIAM J. Control Optim. **26**, 1106–1118 (1988)
10. L.Q. Thuan, M.K. Camlibel, Controllability and stabilizability of a class of continuous piecewise affine dynamical systems. SIAM J. Control Optim. **52**, 1914–1934 (2014)
11. H.L. Trentelman, A.A. Stoorvogel, M. Hautus, *Control Theory for Linear Systems* (Springer, London, 2001)

MIX
Papier aus verantwortungsvollen Quellen
Paper from responsible sources
FSC® C105338

If you have any concerns about our products,
you can contact us on
ProductSafety@springernature.com

In case Publisher is established outside the EU,
the EU authorized representative is:
Springer Nature Customer Service Center GmbH
Europaplatz 3, 69115 Heidelberg, Germany

Printed by Libri Plureos GmbH
in Hamburg, Germany